Burkhardt Wolf

Sea Fortune

Paradigms

Literature and the Human Sciences

Edited by
Rüdiger Campe · Paul Fleming

Volume 10

Burkhardt Wolf

Sea Fortune

Literature and Navigation

Translated by Joel Golb

DE GRUYTER

The translation of this work was funded by Geisteswissenschaften International – Translation Funding for Work in the Humanities and Social Sciences from Germany, a joint initiative of the Fritz Thyssen Foundation, the German Federal Foreign Office, the collecting society VG WORT and the Börsenverein des Deutschen Buchhandels (German Publishers & Booksellers Association).

ISBN 978-3-11-060895-3
e-ISBN (PDF) 978-3-11-061073-4
e-ISBN (EPUB) 978-3-11-060934-9
ISSN 2195-2205

Library of Congress Control Number: 2020939764

Bibliographic information published by the Deutsche Nationalbibliothek
The Deutsche Nationalbibliothek lists this publication in the Deutsche Nationalbibliografie; detailed bibliographic data are available on the Internet at http://dnb.dnb.de.

Cover image: Jan Porcellis: Ship in Storm before Rocky Coast, 1614–1618, oil on canvas, 67 x 35 cm, Stockholm: Hallwyl Museum. Photo: Jens Mohr. By arrangement with the photographer.
Printing and binding: CPI books GmbH, Leck

www.degruyter.com

Contents

Foreword

The bourgeois dreads his fate. Facing this "great mother of danger," he will try to save himself behind a system of care and protection. And nevertheless, "according to the laws of a secret, yet incorruptible mathematics," this "great mother" becomes "more threatening and deadly to the same extent that order believes it has excluded it from itself."[1] For Ernst Jünger, that self-proclaimed heroic nihilist, modern "security culture" was an illusion. In "On Danger," the programmatic foreword to a collection of essays entitled, in English translation, *The Dangerous Moment* (1931), Jünger thus sketches out a little maritime drama containing an exemplary "catastrophe": "humanity's eternal struggle with the elemental nature of the sea," and this "in the temporal form of a shipwreck [*Untergang*]."[2] (Fig. 1) And he systematically collected shipwreck scenarios such as the report of Corréard and Savigny, "Shipwreck of the Medusa." Jünger studied shipwrecks as "doomsday scenarios in miniature."[3] Following their guiding thread, he could paint his doctrine of danger in cinematic chiaroscuro, and with it he could offer his diagnosis of the epoch, declaring its "key date" to be that of the sinking of the *Titanic:* "Here light and shadow luridly collide: the hubris of progress with panic, most extreme comfort with destruction, automatism with catastrophe, appearing as a steering accident." For Jünger, catastrophes inevitably expose those effected to an elementary experience, one distinguishing two sorts of "figures" (*Gestalten*): those enjoying "lesser freedom" from those "who are still attached to primal grounding [*der Urgrund*]," thus seeking nothing less than "a new union of life with danger."[4]

Shipwrecks can serve as an existential metaphor—one that in Jünger's case clearly draws its plausibility from a proto-fascist typology of *Gefahrenmenschen*, men drawn to and embracing "danger." Allied to the "seafarer" is "the artist," together with "the worker" and "the warrior." They all are rooted in what is "elemental," all are adventurers in "the world, which is always dangerous, just as the dead calm sea can hide danger within itself."[5] In evoking a shipwreck, Jünger is less interested in criticizing a concrete form of social or technical disorganization than focusing on an entire culture's fathomless loss: a culture that with its functional systems rests on only putative certainties, on a delusion of "un-

1 Jünger 1993, p. 28; Jünger 2017, p. 33.
2 Jünger 1993, p. 31 (translation modified).
3 Jünger 2019, p. 209.
4 Jünger 1980, pp. 44–46; Jünger 2017, p. 38.
5 Ibid., pp. 32, 34.

https://doi.org/10.1515/9783110610734-001

Luftbild eines italienischen Dampfers im Augenblicke des Unterganges.

Fig. 1: "Aerial photo of an Italian steamer at the moment of its sinking [*Untergang*]," from: Ferdinand Buchholtz (ed.), *Der gefährliche Augenblick. Eine Sammlung von Bildern und Berichten,* Berlin 1931, p. 40 (this volume contains Jünger's essay on danger).

sinkability" and the pseudo-concept of "risk" that is manageable because it can be calculated. In Jünger's writing the nautical gains potential for cultural diagnosis. With the phantom of security, "literature in the old sense" also meets its hour of death, for only "objective reports of experience" can lead to something "of significance."[6] Although Jünger himself refined a correspondingly "cold" view of things, his prose was nonetheless repeatedly marked by a pathos of the elementary. This is shown in what Hans Blumenberg has termed "exaggerations specific to Jünger"[7] such as the "catastrophe": seen objectively, damage at sea only becomes a real catastrophe when it forces entire societies to replace, in Niklas Luhmann's sense, the "form of stability" of its functional system.[8]

6 Jünger 1993, pp. 31 f.

7 Blumenberg 2007, p. 142

8 See Luhmann 2012, pp. 295 ff. (translation modified).

In the end, Jünger's diagnosis is not too different from the conclusions offered by a popular mythology that emerged in the sunken *Titanic*'s wake. This mythology was furnished with worldwide repercussions through an inaugural overseas mass-media network that in April 1912 made an hourly informed "global public" into witnesses of maritime "fate."[9] Since the mobilization of radio and then TV, shipwrecks have always been media events. But these technologies of transmission, to again cite Luhmann, "trivialize...the place from which we see things."[10] We can view Jünger's retrospective dramatization of the *Titanic*'s sinking as a modernization of the existential metaphor that Blumenberg studied in terms of a shipwreck observed by a spectator.[11] But we might also simply call it an example of catachresis, a skewed metaphor. For as a watershed between a culture of security and a culture of danger, what looms, without a doubt, is the Great War. In a manner only all-too obvious temporally and rhetorically, Jünger thus tied that political catastrophe to the old trope of the ship of state. From 1914 onward, his apocalyptic cultural critique was as much common currency as related themes had been a commonplace from time immemorial. Moreover, what Jünger, expert in myth, invoked as the "great mother of danger" was for the adventurer in the "elementary" never simply iron fate, merely *moira* or *fatum*. Rather, as far back as antiquity this mother was honored as *tychē* or *fortuna*—as the goddess of fortune and its vicissitudes.

In Homeric epic and mythology what steered both ships and their fates on the open sea was for the most part the Olympic gods. In Roman discourse, the reference was, more concretely, to a *fortuna redux* or *fortuna gubernatrix*—a goddess of luck at the rudder assigned the task of guiding sea trade and traffic. In late antiquity her cult withered. But once this Fortuna was endowed, as in Renaissance allegories, with the insignias of swelling sails and falling dice, the old goddess emerged as a figurehead for the modern arts of trade and discovery.[12] Earlier simply the addressee of cultic invocation and veneration, she now typified the nautical technologies and knowledge of prospects and danger. And henceforth *fortuna di mare* would serve as a technical term for all the maritime dangers whose calculation meant trading in risk. If we wish to pin down a time and place for the birth of *fortuna di mare*, then it would be thirteenth- and fourteenth-century Northern Italian maritime venues, scenes of mythological-po-

9 See Köster and Lischeid 2000; Rasor 2001; Koldau 2012.
10 Luhmann 2012, p. 88.
11 See Blumenberg 1996.
12 See Patch 1967, p. 102, and Sloterdijk 2014, pp. 818ff. The extent to which Fortuna with the sail is already prefigured in Cicero and in the Middle Ages is debated; the Renaissance seems to have conceived this combination as its own invention. See Warburg 1999, p. 258, n. 52.

etic renaissance and junctures of Early Modern capitalism. The earliest extant reference to a Fortuna conflated with the elementary force of a storm is in a Genoese document of 1242.[13]

A Fortuna who allegorically steers the course of the Ship of Life, who asserts herself in the concrete natural phenomenon of the gale, and who furthermore exerts her influence as a cost factor makes no appearance before the Renaissance.[14] In addition, it seems that starting in late antiquity the sea-wind, the *pneuma* of Christian seafaring, is subject not to chance but rather entirely to divine governance. The wind as an unpredictable quantity, one to be encountered with both heedfulness and daring in the hope of great gain, but which held the threat of bitter loss for the careless or fainthearted, was a discovery made in the context of the thirteenth-century spirit of sea-trading. For this reason, *fortuna di mare* soon symbolized the return of splendid merchant vessels, but also, until the seventeenth century, the remains of shipwrecks chanced upon as flotsam and jetsam—findings from the sinking that at least offer small added value.[15] In the waning Middle Ages, when it came to dangerous long-distance trade a basic distinction was already being made between land and sea: between trade undertaken under a sign of safety and security, *salvi in terra*, on the one hand, and "risk, danger, and hazard of the sea and men" (*rischio pericolo e fortuna di mare e genti*), on the other hand. The sea manifestly contained a menacing potential taking in both elementary forces and the action of other voyagers, extending to piracy.[16] In commercial and insurance contracts, lexicons and encyclopedias, such "fortune" soon shifted into the plural. In Nicolas Aubin's *Dictionnaire de la marine* (1702), for instance, *fortunes de mer* are defined as "those accidents that are caused by tempests & those to which one is subject at sea, such as sinking, running aground, encountering pirates et cetera."[17] In the end the concept covered the totality of perils at sea.

However, from antiquity onward it also stood for the structure of possibilities of a world whose unknown, longed for regions mainly revealed themselves overseas. The Old Italic goddess Fortuna appears to have been a descendent of Egyptian Isis and Greek *tychē*, both endowed with the attribute of the helm as arbiters

---

13 Bartholomaei Scribae Annal. Genuenses ad ann. 1242. See Patch 1967, p. 107.

14 See Reichert 1985, pp. 25 f., note.

15 See Détrée 2005, p. 33.

16 See Stefani 1958, vol. 1, p. 57. On the derivation of piracy from Gr. *peiran*, "to venture, try" (it also means "to attack"), see Schmitt 2003, p. 43.

17 Aubin 1702, p. 427 ("les accidens que cause la tempête, & les autres auxquels on est sujet sur mer, comme d'échoüer, de couler bas d'eau, de rencontrer des pirates &c.").

of fate.[18] Her name was seen as derived from Latin *fors*, "chance," and consequently juxtaposed with providence; from the beginning she was blamed less for misfortune than for unreliability and holding out deceptive promise. The Church Fathers scorned her as a mere delusion with no possible place in God's salvational plan—which could not expel her from popular belief.[19] But Fortuna was not kept alive through superstition alone: encapsulated in her figure was the old discussion, at the latest initiated by Plato and Aristotle, centered on *tychē*, its theological functionalizing and relation to both *moira* and, as we will see, the *automaton*. In cultic and mythological contexts, *tychē* was a living deity, not just an abstraction, allegory, or personification. And because what was imponderable was considered the domain of the sea, in both Hesiod and the Homeric Hymns she is present as a nymph and as the daughter of Okeanos; for Pindar she is part of the world's governing structure, of *kybernēsis*, thus also steering ships to port and on the seas.[20]

One basic reason that as a deity *tychē*, together with her numerous shrines and forms of cultic worship, could endure as long as she did was the lack of a philosophical solution to the problem of chance. Aristotle's doctrine of chance (presented in his *Physics* and the "Interpretation" section of the *Organon*), marking an effort to move the theme from the realm of theological and cosmological speculation to that of philosophy, remained virtually without any real impact for centuries. What did have strong influence was the Aristotelian ethical maxim that *tychē* could not diminish human ethical freedom. For their part, Roman stoics such as Seneca taught that while *fortuna* might distribute this world's parcels of luck and happiness, it could not compromise the *virtus*. Human beings were capable of arresting its power through sagacity and—as Cicero put it—courage, *prudentia* and *fortitudo*, so that *adversa fortuna* could be converted into *prospera fortuna:* a Roman principle that the Renaissance would take to heart under the motto *virtù vince fortuna*. Starting with Alberti's *Libri della famiglia* (1433–1441), Italian merchants seem to agree on one thing in particular: the "great mother of danger" is not simply to be feared and avoided. In the end she will succumb to the *vir*, the truly manly man—after all, she is not only a devouring mother but also simply a woman.[21]

In the context of what Aby Warburg terms this new "stylistic embodiment of worldly energy," reawakened Fortuna had decisive importance for the arts of trade and navigation—something to which the graphic arts in turn bore wit-

18 See Göttlicher 1981a, pp. 27–32, und Göttlicher 1981b, pp. 8 f., 80–84, 219.
19 See Goetz 1996, pp. 75, 81.
20 See Vogt 2011, pp. 96–99.
21 See ibid., pp. 505–507, 527–532, 540.

Fig. 2: Nicoletto da Modena: Fortuna, ca. 1506

ness.[22] Where, for instance, around 1460 Fortuna is transformed into a mast on the coat-of-arms relief of Florence's Rucellai Palace, in an engraving probably created on the occasion of the marriage of Bernardo Rucellai and Nannina Medici in 1461—included on a panel of Warburg's *Mnemosyne* project—the merchant's son follows suit beneath the motto of "the ship of fortune." Henceforth, it seems, he will be spreading out the sails by himself. Chance and ability, *fortuna* and *virtù*, are as it were married in this engraving, as are elementary force and navigation and, in the presented figures, man and woman. Fortuna's effects thus

22 Warburg 1999, p. 240.

no longer appear autonomous but rather as constituting a counterpart or even foil to the clever and courageous mercantile arts—a role informing her emerging three-fold semantic presence in the Romance languages as "chance," "wealth," and "tempest." In sixteenth-century allegories, she would increasingly appear with a globe as her attribute: replacing the old wheel of fortune, which still designated her as mistress of material riches, the globe symbolizes Fortuna's unreliable and uncertain course, her chaotic and arbitrary nature. The instability her role now acquires is shown with special clarity in an engraving of da Modena where she balances herself with one leg on the rudder, the other on a floating globe.[23] The wind, gaining form in the sail's swelling, is here linked back to both the rudder and the rippling waves by means of Fortuna—who thus now finds herself precisely where one incalculable force encounters another and where, in addition, the implement for nautical steering intervenes. (Fig. 2)

This engraving might be brought together with calculations centered on probability and games of chance emerging with Gerolamo Cardano's *Liber de Ludo Aleae* (probably written in 1525 and revised in 1565); these moved Fortuna out of the realm of unpredictability, opening up her "secret yet unerring mathematics." In addition, the engraving evokes both Cardano's solution to cubic and quartic equations and, through the globe, the balls drawn from an urn in the urn model of statistics[24]—the significance Cardano's innovation had for bookkeeping, the insurance business, and the nautical sciences here being self-evident. In any event, Fortuna would also present her double countenance of mythical elementary force and "accountability" in Venice, hence in the place where a "marriage of life with danger" had indeed been arranged: namely, in the annual *Lo Sposalizio* ceremony, when the Doge had a ring sink into the sea. In a copper sculpture by Bernardo Falconi depicting Fortuna standing on one leg upon a globe, the goddess symbolizes the simultaneously mythical and calculable structure of worldly possibility: this *fortuna di mare* holds a movable, rudder-shaped weathervane into the wind. Here the globe—more pronouncedly a world globe than in the da Modena engraving—is held and born by two atlantes. And this superbly balanced figural ensemble was enthroned on top of the Dogana da Mar, the Venetian customs house on the Punta, the promontory at the entrance to the Grand Canal (the sculpture continues to preside on top of the building today). (Fig. 3) Turned toward the sea at Venice's outermost tip, this Fortuna

23 See Fichte 1996, p. 18.

24 See Holländer 1996, pp. 166 f. Alongside pioneering work in gambling odds-calculation and first reflections on the law of large numbers, Gerolamo Cardano (1501–1576) made important contributions to both bookkeeping and designing instruments—including the gimbal known as the Cardanian suspension for ship-lamps and compasses.

looks out toward a horizon of possibility: toward what might befall the city-state in the sense of *prospera fortuna*; what actually meets her in the form of seagoing vessels; and what will come the way of Venetian merchants in payment, as determined by both bills and customs law. For *fortuna di mare,* future arrival and encounter as *ad-venire* is thus also *ad-ventura* in its modern double sense: risky enterprise and marine adventure.

In the following chapters, this double perspective will form the framework for tracing out the fortunes of the Western maritime enterprise as it is manifest in a broad range of techniques and metaphors. For if the goddess of fortune is herself a hybrid formed of image and concept, then the sea voyages under her control can only be observed with double attentiveness to the metaphors and media at work in her figure. Should we choose to not consider the sea solely from the land, then the sea can only reveal itself as an open experimental field between technics and poetics. But in this way, it also opens a horizon of the most varied passages between seafaring and literature. What Jünger dismisses as "literature in the old sense" is an open body of mythological and historiographical texts, others philosophical, or theoretical, or poetic, some of them simple reports and some centered on the marvelous, together with all sorts of writing concerned with practical usage. All these texts have contributed to the great undertaking of describing the sea, that traceless medium par excellence. What was called "hydrography" in the Early Modern period and is now known as "oceanography" was already approached in antiquity by endowing the placelessness of the open sea with loci at once mythological and nautical: in the period broadly marking the inception of Western literature, Homer's *Odyssey* consists of a first extended encounter between writing and seafaring. This epic is in fact a sort of poetic sailing manual, its instructions followed not only by future authors but by the old seamen, cosmographers, and explorers as well. From Homer, then, by way of authors including Dante, Poe, and Melville and extending in to the twentieth century and beyond, in the most varied possible way literature will document the extent to which the sea's vicissitudes are also those of writing.

In the Portugal of the Age of Discovery, *discurso* still meant both the sea route and the course of a narrative. In the end, everything worth narrating, every discovery, odyssey, failed effort has always also been a *discurrere*—a scattering, confusion, even a loss of any sign. In that light, we can see not the classic "ship of poetry" topos as first defining the parameters of maritime literature, but already wily Odysseus, that repeatedly tossed-about seaman nevertheless highly skilled at survival. Strikingly, where the dangerous, tempting sirens—*fortuna di mare* become form—encounter a Homeric Odysseus who evades their grasp, Dante's Odysseus, setting a pattern for modern literature, can no longer escape

Fig. 3: Bernardo Falconi: copper sculpture, *Fortuna di mare* on top of the globe, with rudder-weathervane, 1678

Fortuna, now the shapeless figure of an all-devouring maelstrom. On the threshold of early modernity, we find a realization of what the Greeks already feared in their first maritime sallies: the sea voyage as hubris. When, in their greed or simple, irresistible curiosity about the world, human beings abandon the firm soil constituting their natural sphere, when they transgress the elementary border between land and sea, they place themselves in essential opposition to the cosmological order. The sea becomes an elementary enemy. For that reason they try to mitigate the transgression's results by cultic and religious means—through literal "rites of passage." And so as not to stray hopelessly and forever on the endless, bottomless sea, land-based human beings invent wily cultural techniques such as navigation, meant to restore orientation in the alien zone.

If with its unequalled challenges and dangers seafaring represents a supreme domain of what the German calls *Erfahrungswissenschaft,* practical science or knowledge,[25] it is no less one of political and anthropological self-questioning. The sea serves as a massive reservoir of existential metaphors, unceasingly in motion because of their steady usage. Homeric gales present challenges entirely different from the elementary forces facing captains of modern ships. Most prominently, the topoi of shipwreck metaphorically process an existential and collective groundlessness, so that scenarios such as Lucretius's "shipwreck with spectators" are subject to continuous revision;[26] similar modern existential metaphors attest to a basic loss of grounding on, now, *terra firma* as well (Kafka directly speaks, for instance, of "seasickness on dry land"). Seafaring is in this way accompanied by an enduring renegotiation of elementary borders from which basic distinctions—safety and danger, enlightenment and barbarism, civilization and savagery, art and non-art, and so forth—are derived. Repeatedly, with seafaring we find what Carl Schmitt described as a rupture of "conceptual axes" providing guidance to what he terms the *Landtreter,* the terrestrially bound human being.[27] For this reason, in mythological, religious, political, juridical, and general cultural frameworks seafaring is very much intertwined with scenarios of "grounding" and "ungrounding." Its story is one of disorientation and non-conceptuality, of a knowledge repeatedly divested of locus, washed back into metaphor: in any case a story in which theory and history are repeatedly intertwined anew.

---

25 The Heideggerian term and concept will hover behind my references to "practical knowledge" and "practical science" throughout this book. I will avoid the infelicitous and misleading (although sometimes used) "science of experience" (translator's note).

26 This is likely the reason that circa 75 percent of modern publications concerning shipwrecks are literary, narrative texts—the results of a quantitative analysis of French library holdings between 1600 and 1969. See Zysberg 1999, p. 191.

27 Schmitt 1996, p. 18 (not included in English-lang. transl. by George Schwab, Carl Schmitt, *The Concept of the Political,* Chicago and London 1996).

This is already the case with the emergence and transformation of—to cite Heidegger—different "world pictures,"[28] their "cosmographic" fabrication usually guided by knowledge and non-knowledge of the nautical sciences, its sailing manuals, course information, cartography. Keeping in mind the literal meaning of *poiēsis* as production, from the start seafaring was a poetic undertaking not only in literary respects. Its "world-founding" force revealed itself most spectacularly in adventurous exploration and at times successful, at times fallacious speculation about New Worlds. But it was most enduringly manifest as practically grounded science. Namely, moving past the Pillars of Hercules, its *plus ultra*—"further beyond," turned by Charles V, Holy Roman Emperor and King of Spain, into a personal motto, said to be inscribed on the pillars—gave an impetus to empirical natural research that was more than metaphoric. Even before either the Spanish or English, the Portuguese, setting up mutual feedback between the cosmographs of their territory-managing India House (the *Casa da Índia*) and the nautical vanguard of their India Run (the *Carreira da Índia*), established an empirical, results-based, and in this sense experimental science. In doing so, they catalyzed more than a first form of economic globalization.[29]

That in the voyage around Africa's southern cape, economic maxims eventually suppressed those of science meant the India Run ending up an enterprise that made heavy losses. A new narrative genre now emerged mixing reminiscence and documentation, travel reports and poetic creation, and focused on the countless hardships and shipwrecks of the mostly overladen and poorly captained merchant vessels. The genre was termed *literatura de cordel*, literally "string literature," because the pamphlet-like texts involved were sold on Lisbon's streets on strings. They in fact comprised the first form of global reportage and a pioneering yellow press focused on nautical catastrophes. At the same time, Early Modern *literatura de cordel* was the first narrative genre that—as a counterpart to nautical epics such as Luís de Camões' *Os Lusíadas* (1572)—was dedicated to outrageous fortune alone. For generations, these texts would remain a repository of seafarers' travail, spelling out entire litanies of danger and shipwreck, of disease, hunger, and death, avarice, mutiny, and piracy.[30]

---

28 Throughout this book, I will use the term "world picture" in Heidegger's sense of new procedures in gaining practical knowledge. See Heidegger 2002, pp. 57–85. In this regard and with a view to more recent theory of space see Dünne 2011, pp. 23–30.

29 Pierre Chaunu 1959, p. 34 refers to the *Carreira*, in Braudelian terms, as the first, autonomous "world-economy" (*économie-monde*), which would, however, only become organized on strictly capitalist lines with the Dutch and English maritime undertakings.

30 See Boxer 1984, pp. 155 ff. The reports were collected in 1735 by the Lisbon chronicler Bernardo Gomes de Brito and published in a two-volume selection. A textual corpus that for centuries

In antiquity, the in any event cultically motivated shipwreck anthologies were similar. But what characterizes Early Modern narratives of *mala fortuna* in general, when contrasted with those older texts, is their "empirical" or even "experimental" emphasis.

For example, starting in 1762 William Falconer, himself a sailor who had experienced a number of shipwrecks, published three versions of a poem called "The Shipwreck." Although conceived as a national epic, in a manner untypical of the genre Falconer's verse does not aim at elevating and transfiguring that maritime disaster. Rather, in a style offering something like a nautical equivalent to Virgil's agriculturally centered *Georgics*, Falconer's three cantos deal with the English merchant fleet, and with the minutiae of nautical science.[31] The complementary nature of poetic and empirical knowledge in this work becomes clear in the independent annotation and illustration of nautical matters in its margins. The text is supplied with a map, together with a detailed illustration, accompanied by legends, of the British trading vessel on which Falconer himself experienced the shipwreck his verse describes. But reflecting the wish not only to spell out nautical terms but also show them as an essential element of sailor's practice, he seeks advice in a range of specialized lexicons. Finding them inadequate, he then puts together his own dictionary—one that ended up the undisputed English authority on seafaring for decades.[32]

In both literature and seafaring, shipwrecks stand for the double problem of witnessing, on the one hand, both surety and documentary attestation, on the other hand. These two dimensions constitute a requirement for determining whether and, if so, under what circumstances a shipwreck has taken place, for which definitive proof is rarely if ever available: a situation sailors and merchants have done their best to remedy through the claim to credibility. For millennia the problem of credibility has made necessary specific procedures of nar-

---

was considered mere *literatura de cordel*, material lacking literary merit, then increasingly gained the status of a national classic under the title *História trágico-marítima*. This shift in reception history involved peremptorily incorporating the heterogeneous *relaçãos* into a "historical-tragic" framework and subsuming them to the cultural program of Portugal's John V, aimed at counteracting the nation's faded importance through historical awareness. Brito, at the time certainly speculating about a position in the national historical academy (founded in 1720), thus laid something at the sovereign's feet that previously had been meant for everyone and no one: for a vague form of "critical public."

31 "But I, perplext in labyrinths of art," explains Falconer programmatically, "Anatomize and blazon every part; / Attempt with plaintive numbers to display / The chain'd events in regular array." Falconer 2003, p. 232, B, canto III. 33–36.

32 See ibid., pp. 96 f., and the "Advertisement" for the three versions; in addition Falconer 1780, preface, pp. I–III.

ration, attestation, and simultaneous invention. From the *Odyssey* to Melville's *Moby-Dick* and B. Traven's *Death Ship,* innumerable narratives are located in this zone where distinctions have vanished between myth and poetry, fact and fiction. But both maritime traffic and maritime literature have also been stamped by the problem of credibility in the sense of "credit worthiness." The ancient institution of "sea loans" marked the beginning of long distance trade and thus an early form of capital-based speculation. Beginning in the thirteenth century, seafaring insurance, the first form of any sort of insurance, made it possible to hazard the consequences of *fortuna di mare*'s still unspecified dangers, and thus to create specific "risks" (as it happens, a word derived from *risco*, "cliff") and trade with them on a trading floor for stocks or insurance.[33] As a consequence of this new approach to danger, the medieval ban on interest was ended and a modern credit-economy arrived on the scene. And when it came to both speculative business and establishing technologies of security and prevention, the nautical and mercantile subjugation of maritime *fortuna* became a model for modern "security culture."

Derived from the basic distinction between risk and danger, "accountability" and "experience," a distinction always associated with the difference between land and sea, were two cultures: on the one hand, security and society; on the other hand, hazard and solidarity. The modern pathos of "dangerous life," including its nautical semantics, feeds on just this opposition. But unlike what Jünger among others has suggested, insurance never regained the firm ground long-since lost to the nautical adventurer. With seafaring insurance, trade was nothing less than uprooted on land as well, since that business had shown that not only securely fixed property but also movable and virtual possessions such as ships and credit could serve as a trading base.[34] When not only what is characterized by "fixedness" and "indestructibility" but also what is mobile and imperiled – indeed mobility and danger themselves – becomes a commodity, as Georg Simmel observes, here "as in many other cases, substantive determination has actually been revealed to be functional determinacy."[35] The only

---

33 "Every risk, accident and danger," we read in a commercial agreement of 1310 that anticipates the *risicum* formulas of later policies, "as of God, seas, peoples... shipwrecks... theft, [and] violence in any other instance of accident and danger." (*Omnem risicum, casum et periculum tamquam Dei, maris, gentium...naufragii...robarie, violencia in quodam cujuscumque alterius casus et periculi.*) Cited from Boiteux 1968, p. 64.

34 According to Georg Simmel, the "*principle* of *insurance* has made those objects that are totally lacking in any fixed position in space eligible for mortgage lending." See Simmel 2004, p. 147.

35 Ibid.

thing that can still offer a foothold given the liquidation of both ground and substance, and in a trading context both speculative and liable to fraud, appears to be personal surety. Even more than for insurance companies, which after all work on a basis of mass statistics, it is thus vital for underwriting—the individual insurance for complex risks offered by Lloyd's since the eighteenth century—to be able to distinguish between certainty and speculation, fact and fiction, thus firmly grounding commercial relations in trust. This is likely why authors such as Daniel Defoe and Joseph Conrad understood underwriting as a certification procedure corresponding to verification of literary authorship and, with it, protection of "intellectual property."

Whether on sea routes or in the form of the merchant's booking routines, seafaring is a practical school and test of how one deals with singular, unpredictable events, uncertainty. The ancient view of the sea as a sphere in which things and human beings cannot be defined as living or dead (or as a realm where, following Aristotle, events are assigned neither being nor not-being) also defines it as a space of contingency and potentiality—its structure of possibility reflected not only in *fortuna di mare* but also in literature's own sense of potential. The *Odyssey* presents possible worlds and courses of action, for instance when Circe offers different sailing directions—hence also different directions for narrative—and when during his journey to Hades Odysseus receives a prophecy of death either far out at sea or beyond the sea. For Homer, but even more so for Homer-readers like Joyce, telling stories about the sea means narrating in an experimental way—with potential presents and contingent futures, and even with possible pasts.

But telling stories about *seafaring* means facing up to the ship as what Foucault has described as "the heterotopia par excellence."[36] The ship is a life-space without any natural elements. This is reflected in its downright precarious status as "swimming territory," as a "state-ship" moving within a stateless zone. On board the ship, media, cultural, and literary history have been made *and* written by seafarers. Similarly, elementary processes of technical and political leadership have always been tried out there. From the Early Modern period onward, the ship has served as an experimental laboratory in both pedagogic and disciplinary respects, as we see in the odd history of the captain's rule, mutinies against it, and the repeated efforts at reform. But ships can also be described as mobile "actor-networks" (the term is Bruno Latour's)[37] that can take on completely different structures depending on their goal and commission: Greek cultic

36 Foucault 1986, p. 27.
37 See Latour 2005, *passim*.

ships, for example, have little to do with the large and pompous ships used in ancient Rome; and in the case of modern ships used for exploration, trade, war, whaling, and piracy, the necessary functional systems are clearly different. That the basic model of cybernetics, together with the term itself, goes back to the ancient navigational art of *kybernēsis* is only one of the reasons the ship can be described as a paradigm for that area of knowledge.[38] Steering and regulation processes on a ship are seen as exemplary communicative processes between machines, human beings, and the natural elements. As a target control mechanism, the ship—like Norbert Wiener's servo-mechanical air artillery—continuously opens up contingent futures.

If one particular factor has guided my choice of texts discussed in this book, then it is this cybernetic perspective on the nautical arts. However, to do justice to texts that understand seafaring as a complex field of managing both ships and human beings, of coping with contingency and symbolically discovering the world, we need an approach that tries to mediate between the "two cultures":[39] between technology and nautical science on the one hand and literature and poetics on the other hand. At first glance, it seems that individual textual genres can only take account of particular aspects of seafaring. For example, as in some of Shakespeare's work, plays can present the rules of modern trading and the drama of contingent futures; as in some of Hölderlin's work, lyrics can offer a kind of poetic salvaging of what has been discursively and materially lost; and as already in Homer, both narrative and the epic tradition it is part of display a general affinity with forms of cybernetic processing. Nevertheless, all these literary texts are integrated into a comprehensive network of nautical representation, reporting, accreditation, and writing procedures, just as inversely the writing of literary texts has always contributed to both nautical culture and its cultural history. At certain points, the history of literature and the history of seafaring have as it were been superimposed upon each other, for instance when the *Odyssey* served as a model for post-Homeric sailing manuals, when travel reports from Portugal's India Run justified implementation of a sovereign

38 As a first illustration of feedback control, the basic cybernetic operation, James Clerk Maxwell pointed to the centrifugal governor in James Watt's steam engine; in his own illustration Norbert Wiener then pointed to nautical direction control and engine control: A ship proceeds on a "control path" for whose operation a fixed point such as a celestial body or geographical coordinate needs to be aimed at. Since the ship's course deviates from the first objective because of various disturbing variables (wind, currents, undisciplined crewmembers), that objective is now fed back into the successively measured control variables (the present latitude and longitude), hence varied according to location. See Maxwell 1868; Wiener 1962, pp. 11 f.
39 This term was coined and made famous by C.P. Snow. See Snow 1953, p. 3, *passim*.

regime on board merchant ships, and when Erskine Childers' scenarios of naval warfare prompted actual mobilization.

In the following chapters I hope to track the literary course of seafaring's cultural-technical, medial, and epistemic history. My exploration of a nautical dimension of literary texts is thus interwoven with what in a broad sense can be termed the history of knowledge; readers may also sense certain affinities between the findings unearthed in this way and a kind of marine archeology—an often neglected dimension of the "archeology of knowledge." But what leads to these findings is, consistently, literature: without a doubt only one of many "finding aids" for delving into the historical vicissitudes of ships and the sea. If we wish, for example, to focus on *fortuna di mare* from a primarily art-historical perspective, its pictorial allegories would simply serve as an initial source. This is because a type of painting developed in the Renaissance that was no longer obligated to canonical, mostly poetic motifs but rather emulated the constructive representational and image-creating procedures of nautical charts and instruments. The non-perspectivist and object-free pictorial spaces at work in paintings from this era manifestly inspired Leonardo in his study of waves, eddies, storms, and floods. For its part, the Dutch pictorial tradition took up the sea's atmospheric conditions and contingencies in the "tonal" sea pieces of Jan Porcelli. In these pieces (as shown on the cover illustration) the sky and sea communicate or even blur in a play of light and clouds, waves and foam. Through the low horizon and distant focal point of the paintings, the sea ends up taking hold of viewers, their standpoint shifting into an open realm.[40]

In many of Turner's paintings, we find a move beyond perspective, shape, and representational coherence—a move onto the grounds of possibility of representation itself. Their photographic registering of "the real," of light and its nature, it has been argued, corresponds to Michael Faraday's electromagnetic and graphic experiments:[41] instead of objects Turner paints forces and potencies—instead of visualizable processes only the medium. In this way he arrives at that borderline of visibility that the sea appears to mark for painters. Strikingly, it also signifies a border for poets and writers of fiction—that of the sayable. In this context we are reminded of an observation by Ralf Konersmann: "For 'the sea' does not exist; it no more exists than the unalterable borderline of the shore. There is only this iridescent jetsam of impressions and stories we have

40 See for example Goedde 1989, p. 88.
41 See for example Cohen 2010, p. 131.

of it, washed up over years and centuries."[42] Against this maritime horizon, literature is only jetsam and flotsam, only a gift of *fortuna di mare*. But the hope is that as just such a haphazard entity, it will lead us out into openness, into the possibility that is the sea itself.

42 Konersmann 2003, p. 220.

Acknowledgments

I would like to extend my thanks to De Gruyter for quickly seeing to realization of the publishing project, and to the editors of *Paradigms* for accepting the book into their series.

For its generous support in funding this book's English-language translation, I would like to warmly thank the *Geisteswissenschaften International* translation-funding program. For their steady support, encouragement, and suggestions in the course of my preparation of the book's original German-language version, I offer heartfelt thanks to Steffen Martus, Joseph Vogl and Niels Werber; and for their comments and corrections, to Julian Bauer, Katja Bödeker, Sophie Bunge, Ronald Düker, Lucia Iacomella, Gernot Kamecke, Sara Lehn, Elisabetta Mengaldo, Gloria Meynen, Jörn Münkner, Stefanie Peter, Francesca Raimondi, and Henning Teschke. In addition, I would like to express my special thanks to Joel Golb for his extraordinary efforts in bringing this long translation project to fruition—efforts that included a great deal of valued discussion concerning both the possibilities and inevitable limits of translation. A special word of thanks, as well, to Rafael Jakob for his devoted copy-editing work on the English-language manuscript and its index.

https://doi.org/10.1515/9783110610734-002

Note for the English-language edition

This book was written in 2011 and was published in German in 2013. It has not been substantively revised for the English-language translation, which likewise does not take account of special academic literature appearing after the German-language book's publication: perhaps particulary in the context of maritime matters, such an undertaking could quickly become boundless. Instead, working together with the book's—extraordinarily engaged—translator Joel Golb, we have tried to produce a version of the original text oriented toward an English-speaking and international readership. This does not alter the fact that the book's "global" perspective is also a limited one. It naturally does not try to either do justice to "world literature" in the term's present senses or offer a universal history of seafaring. Rather it thematically addresses the special, at times only loose alliance that seafaring and literary writing formed *in the Western world* from the beginning, an alliance that seems to still be intact. The question of whether this shared "beginning" or "origin" is simply a myth, a retrospective projection, or even one version of colonialist positing is as much a part of the history of this alliance as is the concession that it has always perspectivized and projected onto "the world" according to European standards. The history of Western seafaring (and the writing that cooperates with it) can thus be understood as a history of the "Western gaze"—even if this gaze has never wished to content itself with fixed cultural landmarks but has repeatedly strived to lose itself in the open horizon of the sea.

https://doi.org/10.1515/9783110610734-003

Departure
Topoi of Wiliness: Erring and Returning in the *Odyssey*

To protect himself from the Sirens Odysseus stopped his ears with wax and had himself forged unto the mast of his ship. Naturally any and every traveler before him could have done the same, except those whom the Sirens allured even from a great distance; but it was known to all the world that such things were of no help whatever. The song of the Sirens could pierce through everything, and the longing of those they seduced would have broken far stronger bonds than chains and masts. But Odysseus did not think of that, although he had probably heard of it. He trusted absolutely to his handful of wax and his fathom of chain, and in innocent elation over his little stratagem sailed out to meet the Sirens.

Now the Sirens have a still more fatal weapon than their song, namely their silence. And though admittedly such a thing has never happened, still it is conceivable that someone might possibly have escaped from their singing; but from their silence certainly never. Against the feeling of having triumphed over them by one's own strength, and the consequent exaltation that bears down everything before it, no earthly powers can resist.

And when Odysseus approached them the potent songstresses actually did not sing, whether because they thought that this enemy could be vanquished only by their silence, or because the look of bliss on the face of Odysseus, who was thinking of nothing but his wax and his chains, made them forget their singing.

But Odysseus, if one may so express it, did not hear their silence; he thought they were singing and that he alone did not hear them. For a fleeting moment he saw their throats rising and falling, their breasts lifting, their eyes filled with tears, their lips half parted, but believed that these were accompaniments to the airs which died unheard around him. Soon, however, all this faded from his sight as he fixed his gaze on the distance, the Sirens literally vanished before his resolution, and at the very moment when they were nearest to him he knew of them no longer.

But they—lovelier than ever—stretched their necks and turned, let their awesome hair flutter free in the wind, and freely stretched their claws on the rocks.

They no longer had any desire to allure; all that they wanted was to hold as long as they could the radiance that fell from Odysseus' great eyes.

If the Sirens had possessed consciousness they would have been annihilated at that moment. But they remained as they had been; all that had happened was that Odysseus had escaped them.

A codicil to the foregoing has also been handed down. Odysseus, it is said, was so full of guile, was such a fox, that not even the goddess of fate could pierce his armor. Perhaps he had really noticed, although here the human understanding is beyond its depths, that the Sirens were silent, and held up to them and to the gods the aforementioned pretense merely as a sort of shield.

https://doi.org/10.1515/9783110610734-004

—Franz Kafka, "The Silence of the Sirens", translated by Willa and Edwin Muir, in Franz Kafka, *The Complete Short Stories*, ed. Nahum Glatzer, New York 1971, pp. 473–474 (translation modified)[1]

Kafka's Rescue of the Sirens

Franz Kafka's diary entry of 23 October 1917, which Max Brod would later name "The Silence of the Sirens," has the heading "Proof that inadequate, even childish measures may serve to rescue one from peril." Since it originated at the same time as Kafka's intensive study of Hebrew, this "proof" or demonstration has been compared to Jewish midrash[2]: Talmudic commentary focused on the underlying meaning of a Biblical text, initially developed during the first millennium of Christianity, its form often consisting of a proposition brought together with a canonic citation and the commentary. One author has observed that the goal of midrash is to maintain and shore up the authority of a particular value system, however much artifice this involves. Within the logic of that analogy with midrash, the *Odyssey* would gain the status of a holy text. And in fact, starting in the eighteenth century at the latest, biblical criticism and "Homeric" questions were indissolubly intertwined: in *De sacra poesi hebrærorum* (1753), for example, Bishop Robert Lowth, Oxford Professor of Poetry, instructed readers to read the bible in the way Homer was read, as poetry. Inversely, in his founding of modern Homer philology, the classicist Friedrich August Wolf took up crucial ideas from Johann Gottfried Eichhorn's "Historical-Critical Introduction to the Old Testament."[3]

Up to at least Kafka, the *Odyssey*, that putative "bible of the Greeks,"[4] has served as a pre-text inviting steady interpretation. But in the case of Kafka's text we have a markedly non-scholastic form of interpretation: no textual scholar is at work here, inspecting tradition critically and weighing its variants, sanctioning or annulling them, according to degrees of originality and corruption. Instead, everything appearing to stem from tradition is taken up in one and the same re-narration. The fragments of the putatively transmitted tradition point less to deviations from some certified primal text. Rather, in the course of a "hy-

1 While included in a standard English-language collection of Kafka's complete short stories, this text is actually a diary entry. Citations henceforth without bibliographical reference.
2 Mosès 1987, p. 58.
3 Johann Gottfried Eichhorn, *Historisch-Kritischer Einleitung ins Alte Testament* (1780–83). See Ballabriga 1998, p. 230, Finsler 1912, p. 357, Buschmeier 2008, p. 102.
4 See Burkert 2001, p. 198.

pothetical narrative,"[5] they attest to the impenetrable complexity of the reported event itself: for its *possible* grounds and ramifications. In case here "tradition" is meant to be venerated, then evidently not through first invoking then glossing a canonic citation. The narrator has no recourse whatsoever to Homer's canonic pre-text. This appears lost or at least peripheral. In its place, we have unmistakably corrupt variants.

If we really wish to compare Kafka's text with midrash, then we should understand it as ludic, if not ironic proof that no sacred text can stem from the sirens, or put otherwise: that sacred and epic texts offer radically different forms of transmission. Within this very much stylized (or at least belabored) juxtaposition, a sacred text has, innately, inviolable authority, imposing the obligation of enduring renewal of its value system, so that the inexhaustible meaning of God's word is certified in it and through it. Kafka's text invokes no discourse whose meaning invites plumbing—not even the sirens' discourse attested to in Homer, let alone that delivered by the one and only God. Instead, what Kafka addresses is sensuality and its withdrawal: a mythic song's notorious seductive power and its abrogation; how Odysseus possibly interprets the song's absence; and finally, how the sirens *possibly* interpret Odysseus' *possible* interpretation. Kafka's narrator is no authority, either philological or religious, who can break off the endless play of mutual commentaries to endow it with the meaning of an existing order of values. What precedes this play is only the surfacing or, precisely, absence of that song from which poetry, as the tradition asserts, first came.

This radical difference, as staged by Kafka, between religious and poetic transmission—and thus also between meaning-founding, interpretive-teleological narration on the one hand and wily epic storytelling oriented toward sensual appearance on the other hand[6]–has been connected with the difference between the Semitic consonantal alphabet and the vocalized alphabet of the Greeks: with the possibility to for the first time mark down vowels and syllabic lengths.[7]

---

5 See Schmitz-Emans 2003, p. 363, and in general Menke 1993, pp. 134–162.

6 On this basic distinction see Auerbach 1974, chapter 1, passim. See in this respect Schestag 2009, p. 292 (however solid or not Schestag's assumptions may be concerning the existence and meaning of an "Elohist" biblical strand): "Because in the Homeric epic everything is said and expressed, nothing is absent, precisely the following absence or nothing is absent: that something is not said. What is *entirely absent* from Homer is the appearance of a sign that lacks any explanation. In the report of the Elohists, that sign is *God*."

7 On this controversial supposition and on Auerbach's putative neglect of the writing technologies at the basis of both traditions, see Powell 1997, p. 11, passim.

According to this (to say the least) problematic contradistinction, that Kafka may possibly have here on some level taken up and perhaps attacked, consonantal writing inherently demands contextual and semantic knowledge, and texts such as those of the Hebrews can become holy texts by declaring their own writing-determined need for supplementation as the venerability of an inexhaustible divine word. By contrast, within such an interpretive schema the Greek vocalic alphabet is thought of as a writing system in which what is spoken and even song can be reproduced without knowledge of context and putative meanings.[8] Already on the level of media, interpretive "recourse," the "return" to a pretext whose sacrality is meant to be infinitely confirmed in the interpretive reproduction of authority and value, is opposed to what since Homer serves as poetic "recursion": a return to one's own "origin" that can be considered immemorial and mythical but also simply song. In order for this origin to arrive at language and writing, both must themselves become poetic, hence first of all generate their own origin. Seen in this way, *poiēsis* is "autopoiesis."

If Kafka's text appears to interpret the *Odyssey* in the style of midrash, perhaps, then, only a ruse is involved here: the ruse of irritating one sort of text with the other—and in the end unhinging the opposition between sacred and poetic texts. That his "demonstration" only glosses Homer's episode of the Sirens in a highly contrived way that is also hardly true to the text could thus reflect a method. It is the case that Odysseus "had himself forged unto the mast," Kafka here raising Homer's ante. But his companions are worth no further mention. Rather, this lonely Odysseus lays claim to both chains and wax for himself alone, two devices we can understand as those of modern literature: It is the case that they have no effect on "those whom the Sirens allured even from a great distance," those who consequently could not preserve the distance necessary for aesthetic enjoyment and corresponding medial usage. But at the same time, in the right hands they see to the lonely quiet and receptive passivity of the reading act. They thus simultaneously make possible distance and entrapment: in hearing, even if only silence is perceived here; and in communication, even if this is conveyed as non-communication.[9]

But in Kafka's apocryphal text, these devices appear both all too well-known and ineffective. As if the Sirens know they have become mere literature through repeated finessing, in the face of all proven ruses of listeners or rather readers

---

8 According to Friedrich Kittler because it can note vocalic sounds, syllabic lengths, and, in case its sign is used both ordinally and cardinally, even pitch, the vocal alphabet first generated a knowledge of language and poetry, music and higher mathematics. See Kittler 2006, p. 208.
9 On the paradox of non-communication and the "abysmal openness" of silence, see Döring 2008, p. 63.

they deploy not the weapon of song but the "still more fatal" one of silence. Beyond "human understanding" here is one particular question: whether on the one hand Kafka's Odysseus simply overlooked, failed to hear, and thus left behind the Sirens in excessive pleasure at his own enchantment, hence in a kind of self-forgetting auto-affection; or on the other hand finally knew how to save himself in an act of guile or "pretense"—an act revealing itself as comedy through his suddenly disinterested turn away from the *Seirenes* (the "enchanting or entrapping ones"). In the end the question leads straight to those blind spots, the paradoxes and points of indecision, generally emerging in the thematic realm of (not only) poetic origin.

Otherwise than is the case with Homer, Kafka does not only treat the sound of the Sirens, their voice and singing,[10] which after all limits their seductive power to the radius of their audibility. Beyond that, he speaks of their appearance, which, only seen "fleetingly" by Odysseus before fixing "his gaze on the distance," manifestly feeds fully on the "look" offered by his captivation: at the moment of his turn away from them, they are "lovelier than ever." But in Kafka's parable it is not the sirens—appearing in undetermined number—who possess a true capacity for seduction, but Odysseus's "great eyes" alone, their duality corresponding to the pair of sirens referred to in Homer. In the end it is these "songstresses," now only the enchanting ones by name, who pine for the "radiance" falling from those eyes. What entices is no longer beguiling song but beautiful semblance. And to the extent that this consists of an act of perception alone, in place of a previous constant seductive undertow, what now emerges is labile, merely momentary pretense.

Homer's fabulous creatures know how to seduce not only through their song but also through their promise to disclose their "greater knowledge" (see *Odyssey* 12, 189–191). Kafka's own, silent sirens no longer have such knowledge to impart—or at least they have recognized that their knowledge has little seductive power. Just as the semblance of beauty is tied to Odysseus's rejection, the semblance of truth is tied to his guile. But if Odysseus actually is "such a fox," so infinitely wily that he so to speak drives not only the sirens' language off course but also their silence, then simply describing his enlightening, knowledge-imparting function would be putting it too mildly. He not only recognizes what others consider the sirens' song as their silence, in order to then save himself through simulation of enchanted hearing. Because he is "so full of guile" that

10 Homer 2018, book 12, lines 45 f. Henceforth all *Odyssey* citations will be from this translation by Emily Wilson, with book number followed sequentially by line number(s) both parenthetically in the main text and in the notes.

even former muses like the sirens or an authorial entity like the goddess of fate cannot "pierce his armor," we cannot determine if Odysseus ever really harbored any kind of guile, or rather that was only imputed to him—or if his guile consisted precisely in its endless interpretation. Odysseus counters the strategy of semblance that the sirens follow in their silence with the semblance of a strategy.

Because despite all silence and non-communication, semblance here emerges from an observation of observation, what is at work is a recursive pseudo-process—one not only presenting the emergence of the semblance but that emerges from the presentation as such. Kafka's hypothetical narration constructs a mesh of deceptive maneuvers, of fictions and "as ifs." For readers, it thus initially only marks "*points of transition* in thinking" entrapping them ever more deeply in the text. But where the narration ends up leading is past mere deception and "as ifs," into what theory of fiction terms an "as" structure—into that semblance as reality termed "fictionality," by contrast with the fictive and invented.[11] The turn away from the sirens is a turn from a beautiful song in any case no longer in tune. And tied to this turn away from "the poetic itself" is a turn toward fictionality, to its "inadequate, even childish devices," not only offering protection from the sirens but saving the sirens themselves from their own silence. Namely, their song had not drowned completely but has become a "literary text." To be sure, it no longer attests to omnipotence, but only to conjectural knowledge—that of the truth of semblance. And with that status, it may incite all sorts of silent activities, such as reading—but least of all it will incite song.

Together with Friedrich Kittler and looking back at the great epic song of "Being, as once rising in Greece's dawn," we may choose to consider what remains of "Western literature," as a collection of mere "prose scraps": "a bleak empty writing" entitled "fiction."[12] But with this writing, original song, "song of origin," has not only drowned. In "literary" recursions of the *Odyssey*, recourse to *the* song and *the* "poetic origin" becomes a poetological principle perceiving the possibility of continued backward reference precisely in silence and emptiness. Whether, for instance, the sirens fell into silence or even perished after Odysseus left them behind lies outside the epic horizon, which after all is circumscribed by the hero's progress. But as vase motifs from the sixth century BC indicate, the question was posed not later than with the so-called Pisistratic recension, as an original text was thought to be available.[13] Taking recursive detours, literature becomes a means to rescue the poetic. For just as through its

---

11 Vaihinger 2007, p. 175; Hamburger 1993, pp. 56–59.

12 Kittler 2006, pp. 121, 156.

13 See Heubeck 1988 ff., vol. 2, pp. 128 f.; Burkert 2001, pp. 211 f.

"auto-poiesis," Homeric song salvages "original" poetry, Kafka's prose marks a recursive emergence of Homeric and post-Homeric poetry, albeit only as its falling silent Homer's epic *is* precisely the song whose story it tells—and in the same way literature is silent about the silence it is meant to make speak as interpretation.

In presenting its own interpretation and pushing it toward aporia, Kafka's text closes itself off into a hermetic pseudo-event—but one all the more alluring for that reason. The text maintains that event "as a sort of shield" against all regular interpretation, in order to entice readers to just that.[14] With this metapoetic phrase Kafka's apocryphal transmission in the end thus centers on the mythological, historical, and philological knowledge that seems to be at the disposal of Kafka's narrator and that was always generated from Homeric song. For from the start, "Homer" has served as a father figure or at least a first reference point for Western erudition. But if the song of the sirens forms the poetological center of the *Odyssey*, not only poets and seamen but also textual scholars have had to shield themselves from its powers of temptation. This was also the case in Kafka's time, with questions of Homeric origin still holding scholars in suspense: as we will see, around 1900, speculation was that the birth of poetry and simultaneously that of seafaring may have coincided with the turn to and away from the sirens.

Without a doubt, since antiquity and Early Modern Homer commentators, hardly any scholars have been capable of drawing near to the putative origins of Western poetry in the manner of Odysseus in Kafka's *first* version of the story—"in innocent elation over his little stratagem." Nobody will continue to hear pure song where silence has, in effect, long-since prevailed. For that reason, scholars resort to subterfuge: as Odysseus in Kafkas *last* version of the story, they encounter the Sirens with a "pretense" of enchantment. As fascinated as they show themselves, solitarily wrapped up in their texts as they are, we could almost believe that the Sirens actually are singing and the scholars "alone do not hear them." But they digress, "gaze fixed on the distance," and speak of what is remote and alien, as soon as the Sirens are "nearest to them." Possibly they have "really noticed that the Sirens are silent." For scholars as well, myth's most horrific temptation stems from its silence. Once they succumb to it, a desire for origin may gain strong impetus—an illusory belief to have found the source of poetry, when in reality only their own acts of perception have been in play. Or else, they are then overcome by an "exaltation bearing down everything before it"—namely the certainty of dealing with fiction, with an artistic "water-panto-

14 See Menke 2000, pp. 644 f.

mime" where, it might be said, the Sirens' silence could not go unheard.[15] In order to master this double danger, even scholars need a "shield" of enchantment, allowing observation from both inside and outside, methodological insight and delusion.

Hence if Kafka's text demonstrates anything for scholarship, then it is that nothing is demonstrable here as long as the scholar is not open to total seduction—or else none whatsoever; and that there is no possibility of either complete denial or absolute certainty that something is located "before the text." Scholars only achieve what Odysseus once achieved through such a ruse: to "experience" poetry without succumbing to its myth; to hear the Sirens' song but both assert its silence and themselves before it. And if Kafka's text demonstrates something for literature, then it is that its "childish measures" indeed may serve as rescue—of a tradition whose primal text appears to have always been lost. With its "little stratagem," its entrapment through aesthetic distance and illusory events, literary writing is an exercise not so much of recourse to an origin but rather of recursion consisting of a fruitless search for and simultaneous steady rediscovery of a long-lost beginning. Topoi of transmitted tradition, such as those of the Sirens and Odysseus's ruse, do not attest to a fixed origin but to a wiliness that cannot be overcome: a wiliness involving displacement and dispersal of tradition on the one hand, rescue along circuitous paths on the one hand. The Sirens' message is maintained precisely in the insurmountable distance from an origin, which we can term the song of the Sirens. Kafka demonstrates the impermanence and simultaneous persistence of a tradition. Writing henceforth means, to speak with Blanchot, "to make oneself the echo of what cannot cease speaking—and since it cannot, in order to become its echo I have, in a way, to silence it."[16] Hence the lifebuoy of literature steps alongside the song's sinking. And the possibility of writing opens itself through the song's absence, rather than through revelation of a divine word.

Homer's Poetic Topography

Although linked to fabulous creatures like sirens, poetic enchantment is also a matter of sober practical knowledge. That first and foremost seafaring leads to that enchantment is already clear in Homer's tale. But possibly, the comparison—become a topos after Homer—between poetry and dangerous seafaring re-

15 Kafka 1992, p. 300.
16 Blanchot 1982, p. 27.

veals a recursive commonality: each undertakes a bold passage to its own abyss, its own abysmal locus, in order to repeat a dangerous experience. The sirens' song, writes Blanchot, "was aimed at sailors, men who take risks and feel bold impulses, and it was also a means of navigation: it was a distance, and it revealed the possibility of traveling this distance, of making the song into the movement toward the song, and of making this movement the expression of the greatest desire."[17] Poetry and seafaring had the same origins—at least this is what Homer research claimed around 1900. And what seems to unite the two from the start is their points of reference or topoi.

Homer's *Odyssey* continues to be considered the standard—because most comprehensive—of all pre-Classical sources on Western seafaring. In this text, we find numerous nautical matters terminologically fixed or even documented. For example, *Odyssey* 5, 243–261 offers us a generous (and archeologically confirmed) description of early Greek ship construction, and the first information of any sort on nighttime astronomical navigation, knowledge of rotation of the heavens, and orientation around constellations (lines 272–277).[18] For the description of seafaring experience, for instance of rough seas and storms, and its analogical or metaphoric interpretation, the *Odyssey* remained an obligatory topos-catalog until late antiquity, so that initially rhapsodists, especially the Homerides, then also epic poets, lyric poets, and tragedians presented regular glosses.[19] This commonplace topical material was also collected in the schools of rhetoric as a basis for practice, before Alexandrian librarians brought it together as a poetic heritage for posterity.[20] In the eighteenth century, various philologists were still comparing Homer's "more original" seafaring images with, for instance, Virgil's nautical topoi—this in support of their enthusiasm for rediscovered Homer.[21]

What distinguishes the *Odyssey* vis-à-vis the *Iliad* and its imagery is the existence of numerous loci where "real" and chanted loci are overlayered. For example, Cape Maleas, notorious in antiquity, which as a promontory marked a kind of outer border of geographical describability, precedes the *Odyssey* as a

17 Blanchot 2003, p. 4.

18 On ancient ship building in respect to Homer, see Hausen 1979, pp. 102 f., 125, Göttlicher 1985, p. 47, Mark 2005, pp. 70–96, 185 f. On astronomical navigation in the *Odyssey* see Taylor 1971, pp. 40 f.

19 On maritime images in Homer, especially in respect to emotions, life paths, and political concerns, see Kahlmeyer 1934, pp. 11–45, und Nes 1963, pp. 14, 26, 30. On the "meta-literary use" of this vocabulary, esp in Apollonios of Rhodes, see Rengakos 1994, passim.

20 See Schlaffer 2005, pp. 71, 77.

21 For a prominent example, see Bodmer 1971, pp. 92, 241, 250, 254.

point of divergence between the route of Menelaos and that of Odysseus and thus the narrative course itself.[22] It was no accident that there was a Greek proverb that those who sailed this dangerous cape had to forget their homes.[23] At Cape Maleas with its storms and currents, sailors are not only away from Greek *terra firma* but directly on the high sea, hence driven onto what has no locus. But already in Homer "lose one's place" means lose one's existence, the Cape thus emerging as not only a place of lost locus but also of lost existence. For Odysseus, this loss has become a form of being—the basis for the ambivalence defining both this characterless yet wily and survival-bent hero and the "escape route" traced out by the epic's narrative.

Homer's topoi do not merely refer to a neutral "topography" necessarily serving as the narrative's foundation. Rather, they break new ground in rendering the description of elementary phenomena into a poetical problem. In the case of the *Odyssey*'s descriptions of storms, we find not only recourse to an existing topos but topos-generation. Hence already in the fin-de-siècle, Victor Bérard, classicist, diplomat, and senator in the Jura *département*, pointed to a kind of elementary realism at work in the *Odyssey:* "What we have here is not *the* storm of men of letters, but *a* sea-storm, a storm in the Adriatic."[24] Such topography is not based simply on inspiration by the muses but also, and in particular, on "prosaic fragments"[25]—of the sort Bérard maintained had been conveyed to the *Odyssey*'s author by Phoenician sailors. As even purely philological findings indicate,[26] the *Odyssey* is not rooted on dry land. Consistently, its loci and spatial concepts are rendered dynamic, its objects and personae in a state of movement. The narrative comes to us from the sea, with a view to the coastline: "Interior land only appears as a highly blurred distance."[27]

In a cultural-historical framework, we might tie this poetological observation to a general "deterritorializing" of Greek life marking Homer's starting point: through a sweeping uprooting of the population and the liquidation of traditional landholdings in the Dark Ages, sending a wave of slaves and vagabonds over the entire country; in addition through the establishment of agile mercantile activity, especially on the Asia Minor coast, which developed a maritime trade network and increasingly initiated "colonial" enterprises in—probably still largely

22 See Foulke 1997, pp. 40 – 42.
23 See Braudel 1972, pp. 108 f.
24 Bérard 1902, p. 482. On Homer's verbal shaping of his description of wind, for instance through shifts of tempo and enjambment, see Heubeck 1988 ff., vol. 1, p. 280.
25 Bérard 1971, vol. 1, p. 312.
26 On the shift of style vis-à-vis the *Iliad*, see for example Kahlmeyer 1934, p. IV, passim.
27 Bérard 1971, vol. 3, p. 423.

unknown—western Mediterranean area.[28] In respect to this context, Hegel already speaks not only of a historical "condition of turmoil, of insecurity" at this time, but also maintains a situation-determined instable form of existence on the part of the Hellenes: "The nature of their land brought them to this amphibious existence and allowed them to float upon the waves."[29]

Plato saw a danger of fundamental uprooting, of one's own polis being driven far off course (through wiliness), precisely in this turn from the terrestrial *nomos* and its ontological, ethical, and political commitments: in a turn to "becoming"—to an amphibian existence—that was "oblivious to being," as expressed in piracy, maritime trade, and activity of the Thalassocrats.[30] In any event alienation from and even a state of enmity with the sea was not only articulated in the Classical epoch but already in Homer's time, as testified to in the Odyssey's deterritorialized disposition.[31] Along the path of flight of Odysseus *polytropos*, the highly adaptive, widely traveled and wily hero, *tropus* is intertwined with literary topos, so that each experiences slippage; together they pave the way for odyssey, wandering, steady loss of orientation. This unfolds in a largely undescribed or simply indescribable region: the sea in the unexplored west or simply the open waters, far from any visible landmarks. If we wish to understand the *Odyssey* as a topographical text, it initially seems a paradoxical one, aimed at describing the loci of what has none.

This has, to be sure, not prevented numerous efforts starting already in antiquity, to reconstruct Odysseus's travels corresponding to both the state of geographical knowledge and particular genealogical claims. First Hesiod, then Thucydides and Plato are said to have located his route as between western Italy, North Africa, and the Atlantic estuary.[32] Roman foundational legends claimed the coast of Campania as the Sirens' home. At times the route was shifted to the Black Sea or even the Atlantic.[33] Although the Alexandrian geographer Eratosthenes offered the observation that the *Odyssey*'s original locations would be identified as soon as the shoemaker was found who stitched the bag containing

28 See Braudel 2001, pp. 247 f.

29 Hegel 1914, p. 237.

30 See for example Plato's Atlantis myth in *Critias*, 106a–121c, and *Timaeus*, 17a–27b. See also in this regard Vidal-Naquet 2007, pp. 18 ff.

31 For such an attitude toward the maritime environment see Hesiod, *Work and Days*, verses 236–247, 366 f., 618 ff., and Lesky 1947, pp. 25 f., 33–35.

32 See Wolf and Wolf 1983, pp. 145–147.

33 Ibid.

Aeolus's winds,[34] the "true" route of the wily one remained an object of frequent speculation in the Middle Ages and after.

The series of (still ongoing) efforts at cartographical representation, consistently shored up with all kinds of mythological, cultural, and linguistic theorizing, was opened with Abraham Ortelius's *Ulyssis Errores* (1597). With the "Homerische Welt-Tafel," a folding map attached to his *Odyssey* translation starting in 1802,[35] Johann Heinrich Voß was the first author who avoided simply transferring the poetic topoi to a state-of-the-art map. Rather, he projected them within the configuration, and through indications of semantic distance he saw present in the epic verses themselves (including in their hexameter form). In opposition to advocates of "realism," this marked the emergence of a tradition of "imaginary cartography." For William Gladstone (the later British Prime Minister), not only maps of the underworld episode, hence maps of the "outer circle of the geography of the *Odyssey*," but all Homeric maps were meant to deliberately limit themselves to a "fictitious drawing" of Homer's "pictures of the imagination."[36] There seemed in fact to be good reasons for this, as in the time of the *Odyssey*'s inscription—whatever the rough date actually was—there were no nautical maps, navigation being undertaken on the basis of custom and experience alone.[37]

This notwithstanding, in 1902 Victor Bérard offered an historical and aesthetic counterargument: "The *Odyssey* seems to me to resemble a Phoenician periplus (from Sidon, Carthage, or elsewhere) transposed into Greek verses and poetic legend." For Bérard, "anthropomorphic personification of objects, humanization of natural forces, *Hellenization* of matter"[38] were, on the one hand, solid indications that Homer had already fulfilled the basic requirement for all cartography: a kind of birds-eye perception of landscape formation that —just as the geographer Strabo later called for—makes either a geometric or organic imprint (for instance, Sicily's triangle and the leaf-like Peloponnese, respectively) on views of nature.[39] On the other hand, Bérard disputed the idea that the *Odyssey*'s author had to resort to mere imagination because of the lack of cartographic visualization techniques: "The element of fantasy and the imagination is limited here. Arrangement and logic are the poet's principal con-

34 According to Strabo, *Geographica* I. 2, 15 (C. 22). See also Andreae 1982, p. 10.
35 See the folding map (with drawing by Christoph Friedrich Hellwag) in Homer 1802, vol. 3.
36 Gladstone 1858, I., pp. 219, 223.
37 See Mark 2005, pp. 139 f., 143.
38 Bérard 1902, p. 4.
39 Strabo, *Geographica*, vol. 2. 1, 30 (C. 83). See Wolf 1997, pp. 21, 66.

tribution," precisely when he works on the basis of Phoenician sailing handbooks.[40]

As something like one-dimensional maps, these handbooks were organized along the lines of a list, Bérard's thesis thus having a foundation in classical philology, since what we find in Homer, as Tadeusz Stefan Zieliński speculated at the turn of the twentieth century, is an absence of any representation of simultaneous action.[41] (This forms a contrast to, for instance, the *Argonautica* of Apollonios of Rhodes (circa 250 BC.) which already seems influenced by the first cartographic efforts since the sixth century BC.) To the extent scene and action do not change entirely, Homer's narrative thus consistently traverses a guiding thread of sequential orientation points.[42] For Bérard, whether the places Homer names are real or fictive is simply the wrong question. Instead of posing it, he pursues the orientation procedure from a perspective at once epistemological and aesthetic and thus reconstructs the emergence of a Homeric topography from the continuous balancing of (Semitic) text-transmission, nautical experience (of Greeks and Semites), and creative use of language and writing (here on the part of the Greeks).

Either inserting the *Odyssey* into a realistic topography or depicting its imaginary map are pointless. What counts is gaining a sense of the extent to which *Homer* localized Odysseus and endowed his route with a topography. At the start of a developmental sequence extending from indistinct toponyms in undescribed maritime space to a cartographic visualization of the world (albeit not of the sea itself), two cultures or ethnic entities meet: Greeks (for Bérard in the shape of Homer) and Phoenicians (in the shape of Odysseus); in terms of the order of things, the imaginative loci of the epic and the nautical descriptions in the *periploi:* sailing handbooks whose name already underscores the intent of aiding in a successful "voyage around" the dangerous region of the open or unknown sea. To this end they point to markers when making an approach to land, to special maritime and meteorological phenomena, and to navigable coasts and anchoring berths.

The mid-fourth-century BC handbook entitled *Periplus* by the geographer Skylax the Younger was not the first work of this sort in Greek literature. This honor should be accorded Circe's nautical instructions in the *Odyssey,* allowing the epic hero's long and adventurous return home, his *nostos,* in the first place. Probably such handbooks had circulated for a long time among Egyptians and

40 Bérard 1902, p. 295.
41 See Zielinski 1899–1901.
42 See Grethlein 2009, pp. 275–291. On the relation between narrative composition, catalog, and periplus in Apollonius of Rhodes see Scherer 2006, pp. 134, 198.

Phoenicians before the Greeks founded their own tradition (connected or not with their vocalized alphabet). By the fourth century, most sailors in the region were sufficiently alphabetized to use written sailing instructions. Alongside the nautical information, including specifications of distance in either stages or duration, the *periploi* soon also contained descriptions of loci on land and sea, sanctuaries, and defense works, and finally information on the customs, political circumstances, and trading practices in various areas,[43] the texts thus occasionally developing into regular descriptions of countries in the sailing region. Furthermore, some of the *periploi* authors developed literary ambition: in order to incisively characterize certain types of wind, certain rocks, currents, they had recourse to Homer's poetic topoi, their simple handbooks thus becoming a distinct literary genre—the marine *periegesis* or guidebook, written in hexameters.[44]

For Bérard, travel and adventure stories of whatever sort or epoch are basically poetic recursions of *periploi*,[45] deemed media of narration and fictionality allowing sight and experience "through the eyes of others."[46] And since such experience is not only internal but also unfolds in space, *periploi* are also exemplary media of discovery. When these texts necessarily generate the procedure of bringing loci into existence, they stand at the beginning of both poetry and discovery. Without *periploi*, then, "Homer" (at least the Odyssean Homer) would not have existed, Greek voyages into the unknown not taken place. Bérard concedes that the classical world was aware of this fact, his own work thus merely constituting recourse to "one or two phrases from Strabo." For Strabo—by contrast with Eratosthenes—appreciated the poetic aspects of geography and in addition had argued that Homer had to have used Phoenician *periploi*.[47]

Recursive Modes of Writing

Strabo already touches on what can be termed *recursion* in the context of the nautical-poetic experience presented in the *Odyssey* and its later literary follow-ups. By recursion, I mean a circular movement enabling not only a return to the same but, beyond that, variation through self-referentiality.[48] Recursion

43 See Taylor 1971, p. 57; Hertel 1990, pp. 23f.; Casson 1979, p. 195; and Nordenskiöld 1897, p. 3.
44 See Güngerich 1950, p. 20; Casey 2002, p. 184; Kretschmer 1909, p. 150; and Dilke 1985, p. 143.
45 See Bérard 1971., vol. 3, p. 415.
46 Ibid., vol. 1, p. 309.
47 Bérard 1902, p. 3. On the periplus as a medium of discovery, see Bérard 1971, vol. 4, p. 484.
48 For Odysseus's desire not to repeat what has already been well told, see *Odyssey*, 12, 452f.

is the reapplication of a processing provision (the production and setting into relationship of loci) to a variable that (initially in the form of toponyms, then of *periploi*, and finally of imaginative literature) is already output of this provision. The variable-value (the degree of topographic "poeticity") alters with each iteration—nautical, literary, philological—of this loop, producing difference alongside repetition.[49] Recursions are processes of discovery, because steered by rules they lead to contingency. Precisely putative origins are here repeatedly discovered anew.

The recursive principle is presented with special clarity in the *Odyssey*'s episode of the Sirens. For if Odysseus here draws close to the mythic principle of poetry itself, the "beguiling" (and entrapping) ones promise nothing other than to intone songs from Homer's first epic (*Odyssey* 12, 189 f.). The *Odyssey* is a recursion to the *Iliad,*[50] while the *Iliad* itself emerged from a zone of art that here is articulated in an epic-musical framework and with a prospect of "omniscience." In this origin, song, poetry, and knowledge intersect; Odysseus has not arrived there haphazardly, but through carrying out Circe's sailing instructions (12, 47 ff.). Guile thus not only needs pure presence of mind but the *periploi* as well, or, as an "original" epic needs to claim, the presence of the gods.

In any case, Homer could be considered the starting point for the development of various literary- and intellectual-historical streams not only describing processes of transmission and adoption but also following the principle of recursion. The *Argonautica* of Apollonius of Rhodes, for example, has always been understood as a supplement of the *Odyssey* now, however, opening up the previously unknown eastern part of Greece while also referring back to texts such as the *Periplus* of Skylax. But Circe's sailing instruction already presupposes Jason's adventure (12, 69 f.)—in which, we are reminded in this passage, the Argo (what the Hellenistic scholar Eratosthenes will describe as the very first ship[51]), has participated—and with it, although this is unstated, Odysseus' father Laertes. When, as we read in Apollonios, the Argonauts were threatened with denial of "sweet return" by the "lily-like voice" of the Sirens, Orpheus "overcame" their song with sounds from his "Bistonian lyre"[52]—a contest signifying premature renunciation of song and less guile than has Odysseus. Furthermore, the *Argonautica* reveals recursion to the *Odyssey* in its retrospectively new definition of its origins: the Black Sea instead of the Mediterranean, Jason instead of Odys-

49 On this concept and the recursive recursion, see Ernst 2010, p. 185.

50 See also Kittler 2012, p. 415.

51 See Eratosthenes, *Catasterismorum fragmenta Vaticana* 35; Catullus 64, 11 ff.

52 Apollonius of Rhodes 1912, book 4, lines 893 ff., pp. 355–357.

seus, disturbing sounds instead of song. To this we can add that within Greek tradition the material of the Argonaut legend is older than that of any inherited epic. Because no Greek alphabet or, to say it in perhaps less mystifying fashion, no Greek scribe could save them, the corresponding pre-Homeric songs were lost to time.[53] For its part, through the pre-Homeric material shimmering through its narrative, the *Argonautica* offers us insight into a tradition that may well have furnished a foundation for the *Odyssey:* the tradition of the "helper" fable, recounting stories of rescue—by the hero or by poetry itself.[54]

It is naturally the case that when it comes to substantive and motivic models for the *Odyssey,* researchers have moved past the field of Greek transmission, seeing especially the Egyptian Middle Kingdom "Tale of the Shipwrecked Sailor" as a source of, above all, the Homeric epic's Phaiacian segment (books 6–8). Already in the Egyptian tale, we are offered, in the style of a "minor Odyssey" and as a story in a story in a story, the account of the sole survivor of a failed sea-venture, who, clinging to a "piece of wood," is thrown by a wave onto an island. He is discovered there by the island's godlike ruler—he and his family members have the form of large serpents—who eventually addresses him as follows: "...a ship shall come from your land with sailors, and you shall leave with them and go to your country, and you shall die in your town....if you are strong, and if your heart waits patiently, you shall press your infants to your bosom and embrace your wife. You shall return to your house which is full of all good things, you shall see your land, where you shall dwell in the midst of your kindred."[55]

This narrative is tied to the Egyptian sense of divinity circa 2000 BC. Its eschatological horizon can only be connected to Homeric prophecy with difficulty (although we do have the serpent-omen in Iliad 2, with Virgil picking up on the theme in the consistent, in part eschatological, serpent imagery of the *Aeneid*). But because the text's substantive arrangement is in any case secondary, since historically contingent, Bérard here sees the decisive link with the *Odyssey* in the figure of the *nostos* and its topographical implications. The narrative of homecoming after a marine voyage, he argues, points unmistakable to the use of *periploi*, whose existence in the Middle Kingdom is attested to by various bas-relief inscriptions in Egyptian temples. Seen in this way, the Phoenicians were already eager students of Egyptian navigation before the Greeks made

53 See Montanari 2010, p. 194.

54 See Meuli 1921, pp. 112, 116–118; Danek 1998, p. 252. For a critique of Meuli's argument see Radermacher 1938, pp. 208–217, 224.

55 Flinders Petrie, transl., 1914, p. 43, translation modified. On the context and interpretation see Lanczkowski 1953, pp. 362f., 367–369; Simpson 1984, col. 619f.; Hölscher 1990, p. 110.

Phoenician *periploi* into the starting point of their own poetry.[56] For Bérard, even modern "mirrors of the sea"—at least since the seventeenth century, a term for navigational handbooks[57]—have their source in the Italian navigational manuals, with accompanying charts, known as portolans, which he in turn sees, erroneously, as continuations of the ancient *periploi.*[58] For Bérard, then, in order to read the *Odyssey* properly, it should be compared less with literature of similar contents than with practical nautical literature of whatever epoch. In the end the work is part of a "series of nautical instructions."[59]

This has consequences for the "Homer" question—as grappled with historically, that of the *Odyssey*-poet's originality; for now, not only Greek bards (*aioidoi*) seem given voice with each line and word, but also Phoenician and even Egyptian seafarers. "The poet invents nothing," writes Bérard. "In effect he animates and arranges. Every journey is first of all a string of proper names." And earlier: "Each adventure has its very impetus in a mobilization of toponymy."[60] The *Odyssey*-author's power of imagination is thus no free-floating capacity, rather being initiated on the level of *periploi*, of places and their naming. It is further active in the life given these names in the form of acting persons, in the arrangement of corresponding events, and in the broad narrative disposition, *nostos* and nautical circuit.[61] Even if, as Bérard concedes, toponymy has been subject to scholarly skepticism because of unfounded etymological speculation, it nevertheless does allow a tracking down of loci through their grammatical and lexical elements. (Bérard supported his thesis that the *Odyssey* had its origins in Phoenician nautical handbooks through etymological studies aimed at demonstrating the Hebrew origins of many Homeric toponyms.) Precisely puns,

56 See Bérard 1971, vol. 2, p. 438; vol. 3., p. 35; vol. 4, p. 498.

57 See, for example, T'eerste deel vande Spieghel der zeevaerdt, van de navigatie der Westersche zee, innehoudende alle de custen van Vranckrijck, Spaingen ende 't principaelste deel van Engelandt, in diversche zee caerten begrepen, Leiden 1584. Digitilized at http://bdh-rd.bne.es/details.vm?o=&w=waghenaer&f=&g=load&g=work&lang=es&view=main&s=1. A facsimile edition has been published in the series Theatrum Orbis Terrarum, Amsterdam 1964 (commentary by R. A. Skelton). Also: Der ... Theil dess Spiegels der Seefahrt: von Navigation des Occidentischen Meers, oder der Westseen (...v.d. Noerdschen und Orientischen Schiffahrt), 2 vols., Amsterdam: 1589. Digitalized edition of the Düsseldorf University and State Library, http://digital.ub.uni-duesseldorf.de/urn/urn:nbn:de:hbz:061:1-32777.

58 On portolans see further chapter 3, where I also outline the basic difference between portolans and *periploi*.

59 See Bérard 1902, pp. 56 f.

60 Bérard 1971, vol. 3, p. 414, p. 164.

61 See ibid., vol. 4, p. 490; Bérard 1902, pp. 18 f., 584.

which in Classical Greek still imply a hidden or displaced truth, refer to a linguistic and poetic path of transmission—however winding or devious.[62]

For example on a purely mnemo-technical level, the *Odyssey*'s hexameter could have served to instill certain routines upon toponyms and their sequence. But metrical requirements also contributed to disperse signifiers of real places within the poetic material. For example, Cephalonia appears to have been cut apart on purely poetic grounds, an additional "long island" emerging from the "long *Same*." And even the ethnic designation of Homeric heroes becomes fluid when metrics demands it at one or another place, Walter Burkert thus observing that "the Greeks are named *Argeioi, Danaoi, Achaioi*, depending on whether the beginning, middle caesura, or end of the verse is involved."[63] In the *Odyssey* every Greek is consequently a *polytropos*. Nevertheless, its topoi are so precise that the position of "purely poetically" designated islands has been extrapolated from for instance Calypso's sailing instructions (5, 271 ff.) and the—astronomically calculable—constellations in Homer's time.[64]

Because descriptions of loci from a terrestrial perspective, those like the *periploi* (conceived for "smooth spaces" like the sea and desert[65]), and those emerging in a poetic framework involve different topographical systems, Bérard calls for a study of their relationship as a kind of "topology." If supported by disciplines such as paleontology, linguistics, and anthropology, precisely this topological approach, he argues, could advance to a comprehensive science of human communities, focused on their material and cultural dimensions: to an archeology of cultural origins.[66] "So-called poetical explanations" would by no means suffice to understand how every individual verse of the *Odyssey* is interconnected with a "site detail," how toponymy and topography here meet, and how at such points of encounter we have an overlayering of nautical experience and poetic procedures.[67] "Do we need to be reminded that not only the vision of details but also the groupings and ensembles cannot be explained without turning to the customs and writings of navigators?" asks Bérard, then immediately calling for nothing less than a basic mentality-historiographical shift in research

---

62 Ibid. p. 5; Bérard 1971, vol. 4, p. 500. On punning in Homeric Greek see Heubeck 1988 ff., vol. 3, p. 91.

63 Burkert 2008, pp. 15, 22.

64 See Hennig 1930, pp. 509, 546.

65 On the desert and corresponding Phoenician *periploi* see Bérard 1971, vol 4., pp. 496 f.

66 See Bérard 1902, pp. 8 f., 13, 52.

67 Bérard 1971, vol. 1, p. 335; vol. 4, p. 478.

on Homer: "We need to turn to the sea and the customs of sailors to perceive, beneath the fabulous adventures, the real life that produced them."[68]

In order to identify this "real life" and scrutinize the topological convergence of nautical and poetic realms, at the beginning of the twentieth century Bérard visited the *Odyssey*'s stations to the extent he had managed to localize them; in 1933 his "Odyssey album," including numerous photographs, would posthumously appear.[69] An entire generation of Homer cartographers, representing both "realistic" and "imaginary" schools, would offer no arguments against Bérard's findings, the scholar's eccentric field-research having simply confirmed his original hypothesis that "all the descriptions [in the *Odyssey*] conform to tangible reality, to scientific and experimental truth."[70] Bérard's idea of the simultaneous origin of nautical and poetical experience had in the end, he believed, been experimentally validated. To be sure the experimental media had been limited to charts and photographs (alongside ships and texts, themselves part of the phenomenon being examined). Optical media at hand, Bérard had thus "demonstrated" the interplay between *periplus* and poetry mainly from its imaginary side—on the non-solid basis of maritime perspectives.

The medium of Greek poetry itself, the vocalized alphabet, was not an integral part of Bérard's experimental order. A century later, after Bérard's research had been nearly forgotten, this medium was placed at the center of what was presented as experimental "evidence at original sites"[71] – and not only because acoustic equipment was now available adequate for the relevant frequency-range. Namely, several Homer researchers had increasingly turned from nautical media to the Greek writing system, in order to arrive at new answers to old questions—including those of authorship and originality—regarding the Homeric poems. Starting with the paradox in play since Friedrich August Wolf that with Homer as the "first and greatest of all poets" we also possibly have a poet without the art of writing,[72] researchers (especially starting with the work of Wade-Gery in 1952), have moved toward a more precise dating and functional description of the Greek alphabet.[73]

---

68 Ibid., p. 481.

69 Bérard 1933.

70 Bérard 1902, p. 582. On Bérard's influence on research see Wolf 1983, pp. 173 f.

71 On the effort by Friedrich Kittler and Wolfgang Ernst in April 2004 to demonstrate, at the "Sirens' island" of Gallo Lungo, that upon a passing ship, only vowels and thus simply "song" can be heard, no semantic units, see Kittler 2006, pp. 57 f.

72 See for example Wolf 1908, pp. 92, 127, 138.

73 See Wade-Gery 1952, pp. 9–14, 38–41. On the controversial idea that Arameans introduced the vocalized alphabet, see Malkin 1998, p. 264.

Relatively recently, Barry Powell addressed and concretized the thesis that this alphabet was developed for the specific purpose of recording the Homeric epics and was diffused with them. In Powell's view, there was an "urtext"—albeit one inaccessible to Alexandrian philologists—that Homer dictated to a so-called adapter on the island of Euboea, which around 800 BC was a thriving center of culture and trade. For his part, this adapter may plausibly have been "informed" by a bilingual Semite—perhaps a sailor, to harmonize this scenario with Bérard.[74] According to this theory, transcripts of this urtext initially circulated on Euboea, residents then conveying them as "adventurers" to the Italic west—for example the Pithekoussai settlement.[75] The new writing in a vocalic alphabetic medium, Powell argues, allowed the first comprehensive notation of spoken language. Beyond that, it promoted reflexive awareness of language. The *Odyssey* bears witness to it in the reflexive discursive artful guile that Odysseus accounts for as *utis*, and in a presumed knowledge of the "fictional" impact of speech recited as poetry.[76] Finally, the vocalic alphabet made poetic authorship possible: "poetry" would no longer only be orally composed (as by the *aoidoi*) and memorized through formulae but composed as text by means of topoi, in order to then be memorized by literate rhapsodes.[77]

If we follow both this scenario and Bérard's arguments, then in the *Odyssey* poetic and nautical topography, high literature and practical literature, were still interwoven, before song and "scraps of prose" increasingly separated. To be sure, the verbal arts and seafaring, those cultural techniques that separated Greeks from barbarians like the Cyclops in an elementary way (9, 122–130), were mutually connected from the start: the undescribed because unexplored or distant sea forced a *poiēsis* of loci on sailors just as on poets in their topos-grounded work. It is not only that with "Homer" a new descriptive art including exemplary and recursive topoi had been created, an art that in the future would leave its stamp on, precisely, Greek *periploi*. Because the Homeric epic was simultaneously the medium through which the vocalic alphabet, its vehicle, was

---

74 See Powell 2002, pp. 15, 195 f.; Powell 1991, pp. 66 f. See also Latacz 2003, pp. 23, 26 f.

75 See Powell 1991, pp. 232f. As textual evidence suggests, there was initially no interest in spoken language with this specific syntax and rhythms when it came to fulfilling economic and "public" needs. Still, the hypothesized adaptor would have been interested in Homer's complex metrics, for the sake of trying out a vocalization of the Phoenician alphabet. See ibid., pp. 181 f.

76 See Powell 1997, pp. 28 f.; Ballabriga 1998, pp. 229, 231.

77 Kittler argues that poetry was organized through the emergence of the Greek alphabet. Not first some redactor, but Homer himself, he observes, divided the *Odyssey*'s books in line with the 24 letters of the Greek alphabet, defining the books as something like elementary topoi. See Kittler 2006, p. 120.

spread among the populace, through that poetry precisely sailors and migrants learned to read and write. With that ability, they could increasingly use *periploi* for further excursions and colonizing of the unknown—and they could even write *periploi* themselves.

In this respect as well, the outward voyage was not only a matter for merchants but also for poets. Where the Phoenician *periploi* served to designate known places and open up those still unknown, the vocalic alphabetic notation of syntactically and metrically complex language might have offered an even more flexible and precise means for the determination, relativization, and memorization of topoi. At the same time, the poetry taught how to produce and recursively describe topoi in the first place. It is no accident that in the *Odyssey*'s episode of the Sirens song is presented as a locus of boundless knowledge: here the indescribable is described poetically, namely "auto-poietically" rendered into speech. And perhaps just this poetics of discovery explains why from from Bérard to Blanchot the relation between Phoenicians and Greeks, Odysseus and Homer, has repeatedly been understood as one of recursive identity. Already Bérard speaks of "Homer-Ulysses." And Blanchot still treats the *Odyssey* as a model of topographical-recursive poetics: "To hear the Song of the Sirens, he had to stop being Ulysses and become Homer, but it is only in Homer's narrative that the actual meeting occurs in which Ulysses becomes the one who enters into that relationship with the power of the elements and the voice of the abyss."[78]

The Round Trip's Foundering

Gifted with the art of song and supernatural knowledge, the mythic sirens were once daughters of a muse and a river god. But after they did not intervene in the abduction of their playmate Persephone and were themselves defeated in contest with the muses, they sank to the status of demonic harbingers of Hades.[79] Since then, as Jane Harrison explains, a Siren represents "a kind of evil Muse," "rather of the barren sea than of the clear spring water."[80] When in the *Odyssey* they evoke the grammar, diction, and topoi of the *Iliad*,[81] the Sirens strive to newly occupy or simply replace the locus of the Muses—earlier the very source of poetic inspiration and also their own "origin." At the same time, they in this way promise Odysseus, now more the trickster, a return to the status of Iliadic hero he was

78 Bérard 1902, p. 292; Blanchot 2003, p. 7.

79 See Grant 1987, pp. 374–376; Andresen 1994, vol. 3, p. 2806.

80 Harrison 1882, p. 182.

81 See the detailed analysis in Pucci 1979, pp. 121, 124f.

before being cast off. The landing on the island promises immortality. To be sure, Odysseus chooses another path, homecoming and thus characteristic wiliness allowing him to defeat demonic song as an adventurer, not a warrior, so that song in the end can become poetry. At a distance from the old Muses, the possibility of a new non-heroic epic is opened in this way.[82] Put otherwise, the *Odyssey* founds the epic of another sort of heroism: that of self-rescue in elementary danger.

Said to personify the "dangers of seafaring" or, more ambivalently, *fortuna di mare*,[83] the Sirens, like the *Odyssey*'s other fabulous beings or those of some *periplus*, are affect-laden imaginative entities that according to Bérard signal real danger.[84] Facing them are Odysseus' divine or demonic "helper figures," finding him a saving cloth like Ino-Leucothea or, even more importantly, showing him the correct route like Calypso and Circe—two demigoddesses receiving their instructions from divine figures such as Hermes (5, 28 ff.). But when such decrees from high places do not simply lead to safe harbor but rather straight to the Sirens or even Hades, neither the epic nor Greek seafaring in general can be understood to revolve around safety alone.[85] Both initiate a polymorphic, "polytropic" trial on the perilous but propitious sea.

If the short encounter with the Sirens, those Muses of Hades, is one of the *Odyssey*'s key episodes, then all the more so the *Nekyia* episode, that of Odysseus' voyage to Hades—and this not only because of its length. Rather, the episode marks the most eccentric point, and the turning point, of Odysseus's wanderings. In the *Nekyia*, entrance to the realm of the dead is displaced to Okeanos, the edge of the world beyond the Pillars of Hercules, "the limits of deep-flowing Ocean" (11, 12–13), surrounding the then known Mediterranean world, including its navigable *pontos*, as a circular current. When in the world picture of Anaximander or Hecataeus of Miletus a sea surrounds the earth and not a circular current, Okeanos, the current's personification, thus becoming an ocean,[86] the Odyssean turning point and thus the possibility of any return has disappeared from cosmological view. The high seas have become borderless, hence what in Homer represents a no-man's-land between this world and the afterworld. Those cast away here (like Odysseus already from his grieving family's perspective) are neither dead nor living. What Odysseus has to fear is being swallowed

82 On the religious-historical context to Odysseus' scorning of immortality, see Segal 2004, p. 213.

83 Preller 1860, p. 481. See also Patroni 1891, p. 337: the siren as *divinità del bel tempo.*

84 See Bérard 1902, p. 53. On marine "dangers" and "monsters," see Bérard 1971, vol. 1, p. 494.

85 See Stanford 1954, p. 36; Oswald 1993, p. 73.

86 See Lesky 1947, pp. 58 f., 87.

by the sea without a trace, without a grave, and thus to be lost without a topography of remembrance (5, 311 f.). In the end nameless, in Homer the sea is only "fruitless," an "expanse / of endless salty sea" (5, 83, 100 f.). Historically, Greek timidity vis-à-vis the "Pillars" may explain the fact that with under penalty of drowning the Phoenicians intermittently closed the Strait of Gibraltar and, for the sake of securing their exclusive routes to the ocean, were able to picture a range of insurmountable dangers at sea.[87] But already in Homer the sea itself, in the namelessness and lack of locus threatening to swallow everything, is the ground or "un-ground" of enmity between the Greeks and that same sea.

The wrath of Poseidon, in the *Odyssey* following both the Phaeacian sailors and Odysseus past his homecoming, is only one of countless mythological ciphers for that elementary enmity. Looking back at the Golden Age of land-cultivation, Hesiod saw all social and economic corruption as stemming from the sea; in the Roman period a tradition (linked to the phrase *utinam ne,* "would not" or "let not") developed of seeing all sorts of conceivable evil—war at sea, robbery and pillage, boundless greed—as having first come into the world since Argo, the first of all ships, was built and launched.[88] "Provoked, the sea claims punishment," the chorus in Seneca the Younger's "Medea" still warns.[89] But notably, such provocation lies less in the hubris of certain actions[90] than in—to recall our Kafka text—the "exaltation bearing down everything before it" of seafaring itself. Crossing the coastline upon the sea is an existential transgression because a break in the *nomos,* the "unity of order and orientation"[91] that since Hesiod has meant, as Wolfgang Sellert has put it, "the order of existence ordained by the gods and realized for human beings in tradition and custom."[92] To that extent, the essentially transgressive character of seafaring—after all, a core element of ancient Greek culture—becomes clear. Seen in this light, the relation of the Greeks to the sea is marked not simply by elementary enmity but also by fundamental if not tragic ambivalence: seafaring is, as Dietrich Wachsmuth put it, what "through a unilateral act of 'self-help' preserves the 'ephemeral' human being from sinking [*Untergang*], yet at the same time *in just this way* places human beings in fundamental opposition to the world's divine order."[93]

---

87 See Höckmann 1985, pp. 11, 13, Schulz 2005, p. 49, Roller 2006, pp. 44, 57, Mark 2005, p. 172.

88 See Huxley 1951, pp. 580, 582; Curtius 1973, pp. 465, 479; Heydenreich 1970, p. 32, passim.

89 Seneca the Younger, "Medea," v. 616 (*exigit poenas mare provocatum*).

90 On Odysseus's moral self-elevation to godlike status when he unilaterally condemns and punishes the Cyclops' violation of guest friendship, see Friedrich 1991, pp. 17 f., 26 f.

91 Schmitt 2003, pp. 42, passim.

92 Sellert 1995, p. 7. See also Blumenberg 1996, pp. 7 f.

93 Wachsmuth 1967, pp. 202, 228. On this "ambivalence" see also Heydenreich 1970, p. 11.

Odysseus's last voyage, described only through the augury of Tiresias the seer, is located beyond the epic horizon. In respect to the text's genesis, this might involve the interpolation of an outside chthonic tradition.[94] But on a thematic level, it introduces the atonement that returned adventurer-sailor Odysseus owes angry Poseidon. This atonement has to be removed from the epic action because it would divest Odysseus's wandering wiliness of its basis. For this reason, the tragic dimension of this wiliness thus only emerged in the apocryphal recursions of the *Odyssey*, especially in the lost (putatively Cyrenaic) epic *Telegony:* here the second son resulting from Odysseus' liaison with Circe, Telegonos, set out to find his father, encountered him unknowingly on Ithaca, and accidentally killed him with a venomous stingray spine—action that according to one theory Sophocles would then dramatize in his play "Niptra" ("Footwashing"; only a small fragment survives), through the addition of an Oedipal oracle.[95] In the Odyssey, Tiresias tells Odysseus that following his killing of the suitors, he is to "take a shapely oar / and visit many cities" until he reaches "a land where men know nothing of the sea / and don't use salt to season what they eat," where he is to make a sacrifice to Poseidon, following which he will be able to grow old and die at peace" (11, 120 – 140). But Tiresias also explains in this passage that Odysseus's death will come *ex halos* (11, 136), and depending on whether we understand *ex* in the Greek as meaning "outside" or "from" the sea, we can read Odysseus as fated to die either reconciled with land or else precisely as a seafarer, despite being at home on land, in that latter case enduring the "tragic contradiction of human culture."[96]

When tropes of this sort and thus the nature and status of even his death becomes polyvalent, then it is indeed justified to name Odysseus *polytropos*. The meaning of this word was already controversial in the ancient world—it could refer to a high degree of verbal and intellectual versatility or else to someone struck hard by fate, hence wandering around on land and sea. In the *Odyssey*, scholars have observed, the ambivalence of the term is purposively maintained; it only appears in polyvalent contexts.[97] Already on a purely linguistic level, Odysseus is thus defined as limitlessly adaptable, the basis for his status as the most complex of ancient literary characters, one opening up the possibility of endless recursion.[98] And already for antiquity, this means the following: as

---

94 See Oswald 1993, pp. 76 – 78.

95 See Hartmann 1917, pp. 221 – 223.

96 Wachsmuth 1967, p. 228.

97 See Danek 1998, pp. 33 f.

98 This is the premise of Stanford's detailed study. See Stanford 1954, pp. 6 f.

there is no "true route" in the *Odyssey*, there is no "true Odysseus" to be grasped outside his guileful discourse and maneuvers, beyond semblance as such.[99] Odysseus is the first hero remembered without a "heroic death"—indeed rendered immortal precisely through the deterritorialization of places of memory.[100] Strictly speaking, there can be no conclusion for the *Odyssey* and its *nostos*, for the ceaseless return and simultaneous "de-formation" of its tropes and narratives, because the text consists of just such recursion. For Odysseus, returning once and for all would be being dead, for which reason the topographical ambivalence of *ex halos* only offers one more postponement of death.

Until the end—and even beyond it—the *polytropos* holds himself over water through guile. For this reason, Odysseus's guile can be understood as exemplary for seafaring if not for culture in general: the hero's ritual appeasement of incalculable elementary dangers is in this way invoked to the same extent as is the rationally calculated triage he needs to employ for the passage between Scylla and Charybdis: preferably sacrifice six companions than the entire ship (12, 110). Alongside such knowledge of danger, whether mythological or calculating, it is finally the *technē* of steering a ship that attests to such a "culture of wiliness." "The wind blew straight, the pilots steered" is the basic formula used from the *Odyssey* through the *Aeneid*.[101] With his knowledge of danger, the *kybernētēs* can both fend off and practically use the elementary forces. Adhering to the instructions of the *periploi*—in a mythological framework, the instructions of the gods—is only made possible through his *technē*.[102]

As the vanguard of both seafaring and culture, it is helmsmen who in the epic—from Odysseus's Elpenor to Jason's Tiphys onward to Aeneas's Palinurus—make clear the consequences of culture's tragic contradiction through their death. Only this death guarantees the *nostos*, the hero's homecoming. But because from Homer to Dante joy at return can only be compared with the joy a castaway feels at being rescued,[103] the epic in itself stands for triumph over or

99 On this and the difference between proposition and utterance see Pucci 1987, pp. 16, 20, 122, 128, 150.

100 See Schlesier 2006, p. 108.

101 See for instance *Odyssey*, 9, 78; 12, 152; Virgil, *The Aeneid*, book 3, line 269. All Virgil citations will refer to Virgil 1916; book number of the *Aeneid* followed sequentially by line number(s) both parenthetically in the main text and in the notes.

102 In ancient seafaring the helmsmen were only made responsible for the boat's safety when they were demonstrably experienced—and literate. See Rost 1968, p. 49; Adam 1985, pp. 285, 291.

103 See *Odyssey* 13, 227 ff.; Dante, *Inferno*, canto I, lines 22 f. All Dante citations will use the Robert and Jean Hollander translation available in the online bilingual *Princeton Dante Project*, http://etcweb.princeton.edu/dante/pdp/, with sequential reference to title, canto, or section, together with the line number.

annulment of shipwreck. Just as in Homeric Greek there is no specific expression for a shipwreck,[104] the *Odyssey* together with its (early) recursive *nostoi* follow one and the same reflective schema of *oikonomia:* a rounding of the outward excursion and thus a purposive return of culture to itself.[105] At least for the epic hero, the shipwreck, for now, remains an episode. For the art of surviving it with the art of guile guarantees narrative progress.

This is particularly the case for Latin poetry, whose tradition begins with *Odusia,* the Homer translation by Livius Andronicus.[106] In the *Aeneid,* recursion to the *Odyssey* then became, for the first time, a model for political self-description: the escape route of the expelled Trojan surpasses the course taken by the indomitably wily Greek in that following six "Odyssian" books (the third book recounting the outward voyage in miniature) we now have six "Iliadic" books whose narrative describes a—this time around—enduring colonization in a—this time around—just war. The epic of a wily hero who rather than being a repatriate is now a refugee and conqueror, thus offers Roman principate-ideology its natal legend. Virgil had taken up his work after the sea-battle of Actium (31 BC), in order to cover up memory of the bloody origins of the Pax Augusta—or even put off future civil wars through collective memory centered on a new foundational myth.[107] Not only seafaring but also the epic in this way became an instrument of "thalassocracy."[108] In the case of Rome, such maritime rule went so far that for the sake of containing pirates, the Carthaginians, and all other enemies, starting with Pompeius the Mediterranean was screened with all the rules at work in the art of field-surveying. According to Cicero, Rome since then ruled over this body of water, *mare nostrum,* like a secure and closed harbor.[109] But the empire did not only furnish countless loci to what had none—thus procuring countless starting points for its rule—through a two-dimensional registering of "its" sea. As underscored by Bérard (an employee at the *École supérieure de Marine*, hence here also a specialist), it is particularly powerful Thalassocrats who

---

104 On Homer's *alalēsthai* and *alōmenos* instead of classical *nauagia* see Corvisier 1999, p. 21.
105 On the circular "problematic of *oikonomia*" and the "*odyssean* structure of the economic narrative," see Derrida 1992a, pp. 6 f. On that narrative's composition, see Seidensticker 2008, pp. 17 f., 32.
106 See Schmitzer 2005, p. 34.
107 See Quint 1993, pp. 8, 53. On the Virgilian recursions of the *Odyssey* see ibid., pp. 55 f., 60; Schmitt-Neuerburg 1999, p. 71, passim. On recursion of the *Argonautica* see Nelis 2001, pp. 382, 392.
108 On the concept of thalassocracy, see Thucydides, *History of the Peloponnesian War,* book I, §4 ff., passim; Starr 1989, pp. 5 f.
109 Cicero, *De provinciis consularibus* 31. See also Schulz 2005, pp. 180 f.; Vanoli 2010, pp. 214 f.

had the capacity to transform maritime locations into *lieux de mémoire* for their own narratives of origin.[110]

Virgil presents the procedure at work in this imperial toponymy in the Palinurus episode, which also vouches for the repeal of shipwreck and the heroic return to a primal mythic origin: Neptune demands that "one life shall be given for many,"[111] a sacrifice of the helmsman before Aeneas can reach Italy and found the imperium. After the man falls overboard and the unguided ship heads toward the sirens' reef (5, 864), Aeneas, future terrestrial leader, takes over the vessel, attributing the sacrifice of Palinurus to "mischance," *casus* (5, 869). However, like Odysseus's helmsman Elpenor (*Odyssey* 11, 60 – 78), Palinurus calls on his captain from the underworld to grant him a proper burial: it was in the end not only water that overwhelmed him; washed up half-dead on shore, he was discovered by barbarian coast-dwellers who killed him for his clothing, so that in a sense, a helmsman gone overboard, he has died a doubly dishonorable death. For this reason, the cape where he was stranded is to bear his name "for ever" (*Aeneid* 6, 381). But since Palinuro[112] was always the cape's name in any case, Virgil is here showing us in exemplary form how to inscribe an imperial genealogy into a natural toponymy. Beyond that, in his Augustus epic he transfigures the losses that Rome's older fleet, still under Octavian, had to endure at just that cape, on the way to a thalassocracy.[113]

Once the *Odyssey*'s "polytropism" is integrated into the *Aeneid*'s semantic context of providential imperium, Odysseus's guile becomes mere deceit (see *Aeneid* 6, 529; 9, 602). His virtues are now vices, juxtaposed with upright Roman legal certainty: *pietas* as a kind of generational contract and *fidelitas* as a sort of contractual fidelity.[114] For this reason, from Silver Latinity to that of the Middle Ages, *ulixice* stands idiomatically for "deceitful."[115] But in a scenario of eschatological seafaring involving a recursion of, in this case, the *Argonautica*, Virgil prophesizes that this deceit, and with it the sea itself, will vanish: "A second Tiphys will then arise [as helmsman], and a second Argo to carry chosen heroes," we read in the Fourth Eclogue, concerned with "the birth of the child," who would be interpreted from the Emperor Constantine's Good Friday sermon of 325 AD through Dante as a prophecy of Christ's advent. For with this henceforth *last of all* ships—facing the first of all ships, the Argo—and helmsmen, "even the

---

110 See Bérard 1902, pp. 15, 27.

111 *Aeneid* 5, 815.

112 Also derived from *pailouros*, the headwind at the Cape. See Brenk 1984, p. 777.

113 On Octavian's lustrations there in honor of Neptune in 36 BC, ibid., pp. 790 f.

114 See Albrecht 2006, p. 4, passim; Stanford 1954, pp. 129, 133.

115 See Friedrich 1942, p. 165. On the Odysseus-Aeneas opposition see Cacciari 1998, pp. 63 – 65.

trader will quit the sea, nor will the ships of pine exchange wares; every land will bear all fruits."[116] Here the *nostos*-circle has become a figure of soteriological hope, of the promise that terrestrial wandering and suffering will have an end. Odysseus's own wandering will only continue to stand for a salvational movement in Plotinus: the hero's flight toward home, we here read is an epistrophe: like the soul's path, that of Odysseus's ship describes the observant turn back of the existent to the unity of existence, "the One" of Plotinus. And this odyssey is a movement of ecstasis to the beautiful and finally to the good. For a final time, Odysseus here demonstrates "that for all being its origin is the goal."[117]

At the latest with Dante, a kind of Christian "exit condition" (to cite informatics) is inscribed into the endless cycle of pagan recursion. The circle of *nostos* opens in the twenty-sixth canto of the *Inferno.* In a fiery speech Ulysses convinced his companions to voyage past the Pillars of Hercules instead of taking the path to Ithaca required by salvational economy. As a result, he suffers a definitive death from the sea. Just this drowning of the henceforth irremediably wily hero will make it possible to henceforth not only treat running aground and shipwreck episodically or in terms of tropes, in order to then elevate it into a more powerful, epic, tragic, or salvational, context. Shipwreck will now be a poetic problem, since it concerns the problem of poetics itself. In the end the transgression that Odysseus's "polytropism" represents in light of the Augustinian catalog of sins is also the twofold transgression of all fictional discourse: on the one hand *curiositas*, the ceaseless seeking after pure experience, leading to a neglect of *cura*, of care for the world's harmonious substance and for one's own place in it, for self-limitation in line with creation and, in that framework, a centering of the self; on the other hand sly enchantment and seduction of one's audience, turning into *dolus*—intentional misdirection and false instruction.[118] Consequently, Dante's Ulisse ends in Malebolge, the infernal circle of deceivers, and in the pit for evil counselors. "Adventures" transgressing the zone of established itineraries around 1300 (such as that of the pilgrim's path or, in poetry, that of the courtly epic) posed an irritation to the accepted order of words and things and flouted—both verbally and geographically—the transcendent "being in the world" of everything terrestrial.[119] For this reason they represent fraud against the created order—examples of a God-forgetting and self-forgetting hunger for experience whose consequences will still be born by the "dialectic of the Enlightenment."

116 Virgil, *Eclogues*, pp. 49 ff.
117 Plotinus 1956, pp. 56–63; 245.
118 See Friedrich 1942, pp. 163–166; Blumenberg 1985b, pp. 79, 265 f.; Stierle 2007, p. 125.
119 See Teschke 2008, pp. 2, 11.

Ulisse's aberration from the model of a proper voyage of return to oneself, his disregard for the principle of *nec plus ultra*, and his search for earthly paradise in a southern hemisphere lacking people,[120] drive him to the Mount of Purgatory, where he and his ship fall into a *turbo* or cyclone (*Inferno* 26, 137) before finally being swallowed up by the sea. He only returns when sinking to "being in the world" and that inclusiveness harboring everything terrestrial in Dante's spherical cosmology. But with this shipwreck unfolding "as another pleased," (140), who this "other" might be for Ulisse remains open: still some pagan god of elementary forces or already the God of Christianity. In any case a divinity who "can do all he desires" is as pleased by Ulisse's going under as he is by Dante's ascent.[121] This is the only reason Dante's work can become a "ship that, singing, makes its way" across the sea, *legno che cantando varca* (*Paradiso* 2, 3). In this ascendant manner, looking back and over everything terrestrial, Dante can once again consider Ulisse's *folle volo*, his "mad flight" (*Inferno* 26, 125). Landed in Christian Hell, Dante's Odysseus is thus not only an adventurer who fails to return. He is also the first revenant of that unholy outward wandering that will then repeatedly haunt an Occident (dis)oriented toward experience-based knowledge.

Dante knew no Greek and was only familiar with Odysseus from the *Ilias Latina*.[122] For his Ulisse, he could bear in mind the Genoese brothers Ugolino and Guido Vivaldi (probably more than the medieval tradition focused on Alexander the Great and St. Brendan): the brothers had voyaged past the Gibraltar straits in 1291 to seek a sea route to India; all trace of them was lost on the West African coast, and since then they had fired the imagination of European seafarers.[123] Inversely, Dante's recursion of the *Odyssey* for the modern age, with its fortunate and failed voyages of discovery, offered a proven, downright universal-historical model for self-description.

For not only epics such as Tasso's *La Gerusalemme Liberata* (1581) understood Columbus as a kind of heroic descendent of Ulisse.[124] Columbus actually

---

120 Discussed further later, Aristotelian-Thomistic cosmology plays a decisive role in Dante's cosmology, as described especially in his putative disputation *Questio de acqua et terra*.

121 See Inferno 3 95f.: "dove si puote / ciò che si vuole"; Purgatorio 1, 133: "com'altrui piacque".

122 On the Homer renaissance that developed after Dante, see Kullmann 1992, pp. 353f., 372.

123 On the Vivaldis, including against the backdrop of the failed idea of crusade, and on the possible influence of the *Commedia* on the writing of this expedition's history, see Rogers 1955, esp. pp. 35, 45; Klesczewski 1985, p. 22; Verlinden 1986, pp. 40–47; and Howgego 2003, p. 1080.

124 See cantos 15, 25f. and 30–32 of *Gerusalemme Liberata*.

saw himself as a "bearer of Christ" (*christophorus*) who could always count on providential assistance, even in a shipwreck.[125] However, at the same time he praised himself as a pioneer of unlimited curiosity about the world: he, a simple seaman, had received a divine "spirit of intelligence" allowing him to discover, through *curiositas,* the sciences of geometry and astronomy, classical poetry, and as a final consequence, the world itself. To reinforce this perspective, Columbus —and, commissioned by him, the Carthusian monk Gaspar Gorricio—became adept at the art of typological allegory: in his *Book of Prophecies,* a concordance of biblical topoi, together with recursal to Seneca's "Medea," the *Argonautica* and the *Odyssey,* guaranteed that Columbus could serve both as *polytropos* and nevertheless as godly postfiguration.[126] With him, a typological sequence was carried forward from Homer to Dante and onward to the year 1492.

Since Paul (in Gal. 4, 24), the doctrine of *typos* or *figura* had been closely tied with the procedure of allegoresis—a kind of interpretation with its model in ancient Homer criticism (for instance in writing of Theagenes of Rhegium), but that the Church Fathers severed from poetry, after which they were less concerned with judging its putative meaning than its essential validity or nullity. Already in the case of Lactantius, but most prominently in Augustine, the pagan *fictiones* were placed on the hermeneutic examination bench of salvational revelation. Into the High Middle Ages, poetic discourse could in this way at least receive theologically derived truth-value.[127] Also suggesting itself here was an allegorical interpretation of what was later considered the originary text behind all poetry. In this framework, as a key episode in the *Odyssey,* "Odysseus and the Sirens" became a special touchstone of Patristic interpretive art: the Sirens were seen as both a paradoxical figuration of a figurality that has gone wrong and as beings who demonstrate the dangers of paganism, heresy, and temptation in general. Through an understanding of the mast on Odysseus's ship as a symbol of the cross, the hero's wiliness was also integrated into a Christian reading. Clement of Alexandria, for example, spoke of the constantly endangered ship of faith and a Church whose foundering could only be prevented by distinguishing between Muses and Sirens.[128] Occasionally such allegoresis extended to censure of Odysseus's desire to return home and thus of his insufficient curiosity: as opposed to Dante, an adherence to terrestrial values reflecting greater longing for the wreath of smoke over one's own house than the soul's ascent.[129] A wide

125 See Columbus 1992, pp. 163 ff.
126 See Columbus 1997, pp. 55, 57, 291 (on Tiphys); also Boitani 1994, pp. 46 f., 57.
127 See Stierle 2010, pp. 389–393; Müller 2004, pp. 288 f.
128 See Rahner 1964, pp. 247–270; Wirz 2005, pp. 24, 26.
129 See Blumenberg 1985b, pp. 295 f.

range of Early Modern *Odyssey* recursions fed on this patristic interpretive tradition, including Sebastian Brant's *Ship of Fools* (1494) and Calderón's religious drama "Sorceries of Sin" (after 1635).[130]

Dante was familiar with this hermeneutic tradition: the *Divine Comedy* extends it in a double sense. For in itself continuously producing allegories, it offers their own allegoresis, particularly in dialogic passages. For instance, when in *Purgatory* Dante dreams of being tempted by the same Siren whose song already distracted Ulisse from his (Homeric) odyssey (*Purgatorio* 19, 22), retrospectively different light is cast on Ulisse's transgression. Namely, his rhetoric, the decisive catalyst for voyaging out into the ocean, is equated with the Siren's charm: in the false muses' style, it seduced Ulisse and companions by feigning the Sirens'—in any case fictive—omniscience. The charm Ulisses exerts on himself and his fellow sailors marks the dangerous proximity between inflammatory rhetoric and illuminating prophecy, a constellation Dante knew through the *Retorica* of his teacher Brunetto Latini. Ulisse's infernal appearance is thus that of a double-tongued flame (*Inferno* 26, 52f.); in this way Ulisse represents a sort of antitype to the Biblical prophet Elijah, who having been caught in a whirlwind was transported to heaven in a chariot of fire (*Inferno* 26. 34–39). Dante, by contrast, sees himself in the position of the prophet Elisha—a witness to Elijah's heavenly ascent who now witnesses Ulisse's shipwreck.[131]

Looking back we can describe this in the following poetological terms: in order that poetry ascend again, after antiquity, myth has to go under (for the first time in its entirety). And looking forward, we might put it thus: on its flip side Early Modern poetry becomes the same experience-based knowledge that in scholasticism's Aristotelian-Thomistic doctrine is considered futile confusion on a level of mere singularity. As the protagonists of an epochal transgression, Ulisse and Dante are doppelgangers of one and the same enterprise. This is the case even if the terrestrial topography and topology of Ulisse is only horizontal while that of Dante is moved past vertically—even if Ulisse finally goes under at the Mount of Purgatory while for Dante that is the base for reaching Paradise.[132] Odysseus again becomes, in a higher sense, a *polytropos*, because although divested of the *nostos* guarantee, he also has been freed from the centri-

---

130 See esp. chap. 108 of Sebastian Brant's *Ship of Fools* (*Das Narrenschiff*, 1494). On the *Odyssey* and Sirens in Brant see Gruenter 1966, p. 92; Kemper 1996, p. 84. For a commentary on Calderón's play (together with a partial printing of it), see Flasche 1968, here esp. pp. 10–25.

131 See Rachewiltz 1987, pp. 28, 30, 136; Mazzotta 2008, pp. 328–334. On Elia and Elisa and the biblical whirlwind see 2 Kings 2, 11f., Jer. 23, 19, Job 38, 1.

132 See Barolini 1997, pp. 119, 125–127.

petal undertow of retrospective figural attachment, in order to henceforth bear witness to the centrifugal propensity of Early Modern *Odyssey* recursion.

The author Dante, however, has to first step over the forms of legitimate discourse in order to step onto the path of heaven as a pilgrim. What is represented in Paradise as *trasumanar* (*Paradiso* 1, 70), a "soaring beyond" human terrestrial existence given by God's mercy, is owed to a "trespassing" of sign and language (26, 117): a *curiositas* and presumption that will only be called "poetry" starting with Dante.[133] That in the *Divine Comedy segno*, "sign," repeatedly is rhymed with *legno*, "wood," deciphers this transgression as a type of Adamitic original sin incurred by all poetry, but one aimed metonymically at the ship as the medium for practical or experience-based knowledge.[134] In addition, this rhyme is intertwined with *ingegno* (13, 70 – 72), "talent" or "ingenuity"—the gift of invention from the muses that seduced author Dante to no longer consign the predicate of beauty to the cosmos—that greatest of all artworks—alone. Rather, as God's putative scribe (a *scriba Dei*), or even as a second God (an *alter Deus*), he now unilaterally ventures to create beauty.[135]

In the *Comedy* even Virgil, around 1300 still the "greatest of all poets," has no access to Paradise, because poetry—at least for Dante—simply could not concede this. But in laying out his work as both allegory and allegoresis, in having his poetry—for which no terminology is yet available—thematize its own poetics, and in thus rehabilitating poetic allegory vis-à-vis theological allegory, Dante has compensated for poetry's truth deficit through what later will be called "aesthetic experience." By contrast to theological allegories, poetic allegories, in Dante's sense, cannot "have" metaphysical meaning: they "possess" no vertical complement but rather produce it. On the one hand poetic allegories mark their material, as Dante observes, as "polysemic":[136] they allocate a potentially infinite series of signifieds to the most varied signifiers, which can take in colloquial expressions just as much as, say, historical data. On the other hand, the meaning of this "sea of knowledge" (*Inferno* 8, 7) as divine creation is questioned through such allegoresis—that Virgil (hence poetry itself) presents allegorically. The unlimitedly powerful structure of meaning to which poetry transforms being, finally leads to a horizon of endless ascriptions of meaning.[137] Put otherwise: although structured allegorically, the *Comedy* leads to what Blumenberg terms

133 On the return to *poetari* and *poetria* in Dante's circle see Curtius 1953, pp. 153 f.
134 See Stierle 2007, pp. 225 f.
135 See Schlaffer 2005, pp. 120 f.
136 Epistle 13, to Cangrande della Scala, cited from Raether 1981, p. 285.
137 See ibid., pp. 295 – 297, 305 – 308, und Münchberg 1997, pp. 89, 95 f.

an "explosion of the horizon of possibility."[138] In the *Convivio* Dante himself gave a name to this procedure, which renders allegory into a means of poetic curiosity: parole fittizie,"fictional speech."[139] What in an exemplary manner brings *sapienza* and *poesia,* knowledge and poetry together in "fiction" is a last epic of wiliness.

Literature's "Little Stratagem"

Georg Lukács has referred to Dante's *Divine Comedy* as presenting us with "the perfect immanence of the transcendent."[140] As Lukács indicates, a self-contained, rounded world as in the Homeric epic is here no longer directly present. Still, such a world is indirectly opened for us: through Dante's poetic ascent to heaven, which by contrast with Ulisse's odyssey can be seen as articulating a bifurcation—in mediating between a this-world of confusion and a beyond that for "every lost wanderer" represents "the home that has awaited him since all eternity,"[141] poetry separates the two spheres in an insurmountable way. The rebirth of poetry is thus also a new birth under altered circumstances. In the end poetry has to limit itself to the this-worldly sphere, in order to take up a ceaseless search for its *oikonomia.* With Dante, Lukács writes, we have "a historico-philosophical transition from the pure epic to the novel."[142]

This transition is precisely defined by Ulisse's shipwreck. Through that event, pagan wandering wiliness turns into unsalvageable going under, the circular voyage home together with its *oikonomia* into a voyage outward with no return, allowing only a recursion of measureless wandering. At the same time, the shipwreck, as an expression of "transcendental homelessness,"[143] and as a kind of metaphysical figure for the loss of figurality, marks exactly the point at which new knowledge of the world, of praxis, of representation penetrated into the order of narrative. In Dante the shipwreck has become a representational event marking the precise break-line between the epic and the novel. Where the epic is characterized by closure and totality, the novel first and foremost gains its worldly authority by driving "fissures and rents" into inherited

138 Blumenberg 1985b, p. 299.
139 Dante, *Convivio* (*The Banquet*), book 2, chap. 1, 1–3 (translation modified).
140 Lukács 1971, p. 59.
141 Ibid.
142 Ibid., p. 68.
143 Ibid., pp. 41, 61, 121.

forms.[144] Once shipwreck is a generic feature, the epic, Lukács argues, becomes the novel's concrete, worldly narrative form. These worldly contents initially centers on human life divested of orientation—life for which "neither the goals nor the way leading to them can be directly given." [145] But at the same time, it is centered on the search for an ordering principle capable of grounding all the detours and false tracks, for precisely the novel "seeks, by giving, form, to uncover and construct the concealed totality of life."[146]

Written in the same period when Lukács was working on his *Theory of the Novel,* James Joyce's *Ulysses* (1922) is perhaps the most radical expression of the poetological concept of world-encompassing *discurrere,* of unceasing wandering and seeking, allowing the emergence of an entire series of *possible* organizational principles under the sign of divested epic totality. That unlike historical factography, poetry has recourse to the merely possible here stands at the start of all the novel's narration. As a pedagogic "Nestor," Stephen Dedalus, Joyce's poetic alter ego, places this commonplace at the center of his school lessons. To be sure, when he takes it up—first through Milton's poem of shipwreck, "Lycidas," lastly with a view to Shakespeare (in his wiliness merely "another Ulysses")[147]—this is not through Aristotle's *Poetics.* Instead he has recourse to the *Physics* and *On Interpretation*—to practical and linguistic knowledge.

As Stephan observes, even if something that once was possible never happened, it has still rendered an unlimited horizon of possibility conceivable, an incalculable mesh of ramifications as reality unfolds. When he shifts from the non-specific plural of "possibilities of the possible as possible" to the singular of an "actuality of the possible as possible," he is illuminating two facets of the possible, one physical the other linguistic. For corresponding to the contents of Aristotle's *Physics,* among those possibilities whose sum constitutes a phenomenon's essence, only a single one is actualized.[148] But within the perspective of Aristotle's *On Interpretation*, different modalities of being are at work here—what is real has to always emerge, after all, from a background of potentials. For Aristotle, this background cannot be an object of science. Rendering it into the object of a distinct expressive system—a system moving in a narrative

144 Ibid., p. 60.
145 Ibid.
146 Ibid.
147 Joyce 2000, pp. 30, 250. The "bucolic" elegy "Lycidas" of 1637 was written in memory Milton's friend Edward King, who had perished in the Irish See. Similarly to the Palinurus episode in Virgil, it presents the toponymic procedure by means of the shipwreck: the shepherd Lycidas becomes a "genius of the shore." Milton 1924, p. 63.
148 Joyce 2000, pp. 30, 248. See also Schneider 1986, pp. 54–57.

manner past the horizon of sentence and proposition[149]—would mean actualizing the possible as something indeed possible in its emergence as language. For Joyce, this is the precise place where literature discovers its possibility, this because what it always articulates is no directly present, internally consistent "world" (hence a world capable of mimesis), but rather a space of possibility, of a flexion of modalities of being. "Fictional reality" is nothing other than the verbal actualization of this flexion, or put otherwise: narration opens a topography of the possible—a horizon of possible places, and the space of fictionality is the space of *discurrere* and recursion.

Here reality-effects no longer emerge through recourse to certain referents but in a discourse of circumstances. Everything, whether person or thing, "is absorbed by the circumstances: what surrounds it."[150] Inversely, what is narrated as fiction in this way reveals a hidden totality: the space of possibility of variable circumstances—that of a combinatorial topology, a network or labyrinth of routes in which one has always already become lost without being either an outward voyaging or returning self, and in which the former singularity of the adventure has thus either shrunk or expanded to its everyday nature.[151] For Joyce, such a network and labyrinth is already present in the *Odyssey*. For this reason, his novel also actualizes that epic's two *merely potential* episodes: the passage between the Symplegades, the "wandering rocks," never taken up by Homer because Odysseus follows Circe's alternative route to Scylla and Charybdis; and Odysseus' final outward voyage, which he never begins because the epic narration has reached its end beforehand. The "Wandering Rocks" episode has been described as a labyrinthian Odyssey in miniature distinguished above all by a topology of ramifications and possible worlds.[152] Joyce himself understood all of *Ulysses* as an actualization of the end that Tiresias simply augurs in the *Odyssey* as a possibility. When it comes to Odysseus' homecoming, that figure's prophecy stands in an irrevocable future. However, the choice of potentials tied to the fate of Odysseus' companions and the hero's death is uncharacteristic.[153] Joyce, Richard Ellmann informs us, once indicated that the two prophecies of Teiresias—of the son who will kill him and of arrival in a saltless land—were indeed fulfilled, hence that otherwise than H. Rider Haggard asserted, no two books are missing from the *Odyssey*. That work was simply falsely translated,

149 On the corresponding bifurcation between an Aristotelian approach to propositions and the sentence and a narrative poetics of knowledge around 1600 see Campe 2006a, p. 71.

150 Barthes 2011, p. 52.

151 See Blumenberg 1985b, pp. 80 ff.

152 See Norris 2007, pp. 21, 25, 37–40.

153 See Danek 1998, p. 233.

Odysseus's last voyage thus also being misunderstood. The novelist, Ellmann suggests, thus perhaps intended to present the prophecy's "saltless country" as Ireland.[154] Consequently, the novel *Ulysses* would begin exactly where the Homeric epic has to end.

In respect to Joyce's use of Homer in writing his novel, we not only know the obvious, that his book was based "on the wanderings of Ulysses,"[155] but also that he did comparative research on, especially, Homer, Dante, and the legendary voyage of Brendan, in the process consulting Bérard's work. He appears to have been particularly impressed by his toponymic approach and theory of the *Odyssey*'s Semitic roots—an approach that, entirely in concord with Kafka, problematizes any simple opposition between "Hebrew" legal commentary and "Greek" poetry.[156] And in fact, if, as Bérard argues, Homer "invents nothing" and every episode is only a "staging of toponymy,"[157] then he seems to be referring to a basic principle of Joyce's poetics: in *Ulysses*, narrated actions, but also the forms of narration themselves, have locative character. And just this fusion of places and words renders the novel a recursion of the epic and its own recursions, one centered around not only the motifs at work there but even more so the characteristic procedures. We can thus describe the novel's narrative space not only as an impressively precise topography but also as a scene of *memoria* whose topoi are occupied by various activated images. This represents something like a gliding organizational grid moving between the principle of lists and *periploi* and that of grid and charts. Or we can conceive it as an overlayering of various topological structures—as it were an Einsteinian universe upon which Euclidean or scholastic order has been imposed.[158] In any case the narrated world here founds a space of possibility in face of which historically varying—spatial and verbal—referential systems are utilized, without in this way moving the space of narration toward becoming, once and for all, a sealed off totality. Rather, totality here originates from a return of disparate things.

Victor Bérard's work was not the least of the sources for Joyce's philosophical-historical and poetological concept of repetition. For if the Phoenician iden-

---

154 Ellmann 1959, p. 439.

155 Budgen 1972, p. 15.

156 Ibid. See also citation of Joyce, ibid., p. 174: "And there's a lot to be said for the theory that the *Odyssey* is a Semitic poem." See also Reynolds 1981, pp. 35, 347.

157 Bérard 1971, vol. 3, p. 414.

158 On Joyce's "Memoria" see Reichert 1994, pp. 84, 97; on the ancient "Proteus verse" and its "principle of good ruse," see Hagena 2006, p. 53; on the longitude and latitude of "listener and narrator," see Joyce 2000, p. 870. On the overlayering of topological structures, see Eco 1989, pp. 55 f.

tity already involved a mélange of trading and seafaring, reflecting Mediterranean migration and subsequent assimilation and consistent variation and refinement of a range of cultural techniques, especially writing and seafaring, and if in this way the Phoenicians served as a vanguard in the exploration of new worlds, thus reaching, as tradition has it, the Irish island, then not only the Greeks were themselves a "polytrope" people whose culture emerged from an amalgamation of the most varied elements—but also the Irish. That already on an etymological level Greek topoi were largely of Semitic origin is a thesis of Bérard. In a lecture he delivered in 1907, Joyce maintained that the same was the case for Irish words and places.[159] Seen in this light, Leopold Bloom is a quintessential "wandering Jew" his roots lost along the routes he has taken. From the start, Leopold is both Greek and Semitic: "Jewgreek is greekjew. Extremes meet."[160]

Ulysses demonstrates, indeed *is*, the return of Homeric figures, situations, narrative procedures. Joyce's protagonists are no post-figurations, adaptations, or even copies of original Homeric figures. As if initiated by Homer's *polytropos*, they are all born from that game of wandering wiliness opening up the recursions of a putative primal song. The learned search for an originary route, an originary language and identity, has here wandered into a labyrinth of endless ramifications. It is only in such manifold recapitulations of Homeric origins that a random June 1904 day in Dublin can become, in Hermann Broch's words, a "world daily routine for the epochs"—and Odysseus be reborn, as augured already in Plato, as an *idiotes*, a placid private person.[161] Giambattista Vico's term for this kind of backward movement, which is also a form of creation out of renewed appearances, was *ricorso*, and it is not surprising that already in 1725 Vico declared the "Homeric questions" to be a test case of cyclical cultural poetics.[162] The *ricorso* of Vico-reader Joyce in any case moves forward from the one origin's "disclosure." Origins are always already a second entity, an illusion or fiction, appearing for the sake of steadily returning as such and transforming assessment of the "true" Odysseus and his "true" path into assessment of his wiliness.

Hence what since Homer does not cease not coming to an end, literature's recursion, was slyly driven even further off course by Joyce: onto a sea that is now more a sea of polytrope language than, as in Dante, a sea that constitutes

---

159 On this and relations to the histories of language and migration see Reichert 2008, pp. 78–87.

160 Joyce 2000, p. 622.

161 Broch 1975, p. 64; Plato, *Politeia*, 620.

162 Vico 1948, pp. 289, 293f.

an inexhaustible meaning guaranteed by the beyond.[163] Joyce's novel moves into a chaos of difference drawing its words and places not from a coherent totality (however revoked) but from a continuous "chao-errancy". It develops into a process of productive wandering in the midst of a sea of topoi and associated possible worlds. In *Ulysses*, the Homeric cosmos has become a modern "chaosmos" (both terms were coined by Gilles Deleuze[164]). And narration itself has repeatedly undergone a "sea change,"[165] because it has withdrawn even further from land and its fixed loci to the sea and continuous creation of words: because, as described by Bérard and formulated programmatically by Ezra Pound, it has taken up the perspective of the *periplus*, "not as land looks on a map / but as sea bord seen by men sailing."[166]

Joyce insisted that the "modern writer" had to above all be an "adventurer": "In other words," he explained, "we must write dangerously: everything is inclined to flux and change nowadays and modern literature, to be valid, must express that flux."[167] The adventure of literature, its necessary self-endangerment, consists of a continuous protean shift of its forms and topoi, of a "chao-wandering" needing to venture upon language in a fully fluid, chaotic state while nevertheless shielding itself from it. In this sense, the allegory of writing as a dangerous marine voyage is always implied in Joyce: no longer as a voyage from and to home within the delimited order of classical literature but as a chaotic odyssey. This leads, as in the case of the sailor-raconteur D. B. Murphy in the "Eumaeus" episode of *Ulysses*, past the Cape of Good Hope and around the world. Murphy's wife, we learn, has been waiting for seven years for him to come home. Emerging from ceaseless nautical yarn, Murphy's story is unverified because even the sailor's certificate of hire is merely "literature." But with his fabulous tales he shows in exemplary form how the sea, which is omnipresent and must thus be crossed, can at least be "diddled"—in order to in fact come home.[168] But in *Ulysses* we not only encounter the sea of language itself as an "originary" endangering of literature—also what may be found "before the text and language" has such a function. When in the "Sirens" episode Bloom, and with him the entire male world, experiences seduction through singing and silent barmaids behind

163 See Hagena 2006, p. 20: "If for Odysseus the sea is a labyrinth and language the labyrinth of *Ulysses*, then language is the sea."
164 Deleuze 1994, p. 57, passim.
165 Joyce 2000, p. 63.
166 Pound 1996, p. 324. – See also Feshbach 1985, esp. p. 333; Hesse 1969, pp. 38f.
167 Power 1974, pp. 95, 110.
168 Joyce 2000, p. 719: "Mr Bloom could easily picture his advent on this scene, the homecoming to the mariner's roadside shieling after having diddled Davy Jones."

a counter, the text runs the risk of losing itself in its formal musical structure or in mere facsimiles of sounds and noises.[169]

That this episode (whether organized fugally or in terms of leitmotifs) consists of a musical-artistic notational system is, looked at closely,[170] simply a fiction—a kind of deceptive maneuver to dissemble the limits and menaces of literature. In the end the episode is located between the "ineluctable modality of the audible" and the "ineluctable modality of the visible"[171]—a partition whose bridgeover musical notational systems can, at best, simulate. It is only in its literalness that, literary language limits itself to sequentiality, as Lessing observed; rather, literature always finds itself located *between* sequentiality and simultaneity[172]—between the one-dimensional or symbolic register of the *periplus* and the two-dimensional or imaginary register of the chart. The element of enchanting discourse and song that cannot be absorbed into notation is the "in between" of its "supra-segmentalia." But above all it is the acoustic borderline values of language and song and, in the end, physiological facets of perception, that define the hearing of what can and even cannot be heard. That the sea already sounds in the organism; that the "flux and change" dangerous to literature is always already heard;[173] that no ear can seal itself off from this hearing and the Sirens' enchantment is from the start auto-affective: in Joyce the figure of the seashell confirms all of this, as an allegory of sounding and hearing at once.[174] Consequently, where poetry describes its immemorial endangerment as mythical, literary prose describes it as entirely real.

Extending onto Joyce, those threatened by the Sirens appear as seafarers, scholars, and, say, "canvassers" like Leopold Bloom—spelled out, men and their authorship. The last "little stratagems" of literature serve for their rescue. When Kafka wrote his text on Odysseus and the Sirens on 23 October 1917 in Zürau, a few days after his last letter to Felice Bauer, this scene of writing was tied to a turn away from the "muse" of his previously most productive writing phase, but at the same time to a deflection of her marriage promise and a

169 Joyce 2000, pp. 328 – 376, 359.
170 On the merely simulative character of Joyce's putatively musical writing, whose art of form and notation is deliberatively misleading, see Wood 2007, pp. 68 – 74.
171 Joyce 2000, p. 45.
172 Ibid.
173 According to Bloom, male Sirens, namely tenors, can develop their erotic power through tonal frequencies that catalyze chemical processes in the female body, leading first to "flow," then to "flowers"; see ibid., pp. 353 and Hagena 2006, pp. 120 f.
174 Joyce 2000, p. 363.

kind of reversal that retrospectively makes Kafka appear the actual seducer.[175] Considered in this way as well, the episode described in the "siren" fragment, and which can be read as an episode in Kafka's life, again treats that ruse allowing the lover to become an author, Odysseus to become Homer. And writing again means slyly shifting, in Blanchot's words, "from the first to the third person, so that what happens to me happens to no one."[176] In any case, precisely when Sirens are in play, the "dangerous writing" that Joyce called for involves first and foremost dangers posed to literature itself: whether by women intending entrapment through marriage; or as something pre-symbolic that always already undermines signification as mere noise.

For the Greeks, not only mythologically symbolized dangers were "sirens" but all imaginable things that emitted tones and noises, screeching and howling, and thus signaled danger.[177] Since around 1900 such signal techniques prevailed on water and land, the field of writing treating Sirens (along whatever detours) had to expand correspondingly. Of interest for such writing were not only techniques of the sort very opening signaling danger, for example the modern "sirens" designed by de la Tour and Helmholtz.[178] When from now on technologies of communication in general could be considered fallen muses, this posed an entirely new challenge for writing in the *Odyssey*'s wake: at stake beforehand was symbolizing elementary phenomena in a way neither limited to the merely imaginary nor claiming to describe the locus of the real; at stake henceforth would be coming up with new "little stratagems"—running alongside aesthetic distance and fictional experience—to stand up to the *poiēsis* of technical media. Put otherwise: the Homeric gods may still have delivered danger to their heroes for the sake of initial narration; Odysseus may have warded off the danger of death by narrating, in a recursive sleight of hand, his previous adventures and finessing of mortal danger; and finally language may have slyly saved itself in writing, kept itself alive, through recursion, within what Foucault terms this "virtual space of self-representation and reduplication." But now writ-

175 See Döring 2008, pp. 62, 64.

176 Blanchot 1982, p. 32.

177 See Rachewiltz 1987, p. 58.

178 Helmholtz constructed his "double siren" in 1856, on the basis of Charles Cagniard de la Tour's "siren" (1819) that made possible a production of guidable noises without an oscillating sound-box, through the clocked delivery of air-blasts. Helmhotz here made use of his physiological research on tone perceptions, in the process directly referring to Plato's and the Pythagorians' myth of the Sirens. In fact, their own cosmological doctrine of spheres centered on the connection between quantifiable oscillation of air and audible harmonic intervals. See Welsch 2004, pp. 57–60, 65–67, 73 f.

ing "in our day has moved indefinitely close to its source, to this disquieting sound which announces from the depths of language—once we attend to it—the source against which we seek refuge and toward which we address ourselves. Like Kafka's beast, language now listens from the bottom of its burrow to this inevitable and growing noise."[179]

Texts such as Joyce's and Kafka's do no simply drown in the noise emanating from various channels. Rather, as it is put in Kafka's "Pontus letter" to Felice, they make audible "a sad, mighty, wordless song" that is "impossible for human voices."[180] Evidently such literature no longer is nourished by muses' mouths, nor from any meaning that scholars can interpret.

But like earlier *Odyssey* recursions, it still has recourse to that *technē* meant to produce loci in what has none and render them into speech. If finally it ascribed its own mythic origin to technical media—media such as the telephone and radio that may have induced a general delocalization of commerce, indeed the collapse of the elementary difference between land and sea—this is perhaps merely a ruse for self-rescue. And it is perhaps an opportunity, as well, once again to be slyly driven off course, into noise and the placeless, to be driven once again out to sea.

179 Foucault 1977, pp. 56, 60.
180 Kafka 1973, p. 166. See also Siegert 2007, pp. 28 f.

Chapter 1
Rites of Passage: Cults and Religions of Fortunate Seafaring

Religions of the Seaways

In the West from the beginning, seafaring has been considered a transgression: a violation of elementary borders. As soon as human beings venture onto the seas, leaving behind terra firma they also abandon its laws and obligations, its order and safety. A passage leading through the Other of the *nomos* and of culture—through, in Hans Blumenberg's words, a "sphere of the unreckonable and lawless, in which it is difficult to get one's bearings," indeed an abyss of unknowable depth and heavy danger, has to be viewed as an existential wager.[1] What here endows such an intense sense of precariousness to the anticipation, circumnavigating, and coping with threat attached to most marine dangers are associated qualities of latency, shapelessness, and suddenness. Forming a precondition for virtually any sea excursion are certain schemata of attentiveness and certain routines of action: types of perception and usage meant to open a path to damage-free, successful, and therefore "fortunate" seafaring. As a threatening environment *par excellence*, the sea is in this way the source of idiosyncratic cultic and religious forms; these are characterized by a specific sense of danger while at the same time promising or even guaranteeing rescue and salvation. To the extent not only various cultic practices of avoidance, mitigation, and precaution emerged from the elementary anxiety at imminent drowning but also many institutions for safety and insurance, the modern "culture of security" is also rooted in the danger-zone of the sea.

Albin Lesky has given us a theory of primeval alienation from the sea, or even categorical enmity toward it, on the part of the Balkan-stemming Greeks —the alienation already manifest in a lack or borrowing of maritime and nautical terms. A further sign of it, Lesky argues, is Hesiod's praise, since become a topos, for a Golden Age fulfilled in agriculture and a firmly terrestrial *nomos:* an age that had not yet known the misappropriation of logs for seafaring and the corruption of maritime transport and trading. In addition, Lesky refers to fabulous lore such as that devoted to the Cretan despot Minos, to whom the Athenians had to pay tribute on account of their poor nautical skills, and to Plato's

1 Blumenberg 1996, p. 8.

https://doi.org/10.1515/9783110610734-005

political skepticism about all marine undertakings.[2] To be sure, a large amount of literary, cultic, and mythological evidence points to a more complex ancient Greek attitude toward the sea. For its part, as a sphere of the "ambivalent," the Greek sea displays a religious structure. For example, *Okeanos* is not only considered a circular stream of water surrounding the *oikumenē*, the inhabited human world on firm land, but the generator of everything living (*Il.* XIV, 246), also present beyond the salt tides. For this reason, temples such as the Acropolis possess a "little piece of the sea," so that the shrine together with its bloody and non-bloody cults can thrive.[3] Nevertheless handing one's existence over entirely to the sea, possibly founding an alien existential form, community, culture upon it, would have been deemed hubris.

As Blumenberg observes, while human beings manage their life and institutions on firm land, they try to comprehend the "movement of their existence" in terms of seafaring and its hazards.[4] If the ancient sea represents the locus of liquidation, of an impure mixing of the animal, human, and divine, then ritual systems such as sacrifice here have a more urgent presence than anywhere else. Namely, they mark and confirm social and existential boundaries, in order to at the same time open them up for regulated transgression. With transitions and changes being as vitally important for social life as a firm and secure topography, confirming the vitality and very existence of the community, and of culture itself, depends on rites of passage such as the sacrifice. These are structured along the lines of a voyage, especially a sea voyage with its transgression of elementary limits. Hence as Arnold van Gennep has shown, in a downright exemplary way moving on board and then back on land is tied to rites of separation and affiliation: a framework containing—just like sacrificial ceremonies—threshold, transition, and transformation phases. Ship and crew are here sent into a state of "extimity," of simultaneous inclusion and exclusion.[5]

Ancient Greek rights of passage are marked by integration of the natural environment in both a spatial and temporal sense. For one thing, cultic calendars reflect the year's periods of uncertainty; for another thing, in their coming of age ceremonies Athenian ephebes, for example, enact their oath by naming local riv-

---

2 See Lesky 1947, pp. 8, 14, 25 f, 33 – 35. The ancient Greek cult of the helmsman, the *Kybernesia*, was derived from this tribute. On this tradition and its critique see Mommsen 1864, pp. 269 f., and Wachsmuth 1967, pp. 385, 390.

3 Burkert 1997, p. 177.

4 Blumenberg 1996, p. 7.

5 See Gennep 1960, pp. 23, 184. On the middle, liminal phase see Turner 2009, pp. 359 f. On consecrated ship models as attesting to ancient rites of passage see Göttlicher 1978, pp. 5 – 9.

ers, springs, and seas.[6] Since on account of the regional topography and relief, the Mediterranean winds arrive with strict regularity. They can thus be addressed directly, various wind cults emerging for the sake of safe sea passage—the later personification of the wind as *fortuna di mare* here has its starting point. And because in the epochs of navigation by sight and stars alone, islands and especially promontories were vitally important reference points and nodes of maritime traffic, they were considered sacred or even to be *arae*, natural altars.[7] They were venerated either at on-site shrines or through prayer-formulae and a sacrifice while passing through.[8]

As already clear at this point, the Greeks are adept at a "pathway religion" counting on secure divine accompaniment for a good voyage, the *euploia*.[9] The appeal is first directed not at the "Soter" divinities, like the Dioscuri only appealed to in moments of extreme danger, but at a *theos pompimos*, a guiding and protective god meant to put off emergencies from the start. To gain the god's favor, there is a resorting to even drastic measures, for instance bloody or indeed human sacrifice. Hence Karl Meuli argued that the great seafaring epics consistently refer back to "mythic tales of assistance," then this is no less the case with traditional sacrificial cults.[10] Regulated consecration ceremonies and "fine offerings" (*Od.* IV, 473) are good and obligatory usage in the *Odyssey*, for as Telemachus is taught, only then will the gods "restore your longed-for homeward path" (*Od.* IV, 480; after Odysseus's companions have slaughtered the cattle of Helios against divine command, thus violating the basic provisions of ritual sacrifice, they permanently lose two things: the "good path" and, beyond that, the locus of a ritually subsumed death, the locus of memory: they are devoured like animals).[11] In the *Argonautica*, the gods are described as arbiters of fate who promise, for example, "to point out and show me the paths of the sea, if by sacrifice to [Apollo] I should begin my venture...."[12]

---

6 See Horden / Purcell 2000, p. 411.

7 For an ethnological perspective see Beck 1985, p. 116: "Off every large cape, headland, point or bluff, at the entrance to any body of water like the Baltic Sea, the Mediterranean or the Black Sea, one is likely to encounter strong tidal currents and confused seas. To the unsophisticated mind that knew nothing of tidal streams, ocean currents and the rest, these disturbed waters were the result of the action of subaqueous monsters. In order to propitiate them, some kind of ritual varying from human sacrifice to prayer had to be performed. A successful crossing led to festive activity."

8 See Wachsmuth 1967, pp. 115, 394–417.

9 Ibid., p. 78.

10 See Meuli 1921, p. 112, passim.

11 See Detienne / Vernant 1983, p. 248.

12 Apollonius of Rhodes 1912, p. 27.

It is the case that the most prominent Greek sea-gods, who according to Hesiod determine "the outcome of good things and bad things alike,"[13] appear less a cultic product than that of the epic and theogony. Where the Roman sea-gods are seen as derivations of their Greek counterparts, thus standing in contrast to river gods reflecting native tradition, the Greek sea-gods—Poseidon and his wife Amphitrite, the Tritons and Nereids, Proteus, Phorkys, and the "Old Man of the Sea," Nereus—emerged from highly varied contexts. Alongside Olympian gods and goddesses such as Apollo, Artemis, and Athena, a wealth of lower and possibly pre-Hellenic counterparts swim about in the Greek sea. We find for instance minor deities such as Proteus, capable of changing shape when needed but, once captured, revealing the future contingencies of the sea and of life, together with local gods of the sea and harbor. We even find wine-god Dionysus, who was said to have transformed sacrilegious seamen into dolphins. The sea thus even deterritorializes the world of the gods. But what we now have as testimony moving past such mythographic material is tied to the shrines and altars founded in honor of the Nereids und wind deities after having received assistance.

It is also tied to Poseidon, great god of the sea, ally of the Achaians—but also enemy of Odysseus. Poseidon is anything but a prototypical sea god. Named "earth shaker," he has been attributed to the Mycenaean populace, still turned away from the sea.[14] In addition, Poseidon is no personification of the sea but only the ruler of its elementary forces. As an originally terrestrial divinity he is associated with springs and horses—hence the cult of *Poseidon Petraios*, Rock Poseidon, from which the first horse sprang after Poseidon spilled his seed upon it. At sea the horses could be made of foam or sometimes take the form of living ships. In Euripides' "Hippolytos," Poseidon also produces a bull from the sea-depths, its absent return to its source then spelling doom for Cretan Minos.[15] But what finally rendered Poseidon into the dominant sea divinity was the associated cult of sacrifice by drowning (of bulls and horses)—an act that has been described as sympathetic magic and as the least specific type of maritime sacrifice in its independence from any image of a god. Within the mythological and poetic imagination, the sacrifice to Poseidon, this borderline form between magical practice and institutionalized cult, was projected back onto the origin of human sacrifice, as if with this extreme of abandon rescue from maritime distress was meant to be nothing less than forced. Figures such

13 Hesiod 1988, p. 57.

14 See Larson 2007, pp. 68 f. On the chthonic origins of Poseidon and their consequence for the *Odyssey*'s non-epic, "tragic" end see the previous chapter.

15 See ibid., pp. 68 f., und Lesky 1947, pp. 95–98.

as Iphigenia or even Absyrtus (from the *Argonautica*'s ambit) have thus been understood as poetic sublimations of this "primitive" sacrifice.[16]

By interpreting sacral actions in an "economic" sense, we can also generally understand sacrifice at sea as a kind of fare paid according to urgency and danger, extending from human sacrifice to that of bulls and other animals and onward to hair, blood, supplicatory, votive and other offerings, and then further on to retroactive or precautionary sacrifices of ships. As we read in the *Odyssey*, this price is owed the gods.[17] However, such sacrifice also imposes a duty on the deity as a kind of contract: a duty addressed with special clarity in the context of Roman seafaring as *creditum* and *depositum*. Already on grounds of the effort involved, regular altar offerings in difficult episodes at sea would have been very difficult. But in normal sailing circumstances, it was entirely usual to use the ship—the *Naus*—as a temple—a *naos*, the ship, as a "heterotopia," for its part being divided up in line with sacral topology: especially rudder and keel were considered holy, because of their permanent location in the water and a good voyage depending on them first and foremost. Beyond this, as a result of ritual appeal to a god or goddess, the *epiklēsis*, there was assumption of subsequent theophany, divine accompaniment on board. And finally, nautical altars allowed a specifying of sacral zones and thus offering sacrifices in a strictly ceremonial framework. Probably the stern was the preferred place for this, and probably on longer voyages entire hecatombs of sacrificial animals were placed on board.[18]

In an event the social order of maritime traffic was mirrored in the ritual order: in the Athenian, Roman, and Byzantine fleets, officials specialized in sacrifice sometimes accompanied voyages. When this was not the case, the sacral actions were seen to by the ship master or, in cases of indisposition and absence, by the first deputy, the *pronauklēros*. But elementary distinctions, not only those of a social nature, were introduced and reinforced through sea sacrifice. As soon as it was located in seawater, the ship belonged in a "thalassic" realm. But it was only separated from the homeland's soil, from its dominion and local divinities, after successful ritual taking down of the ropes, once terra firma was no longer visible, once one was so to speak lost between the two "oceans" of the heavens and the seas and, in addition, the discharging sacrifice had been sunk. Just as sacrificial substitutions were sometimes proffered, the rites of separation could also be temporally and spatially pushed forward. The voyage's *prooimion*

16 See Robertson 1997, pp. 85, 88, 93, 96 f.

17 See *Od.* IV, 352, 472 f. The "Tale of the Shipwrecked Sailor," discussed in the previous chapter, could be considered a model for the *Odyssey* in this respect and in relation to the fear of drowning in the sea and thus having no burial. See Fabre 2005, pp. 191, 195.

18 See Wachsmuth 1967, pp. 157, 324, 342 ff., 367 f., 448 f.

could thus be shifted to the evening beforehand or to one's own doorstep. Likewise, whether a voyage should actually take place was explored with the help of *omina*—mantic and divinatory practices, hence through translation of divine will capable of influencing not only embarkation but landing.

As an elementary sphere of ambivalent hierophanic experiences, the sea was suffused with taboos, with a system of proscriptions and avoidances. That ships gouged through this *nomos*-free space like ploughs and, in addition, angered the sea with oar strokes, was feared to be a provocation of the sea, so that appeasing apotropaic objects were placed on the bow. Or else the ship was treated as a living being endowed with divine names and meant—as made clear in eyes painted upon it and its fishlike form—to undergo a kind of mimicry of the sea and its creatures.[19] Finally, both the ship and its crew had to undergo cathartic rites extending from purity sacrifices for individual ships to, in ancient Rome, *lustratio classis*, with its ceremonies of purification for the entire fleet. Impurity was understood as a guarantee of shipwreck, for which reason corpses and sexually tainted persons were taboo. The success or failure of a voyage could emerge as a divine test of a passenger's purity: before the court, accused men such as Antiphon and Andokides would present successful sea-voyages as proof of innocence,[20] and the presence of a pious and just person, a *eusebēs*, served as something like divine maritime insurance. Inversely, presence on board of someone tainted, an *asebēs*, was seen as grounds for distress at sea. According to the specific social and nautical situation, a mere passenger could be selected as a *pharmakos* or sacrificial scapegoat.

In this context, we can understand the Biblical story of Jonah as a both a synopsis and new interpretation of nautical sacral ideas cultivated into Hellenistic times. Burdened with guilt through his flight from his God, Jonah brings dire misfortune down upon the ship he hopes will convey him to Tarsis, located at the extreme western end of the *oikumenē*. On his account, God ends up sending "a great wind into the sea, and there was a mighty tempest in the sea, so that the ship was like to be broken."[21] Alarmed, the sailors implore their gods and toss various objects overboard. When nothing helps they call on Jonah to pray to his own god, and the crew casts lots to find out whether someone akin to an *asebēs* has possibly come on board. Identified in this way and confronted, Jonah readily acknowledges he is the cause of the ship's troubles and advises the crew to toss him aboard; at first, however, they appeal to Jonah's god not

19 See ibid., pp. 202, 234 f., 277; Göttlicher 1992, pp. 106–109.
20 See Antiphon's plea in Sasson 1990, p. 91.
21 Jonah 1, 4, King James Version.

to "lay...upon us innocent blood" (1, 14). Jonah only goes overboard when all despairing nautical maneuvers prove fruitless. Then the sea subsides and, motivated by a mix of dread and gratitude, the men now offer sacrifice and make vows to Jonah's god. We also have a retelling of the story in the text known as the "Chapter of Rabbi Eliezer," where the effectiveness of Jonah's tossing overboard, a last means against the wind cast forth by God, is initially tested by the sailors, "men of the seventy languages": Jonah is very briefly thrown into the sea, the quieting of the gale observed, together and its strengthening when he is taken back on board. Only then is the *asebēs* finally consigned to the waves.[22]

It is clear that Jonah's actual maritime passage, his being swallowed and then spit out by a big fish, follows the basic pattern of all initiation rites—and also all myths of coming back from the depths of the sea.[23] Jonah's prayer in the fish's belly here simultaneously signifies the greatest distance from God and closeness to Him, in this way marking the middle of the fully symmetrically structured story of Jonah.[24] All told, Jonah enters into divinely mandated homecoming, in the course of which he ends up traveling eastward, to Ninive, despite his initial resistance.[25] The rejection of human sacrifice and bloody sacrificial acts in general is clear here, and the popular practice of drawing lots, cleromancy, is present. In addition, survival of a shipwrecked man is explicitly pointed to as a sign of election. But what is most important is the monotheistic presentation of the scenario of voyage and engulfment. In the Hebrew Bible's Jonah story, God is, of course, the creator of Heaven and of both the earth and sea, and ruler of all the elements. Even a maritime monster is disempowered into His instrument. God causes fear and is incorruptible, but in His mercy attests to universalism —precisely for the sailors—despite his name being Jahwe.[26] Whether or not one is inclined to understand the Jonah narrative as a kind of bridge between "pagan" "path"-centered religions and the Christian concept of the life journey, alongside the motif of transformation and with it that of resurrection this universalism remains the decisive figural moment in the Jonah-story's plot.[27] And not

---

22 On this variant see Limburg 1993, pp. 105–107, and Steffen 1994, p. 22.

23 See Steffen 1963, pp. 44 f., 88 f., 96; Gerhards 2003, pp. 225, 230.

24 On the structure of the book and on Jonah's prayer, which presents the fish as a kind of "temple," see Kratz 1979, p. 185, Craig 1994, pp. 82, 121 f., Gerhards 2003, p. 227. For a lengthy and detailed analysis of the text, including an excellent study of its status as a literary work, see Wolff 2003; see also Uriel Simon's detailed study, specifically offering an alternative to the usual Christian interpretive framework, Simon 1994.

25 See Kratz 1979, pp. 161, 175.

26 See Sasson 1990, p. 341, Salters 1994, pp. 53 f., Gerhards 2003, pp. 246 f.

27 On the Jonah typology in the St. Matthew gospel and Luther, see Steffen 1994, pp. 93, 107–110.

only the Roman poets, but also early Christian literature, moved the sea and seafaring into an eschatological and even apocalyptic horizon. Already because the sea is all-encompassing, its conquest or at least the management of its dangers was meant to benefit everyone. All persons were meant to become participants in one and the same life journey with equal status.

Paul and Universal Rescue

The "coming child," Virgil wrote in his Fourth Eclogue, will see to it that "even the trader will quit the sea, nor will the ship of pine exchange wares; every land will bear all fruits."[28] It is the case that the early Christians longed for the arrival of a helmsman who would do away with the sea and its unpredictability, with streams of human beings and ware, indeed with the Roman thalassocracy and finally the secular empire itself. In the interim, however, they made zealous use of imperial ships and shipping routes—namely as means for missionary work. Maritime travel was naturally a precondition for Paul's communication system, consisting as it did of an increasingly narrow network of enduring epistolary contact with individuals located overseas. This thirteenth apostle could only offer himself as a witness of his calling to preach. He could only gain authority by authorizing himself in his letters—through appeal to the "God in me"—to speak "in the name of all." The possibility of his theology of spirit and universal promulgation thus rested less on "law" and more on "true scripture," together with the secure delivery of letters, distance notwithstanding.[29] And since these letters also represented the first written evidence of Christianity, the dangers tied to their shipping were also dangers for the faith's universalization. Shipwrecks endangered not only goods and people but—even worse—the good news. This alone explains the drama attached to Pauline accidents at sea.

In the Second Epistle to the Corinthians, Paul indicates that "thrice I suffered shipwreck, a night and a day I have been in the deep," and among all the perils he has encountered along his mission's paths, he singles out the "perils in the sea" (2 Cor. 11, 25 f., King James Version). To be sure, we here do not have seafarers' heroism in the Greco-Roman style: "If I must needs glory, I will glory of the things which concern mine infirmities" says Paul (ibid., 30), here praising only his good connections with God.[30] Acts 27, the report of

28 Virgil, 1916, p. 35.
29 See Siegert 2003, pp. 53 – 56.
30 See Badiou 2003, p. 67.

Paul's voyage by sea to Rome, can then build on these "authentic" references to shipwreck. It is to be sure unclear whether Luke the Evangelist, traditional author of Acts, is identical with Paul's like-named companion: the text never explicitly refers to Paul's epistle and was perhaps written a generation later. It is also unclear why the narrator sometimes speaks in the first person: whether this is meant to certify personal witnessing or rather third-party reports were simply integrated without additional editing. Finally the report of Paul's sea voyage clearly borrows a number of concepts, motifs, and narrative elements from the *Odyssey* in particular, as well as making use of the general toponymic and fictional narrative procedures of the *periploi.*[31] In any case what we have here is a description, speedily canonized as authentic, of Paul's last voyage, and thus a key episode in the first exegesis of Paul, which is after all what Acts represents.

The episode's starting point is Paul's arrest and appeal to the emperor. In making use of this right enjoyed by all Roman citizens, Paul prevents his possibly mortally dangerous (and thus mission-imperiling) transfer to Jerusalem for trial. At the same time, he in this way delays his premature liberation, for once it has been issued, the appeal obliges the authorities to bring Paul to Rome. Consequently, he is to be shipped to Italy, the prisoner of an imperial cohort. In Lycia, the crew switches to an Alexandrian grain-freighter, whose voyage to Crete becomes stalled: The autumnal equinox has been reached and October soon follows, hence a time of the year when seafaring actually stops on account of turbulence of wind and waves—but even more so on account of the overcast heavens, rendering astronomical navigation impossible, the voyage a big gamble.

In the city of Lasea Paul thus issues a warning: "Sirs, I perceive that this voyage will be with hurt and much damage, not only of the lading and ship, but also of our lives." "Nevertheless," we read further, "the centurion believed the master and the owner of the ship, more than those things which were spoken by Paul" (Acts 27, 10 f.). Both the ship's helmsman and captain owner (the *nauklēros*) thus insist on voyaging further, since unintended stopovers involve immense costs.[32] But Paul's warning having been brushed aside, the ship is caught up in an *eurakylōn*, a cyclone with ever-changing direction.[33] Tossed about by the winds and

31 See Macdonald 1999, pp. 88 – 107, Kratz 1979, p. 336, Cornils 2006, pp. 237, 248. On the *Odyssey*'s "toponymic" and "fictional" narration see previous chapter; on ideas circulating between the eighteenth and twentieth centuries concerning Luke's authorship, see Guillaumier 1992, pp. 58 – 65.

32 See Breusing 1886, p. 161.

33 See Zmijewski 1994, pp. 860 f., und Kettenbach 1997, pp. 55, 209.

with a soaking, swelling load of grain on board, the ship is girded with strong ropes to keep it from falling apart.[34] The situation not improving, for the sake of the ship's stability the crew is forced to throw things overboard—first crockery, then equipment, and finally the load of grain. Hence through maritime *fortuna*, the governing power of higher authorities in general here seems suspended—and, in particular, the *annona:* both the divine Roman personification of the grain supply and, through this, the theophanic symbol of universal imperial care.[35]

At this point of crisis, however, Paul again speaks up and announces the good news to be delivered by an angel: of imminent rescue not of the ship and its cargo but of its crew. He calls on everyone to fortify themselves with a shared meal. And he exhorts the soldiers to keep those sailors on board who seem about to abscond on the lifeboat—for only together can *everyone* be saved (27, 31). When the ship approaches unknown land, the Roman soldiers consider killing their prisoners before they can use the chance to flee, thus trying to continue imparting absolute validity to imperial law in a state of emergency. The captain forbids this, instead organizing a beaching safe for swimmers, and with their help for non-swimmers as well.

"And so it came to pass, that they escaped all safe to land"—namely "two hundred threescore and sixteen souls" (27, 44 and 37). Transparent here is the providential dispensation for Paul to reach Rome, so that with the appeal to the emperor there is also empire-wide promulgation of the word. Here the church's later political measures to build up the Christian community, maintain its communal cohesion, and finally see to everyone's salvation are here already manifest. Two hundred and seventy-six saved souls are mentioned, and this precise count is unique in the Gospels, as if a painstaking merchant's accounting of loss of ship and freight has become a registry of salvational events. Paul does appear to misconstrue the sailors' intention—they are actually not trying to flee but simply to save the in every respect unreliable lifeboat.[36] But when he

---

34 On this girding technique see Köster 1923, p. 140.

35 The *annona*, the empire's universally gratis basic provision of grain, mainly from Egypt, was first tried out under Gaius Gracchus, then fully executed starting with Claudius. For reasons of both political stability and propaganda, it was meant to be guaranteed at any price. To this end, into the Early Modern period the Roman authorities organized gigantic transportation systems with huge grain freighters and employment of countless private shipping enterprises. Commercial speculation with lack of grain was, at least nominally, forbidden through strict sanctions. See Rost 1968, p. 95, Casson 1979, p. 195, Göttlicher 1981b, pp. 146–149, 158, Garnsey 1983, pp. 123–128, Huber 2002, p. 175.

36 See Kettenbach 1997, pp. 27–29.

has the line to the boat cut, he is sharpening the distress at sea out of, as it were, salvational strategic cunning. That the description of both nautical procedures—from navigational maneuvers to "girdling" to beaching—and maritime legal matters—for instance jettisoning in line with the *Lex Rhodia de iactu* and the procedure for majority decision in questions of ship guidance—reveal a high degree of knowledge, for the sake of promptly giving them a parabolic turn, is unique in ancient literature.[37] In this narrative of miraculous rescue, in a downright technical manner, seafaring is endowed with a salvational sense.

Between Two Seas

Precisely the locus of Paul's beaching has significance for this salvational meaning. For this locus, where individual passengers and crew members experience miraculous rescue at the end of a voyage lacking orientation in the midst of all-encompassing darkness, only appears a saving natural harbor, a bay, at first view. It emerges as a *topos dithalassos*, "a place where two seas met" (27, 41). In exegesis based on a passage of Strabo, the reference is to a "promontory," "strait" and "small sea-strait," "channel," "sandbank," and "outer ground"; traditionally it is located at Malta, later at Mljet, and finally at Cephalonia.[38] But what is decisive here, already because Greek has its own expressions for all these significations,[39] is less the identification of a geographic or hydrographic place as the structural determination of an "event locus": the interface between two masses that can be named "sea." Because in Paul's report the skies are heavily overcast, astronomical navigation impossible, indeed blurred with the water, the ship is initially lost in absolute placelessness. What renders the beaching an event is the fact that the "place where two seas met" denotes not merely some place but the very possibility of self-localization. Namely, what here can be experienced as something visualized is a form of basic articulation: the creation of a place out of a dark horizon of expectation in which heaven and sea are still intermingled.

This unexpected event realizes more than the complex meaning of *contingere:* to touch on or border on something; to reach a goal; to befall; to turn out well. Beyond that, it rests on the ground—or rather, the groundlessness—of semiotic-logical processes. It is no accident that modern semioticians such

---

37 See Praeder 1984, p. 701.
38 On the *topos dithalassos* see Breusing 1886, p. 202, Balmer 1905, p. 415, Brannigan 1933, p. 186, Warnecke 2000, p. 69. On the controversial location on Cephallenia see ibid., p. 64.
39 See Gilchrist 1996, pp. 42, 45.

as Saussure describe the interplay of conceptual and phonetic entities as an encounter between two seas. In this context, Saussure refers to the "somewhat mysterious fact" that "'thought-sound' implies division, and that language works out its units while taking shape between two shapeless masses. Visualize the air in contact with a sheet of water; if the atmospheric pressure changes, the surface of the water will be broken up into a series of divisions, waves; the waves resemble the union or coupling of thought with phonic substance."[40] The Pauline sea-drama does not offer a simple representation. Rather, it presents the scene of a—for non-believers—"mysterious" articulation that even at first glance can be understood as a recursion of the creation story from Genesis. Correspondingly, the rescue from maritime distress resembles not only an individual new birth but the birth of a new cosmos—so that the Pauline shipwreck marks the decline of the pagan and rise of the Christian world.

With this articulation "between two seas," an ontological and representational-logical horizon is opened that is not limited to strictly theological questions. Initially it concerns the relationship between the infinite and the finite, potentiality and actualization, a relationship whose treatment in the framework of aquatic metaphors was inaugurated not only by Hebrew cosmogony: throughout Near Eastern creation mythologies, we find recourse to a world without signs and images emerging from chaos conceived as a dissolution of the primal element water and as a separation of the primal oceans.[41] In order to not become lost in this original de-differentation always already and repeatedly threatened by the sea, Greek sailors, once land and its names could no longer be seen and sea and sky seemed to meld, threw an offering into the water—a rite of passage establishing a basal difference postponing a swallowing up by the watery void.[42] For their part, Christian seafarers would make their way through the sea's chaos by orientation around *stella maris:* the "star of Mary" (Venus, Sirius, or the North Star) offered a connection between the heavens and the sea in that Mary, "Queen of Heaven," was simply a Latinization of Miriam or *mar iâm*, the "drop of the sea," which was sublimated from *stilla*, "drop," into *stella*, "star."[43]

If, as Arnold van Gennep argues, passages always take place between two worlds,[44] it cannot be surprising that the *topos dithalassos*, the "place where

40 Saussure 1959, pp. 111–112. On the static nature of this articulation and its dynamization in Lacan and in the linguistics of Deleuze / Guattari (grounded in the ideas of Louis Hjelmslev) see Berressem 2004, pp. 231f.

41 See Ramras-Rauch 1985, p. 142, Detel 1988, p. 46, and Böhme 1996, pp. 33f.

42 On this rite see Wachsmuth 1967, pp. 321f.

43 On this derivation, going back to St. Jerome, see Hönig 2000, pp. 52f.

44 Gennep 1960, pp. 20f., passim.

two seas met" became, on its part, a topos to describe the sea: from Virgil's poetic *periplus*, the *Aeneid*, to the founding of the modern seafaring report by Hakluyt and Purchas, the "sky on all sides and on all sides sea" (*Aeneid* 3, 193) serves as a topos of "primarity," through which the fundamental articulation of navigation, language, and poetry has to first constitute itself. "The Text it selfe is a Sea," is the way Samuel Purchas put it in 1625—the writing tool thus being the counterpiece to the ship.[45] From the Renaissance onward painting has (de)figured the null-point and beginning of all symbolic and pictorial representation—but also the horizon of perception and subjective "imagination" in general—in the mutual engulfing of sky and ocean, clouds and waves.[46] Correspondingly, modern oceanography describes the scene of a primary articulation as the communication between the heavens, its winds and its magnetism on the one hand, the sea, its waves, and its streams, on the other hand—a scene that, to be sure, is always threatened with destabilization by atmospheric disturbance.[47] In both cases, however, *topos dithalassos* designates the point of an elementary confrontation.

The elementary event settled for Paul between two seas is initially more closely tied to creation concepts found in Genesis than to an allegedly genuinely Christian concept of universal salvation. As suggested, a scenario of universalistic salvation is clearly identifiable in the story of Jonah. This notwithstanding, within a certain strategic perspective—that, for example of the Gnostic Marcion—the Jewish God of creation here at work is a God who, as primal ground and abyss, sees to an immemorial articulation but distinguishes Himself through an immeasurable and erring omnipotence, through threat and danger, and for that reason has to produce elementary fear. The "benevolent God" Christians consider their own, on the other hand, the God who has his only authentic apostle in Paul, as Marcion says, is the addressee of hopes for salvational-theological expectation. To remain within that strategic perspective: if the God of creation is the sole spectator of the shipwreck treated in the Hebrew Bible, the Christian God is anything but, to cite Karl Barth, an "observer, a non-participator, a

---

45 Purchas 1905, p. 57. On Virgil's topos in Purchas, see ibid., p. 56.

46 On Leonardo's maritime studies see Mertens 1987, p. 45, and Vinci 1996, pp. 21, 35, 40. On Dürer see Fricke 2010, pp. 42, 52f. On Turner see Nova 2010, pp. 86–89. On the perspectively framed sea and its churning "abyss," see Kant 2000, pp. 152f.

47 On the "tragic dialogues of those two Oceans, of air and water," see Michelet 1861, p. 26. On the "magnetic sea," see Maury 1861, p. 151. On the interplay between magnetism and currents as well as wind and waves and the "disarticulation" or "de-differentiation" of both in the seastorm, see Munk 1992, pp. 443–446.

non-sufferer."[48] Whether a messianic helmsman or spirit become immanent, He is present precisely in the abysmal moments of most extreme human distress, founding the alliance of fearlessness that Paul announces—a relational form involving human beings and God but likewise involving the relation between human beings themselves, hence their community. Hence the Pauline community in and with Christ does not first present itself in that fundamental articulation between two seas but rather in an act of self-help, as unauthorized as it is a manifestation of solidarity, through which the ship's crew, as a body, saves itself in the moment of shipwreck.

With this act, something is presented that can—despite scenes such as a last shared meal[49]—only very imprecisely be described as a repeated Christian foundational event, especially since the idea of the "Christian" is not found in Paul. More precise would be reference to a "messianic" event that, as Jacob Taubes indicates (in doing so, he himself articulates that strategic perspective), signifies "the establishment and legitimation of a new people of God"—a community of salvation rather than one limited to an exclusive "community of solidarity" and "kinship of promise" called Israel, but takes in Jews, non-Jews, and Christian Jews.[50] If, then, a foundational act takes place here, it is the foundation of universalism out of the solidarity of the *klēsis*, the "messianic announcement" that, aside from all legal and worldly identity, generates the messianic community of the *ekklēsia* in an instance.[51] This event is not "once and for all" but rather circumscribes a horizon expressed in the pronoun "we"[52] and later in the name "Christ." Even Paul's declaration that "there shall not an hair fall from the head of any of you" (Acts 27, 34) is less a genuinely Christian salvational assurance (compare Luke 21, 18) than a specification of locus for what is possible through a universalizable momentary event: everyone being saved.

The crew on Paul's ship does not consist of representatives or allegories of ethnic or legally circumscribed groups. It personifies a universalistic calling, for as long as a contingent situation (such as beaching) is "commensurable" with a truth (such as universalism), "anonymous individuals" always become "vectors of humanity as a whole."[53] What in this context is designated through

48 Barth 1968, p. 335. See also Badiou 2003, pp. 34f., 69.

49 On this "communal meal" of the Christians, which can be defined neither as mere "breaking bread" nor as a regular "Eucherist," see Schneider 1982, pp. 395f.

50 Taubes 2003, p. 42.

51 Agamben 2005b, pp. 59f., passim.

52 On the narratological analysis of the pronoun for the Pauline "narrational and mnemonic" community, see Cornils 2006, pp. 20, 237, 248, and Guillaumier 1992, pp. 62f., 70.

53 Badiou 2003 p. 20. For the following, see ibid., p. 81, and Agamben 2005b pp. 54f.

the term "vectors" denotes something pointing to the possibility of a momentary event neither absorbed into a totality—for example the assembly of Catholics—nor existing in the singular, as with the Jewish covenant—and that instead enables a new communal bond. Such a grouping—what Taubes refers to at one point in his German-language text as a *Ver*-Bund[54]—is not reconcilable with a law, whether that of the Jewish Torah, of the Greek cosmos, or of the Roman world. Namely, the law preserves the father's transcendence and is thus a cipher for finitude. The momentary event, to the contrary represents the possibility for stepping past finitude, through which it simply ignores the law as a political—either theocratic or imperial—regulative power and as a fundamental separation or allocation.

In view of the shipwreck situation and its (differential) repetition from Homer to Paul, we might also say with Alain Badiou: Greek thinking, as articulated in both epics and cosmology, signifies the "(philosophical) desire to occupy the place allotted to you in adequate fashion, an allocation of places whose principle thought can recover." "For Paul," Badiou continues, "the Christ-event, which shears and undoes the cosmic totality, is precisely what indicates the vanity of places."[55] If the meaning of both the Greek helmsman, the *kybernētēs*, and Roman *gubernatio* is continuous or definitive on-site location, the Pauline event fulfils itself in a diminishing of locus[56]—or in cosmic rupture.[57] Or to speak with Karl Barth: it fulfils itself in a place that "is no place; for it is the 'Moment'... when men surrender themselves and all that they are to God."[58] For this reason the Pauline event is not tied to an original locus, of the sort seems to be present in the singular articulation between two seas.[59] Rather, this momentary event disarticulates this origin by having the immemorial creation only emerge in the universalistic event of everyone's salvation, thus rendering the origin only possible in its repetition. Relevant here is Saussure's insight that living articulation is dynamic, an overlayering of anticipative and retroactive processes of

---

54 Taubes 2003, p. 146.

55 Badiou 2003 p. 56.

56 We can observe the narrative topography's disorientation through, already, the wind, in Paul's period the natural indicator of the heavenly directions. The wind's proper name, *Eurakylon* or *Euraquilo*, is "a Greek-Latin mixed form used in Hellenistic maritime language (from Greek *euros* = east-southeast wind and Latin *aquilo* = north wind), which is inherently contradictory." Roloff 1988, p. 362.

57 Peterson 1950, p. 79 proposes that Paul's symbolism stems from a now lost apocalyptic book and the expectation of a fundamental cosmic break.

58 Barth 1968 p. 110.

59 For the effort to archeologically pin down the locus of Paul's beaching, see Meinardus 1988, p. 205, Hirschfeld 1990, p. 25, Hellenkemper 1994, p. 161, Gilchrist 1996, pp. 29–51.

enunciation and thinking. Like speech events, faith events are a sustained vortex of signs—an enduring whirlwind of traditional topoi (of both navigation and the epic), but also the enduring "symbolic" assemblage of what has been smashed or sundered.

Instituting Salvation

For Paul, tradition does not primarily involve instituting memory or preservation. It is for its part an event—the framework being first and foremost the relation to both pagan and biblical topoi. On the one hand we find retention of a Lucan-like report, with concepts and images from the *Odyssey*, especially from the shipwreck scenes of the fifth and twelfth books. But mythological figures such as the sea-god Ino, Odysseus's helper in distress, become Christian angels. Storms at sea are no longer divine acts of punishment but have natural-contingent causes whose power is then overcome by the one God; and finally, at work here is a belief-based hope alien to Homer. This apostolic episode is particularly rich in typological references: the creation story and Noah's arc, Moses's founding of an ethnos he rescues by parting the waters, the story of Jonah—all have their antitype in Paul's last journey; and older promises of a sea-journey's end can be heard here as well. But the Pauline event does not only consist of fulfillment of traditional typological episodes. Its messianic or political-theological vector lies in the typological relationship itself—in the event's capacity for repetition and regressivity, which generates the universality of a truth in the first place.[60]

Because this "true event" founds *salvational* history, which is to say the possibility of historical fulfillment, it is no historical or tradition-grounded event: Paul's shipwreck does not have the status of a *canonical* paradigm in salvational history.[61] Rather, it is a paradigm of mercy. We might here speak of something like what Badiou terms a "universal multiplicity" whose possibility is stipulated by the momentary event, and of a state of surplus that is mercy itself, in the horizon of history and tradition.[62] *Charis*, grace, which as an act and concept stems from the legal world of slavery and legal judgment,[63] as it were tears the law from the finitude of history and tradition. By offering a prospect of liberation and freedom for all, regardless of origin and accomplishment, it gives, as for instance

60 See Badiou 2003, pp. 81f. and Agamben 2005b, p. 74.
61 See Göttlicher 1999, p. 250.
62 Badiou 2003, pp. 45, 63.
63 See Feneberg 1992, p. 157.

Badiou has seen it, everyone a universal law of love along a pathway different than writing, thus making possible for everyone, without distinction—or rather, with respect for the difference between each person—subjectivity in faith. This merciful momentary event renders love into the only command, thus placing it above faith and hope.[64] Namely faith (as articulated by Paul on the foundering ship) points to universal truth; hope designates perseverance and faithfulness along this path (what the ship's crew demonstrates while being en route without any instruments); but love circumscribes the universal impact of love itself (an impact demonstrated in the rescue of everyone on the ship). Love specifies the dimensions for understanding what belief is capable of and what hope makes possible against the law of reality.[65]

The event of mercy-grounded rescue is a moment of salvation that is necessarily missed with an assimilation into tradition or history or a monopolization as institutional founding. On the one hand, in its own self-understanding, the Church is, as Karl Barth has put it, "[t]o a greater or lesser extent...a vigorous and extensive attempt to humanize the divine, to bring it within the sphere of the world of time and things, and to make it a practical 'something,' for the benefit of those who cannot live with the Living God and yet cannot live without God."[66] On the other hand, as soon as the Church takes on administrating salvation and begins to propagate the idea of enduringly supplied grace and thus guaranteed security, it loses its eventfullness. The dynamic between institution and event reveals itself in all its sharpness in the case of the Church: this institution only exists with recourse to an instituting event that for its part depends on the possibility of representation and retroactive actualizing. The Church can by no means be derived from a symbolic sphere but cannot exist without it. In such an institution, there is an intertwining of what Cornelius Castoriadis describes as "a functional component and an imaginary component... combined in variable proportions and relations" while at the same time being nothing but the "form, rule and condition of what is not yet."[67] Institutions are not the direct result of events—for instance of universalism's founding—but at best the perpetuated effort to grasp such events recursively and actualize them along with the possibilities they enclose. However, an institution becomes the reactive result of a result when it strives to grasp the possibilities of an event once and for all in a past totality, a singular foundation, or a legacy.

64 See Taubes 2003, pp. 52f., 55.
65 See Badiou 2003, pp. 89f., 93f., 96.
66 Barth 1968, p. 332.
67 Castoriadis 1987, pp. 138, 219.

Even in its regressiveness, the momentary event itself preserves the potency of non-being. If in the case of Paul, the event takes place, as Taubes suggests, in the framework of "political theology," being thus directed against the exclusive Jewish covenant, and especially against the massive constitutive power of the Roman Empire, it by no means establishes the metastructure of a new universal grouping. It simply circumscribes the realm of its possibilities, which also comprises the possibility of non-existence. The momentary event thus offers no guarantee of salvation or a dispensing with existing constitutions. Rather, it articulates a problem, thus opening the possibility of a solution. What Paul names *klēsis*, calling, organizes fidelity to the event. In this context Agamben speaks of an "exigency" that as a specific messianic modality appears alongside the categories of possibility, impossibility, necessity, contingency: this exigency "consists in a relation between what is or has been and its possibility. It does not precede reality; rather, it follows it."[68]

In the account offered in Acts, the captain, helmsman, and ship master initially show no fidelity to Paul following his warning. But when the situation becomes dire they do so, allowing everyone's rescue by independently doing his bidding.[69] With a view to this shipwreck situation, we might understand the messianic possibility in terms not only of the *kybernēsis* but of modern cybernetics as well: within this schema, Paul first frees the faithful from the "fixed setpoint control" of Jewish ritual law, replacing it with a non-deterministic teleology. This not only allows a cybernetic modeling of what we might term coincidence or contingency. A rule-led relation to freedom allowing discipleship for everyone and nevertheless only insisting on an opening toward possibilities can be described as "follow-up control" whose "command variable" is then the presence of Christ in the spirit. In this respect, Hans Reinhard Rapp observes as follows: "Adaptation to the various peoples and their social and religious pre-understanding, indispensable for the Christian mission, was only made possible in this way—the adaptation not, however, leading to a lack of purpose."[70]

This idea of *kybernēsis*, both universalistic and open, was immensely important for early Christianity and its eschatological horizon of expectation. It is even arguable that through divine charter and pneumatic enthusiasm, Paul engaged in nothing less than a hijacking of the world-imperial power's cargo ship, image, and credentials. After all, he prompts the soldiers to abandon the lifeboat and the sailors to a beaching that is nautically equally nonsensical in that an or-

---

68 Agamben 2005b, p. 39.

69 On the "unrealistic" prominence of Paul the prisoner see Roloff 1988, pp. 358, 361.

70 Rapp 1967, p. 99.

derly anchoring would be very possible.[71] Before the age of Christian ecclesiastical hegemony, missionary work thus amounted to piracy, in the literal sense of *peiran*, to test, try, risk. To be sure, as soon as it was institutionalized, the Church did not so much orient itself around cybernetic control processes and practice audacious forms of piracy as seek firm representational forms. It did understand itself as a ship, but in its real organization, its didactics, and its narration,[72] it exposed itself less and less to the experiences and events leading through a continuous vortex of signs. Rather, it increasingly limited the strict observance conveyed by *religare* to the original, nautical sense of binding fast, mooring.[73]

When in what Hugo Rahner has termed their "nautical theology," the Church Fathers visualized the constitution of this salvational institution, having been founded once and for all, as an ecclesiastical ship, they did so in a firmly allegorical framework.[74] This extended from the yardarm (called "antenna" and something like the reception instrument for heavenly messages) to the mast (the cross of Jesus) and onward to the rudder. Patristics understood faith-enlightened feeling and understanding (*sensus*) as helmsman, guided by Christ himself, of the Church's destiny-grounded community. Christ was here seen as entering into the presiding providence of helmsman and father through the winds of the pneuma located in the divine *logos*, the final authority of governance and rule.[75] Where in Paul's First Epistle to the Corintheans (12, 28), *kybernēsis*, governance, would still be understood as an eventful and merciful divine gift, it was now handed over, with finality, to the pastoral office.[76] Since then, the cybernetic calling has been considered either a regulated vocation of congregational leadership—belief here being governed less in the horizon of its possibilities than in that of its maintenance—or, extending to Justinian's idea of governance and beyond, as what Peter Classen refers to as an "imperial administration commissioned by a higher power,"[77] hence as a straightforward *calling* to power.

---

71 In Acts 27, 5 the voyage moves through "Lycia," a center for eastern Aegian piracy, before the angel conveys this message to Paul: "lo, God hath given thee all them that sail with thee" (27, 24). See Wehnert 1990, pp. 92f., Zmijewski 1994, p. 862, Rapske 1994, p. 28, Kettenbach 1997, pp. 27–29, 37, 209, 215–220.

72 On the "narratological exegesis" of the Pauline "event," see Eisen 2006, p. 130.

73 On this "original" nautical sense see Kettenbach 1994, p. 433.

74 Rahner 1964, p. 268.

75 See ibid., pp. 365, 328, 332.

76 See Goldammer 1950, pp. 235–237. According to Foucault, Christian *kybernēsis* outdoes its Greek counterpart in that it involves leading not only things (especially states, enterprises, and ships) but henceforth human being. Thus seen, Greek navigational *technē* here comes together with the pastorate. See Foucault 2007, pp. 96–99; 122f., 165ff.

77 Classen 1972, p. 9.

Once the apostolic missionary project on a maritime medium was completed, the Church thus finding itself on firm, indeed unshakable, ground, then the sea was meant to simply vanish, entirely in line with the apocalypse. For preferably everyone, including especially sailors, would be pinned down to the true faith once and for all. Repeated probation in the matter of faith, love, hope can hardly be in the interest of an already established, ecumenically oriented salvational institution, and the same goes for experimenting with what we might call a culture of self-help and indeed solidarity—as already manifest in both the book of Jonah and the Greek "religion of the seaway." But such a culture is mandatory precisely in voyages on the high sea. For this reason, the Church, come to power but powerless vis-à-vis the sea, considers ships as breading grounds for pagan and wildly superstitious, blasphemous beliefs, cults, and ways of life otherwise overcome. It was the reason, as well, for the Church considering sailors, if not exactly what the Spanish humanist Juan Luis Vives termed "scum of the sea,"[78] then at least as unreliable characters. And it was the reason for entering into an unceasing struggle against transgressions and errors of faith precisely in this maritime locus—as if the sea were a wild zone or no-man's-land. Seen in this way, the *navis ecclesiae militantis*—the ship of the church engaged in worldly battle—was more than mere allegory.[79] For precisely on those ships undertaking voyages made under a highly Christian sign—voyage such as those of the Spanish fleet—religiosity served as a means of discipline and authority-production.[80]

In actuality, the old nautical sacral practices had hardly vanished under Christian auspices and in the age of established ecclesiastical observance. We have a fourth-century report, for example, which is to say from a period when mass conversion had necessitated compromises with forms of popular piety, of a nautical practice of using the consecrated Eucharistic host as a kind of Christianized amulet, what Franz Joseph Dölger terms a "means of protection" and "assurance against the dangers of seafaring".[81] Even Neapolitan votive pictures from the sixteenth century record the sacrifice and apotropaic use of a white sheep. In addition, we have documentary evidence of various modern practices for invoking and cursing the elements and inhabitants of the sea—in the case of Portuguese sailors, for instance, exorcistic rituals for driving out the devil from

78 See Boxer 1984, p. 110.
79 On the engraving of Pieter van der Borcht (1545–1608) with the title *Navis ecclesiae militantis*, see Göttlicher 1999, p. 168.
80 On this, on the Christianized rites at departure, in maritime distress, and on arrival, and also on the inquisitorial inspection of ships, see Pérez-Mallaína 1998, pp. 237 f., 241, 245.
81 Dölger 1976, p. 247.

the sea, or the propensity of Spanish sailors to toss reliquaries into the depths.[82]And finally, starting with the epoch of globalized seafaring novel rites of passage developed, for example "equatorial baptism," the line-crossing ceremony already documented for 1529 and still being practiced in the early twentieth century.[83] Because precisely the cultic moment appears to count as a universal feature of seafaring, not only the Mediterranean and Atlantic nautical cultures but also putatively "primitive" or "non-historical" societies have become the object of observation for modern ethnology and history of religion.[84]

The Sea of Dread

In the Early Modern period, which saw the initiation of the first comprehensive organization of maritime transport, nautical prayer books served in part to encourage a trust in providence meant to render superfluous magical practices and superstition. God "guides all things through His veneration-worthy providence," writes the Reform pastor Théophile Louis Barbauld in 1688 in his *Prières pour ceux qui voyagent sur la mer.* For Barbauld, God is the master "of all unfortunate and adverse events."[85] Under this portent the book of Jonah was of special interest for Protestantism—for after all, the Jonah story made clear what providence meant for Christian seafaring and Christianity in general. For this reason, an exegesis of Jonah such as Calvin's offered an ambitious seafaring nation such as England the guiding principles of proper faith at sea. Already both in his institutional doctrine and approach to predestination, Calvin excludes any unauthorized historical-critical rationalizing about Holy Scripture and its providential consequences: "Let us, I say, permit the Christian man to open his heart and his ears to all the discourses addressed to him by God," we read in an English translation of the *Constitutio*, "only with this moderation, that as soon as the Lord closes his sacred mouth, he shall also desist from further inquiry."[86]

---

82 See Delumeau 1984, pp. 58, 63.

83 On the course of this initiation rite, in which even the higher-placed crew participated, see Beck 1985, p. 117.

84 On the role of magic and the "social fact" of the cult of sacrifice in shipbuilding and shipwrecks, together with the function of ring-exchange ("kula") in Melanesian seafaring, see the detailed discussion in Malinowski 2014, passim.

85 Cited from Delumeau 1989, p. 429.

86 Calvin 1843, p. 143 = III. 21. On the disciplinary impact and function of Calvin' Jonah sermon, see Sherwood 2000, pp. 33–38.

A problem traditionally understood as treated in the Jonah book is that the prophet finds no sense in pronouncing a prophecy that God will not fulfil; Calvin interprets the problem in the framework of the doctrine of providence—God can change his will at any time. Precisely what appears to be caprice or coincidence reveals God's hidden providence. Correspondingly, the storm buffeting Jonah's ship, like any accident at sea, is not due to "chaunce," but rather is divinely sent, for "nothing can come to passe, unlesse from thence it be both fore seene and decreed."[87] Furthermore, Calvin argues, God never lacks reasons to test every individual sinner, even when the sinner presents himself as innocent. The encounter with putative chance, whether maritime or as drawing lots, leads to necessary confession of guilt. "Let everie of us acknowledge our sinnes, when they be reproved," demands preacher Calvin, "even although chyldren be our Judges, or if any of the most bale or contemptible ryseth against us, let us suffer it patientlie, and let us know this kind of Controwlers happeneth unto us by the providence of God." Jonas deserves to be swallowed up in the "bellie of hell," "terrible punishment," and "continuall torment...more greevous unto him then a hundred deathes."[88] But even if daring seafaring (or life's journey in general) inevitably is accompanied by countless dangers, firm believers can count on their delivery—and consider Jonah's miraculous rescue as an "image of our resurrection."[89]

Against this backdrop, it cannot be surprising that the first English-language anthology of shipwreck reports, *Mr. James Janeway's Legacy to His Friends, Containing Twenty-Seven Famous Instances of God's Providence In and About Sea-Dangers and Deliverances* (1675), is fully attached to Calvin's doctrine of predestination, while also having the Jonah story as a narrative and interpretive model. If hungry mariners, off course, Odysseus-like, somewhere in the waters, here see "nothing but the Belly of destruction opening for them," the "finger of God" has a fish appear and throws it for unexpected satiation of all on board, then seeing to it that their route crosses with that of another, rescuing ship. In tales of providence of this sort, rescue and mercy are consciously put off to the last moment: "when we think of nothing but sinking and perishing, then doth God think of saving and delivering." Seafaring, then, is an unparalleled school of faith, precisely on account of its indissoluble dangers: "let going to Sea, being in storms at Sea, being brought to extremities at Sea, learn you then to pray."[90]

87 Calvin 1578, p. 9v.
88 Ibid., pp. 21 v, 31 r, 33 v, 34 v.
89 Ibid., pp. 34v, 42r. – See also ibid., p. 10r.
90 Janeway 1680, pp. 2, 5, 133.

Through the enduring encounters seafaring makes possible, but also through sailors' characteristic persistence, a regular competition between outdated rituals and sacral and superstitious beliefs of the most varied provenance prevails on the high seas. Christian seafaring is marked to considerable extent by an effort to absorb these heterogeneous elements into a unifying matrix of true faith. To this end, a foundational narrative document such as Acts is useful. Since Paul, stories of voyaging and conversion are a firm ingredient of hagiography, so that Christian saints—perhaps most prominently Brendan[91]—would gradually replace the ancient maritime divinities. Even more than the emergency deployment of holy water, their relics served to magically placate the elements. Furthermore, there are countless nautical prayer-formulas and on-board cults, chapels, and processions. In all of these, antique and local survivals corresponding to Christian cultic rituals were elevated and sublimated.

We here are offered witness to the omnipresent anxiety seizing sailors in face of the sea. The genetic theory of religion deriving cults and sacrifice from dread of the waters does not originate in either Early Modern observance (exemplarily in Calvin) or enlightened cultural scholarship (as with Herder) but in the High Middle Ages, for example the writing of Josephus Iscanus.[92] With a view to nautical sacral practices and their persistence into Early Modernism, Dietrich Wachsmuth has spoken of a "tragic contradiction within human culture," one in which on the one hand an "act of 'self-help' preserves the 'ephemeral' human being from going under," while on the other hand the human being "is placed *through this* in basic opposition to the divine world order."[93] Spoken cosmogonically, this opposition was originally absent—whether we are speaking of a Golden Age or sea-distant paradise. But the "origin" and its depictions are only rooted in loss, separation, and contradiction, for which the sea stands, its endurance signifying "culture." The sea, then, stands in exemplary way for a "culture of failure"—for a dynamic of exposure to loss and danger together with initiation of self-help measures. In establishing a "total social fact,"[94] that culture's cultic practices, its rites of passage, and there especially its rites of sacrifice, respond

---

91 On the Brendan tradition and its dependence on the *Odyssey*, especially the *Vita Pauli*, see Pietrzik 1999, pp. 53–55, Strijbosch 2006, pp. 1–7, Gerds 2009, p. 28.

92 See Calvin 1578, p. 11r.: "Feare compelleth unwylling men to come unto God. Wickedlie truely, sayth one, that feare is the cause of Religion, and was the originall cause why men thought that there were Gods." See also Herder 2002, pp. 21, 26: "Data lies in everything explaining the first mythological age....a genetic explanation of the wonderful and adventurous." On Josephus Iscanus see Heydenreich 1970, p. 77.

93 Wachsmuth 1967, p. 228.

94 See Hubert 1968, pp. 306 f.; Durkheim 1976, pp. 72 f.

to the threatening loss and fear of it. The living beings, but also the objects, proffered or cast off, destroyed in the process are delegates of, belong to, the collective. Namely, they establish a transition from the *Ding* to the *thing*, from the cast off object to *cosa nostra*.[95]

In the framework of religious history, we might say that the sea has from time immemorial been what Jean Delumeau calls the "locus of anxiety, par excellence," and sailors have always been both the most pious and impious of persons.[96] Especially clearly in rituals such sacrifice, their religions serve to mitigate and defend against *phobos*,[97] the menacing moment that in a cult such as the sacrifice by sinking is still largely undefined, but is now addressable in the form of a feared object. Sacral practices are meant to bridge over the abyss and reach the "great Other" and its will. Without the certainty of monotheistic faith, who this Other is and what it can want are open questions.[98] The only thing that is fully clear is that the abyss does not so much reveal itself within society as yawn at its edges, and that in the "dread at," something expresses itself that is only determined—spatially or temporally—by this "at."

The sea itself is this abyss and this "of." A ritual practice such as sacrifice stands surety for a consistent Other being present in its elementary status, however obscure: an Other that has to be feared, that can be placated or else cannot be.[99] On the one hand phantasms of the sort characterizing maritime religion constitute a kind of shield covering the abyss, rendering it bearable in the first place and allowing it to be dealt with. For as it would be "horrible", as a text appearing in 1600 on threats from the sea put it, to imagine all the mortal dangers and possibilities of loss tied to a sea voyage.[100] On the other hand, while not capable of depiction or designation as feared objects, the quasi-objects and quasi-causes in play here, progeny of the "of" or else its opening to the abyss, are bound to the imaginary and the chain of the symbolic—in order to incontestably "insist" within that locus. As objects included in exclusion, they signal that the anxiety or dread is no deception. "The water *observes you*," reads an ancient

95 See Lacan 1992, pp. 43 ff.
96 Delumeau 1984, p. 51. – See also Fabre 2005, p. 191.
97 See Böhme 2009, pp. 157–162, 174–176.
98 On sacrifice as healing of the rift or abyss opened up in the other, see Žižek 1991, esp. pp. 58 ff. On monotheism vis-à-vis polytheism exemplified by "Islamic seafaring," see Barthold 1929, pp. 40, 43.
99 See Lacan 2006, pp. 740 f.
100 Cited from: Delumeau 1984, p. 56.

Greek command for caution that points by way of "the gaze," a prototypically non-depictable and non-signifiable object, toward the "great Other."[101]

This hides itself in the sea, and its will is as unfathomable as the sea itself. But the religious human being tries to address it and its will through cultic practices such as sacrifice. Mantic sacrifices, for example, are meant to fathom the will of the great Other, thus also anticipating the future. For their part, things or living beings tossed overboard and into the sea's abyss, thus being "abjected," are tools for such fathoming in a literal way.[102] The seafarer enacting a ritual renders himself existentially dependent on the great Other, as shown already in the tossing away of his possessions. At the same time, he is in this way rendered seducible, indebted, even at the price of manifest profligacy. Hence in antiquity even pirates owed the gods a tenth of their gains.[103]

Christian seafaring tried to do away with such cults, grounded as they are in the "superstitious" belief in a devouring great Other, whether as the formless sea or as a monster of the depths. Instead, faith was meant to center on divinely blessed rulers who, like Herakles, triumph and rule over the monstrous sea, and in the process secure seafaring by vouching for their own safety.[104] In the end, what gave impetus to reformation of the nautical mentality was dread at dread (and wild forms of its overcoming). By contrast with pagan, putatively primitive sacral practices, Christian seafaring avoided material loss, let alone loss of life, in its place founding a new cult: that of safety and the soul's salvation. This involved not only what religious historians have been wont to describe as an interiorizing or sublimizing process: here sacrificial beings, be they animals or tainted persons were personified and saved, and at the same time it inserted a premise of objectivity into what had been the province of sacral practices in pre-Christian times.

What prompts anxiety first and foremost, and what especially prompts it among seafarers, is "the real," lying behind the regularly returning signs and manifestations of the sea and seafaring, its cycles and routines. Lying behind the *automaton*, *tychē* had menacing import, as chance become fate. The Greeks tried to gain a sense of *tychē* through oracles and omens, mantics and divination.[105] But finally the ideas tied to chance and fate stemmed from the cult of sac-

101 Wachsmuth 1967, p. 287. On anxiety as a "signal" of the real, see Lacan 2014, pp. 160 ff.

102 On Lacan's pun of *je* and *jeter*, thrownness and throwing, see ibid., p. 7.

103 See Höckmann 1985, p. 158.

104 On the Early Modern ruler as Heracles and the relationship between *sûreté* and *sécurité* see Delumeau 1989, pp. 11 f., 24, 524. On the Heracles-Hesione saga see Gerhard 2003, p. 230.

105 On the relationship between Greek augury, seafaring, and fear, see Minois 1998, pp. 72, 81, 109–111, and Bowden 2005, pp. 122, 158.

rifice, its partition and distribution of life and value, divine and social participation: the most important gods of fate, Moira and Aisa signify "portions"; Ker, the fateful god of death, is derived from *keirō*, "to cut"; and demons could intercede into the course of events at any time as an "allocator."[106] The Roman pendant of *tychē*, Fortuna, derived from *ferre*, "bring," and thus related to what will come, was itself elevated to a deity that alongside amorous matters, luck in battle, and affairs of state would also be tied to dangerous or auspicious seafaring.[107] Hence together with a globe, a wheel, and a sacrificial offering cup, a rudder was one of the attributes granted Fortuna, so that soon the cult of the maritime driver of fate emerged, Fortuna Gubernatrix. Notably, Fortuna handed out her gifts in the same capricious way dramatized on a daily basis in the course of the sometimes fortunate, sometimes failing sea voyage. Supplied with a rudder, the only nautical attribute to become an imperial insignia, this Fortuna stood closer to the Goddess Annona than to the Early Modern merchant's goddess of fortune: less skill in trade than imperial provision was her main concern.[108]

When it came to maritime loss and danger, already in the Hellenistic and Roman periods there was more reliance on the law than on either the goddess of fortune or the protection offered by sacral practice. Otherwise than nineteenth-century legal history of antiquity generally maintained, the ancient world was largely unfamiliar with the legal concept of *casus fortuitus*, coincidence or coincidental misfortune. Such events were simply part of Fortuna's domain, not that of law. *Casus* was understood in the sense of "negligent behavior," so that when suitable forms of self-help were at stake, not so much contingency and its apprehension inherently stood at the center but rather appropriate behavior in maritime distress and the legitimate ways to deal with what was unpredictable.[109] In order to allocate damages, the legal institutions of *jactus* (law governing jettison) and *nautikon daneion* (law governing maritime credit), took their place alongside the institutions of cult and sacrifice.

Codified in the *Lex Rhodia de Iactu*, common law regarding loss suffered during navigation provided for a community formed of all those exposed to the danger of loss at sea: those directly damaged, the shippers, and the owners of salvaged cargo.[110] What would be practiced into the modern period as "average" (or as *Große Haverei*) preserved its earlier context in the technical concept

---

106 See Bremmer 2007, p. 138.

107 See Patch 1967, pp. 89, 144f.

108 See Göttlicher 1981a, pp. 29, 31, and Göttlicher 1981b, p. 219.

109 See Meder 1993, p. 22.

110 On the reports of Livius, Sueton, and Cicero concerning occasional, official, and private arrangements on maritime insurance for goods and persons, see Vance 1908, pp. 1–3, 6.

of "sacrifice." A community-born "sacrifice" is defined as what in maritime distress cannot be preserved by giving up something else, but what itself preserves other material—the ship, the cargo, or the crew. To be compensated communally (through a share of received goods), as already stipulated in the *Lex Rhodia de Iactu* but still adhered to at present, the "sacrifice" must be voluntary, necessary, and successful. The success has to result from its rescuing or preserving effect. The necessity indicates real or at least evident urgency, traditionally verified not only by the shipmaster but the ship's entire council. And the voluntary quality of the "sacrifice" lies in its being the result of a conscious decision to jettison goods and not of force majeure, an act of God such as a mast breaking in a storm.[111]

These regulations establish solidarity, because they distribute damages justly, prescribing a kind of rite of passage as it were spontaneously, when an emergency is stipulated, and fix the distribution retroactively. Similar here to the previous sacral practices, the regulations are thus of a guiding, defensive nature. By contrast with the case in the old cults, they contain no anticipative or speculative moment. That was rather the concern of the *nautikon daneion*. Corresponding to mercantile regulations, the *emporikoi nomoi*, the institution of maritime credit allowed a long-distance trader (an *emporos*) to take funds from a ship-owner (a *nauklēros*) or shipmaster (a *nauta*) for major voyages without risking ruin in the case of total loss of cargo. For if the ship and thus the whole enterprise went under, then the repayment duty stipulated in the loan contract (the *syngraphē*) was also extinguished. Whoever agreed on the opposite side of such a speculative investment had a right to an interest occasionally as high as 30 percent. As a rule, debtors with such credit could continue to do business even when cash was short, together with being able to lay claim to a certain form of insurance protection upon surviving an accident, while prosperous creditors—already because the high interest they charged was morally justified—could productively invest their capital.[112]

This practice, documented in most detail for antiquity by forensic speeches of Demosthenes (on the recovery of overdue loan repayments) was commonplace into the High Middle Ages. Maritime loans became increasingly less worthwhile after Pope Gregory IX radically limited interest income through his usury ban of 1230. In order that capital that had become mobile in the High Middle Ages could

---

111 On the *Lex Rhodia de iactu*, see Ashburner 1909, pp. xcf., cclii, ccliv–cclvi. On the corresponding maritime regulations from the eighteenth to the twentieth centuries, see Falconer 1780, art. "Average," Dana 1856, pp. 293–328, esp. pp. 298, 316, and Kreutziger 1966, pp. 38–41. On acts of God, see Spooner 1983, p. 11.

112 See Millett 1983, pp. 36, 42–51.

nonetheless be used in long-distance trade, *foenus nauticum*, the bottomry bond, was considered on the one hand to be an interest-free loan transaction, on the other hand to involve a contract for remunerated assumption of dangers.[113] Finally, the church gave its blessing to a practice in which dangers were understood, for the first time, as a "risk" (derived from *risco*, "cliff") and this as an article of commerce, through which, however, *securitas*, previously the religiously guaranteed securing of the soul's salvation, became a matter of *assicurazione*, insurance against material loss. It is not surprising that Calvin would be the first to declare the ban on interest not obligatory. For Calvin would mark a turning point in the Western sense of economy and providence, but also in the modeling of accident and contingency. His doctrine of predestination as hidden divine will formulates our still-unknowing state as a mandate for both economic success and regulated social provision.[114] Religious calling informs the ethos of its secular counterpart, the vocation, whose economic balance-sheet is still meant to be read in salvational terms. From that time onward, "hope" is not merely a theological but a mercantile concept, expectation of profit having become calculable along with risk. Now operationable in insurance-technical terms, hope is tied to shares and allocation, while universalism and security appear subsumed into a monetary medium.[115]

For the history of seafaring and salvational assurance, maritime insurance was in any event a pioneering project in that it for the first time took on all the dangers potentially facing ships and cargo during a voyage—"every danger," we read in a Pisan policy of 1383, "sent by God or coming from the sea or from human beings and every chance event or danger or stroke of fate and every misfortune that could come about in any sort of way."[116] Under a regime of provision against dangers, a regime replacing cultic practice that founded community through wasteful sacrifice with a culture of security and compensation, hope involved not so much rescuing as insuring everyone and everything. In his manuscript on public insurance companies, Leibniz thus declares the ancient institution of jettisoning to be the basic principle of all insurance: "For just as the *Lege Rhodia de Iactu* wisely saw to recompense being paid from shared costs for ware tossed out to relieve a ship, the entire republic should be considered a sort of ship enduring much weather and misfortune, so that it would be inequitable for only few to be affected by the misfortune while the others escape free."[117]

113 Discussed in more detail in following chapter.
114 See Campe 2012, pp. 23 ff.
115 See Badiou 2003, pp. 96 f.
116 Cited from: Liebig 1914, pp. 52 f.
117 Leibniz 1923 ff., IV. 3., p. 424.

Suggesting itself in this context is a developmental schema for the history of mentalities in respect to confronting boundary-transgression and corresponding dangers: a schema that with the ideas of *taboo, sin,* and *risk* distinguishes magical-cultic, moral-conscientious, and rational-probabilistic approaches with corresponding collectively sanctioned options for action.[118] Aligned with this schema is an understanding of the policing to which mariners were subjected starting with the Early Modern reorganization of the fleets as a kind of reformation of still remaining, wild and superstitious cultic practices. As for instance the monumental *Hydrographie* of Père Fournier (1643) shows, the disciplining drew on the strict piety of the mariners (*gens de mer*), their Christian form of life and clear understanding of authority.[119] Through all-encompassing demands and regimentation imposed upon them, the ship would now become, in words of Ralf Lisch going back to Erving Goffman, a "total social institution," an isolated sphere, removed from the profane realm yet totally subsumed to the commonweal's concerns.[120] And the fact that fear of the dangerous influence of a putative *asebēs* is at best a remainder of seafaring's "superstitious delusion" (Kant's—obscenely shaded—*Afterglaube*),[121] is self-evident for the Enlightenment: Kant, for example, demands replacement, precisely in the case of seamen, of the cult of "sacrifice" by a moral culture that "sacrifices," even in the worst emergency, everything that is merely contingent, including one's own life, to absolute duty.[122]

It is nevertheless unclear whether the significations and functions that cultic sacrifice and sacral practice once had for the social test case of maritime commerce found their true continuum in processes of interiorization and objectification. In the technical framework of insurance, force majeure—be it in the form of paralyzing dead calm or of a raging form—is uninsurable. It is no risk but rather a danger, thus remaining an act of God, itself addressable through hope, prayer, or other sacral practices.[123] And that what is termed "sacrifice" in insurance law concerns more than property law is shown by various modern cases involving seafaring. For example, into the nineteenth century slaves, whose legal status

118 See Wiedemann 1993, pp. 45–50.

119 See Fournier 1667, esp. *Livre vingtiesme,* "De la Deuotion & Pité des gens de Mer."

120 See Lisch 1976, pp. 11–14, 56.

121 See Kant 2009, p. 215.

122 Kant's reference to mariners' concerned behavior in maritime distress: "but if it is said of someone who, in order to preserve his own life, pushes another survivor of a shipwreck from his plank, that he has a right to do so by his (physical) necessity, that is quite false. For to preserve my life is only a conditional duty (if it can be done without a crime); but not to take the life of another who is committing no offense against me and does not even lead me into the danger of losing my life is an unconditional duty." Kant 1966, p. 299.

123 On acts of God in antique liability law, see Brecht 1962, pp. 145f.

—person or property—was in any case always controversial, would be thrown overboard in maritime emergencies, compensation claims then being made for the "sacrifice" of alleged property.[124] Even at the start of the twentieth century, there were cases involving situations of maritime distress and hunger in which individual crew members were killed and eaten, those accused of murder and cannibalism claiming exoneration on grounds of the ancient seafaring ritual of casting lots to choose a sacrificial victim.[125] In such cases procedures of maritime law were consistently overwhelmed, then reformed, as a culture of security had here been overlayered with cultic elements.

Even if placed under a sign of strict accuracy and objectivity or shifted into the state of claimed emergency law: repeatedly, maritime peril has compelled wild or outdated rites of passage and asserted the cultic and sacral moment of seafaring, against all legalizing, capitalizing, and reforming. Perhaps one thing in particular explains the stubborn persistence into Early Modernism and well beyond of seafaring customs and rituals attesting less to piety than—to speak with Jacques Lacan—to a special form of nautical surplus pleasure: the continued existence, despite all mechanization of sea passages and insurance of imperiled assets, of a "community of danger" that always has to consider failure. Integrating this community into the society of terra firma, with all its *nomoi*, its regulations, duties, forms of commerce, would mean conferring a social structure of security, care, and participation on its members—hence what ethnologists see as preformed in the archaic system of the "gift," of regulated living with and for one another.[126] Within the Early Modern regime of maritime and globally circulating capital, mere sailors were, however, always limited to a system of living from one another—hence of sacrifice. As lower "workers on the sea," they hardly participated in the command of their voyages and resulting gains. They had little to lose when it came to fixed tenets, accumulating capital, inalienable rights. Their domain was and remained the sea of dread; their hope was for a fortunate voyage.

---

124 On the ancient exclusion of slaves from the *Lex Rhodia de iactu* see Ashburner 1909, p. ccliii. On the Kantian humanity-grounded argument against "jettisoning" slaves, see Schiebe 1840, p. 115. On insurance of slaves as transport insurance, see Ebel 1963, pp. 208–215.

125 On the controversial insurability and "sacrifice" of slaves, together with the putative justification of cannibalism though "randomization," see Armstrong 2004, pp. 168–182. On providential rescue from cannibalism in situation of hunger and distress at sea, see Janeway 1680, pp. 4, 7.

126 See Mauss 1966, pp. 63 ff.

Chapter 2
In Search of Fortune: The Birth of Insurance from Maritime Danger

Aventiure and Fortuna

The pecuniary significance of liquidity—that financial attribute already attached to the sea since its earliest damnation by "cultural criticism" only emerged in the Early Modern period. Money could only become a medium for "absolute," because global, connection once modern capitalism ventured onto the sea[1]—according to Hegel, the "supreme medium of communication." To be sure, in order to deploy itself within "the element of fluidity," trade had to simultaneously expose itself to "danger and destruction.[2] To again cite Hans Blumenberg, incalculability, lawlessness, and the sea's resistance to orientation—its third demonic moment—inevitably stamp sea-trade, however "calculably" it can be organized. The sea, that abyss of dangers, has been the source, and this from seafaring's start, not only of the metaphorics of life's journey and *homo viator*, but also of the Early Modern concept of "adventure."

Let us use the German-speaking lands as our example. Middle High German *aventiure* (derived from Latin *advenire*) was initially the key term in a court context, simultaneously denoting a narrative art centered on both the challenge and preservation of honor, *êre*, and knighthood, *rîterschaft*. Then, beginning in the Late Medieval period, it increasingly came to designate a daring venture, a danger, or, even more specifically, *risk* in the area of mercantile activity.[3] Where the pilgrim, aware of his limited time on earth, turned from earthly activity and its perils, and the knight—already because of the ethos of his estate—tried to defy fate in some extraordinary trial, the aspiring mercantile middle class would strive for stable prosperity—this achieved through worldly activity shielded against chance and damage. *Aventiure* had now become a business matter. As such, it signified either the fortune or misfortune of what the German term *Handeln* captures in its double sense: trade and action. Hence just as starting in 1443

1 On the cogency of the term "capitalism"—only coined in the nineteenth century—and on the economic-historical linkage of capitalism's takeoff with Early Modern maritime trade, see Braudel 1985, p. 49 f.; Braudel 1972, pp. 510 f; Wallerstein 1998, pp. 45 f., 59 f., Leidinger 2008, pp. 22, 47 f., 51 f., Le Goff 2012, pp. 142 ff.

2 Hegel 1991, p. 268.

3 See Keller 2009, pp. 124 f.

https://doi.org/10.1515/9783110610734-006

English merchants who had taken up maritime textile trading with Iceland and Scandinavia, the Netherlands and Southern Europe called themselves "merchant adventurers," in the German-speaking lands market production with uncertain demand was now termed *aventiurehandel.*[4] This activity would soon be carried out through the *commenda* money-lending system, in which at their own risk creditors placed their capital with traveling traders in order to gain a portion of the eventual total profit—should the enterprise succeed. In this way, on the one hand merchants fixed in different locations could be interconnected: instead of personally accompanying their business they could manage it from their trading station. On the other hand, this was tied to recognition of their adventurous courage, their willingness to bear the risk of delayed or no repayment—recognition that in the longer term led to a precise determination of capital-loss risk (*periculum sortis*), to its offsetting as legal "interest," and finally to contractual partnerships and sanctioning of promissory notes.[5]

In the early sixteenth century, both what clearly owed its value, not to established custom or official certification, but rather to trading and market processes (for instance jewelry) and something derived from happy fortune (for instance unexpected earnings or possessions) were termed *Abenteuer* in the German-speaking areas.[6] In this context, Early Modern High German *auventura*, in the signification of commercial uncertainty or haphazardness, would establish itself as a synonym for *risigo*, together with the terms *angst* and *anxst*. At the same time, *auventura* would refer to a sphere of lawlessness, incalculability, and disorder, as feared outside urban walls or terra firma.[7] It is thus unsurprising that in the eighteenth century, when the earlier adventurers wished to be understood, more than ever, as consolidated entrepreneurs, the previous honorific *Aventurier* was mainly awarded pirates—after all, as a lexicon of the period explains, lacking "fixed abode," they enriched themselves for the most part through simple "plunder."[8]

With the Early Modern concept of adventure being tied first and foremost to the fortune or misfortune encountered by a daring or at least uncertain undertaking, it was only natural to bring such *adventura* together with old goddess Fortuna and understand it as chance, or coincidence, or what befalls the seeker of

---

4 See Bonß 1995, pp. 123 f.

5 See Nerlich 1997, pp. 276, 297, 305 f.

6 See Welzig 1969, pp. 439 f.

7 See Kuske 1949, pp. 547–550, und Rammstedt 1992, col. 1046.

8 *Allgemeines historisches Lexikon*, Leipzig 1722, vol. 1, p. 265, cited in Fohrmann 1981, p. 183.

fortune.[9] In literature, the manner in which an increasingly established art of trade became aligned with a gradually pragmatized understanding of fate, providence, and chance was initially reflected on in Late Medieval and Renaissance Italy: source of both the revolution in maritime trade and poetry's renewal. Dante, for example, still very clearly subsumed Fortuna to providence. As a pagan goddess, Fortuna could only persist due to a lack of a Christian equivalent in everyday understanding;[10] she was now meant to symbolize the structure of possibilities within the terrestrial realm, and at best stand for divine governance. But speculation on her favor was strictly off limits—on pain of seduction by the sirens' song of worldliness, that unfruitful desire bearing its most unnatural fruit in the "excess," *dismisura* (*Inferno*, XVI. 74), of usury.[11]

In the late medieval Italian *novella*, the merchant's speculation on Fortuna is, to be sure, no longer simply rejected. But it remains associated with an always menacing possibility of boundless loss, and in addition with the need to secure one's fortune in good time. In the "Landolfo Rufolo" tale in Boccaccio's *Decameron* (c. 1350), Landolfo wishes to see his in any case huge fortune doubled and thus invests all he has in a shipment of goods to Cyprus, where, however, it can only be sold at cutthroat prices because of the local market situation. Suddenly impoverished, he turns to piracy and thus gains new riches. But first caught up in a storm, then—at harbor—robbed of his wealth by other merchants, fate has it that, having become a castaway because of another storm, he chances on a casket full of jewels. Landolfo can now retire from trading, and especially from trading at sea.

This sort of *aventiure* at sea is consistently centered on the occasion of *adventura*, the involuntary arrival of profit or loss. It also presents us, in the process, with the great difficulty in distinguishing between business and robbery, enterprise and perdition. And in the end, it thus only thematically addresses the challenge poised by—or indeed the avoidance of—all-powerful Fortuna, rather than spelling out any options for acting in face of the contingent. At most, it demonstrates the unfathomable nature of providential decree—or put allegorically, the steady process of up and down, the continuous turning of Fortuna's wheel. In Brant's *Ship of Fools*, Fortune—the overestimation of its reliability, and with it of one's own privilege; venturing onto sea and underestimating

9 On the similarity between *ventura* personification and *fortuna* allegory, and on the conceptual convergence between *adventura* and *occasio* in the sense of "opportunity" and "chance," see Patch 1967, pp. 39 f.
10 On the Christian denial or adaptation of the goddess of fate, see Gregersen 2003, p. 362.
11 See *Inf.* VII, 61–96; Patch 1914, p. 20; Rachewiltz 1987, p. 141; Münchberg 2005, p. 160; Ruud 2008, pp. 53, 58.

"luck and weather" (hence *fortuna di mare*)—still functions as the most lurid image of the foolish search for happiness, and with it of a sinfulness leading directly into the abyss.[12] No less than a turn away form the world is a preferable option, meaning avoidance of the sea and all its commerce: the perspective, it would seem, of the southern German-speaking lands at the time. However, within the perspective of a doubly adventurous experience of the world and a correspondingly facts-based narration, the sea becomes—in this region as well—the medium of a new art of trade. For in what was probably the first of all German prose novels, the anonymous *Fortunatus* (1509), the key turning points of the eponymous hero's life history are tied to seafaring, its perils and their overcoming.

Probably written in Augsburg, which since the mid-fifteenth century was the center of German trade and thus a collection-point for global empirical knowledge, this *Volksbuch* is arranged like a kind of *hystoria:* a heterogeneous narrative genre developing didactic material from things both true and invented. Supplied with elements from fables and legends, chronicles and novellas (including motifs from the *Gesta Romanorum*, the German folktale, and the *Decameron,* it also contains traces of the travel reports and numerous treatises on trade circulating at the time. Fortunatus personifies the type of the new adventurer: after leaving his impoverished parents' house in Cyprus, because "there's a lot of happiness in the world" (*noch vil glüks in dieser welt*) to be discovered,[13] this fortunate, fortune-seeking young man, following various detours and false paths, enters a "wild wood." Here "Lady Fortune" (the *junckfraw des glücks*) appears before him; rather than corresponding to the allegorical figure of the German *tychē*-tradition—for example the *Saelde* or "Lady World"—she is similar to the Romance world's Fortuna, her mercantile and "accounting" connotations included.[14] The fortune she grants Fortunatus in the form of gold pieces form an inexhaustible "purse" (*seckel*) allowing him not only intermittent wandering into the unknown but travel routes and seafaring expeditions chosen as he pleases—and finally development of a far-reaching trading network, purchase

---

12 See Brant 2002, p. 88f., 134f., 415–417 = chaps. 23., 37., 109. See also Gruenter 1966, p. 92; Kemper 1996, p. 79.

13 *Fortunatus,* transl. Michael Aldane (first complete translation of the *editio princeps* [1509] into English, with original illustrations). Only in digital form at: http://www.michaelhaldane.com/Fortunatus%20Illustrated.pdf, p. 3; for the German original see: Roloff 2007, p. 8.

14 *Fortunatus,* transl. Aldane, p. 35; Roloff 2007, p. 46. See Wiemann 1970, pp. 62f., 66f.

of land at home, and—with the advent of risky curiosity about the world—a voyage to India.[15]

The *aventiure* as a wandering into terrestrial labyrinths, leading in the end to successful testing, can only find its continuation in the Early Modern adventure when the basic premise for middle-class mercantile activity is fulfilled, when money is available as Alfred Sohn-Rethel's titular "bare coin of the a priori."[16] For worldly experience, fortune is thus needed in the form of inexhaustible assets. For this reason, narration in *Fortunatus* will now be novel-like and full of "world substance," corresponding as it does—in the fashion of itineraries, with their indications of stations and distances—with the experience of the world, "written down," *erschriben,* in the style of a travel diary and its descriptions of lands and peoples.[17] But the worldly experience and recording always depends on the deployment, the increase and circulation, of the assets—something like the starting capital of all Early Modern knowledge of the world. Global exploration thus unfolds in a general context of mercantile bookkeeping; continuously observing what can be experienced, taking its inventory, involves a process of exploration oriented towards the horizon of possible future experiences, indeed possible worlds—and at the same time calls for an economizing and efficient use of time. Consequently, we can understand *Fortunatus* as not only attached to a generic tradition of commercial and instructional handbooks circulating in upper Italy and soon also north of the Alps starting in the early fourteenth century, meant to help merchants achieve bookkeeping and arithmetic competence, the art of trading, practical knowledge of the world at large.[18] Rather, it also should be read in the context of various published itineraries, especially the travel descriptions of Johann Tucher and Sebald Rieter the Younger.[19] And last but not least, starting with the Portuguese seafaring expeditions, and with the hope tied to them of forming an alliance in Ethiopia with the mythic

---

15 See *Fortunatus,* transl. Aldane, p. 83: "To his wife Cassandra: 'I am not going away for the sake of pleasure, or luxury, or material profit...I have seen half the world, and now I want to see the other half, even if it costs me my life. I cannot clear my head of this wish. So give me your consent, for no one can prevent this but God and Death.'" Roloff 2007, p. 99: "'ich zeüch nitt auß umb wollust / wolleben / noch umb güt zu gewinnen. ich hab das halb tayl der welt gesehen. so will ich das ander tayl auch besehen Unnd soltte ich mein leben darumb verlieren [...] das mag niemant wenden / dann got und der tod.'".

16 See Sohn-Rethel 1990.

17 *Fortunatus,* transl Aldane, p. 66; Roloff 2007, p. 80.

18 On this genre's form and tradition see Dotson 1994, pp. 13–26; Friedmann 2000, p. 394. On the bookkeeping economization of time, meant to prevent accumulation of inactive capital through a precise fixing of the duration of sea trade, see Tenenti 1990, pp. 227f.

19 See Raitz 1984, p. 105; Roloff 2007, pp. 247–251 (appendix).

Christian patriarch Prester John, it occupies the eschatological horizon of expectations always characterizing Early Modern capitalism, in its expansive, maritime, and global variants.[20]

The perils Fortunatus will henceforth encounter on his adventurous path are mainly of the sort concerning the mercantile system. Mythic threats such as those once marking the knightly *aventiure* (in the form, for example, of fabulous creatures) here no longer play a role, the sea's own role now being defined by the indistinct and uncertain. As an elementary uncertainty, *fortuna di mare* is henceforth only mentioned when praying for *gütt wetter* for an East India voyage.[21] Natural dangers have here lost their abysmal quality, as we perhaps see most clearly in the *Fortunatus* episode concerning St. Patrick's "purgatory" or cave: where previously people fantasized about a mystery, even the entry to the underworld and, presumably, absolution from all sins, now we have the emptiness of a ramified cave system—danger consisting only in the possibility to get lost and thus lose economically useful time.[22] With one's lifetime or even life itself become a temporal resource, the real danger to Fortunatus's path forward would be abuse of his mercantile ethos, of the form of intersubjective commerce to which he owes his career: contractual reliability, the presence of trust in trading reciprocity and the obligations held to the "symbolically generalized communicative medium" (the phrase is Luhmann's[23]) par excellence, money.[24] That in conflicts with the feudally structured world, threats take the form of indissoluble differences of status, legal asymmetries, and the capricious exercise of torture, is something Fortunatus learns at the very start of his life as a merchant. And that a phlegmatic temperament or even one marked by too much restlessness and amorousness, or by either rapacity or inadequate discretion, must mean the end of one's own fortune, is spelled out in the quick bankruptcy suffered by Fortunatus's wayward sons and heirs, Ampedo and Andolosia.[25]

When it comes to such dangers *Fortunatus* reveals an unusual realism and nearly textbook precision. On the other hand, the novel is rather mystifying in its description of a money economy as such. In fact, the lack of a grasp of the self-aggrandizement of capital, of its detachment from utility-value and annulment of the estate system, is the basis for the narrative's central magical ele-

---

20 Raitz 1984, pp. 105f.

21 Ibid., p. 100.

22 See *Fortunatus*, transl. Aldane, pp. 48f. See also Müller 2008, p. 447.

23 Luhmann 2013, p. 180.

24 On the violence-deflecting impact of money as a medium, see Luhmann 1996, pp. 253, 259f.

25 See the episode with the count and the career of Fortunatus's son after his death, *Fortunatus*, transl. Aldane, pp. 103ff.

ment: the money-purse, presenting the phallic character and wondrous generative power of money. The novel points to *Fortunatus* as not only attesting to the impetus of rationalization and calculation within Early Modern experience—but also, at the same time, to the compensatory function of Early Modern imaginative literature. That literature is the locus where "what withdraws from the construction of meaning is [nevertheless] retained," reflected on formally and substantively.[26] In respect to motif, this leads to an identification of sexual with financial potency and between the natural sequence of generations and capital: we thus see Fortunatus already, although not yet a "capitalist," distinguishing himself in a courtly context, "thrusting before the lords and ladies"[27] before taking flight because of a threat of castration; being obliged, once furnished by Fortuna with the fortune-bringing purse, to annually endow a marriageable maiden in the manner of the charitable medieval "mounts of piety"; having his purse cut away from him in Constantinople, so that he fears he has been "deprived...of virtue"; and his son Andolosia being rendered drunk and insensible during the night of an anticipated amorous adventure and having the purse stolen from him once and for all, so that he has overslept both "fortune and felicity."[28]

Here money has become what Walter Raitz calls an "absolute mover,"[29] the catalyst of all the novel's action and striving, while at the same time serving as a mediating vehicle for all worldly experience and options for such action, hence for the narrative process itself. If at first Fortuna leaves it to Fortunatus's discretion to choose between the virtues of "Wisdom; Riches; Strength; Health; Beauty; and Long Life,"[30] eventually the choice of riches will reveal itself as opting for choice itself, for only money can allow options both in the technical trading and more general sense of action; it thus sets up an option-tied future, consequently one with risks that invite daring rather than simply being full of perils. The money-choice renders the adventurer into a *homo oeconomicus*. And what opens the space of experience and narration is, as Joseph Vogl observes, "less adventure, which, befalls you and is overcome, than risk, which you intentionally enter into and anticipate."[31] When, drawing the narrative's moral, the epilogue

---

26 Müller 2008, p. 436.

27 Roloff 2007, p. 13 (this phrase omitted from *Fortunatus*, transl. Aldane).

28 *Fortunatus*, transl. Aldane, pp. 6, 10 ff., 36, 56, 118 f. See also Hörisch 1998, p. 114. On the Early Modern social-provision and endowment funds known as the *montes pietates*, which as an opportunity for speculation would at the same time contribute to dissolution of the ban on usury, see Braun 1963, pp. 21 f., 27 f., 41–43; Koch 1968, p. 41.

29 Raitz 1984, p. 29.

30 *Fortunatus*, transl. Aldane, p. 35.

31 Vogl 2002, p. 180.

of *Fortunatus*—very likely not written by the author, and already omitted from the second, 1518 edition[32]—concludes by stating that "Lady Fortune...has been hunted from our lands, and is to be found in this world no longer,"[33] then this at the same time sets a seal on the vanishing of any decisionary option concerning the existence or non-existence of capitalism.

Once opted for, choices can only be made within its increasingly expanding boundaries, legal security being expected to that end—as evidenced in *Fortunatus* especially in trading-metropolis Venice.[34] But the mysterious origin of capital recedes to the same degree it expresses itself as regular repetition compulsion: in investing his accumulated fortune in real estate on his homeland, Cyprus, Fortunatus makes use of financial difficulties suffered by the Count of Ligorno to properly dispossess him; and even when he is received with all honors by the sultan in Alexandria and initiated into the secret of his magic hat, he feels justified in fleeing with it to Cyprus, in order to then never give it back: after all, in trying on the hat, which transports whoever puts it on wherever he wishes in his thoughts, he almost landed in the sea, thus falling into a "fear" "I shall remember until my dying day."[35] Resting on "original accumulation," Early Modern capitalism demand an ability to seize the moment. This constitutes, to cite from Walter Raitz, "the real substrate of the allegory of fortune."[36]

Fortunatus thus reflects the evolution of trading and financial capitalism and the role it played in the equipping, organizing, and expanding of gradually globalized commercial maritime commerce. In the end the Augsburg merchant houses generated their fortune and their power (which would become unmistakable with their electoral assistance for Charles V at the latest) to a significant degree from East India trade, before eventually concentrating on banking and turning to real-estate investment.[37] Against this background, the Sultan's magic hat in *Fortunatus* has been interpreted as a mystification of double bookkeeping: that technique of rationalization, introduced in fourteenth-century northern Italy, which through inventories, dating, and balancing of all business operations

---

32 See *Fortunatus,* transl. Aldane, p. 166.

33 Ibid.

34 Ibid., p. 67: "Here, there are many rich people; here, you may open your money to view." "hye tarffestu dich auch lassen mercken das du gelt habest," Roloff 2007, p. 81.

35 *Fortunatus,* transl. Aldane, p. 97.

36 Raitz 1984, p. 27. On the allegorical proximity between *fortuna* and *occasio* see Patch 1967, p. 116.

37 On the ascent of the Fuggers, their long-distance trading and dispatch, and their agitation against the usury ban see Fugger 1995, pp. 112–117.

offers a kinds of panoptic view of all trading events, and in this ubiquity made possible, among other things, control of partners and commissioned agents.[38]

But not only this already long-established, calculative procedure, together with its displacing impact, is reflected in the episode of the magic hat. The sea route to East India was only opened up a short time before *Fortunatus*'s writing. And with this trading route, which first attracted financing from Augsburg in 1505, the entire world became a sphere for (concrete) trading and (abstract) circulation of capital: a development tied to the decline of the old Mediterranean trading places. Alexandria would now be absent as a relay-station for East Indian trade, for which reason the merchants and authorities there no longer needed to be treated like equal trading partners. Venice reacted to the loss of its spice-trading monopoly by founding a commission whose proposals included alliance with the Sultan of Egypt so that he would then check Portuguese activities, to the extent possible.[39] In any event, fortune was no longer to be sought on trading routes headed east. Henceforth the ocean would be plied for the sake of making capital from the incalculable—through a consolidation of seafaring routes and routes of trade.

Risky Partnership

Otherwise than beforehand, in the Early Modern period there is a sense Fortuna can be banked on—that she is more than a moody goddess of unfathomable fate. It is the case that in the form of *fortuna secunda* and *fortuna adversa* she continues to present herself with the double face of fortune, understood as simple chance: on the light side, gain or enjoyment, say of riches or of love; on the dark side, perils from which pain or loss are the expected outcome. But Fortuna no longer presides over a fate to which individuals are simply delivered—for better or worse. Henceforth she stands for a *need* for fortune, a need constituting the basis for courage to take their own fate—or better: a shared future—in hand. She has become an allegory of calculated daring. As such, she attests to the new conviction that fortune is to be achieved by consciously and, if possible, together accepting the threat of loss and pain.

---

38 On the rationalization and control function of double-entry bookkeeping see Lane 1987, pp. 177, 179 f., 190 f.

39 See Stefani 1958, I., p. 96.

This is evident on the field comprised of both intimate and commercial partnership—a field that, albeit only partially, has opened up to uniquely favored Fortunatus, that pioneer and champion of a—still heroic—capitalism circa 1500. Shakespeare, who possibly was familiar with the material of *Fortunatus*, including the motif of fateful choice, through its English adaptation, *The First Part of Fortunatus* (1596),[40] offers a contrast to this by staging the adventure of partnership in "The Merchant of Venice" (1596/97). Shakespeare's statement of the guiding principle of all pursuits of fortune is equally applicable to love and commerce: "They lose it that do buy it with much care."[41] In both realms, partnerships are defined by peril.[42] As such, they promise protection and security. But to the same extent, they are grounded in an acceptance of insecurity and uncertainty, for the sake of gain or pleasure. In both realms, certain risks are voluntarily undergone in order to open chances for the future. Furthermore, in the modern period partnerships are accompanied by increasing risk—meant to be understood as a (second) vanguard of the arts of trade and human action. For the partners are driven on by long-term turmoil, pressing needs, expansive interests; with their mutual promises, obligations, and credit, they act in anticipation of an uncertain future; and when they support each other with forms of speaking and commerce bridging all sorts of space and time, we can speak of a regular "drama" of Early Modern striving for fortune.

In the commercial metropolis of London around 1600, it was actually customary to contemplate one's own art of trade and manner of acting as a drama. For example, as a "city comedy," Ben Jonson's, George Chapman's, and John Marston's "Eastward Ho!" (1605) presented a panorama of the instable trading world in the England of James I, with the maritime portrayed as a benefactor of the common weal, the clerk as a usurer, hence as the actual enemy of city, court, and country. With the help of mere "gossip" and under the alias of "security," the clerk sells ""excellent uncertainties," consequently an illusion of protection at the seafarer's cost; he is driven by "slavish avarice" and must finally offer

40 This adaptation served as a model for Thomas Dekker's "Pleasant Comedy of Old Fortunatus" (1600), a play that may have also been influenced by Shakespeare's "Merchant of Venice." See Reichert 1985, pp. 40 – 43" Hörisch 1998, p. 268.

41 Shakespeare, "Merchant of Venice," 1, 1, 80. All references to "Merchant" in this chapter, both in the main text parenthetically and in the notes, as well as to other Shakespeare plays throughout the book, will indicate act, scene, and line numbers sequentially.

42 On the modern definition of insurance as a *Gefahrengemeinschaft*—"community of danger"—and association of similarly endangered see Büchner 1976, p. 111.

the most abject apologies before the collective.[43] Seen in this light, "The Merchant of Venice" is one of a number of plays examining the question of the quest for fortune as one centered on *both* amorous and commercial trade. But although or precisely because it mirrors Shakespeare's present in Early Modern Venice, it possesses special historical acuity: as the leading Mediterranean maritime and trading power, Venice was both model and precursor of an Elizabethan England where at least equally immense capital was being expended, for the sake of then investing and augmenting it through maritime overseas trade. Starting with the Middle Ages' waning, a range of forms of commercial partnerships were maintained in Venice, influencing European trade in general. Initially formed by well-established and prosperous families, these partnerships were increasingly granted legal security by the authorities in respect to both transport and business activities at home and abroad. Sea trade was organized within this commercial framework, in the form of loans, the above-mentioned *commenda* contracts, and other alliances of work and capital.[44]

Several key fourteenth-century developments were the foundation for success of this Early Modern trading and finance-capitalism. The import of paper and of Indo-Arabic ciphers and, based on this, the invention of double bookkeeping prompted a *rivoluzione commerciale*. It not only made possible the cooperative creation of credit but also allowed the former itinerant trader, the *mercante viaggiatore*, to take care of his business through agents, from his office, and thus as a *mercante residente*.[45] At the same time, a nautical revolution unfolded—the introduction of new sorts of ship, widespread use of the stern rudder, improved tackles, compasses, and portolan charts now allowed regulated and largely reliable shipping commerce even in winter. This commerce created the infrastructure for steady expansion of overseas trade and the profitable deployment of suitable financial means.[46]

While the customary initial formula in the account books used by merchants was "in the name of God and good adventure" (*col nome di Dio e di buonaventura*), or "in the name of God and good profit" (*in nome di Dio e di guadagno*),[47] the reference here was less to the Christian divinity and His providence and more to Fortuna and her promise. In trading metropolises such as Venice, it was nat-

43 Jonson 1994, pp. 35, 44, 57. On shipwreck and insurance fraud in Jacobean "city comedy" see Zwierlein 2004, pp. 79–82, 92f.
44 See Lane 1944, pp. 85–96.
45 See Ifrah 1993, pp. 528–538; Seife 2000, pp. 90f.; North 2009, pp. 28–32.
46 See Lane 1963, pp. 606–611; Lane 1973, pp. 119–122, 132; Lane 1987, IX: pp. 233–242; and Rösch 2000, pp. 100–107.
47 Cited from Nerlich 1997, p. 322; Nehlsen-von Stryk 1989, p. 200.

ural for this goddess to manifest herself primarily in fortunate or unfortunate maritime enterprises—in *fortuna di mare. Lo Spozalizio*, that Venetian rite of passage, already pointed to this: the duke here annually married the republic to the sea through a ring thrown into the water, in this way not only ceremonially renewing their partnership but also incorporating the sea, excluded from all ownership since antiquity, into civil law. But it was also attested to in the contractual formulae now defining the maritime venture in the context of commercial partnerships as well: "risk and fortune" and "risk or fortune."[48]

Shakespeare's "Merchant" takes up the theme of risky partnership in a double plot. The play is located in both Venice's mercantile world and in an idyllic country estate named Belmont where the nobleman Bassanio wishes to court beautiful and wealthy Portia. Because he is in debt, he borrows money for this "pilgrimage" (1, 1, 127) from his friend Antonio, a melancholy merchant whose capital is currently not available because scattered on several trading shops on various seas. Antonio will naturally grant Bassanio the loan interest-free, assuring him that "My purse, my person, my extremest means / Lie all unlock'd to your occasions" (1, 1, 145 f.). And, in a plot as iconic as it is notorious in Western literature, in order to lend the money, he goes so far as to pledge his "person"—which is to say a pound of his flesh—to the Jewish moneylender Shylock, in the event he cannot repay the interest-free loan of 3,000 ducates within three months. Antonio sees no problem with this, expecting three times as much to be available with the return of his ships. For this reason alone, he does not hide his contempt for the moneylender.

Supplied with Antonio's loan, Bassanio now travels to Belmont for the courtship process specified by Portia's father in his will. Like two failed suitors before him, he has to choose between three caskets, gold, silver and lead, each bearing a different maxim of fortune in partnership. If he decides for the false option, he will never be allowed to court another woman. His fate is thus closely bound up with the decision. Counter-intuitively, he chooses the correct, lead casket, but receives word of Antonio's misfortune immediately after: news has reached Venice that his ships were wrecked; and Shylock is demanding timely repayment—or bloody fulfillment of the agreed-on bond. In the ensuing court hearing, Shylock insists on honoring the contract, the Doge fruitlessly asking him for mercy. Then Portia, disguised as a young solicitor, takes over the judge's office. She confirms the contract's validity, but after Shylock again insists on his rights, she interprets the contract literally: the moneylender has a right to precisely the agreed on flesh, nothing more or less—and in addition not a drop of blood. But since as

48 See Villain-Gandossi 2001, p. 60. On *Lo Spozalizio* see also Heller-Roazen 2009, p. 120.

a "stranger" Shylock is manifestly striving after the life of a Venetian citizen, she condemns him to loss of his fortune, also placing *his* life at the disposal of Ducal mercy. Finally, Antonio sees to the Jewish moneylender's forced conversion and claims administration of his fortune, before—last but not least—Portia announces the providential return of the seemingly lost ships.

Consequently, fortune and misfortune are here not simply subsumed to the unfathomable moods of Fortuna. Rather, they are treated in a complex way—through an enmeshing within the plot of bond and the second plot of casket. Already through the fact that in the play a loan is required for Portia's courtship, love and business are inseparable, while inversely the credit relation with despised Shylock only comes about because of the love-dealings. *Fortune* here means aptitude but also earthly riches that include both amorous and mercantile acquisition. This ambivalence of economic and affective realms, of commercial and personal valuation, is conveyed in "Merchant" through terms such as *worth* and *dear*, just as *business* is used for both business and courtship. And when Bassanio accompanies his fortunate choice of casket and fortunate courtship with the words, referring to an inscribed scroll he offers Portia, "I come by note to give, and to receive" (3, 2, 1510), he is speaking like a debtor whose timely repayment is about to be confirmed by his creditor. Portia, who, addressing Bassanio, insists on freeing his friend and creditor Antonio from his debt by paying many times its amount, logically converts such monetary value into the price of love: "Since you are dear bought, I will love you dear." (3, 2, 1691) Furthermore, in the case of both Portia and Bassanio personal quality is expressed in material fortune, amorous capacity in economic competence: just as Portia represents not only a good catch but a young woman with noble character, Bassanio is less a dowry chaser than a gallant nobleman who knows how to mutually convert the pecuniary and amorous sides of partnership.

Forming a contrast to this in Shakespeare's play is the Jew Shylock, an unfortunate figure *par excellence*. Namely, were the play's import limited to an antisemitic resentment that, always furnished topoi for a basic critique of capitalism, the "worth" of Shylock's character would need to be constituted in as clear-cut manner as is the moral condemnation of charging interest. But what the drama describes is, on the one hand, Shylock's successive isolation and a resulting embitterment, and, on the other hand, a fundamental contradiction between economic practice and doctrine: a contradiction from which Shylocks ostracism and eventual disfranchisement emerges. Antonio lends his money at no interest and thus lowers the interest rate, Shylock's "well-won thrift / Which he calls interest" (1, 3, 371 f.). He scorns the Jews as a usurious nation that lays claim—as an enduring basis of existence, indeed as its well-earned profit—to what "the hand of heaven" (1, 3, 420) at any time can give or take. But for the sake of Bassanio's courtship, he is even willing to borrow from Shylock—albeit not without discrediting him as a partner and directly declaring him an "enemy" (1, 3, 462). In the concluding court proceeding, after Shylock is robbed of his—speaking in a strict legal sense—legitimate rights and fortune, Antonio for his part lays claim to interest at the moneylender's cost (see 4, 1, 2330). In the end, Shylock, the Jew, personifies what is standard procedure in the trading metropolis but is nevertheless condemned—what for the house-keeping subject is good practice but for the moral subject bad, and what for that reason is included in the name of good *aventura* and good profit but excluded in the name of God.

When Antonio reproaches Shylock with being incapable of friendship and partnership "for when did friendship take / A breed for barren metal [= interest] of his friend?" (1, 3, 460 f.), he is taking up a polemic against usury carried forward in the West from Aristotle into Scholasticism and the Early Modern period. In the *Politics,* Aristotle defines "Chrematistics" as an "art of acquisition" equal to an "art of wealth-getting" that "has...suggested the notion that riches and property have no limits."[49] Wealth in accordance with nature emerges from *oikonomia* understood as the entirety of a household. But money acquisition rests on trade and sales, and like every art has no natural limitations in its activity—rather setting itself as the highest telos, thus denigrating real fortune. For Aristotle, chrematistic's most extreme excesses are arrived at in the usury business. Usury, we read, is "most hated...and with the greatest reason,"

> making a gain out of money itself, and not from the natural object of it. For money was intended to be used in exchange, but not to increase at interest. And this term "interest" [*tokos,* lit. "offspring"] which means the birth of money from money, is applied to the breeding of money because the offspring resembles the parent. Wherefore of all modes of getting wealth this is the most unnatural.[50]

On account of this conflict with nature and true fruitfulness, but also because it violates the imperative of justice and loving one's neighbor, usury was condemned in the New Testament, the Patristic writings, and the Church councils. It was considered a sin, even the soul's death. "Lend, hoping for nothing again," said Luke (6:35; King James version), from which already Ambrose, Bishop of Milan, developed the following definition: "Usury is receiving more than one has given."[51] Interest, which seemed useless even for exchanging goods let alone producing them, and in addition violated the principle of "just

49 Aristotle 1941, 1257b (p. 1141). Henceforth all citations of Aristotle are from this edition.
50 Ibid., 1258b (p. 1141)
51 Cited from Le Goff 1988, p. 26.

price," attested to avarice. Simply propped up on the work of others, it also offended the divine plan of creation in a double sense. For in the Christian Middle Ages, work was considered necessary expiation for the Fall. But at the same time, it was understood not only as punishment but also, increasingly, as a divinely granted opportunity to recuperate dignity through productive cooperation in the work of creation. Usury—which in this period not infrequently was deemed the same as charging any interest—defied both interacting conceptions. Beyond this, seen from the perspective of natural law, trading with someone else's property—and possibly against the owner's will—was nothing but theft. And in that credit was based on deadlines, it involved theft of what belonged to God, not human beings: time.[52]

This theological critique of usury initially developed independently of Aristotle; his money-theory served more to reinforce that critique than to establish it. Starting with the Synode of Pavia in 850 and continuing to the late twelfth century, the ecclesiastical arguments gradually gained shape. Citing words of Jesus (Pope Urban III was the first to do so) and drawing ontological distinctions (the approach of Thomas Aquinus in particular), practical measures were formulated such as precisely limiting interest and offering compensation to those affected by usury. But there was also an increasingly differentiated assessment of money lending and its economic function, also at work in business procedures such as purchasing credit. Especially the *societas*, as an established form of alliance in which two or more connected parties invest their money for the sake of their working toward a shared goal, was discussed in terms of one all-deciding question: whether the possibilities of loss and gain were distributed asymmetrically or justly, hence whether a usurious or legitimate commercial partnership was present.[53]

Around 1600 England, its commercial center London included, was in a state of upheaval, moving from an economy based on an "entire household" toward one based on chrematistics: a situation manifestly reflected in "Merchant" in the opposing economies of Venice (unmistakable in the risk-taking not only of Shylock but also of Antonio and Bassanio) and Belmont (presented clearly in Portia's sovereign economic control of her surroundings). The shift was initiated in the later sixteenth century—notwithstanding the Puritanical preference for a limited household in line with Xenophon's conception in the *Oikonomikos*—by none other than Queen Elizabeth herself. The unmarried and childless regent was concerned less with measures for genealogical reproduction than with pro-

52 See ibid., pp. 39 ff.

53 See Blumhardt 1911, pp. 74 f.; Noonan 1957, pp. 12, 16–20, 53–55, 133 f.

moting commerce and world trade, so that she saw to the purchase of a great deal of high-interest credit; accordingly, Elizabeth legalized credit at interest in 1571, under the pressure of maritime enterprises demanding ever more capital.[54] There was a ten percent top interest rate, mostly used to its fullest. But the process continued to be condemned in the framework of a confessionally coded critique of usury. In his *Discourse upon Usury by Way of Dialogue and Orations* (1572), Thomas Wilson thus states that "lending for gayne is a chiefe branch of covetousness" and the usurer is "a greedy gayner for him selfe, seeking his own welfare upon good assurance, without any care at all what becometh of his neighbour, grawing him unmercyfully, to satisfie his own wrechted and most greedy hunger."[55] In this context, it was an easy matter to draw on capital loans as abstract business transactions with no negative moral implications for the debtor—while at the same time to tie despised interest-taking to greed, mercilessness, and evil character. After their expulsion by Edward I in 1290, there were barely any Jews in Elizabethan England. Nevertheless, within the play's imaginative structure, professional moneylenders are defined as Jews; it is as easy to obtain credit from these moneylenders as it is to discredit them.

If—as is manifest in Shakespeare's play—the value of property and that of personality mirror one another, then within Europe's medieval and Early Modern society as reflected in "Merchant," Jews were often viewed as not due any larger amount of wealth. Notwithstanding the creditor's legal power in this period in respect to liens and mortgages, if Jewish assets were needed to, for example, solve severe royal money problems, the assets could be confiscated by royal decree—a constellation that seems reflected, first in Shylock's absolute power over Antonio's life and death, and in the end, however, in Shylock's dispossession.

Christian nobility, on the other hand, embodied a value which involved the personal no less than the pecuniary. If noblemen were already credit-worthy on the basis of reputation and social esteem, it was also the case, in Georg Simmel's words, that "it is not necessary to be a gentleman in order to obtain credit, but rather that whoever demands credit is a gentleman"[56] If Antonio's noble-mindedness appears to show itself in doing the absolute utmost for his friend through financial *and* personal credit-worthiness, Shylock does not baulk to translate Venice's social and moral code into its plaintext: "he is a good man" here means only that Antonio is solvent, "that he is sufficient" (1, 3, 339 f.).

---

54 See Damlos-Kinzel 2000, pp. 116–127.

55 Cited from: Cerasano 2004, p. 42.

56 Simmel 2004, p. 485.

In openly stating that personal qualities and liquid capital are convertible, Shylock lays claim, as a Jewish creditor—*a fortiori*—to the credibility that, as presented in "Merchant," for Christians can be deduced from simply need of credit. That the claim is publicly rejected not only reveals a confessional asymmetry in forms of social commerce in Early Modern finance-capitalism, but also the fundamental catachresis of its self-description: where personal reputation is the condition for commercial partnership, and inversely asset value is the condition for social esteem, in both the societal and economic spheres credit on the part of the debtor is indispensable. But the imbalance of this socio-economic analogy quickly becomes visible on the side of the debtor: "social credit" is indeed granted, but charging interest is not conceivable as even when increased "repayment" takes place here, it is necessary to speak of a "gift," as denoted in the context of love or friendship. In the credit-based economy, however, paying interest is necessary for the reliable and steady capital lending that keeps commercial activity afloat. Hence one "unnatural" fact in particular underscores the distinction between the credit systems characterizing societal and economic spheres: that we do not only work with money and pave the way to fruitful social relations with it, but rather, to the extent money penetrates social life, it itself must "work" and—albeit unfruitfully—"multiply"."

Shakespeare's Usurer of Venice

If everywhere comparisons are made between things as incommensurable as money and personal qualities, then it would be inconsistent to exclude the comparison, on the basis of "essential differences" between confessions, at points where it collides with the principles of Christian anthropology. In taking the concept of "personal worth" and that of the "bond" literally, Shylock is being more consistent in this respect: he designates a pound of Antonio's flesh as a pledge, recognizing that "if it will feed nothing else," which is to say its negligible exchange value notwithstanding, "it will feed my revenge." (3, 1, 1288) In addition, he understands the "bond" or contract in the sense of *contrahere*, binding Antonio and opposing this elementary dependence to the economic and moral fiction of free and equal exchange.[57] He here turns the operation that turns fatal to him

57 On the monetary effect of "transformation of moral networks by the intrusion of the impersonal," hence on the economically unorthodox founding of the economy on credit instead of on free but indebting exchange, see Graeber 2011, pp. 21 ff., 43 ff., 332 f., passim.

as a Jew against his Christian business partner, rendering what actually is merely a function of calculation again into substance. Namely, the phenomenon that furnished sixteenth-century economic practice with what Werner Sombart described as the elements of "de-concretization, de-naturalization, and de-personalization"[58] (of reification, real abstraction, and value-mobilization), elements later attributed to the modern economy in general, was standardized and extensive money-circulation. Beyond that, it was the introduction of money substitutes in the credit business, from personal certificates of debt to fungible bearer-securities. "The money economy differentiates...possession and personality," which become "independent opposites," observes Simmel.[59] But precisely this independence allows economic management with both in a homologous way, so that credit and cash can increase each other, just as "the fundamental emotion of love can manifest itself sensually and spiritually in such a way that these two forms strengthen each other."[60]

Just as love is most alive when daring and high stakes are present, commercial capital is augmented when someone risks investing it for use of others. This is especially the case with the risk-taking tied to credit, although in the context of professionally impersonal operations and larger amounts of capital, good repute and the trust of partnership are no longer a sufficient safeguard. In modern business relations, assuming such risk cannot, in the long term, be a personal benefit but must be a service, consequently something needing compensation—if only, in the worst case, to dispose over sufficient capital reserves. That this understanding can very much be theologically grounded would become clear with Calvinism at the latest. No longer tied to canon law, the Reformation could finally largely ignore the scholastic argumentation against interest and develop its own doctrine, which chiefly followed the "Golden Rule": charging interest would only be considered a sin when inflicting disadvantage on one's neighbor. The sin was now meant to be defined not in terms of value derived from mutual gain, *lucrum ex mutuo*, but rather of value derived from another's loss, *lucrum ex damno alieno*. And, last but not least, since exact cost accounting and a deep understanding of all economic connections was beyond the capacity of both ordinary laypeople and clergymen, one's own conscience was meant to serve as guide. Calvinism was far from pleading for unfettered freedom to charge interest—ideally the world would be free of it. But there was now unceremonious recognition

58 Sombart 1987, III., p. 222. For an analysis of reification, esp. with a view to the connection between commodity and conceptual form created by the finance-economy, see Sohn-Rethel 1971, esp. pp. 30, 72, 90, 110.
59 Simmel 2004,, p. 335.
60 Ibid., p. 193.

that when it came to a shared economy, interest had functions as indispensable as, say, double bookkeeping. For this reason, and also because individual economic success was understood as a sign of election, interest was not to be damned wholesale.[61]

Being able to calculate absences, which is to say loaned capital, dispensing credit on an impersonal basis, assessing the risks tied to that allocation, and receiving compensation through a "fair price" as interest—all these operations, indispensable for the acceleration of capital turnover, for the investment-related utilization of savings, and for the dynamization, future-orientation, and general expansion of capitalism, were sometimes taken over by moneylenders, but were placed on a stable foundation by a modern banking sector. Banks emerged from the problem of interest, depersonalizing it and thus professionalizing a response to it. Banks deal with promises of payment, hence with both money *and* time, or put otherwise with the danger of not receiving lent capital back. Because they calculate this danger and its costs, they create "risks." And because they continuously monitor the agreed-on payment circulation, they engage in danger-prevention. This regulation of a highly complex process of simultaneous payments and non-payments—the coordination of different temporal horizons—is the "risk management" that banks engage in to maintain the capitalist economy, compensating themselves for this activity with interest. A necessary condition for both collecting and paying interest, and thus being able to support a wide range of enterprises with their specific orientation toward fortune and the future, is in any event the existence of deadlines.[62]

Shylock's mercilessness, revealed when Antonio cannot honor his deadline, consequently also shows foresight. Usurers were the forerunners of bankers, hence what Jacques Le Goff has termed the "instigators of capitalism...merchants of the future, sellers of time."[63] It is on their account that Shakespeare can stage his drama of the capitalist pursuit of fortune. For what would soon proceed as purely "calculable" business with risks and futures in the banks' booking routines could here be presented as contract and promise—as a dramatic entanglement between faithful and treacherous characters, those who are noble and those who are greedy. Shakespeare's "usurer of Venice" is a precarious transitional figure in that he already claims the interest that capitalism will henceforth require; in that he only has to set the interest so repellently high because it is not yet equitable practice in commercial life; and in that this offensiveness is attrib-

61 See Noonan 1957, pp. 365–367.
62 See Baecker 2008, p. 186.
63 Le Goff 1988, p. 93.

uted to him as an essential feature, the business risk for which interest, after all, stands thus becoming a personal danger. Shylock's fate attests to the crisis to which the early capitalist shift to risk-based business and a credit economy stands. For this reason, he seems to be socially, legally, and religiously ostracized at once—albeit for very different reasons. For example, within the late medieval Christian behavioral codex, no longer feudal pride, *superbia*, but middle-class avarice, *avaritia*, counted as one of the worst of sins. Antonio's pride thus appears pardonable, Shylock's avarice downright devilish. And it is only because the ban on the practice of usury by Christians was void when enemies were involved that Antonio, following declaration of enmity, can take interest from Shylock.[64] For his part, already because he demands interest, Shylock is deemed "greedy to confound a man." (3, 2, 1653)

When the Duke of Venice asserts that Shylock is "empty / From any dram of mercy" (4, 1, 1936 f.), on one, important, often neglected level, the accusation is not only evidently a culturally overdetermined collective statement about Jewry, but is also aimed at capitalism, its credit system, and its deadlines—adhering to which is not only the prerequisite of Shylock's existence but of the modern economy in general. But where the duke requests mercy—renunciation of Antonio's contractual penalty and dispensing with a part of the unhappy man's debt—and where Portia instructs him as to transcendent justice and the providential gift of mercy (a gift that all legal subjects need in the end), Shylock, without affect, refers to his solid, contractually secured rights. Why on earth should a Jew suddenly practice "Christian" mercy after even eligibility for that quality has been denied him? Portia's juridical chess-move, robbing Shylock of all his claims and even his fortune, is, looked at objectively, simple fraud: for example gaining something one has a contractual claim to must be legally secured, for which reason spilled blood or an unnoticeable excess or shortfall of agreed-on weight cannot be a basis for contractual non-fulfillment.

We might thus understand Portia's judicial machination as an action—allowed the nobility alone—on her own behalf. Antonio's rescue is, after all, the only way she can save her promised marriage and future domestic peace.[65] We might also see Portia as an embodiment of Fortuna herself, using unbounded,

64 See ibid., pp. 9 f., 22.

65 On this, on the actually only "substantial" partnership between Antonio and Bassanio (hence a partnership that absolutely must be annulled by Portia), and on the *sponsalia per verba de futuro*, hence the initially still only virtual marriage between Portia and Bassanio, see Shell 1982, pp. 60, 74 f.

sly, and artful weapons (this time, rhetoric) to shape fate according to her will; in the process, this reading would have it, she perverts, in pagan fashion, the theological concept of mercy, revealing herself, when she takes Shylock's fortune but not his life, as a sadistic goddess of fortune to boot.[66] However, it is in fact clear that Portia's machinations are aimed at preserving the legal certainty vital to Venice's trade, while nevertheless placing a check on the bloody excesses—unfolding *ad extremum* precisely in Venice—of capitalism.[67] From the perspective of natural law, the principle of *pacta sunt servanda* represents a special case of the imperative of the subject's inviolability—in "Merchant" not only that of Antonio but of Shylock. For this reason, Portia's intervention consists not only in Shylock's outwitting and disfranchisement. At the same time, it reveals concern with equity as an English legal institution: the law's adaptation to the particular case, while nonetheless adhering to principle. Already within Roman law, *aequitas* had a corrective function for civil law, especially when there were legal lacunae or when strict interpretation of the existing legal clauses appeared to clash with the collective sense of justice.[68] The Elizabethans were either directly familiar with the first conception of this legal institution in Aristotle's *Nicomachean Ethics* or else as mediated by Thomas Aquinas, Jean Bodin, and Richard Hooker. The desired milder moment in the legal process was reflected in the other name given the Chanceries with their corrective function: Courts of Conscience. From that perspective, we can understand Portia's adjudication as involving an appeal to conscience.[69] To be sure, for Aristotle equitability is only practiced by the individual who "is no stickler for his rights in a bad sense but tends to take less than his share though he has the law on his side."[70] And in this Aristotelian light, as a—by no means disinterested—judge, Portia does not act any more equitably than do Shylock or Antonio.

Portia only produces equitability to the extent that she takes the contract literally and supplements it through other "laws of Venice" 4, 1, 2257). She oversteps law to preserve its validity.[71] Extrapolating from this, we can also note the following: because a risky partnership has here led to a singular legal prob-

---

66 Antonio on Fortuna: "it is still her use / To let the wretched man outlive his wealth, / To view with hollow eye and wrinkled brow / An age of poverty" (4, 1, 2213ff.) See Reichert 1985, pp. 89–94.

67 See Darcy 2003, p. 189.

68 On the legal problem of equity see Maye 2006, esp. pp. 58–61. On the chancery courts see Vismann 2000, pp. 52f.

69 See Schoeck 1999, pp. 225f.

70 Aristotle 1941, 1138a (p. 1020).

71 See Spencer 2003, pp. 152f.

lem, law and its guiding code, centered around the "just"/ "unjust" dichotomy, is here undermined by a form of "just injustice." For if the code can no longer process a case of risk, and if the code is in the end applied to law itself, a decision looms that is no longer merely juridical but has fundamental import. This alone is the reason for the shift in "Merchant" to the transcendent level of mercy and sovereignty. Shakespeare's judicial scenario thus shows that, for the sake of not simply declaring contracts such as Shylock's invalid because their stipulations do not (cannot) account for all possible future circumstances, new sorts of law and contract are needed to adequately address the risk-orientation of the Early Modern art of trade. Running alongside the legal principle's structure, conditional contracts, for example, introduce a specific narrative order of knowledge into juridical discourse: a singular and contingent event's inner and outer circumstances—themselves deciding whether a contract's conditions of validity have been fulfilled—can only be described in this way.[72] In the end, then, what is at stake in the process being staged here is not only a legal conflict (even one meant mainly as an example) or a troubling case of antisemitic miscarriage of justice. Rather, the decision concerning what law can ever accomplish in the case of risky partnerships centers on the shift from unconditional responsibility to conditional solidarity, and from individual and moral guilt to collective and financial indebtedness.

"Mercy" is demanded of Shylock because Antonio is "within his danger" (4, 1, 2119). He is expected to desist from both personal revenge and compensating for his own financial damage, incurred by the delayed repayment, through physical harm to the opposing party. Economically assessing body parts is grounded in legal codices of the early and later Middle Ages that stipulated specific compensation for specified injuries, clearly for the sake of dispensing with revenge. Until the Early Modern period, this conversion of corporeality into economics was as unusual as the basic distinction between civil and penal law self-evident; but in the Elizabethan age it resurfaced—in the form of compensation funds for injured mariners (for instance the Chatham Chest that Sir John Hawkins established in 1590).[73] To be sure, the inverse conclusion of specific physical injury due according to an amount of lost money-value has always been forbidden. Furthermore, with Roman law forbidding any money-value being placed on human

---

72 On the narrative as opposed to the propositional constitution of the event within the Early Modern episteme and the corresponding "structure of apprehension," see Campe 2006a, pp. 70–73.

73 See Wilson 2003, p. 21.

life into the early Modern period, the inverse movement from specific money-value to a human life was all the more inadmissible.[74] Shylock is not allowed access to Antonio's pledge for both these reasons: something for which above all Portia's ban on blood-shedding to satisfy a debt stands surety. That the usurer nevertheless insists on his rights presents the mercilessness of the new economy in a double manner: he does not respect the border (marked by blood) between natural fruitfulness (the procreative tie between Portia and Bassanio) and its unnatural counterpart (Shylock's usurious demand). Until the end, he baulks at introducing the creaturely into his economic calculus.[75]

What in the Early Modern period makes possible the conversion of corporeal damage into financial value is the gradually developing idea of something like "human capital." Delinquent curtailment of or grave injury to it may be criminally prosecuted. But in Antonio's case there is no punishable behavior; rather, we have a venture in its double sense: that trader has invested his fortune in a daring enterprise while also placing his person in danger, namely in reach of Shylock's bloody demand. The Duke of Venice, the ultimate legal authority, thus decides that indeed compensation but by no means punishment, let alone capital punishment, will follow from Antonio's venture. For the sake of justice, in that Antonio's representatives offer Shylock an augmented repayment, there is an acknowledgment that with his credit, he himself has taken over the danger of losing his own capital, so that he merits an allowance. In return, he may no longer be a usurer in the sense of the "greedy gayner" referred to in Wilson's above-cited description. Shylock is, then, forced into solidarity, but for his acceptance of danger he is allowed a bonus. If he scorns this solidarity, he is a usurer who has to fear both the confiscation of his fortune and for his life. With the bond-centered plot, we consequently find not only Jewish Shylock forcibly converted to Christianity, but, just as much, merely selfish "assurance" converted into the solidarity of insurance.

Within the argumentative framework of "Merchant," this appeal to solidarity would seem grounded in a Christian imperative of loving one's neighbor, to the same extent that it is put forward against usury. But finally, a double maxim is connected to the appeal: not only to generate communal security, but at the same time to embark on individual (ad-)ventures quite often tied to business involving credit. Security through solidarity is not the "other" of that adventure; rather, they designate two sides of the same new orientation, so that increased

74 Early Modern Venice nevertheless knew a special form of life insurance: insurance on credit taken out on the life of a debtor. Ship captains and sailors were excluded from this, probably because of various cases of fraud. See Stefani 1958, I., pp. 119f., 123.

75 See Weigel 2004b, pp. 76–78.

security signifies no diminishing but rather an augmentation of risk. And if the order of the day is to no longer imagine oneself (spiritually) safe before a providential salvational God, in order to calmly hazard all sorts of dangers, but rather henceforth to consciously engage in risk while simultaneously collaborating with others to establish (material) security, then the previously ritual-theological assurance has now become an insurance grounded in mercantile solidarity. In this respect, Lucien Febvre has spoken of a "transfer of the heavens to earth" and, in the specific context of seafaring, Alain Cabantous of "the heavens in the sea."[76] This might suggest understanding the new orientation to security *and* risk as the result of a secularization process. But this development was prepared not outside but by way of a body of theological-canonical conflicts and debates. This was the context for a first effort to mark out the borders for determining the calculability of chance and thus the distinction between divine providence and earthly contingency. And it was the context for initial discussions of the extent to which dangers, in their contingency, can be calculated to any extent whatsoever, and the extent to which such calculations can be socially useful and lead to a just price.[77]

The Drama of Danger

By around 1600, what previously would have been considered a presumptuous challenge to Fortuna and providence had become something nearly self-evident: the imperative to love the venture. This is already manifest in "Merchant" in the strict arrangement of the casket-plot. The last will of Portia's deceased father renders his daughter's choice of casket into a "lottery of my destiny" and a "hazard" for her suitors (2, 1, 529, 561). "Who chooseth me, must give and hazard all he hath" (2, 9, 995) is the motto of the lead casket opening true love. The motto prescribes not only a daring choice but the choice of daring itself, an *amor fati* that as, in the Early Modern sense, a free choice of partner is the same as *amor fortunae*. Instead of re-assuring itself through the parental option, and instead of yielding to the heroic wooing of a knightly adventurer (the Prince of Morocco) or the meritocratic claims of a solid gallant (the Prince of Arragon), love now solely follows the criterion of wholehearted commitment. Henceforth the insecurity of all partnership, all "double contingency," will be reflected in a characteristic semantics and rhetoric of hazard.

76 Febvre 1956, p. 245; Cabantous 1990.
77 See Ceccarelli 2001, pp. 612, 630 f.

Without a doubt, amorous striving for fortune continues to mirror its counterpart in business. But it no longer can be translated into a fully objective calculus of capital and costs. "In contrast to what is true of one's interests, it is impossible in love to calculate the costs or weigh up the accounts," writes Niklas Luhmann, "because both one's profits and one's losses are enjoyed; indeed, they serve to make one aware of love and to keep it alive."[78] Partnerships settled into both the field of business and the field of love continue to be interrelated, but are now mutually independent: never before have there been more and better possibilities for entering into partnerships that, on the one hand, are impersonal, purely practical, and calculable, and, on the other hand, are personal and intensified through experience. Preceding each of these forms of partnership is, however, a readiness for daring—since then, the only opening to a fortunate future. In this new context, what ties partners together is not certainty or safety, but rather hazard in love, risk in business.

In Shakespeare's "Merchant," love forms the bridge between the simple venture and its implementation as a speculative business. In actuality, Antonio borrows money from Shylock not for the sake of his own trading activities, but for that of an adventurer's love life. Consequently, the adventure, whose etymological grounding in *ad-ventura* signifies the advent of fortune, is here not primarily concerned with the sort of speculative partnership that in the seventeenth century's business parlance was called "adherence to a venture." But the play's amorous material is already guided by a behavioral orientation toward daring that will give impetus to business partnerships starting with Shakespeare's age. Indeed, in the intersection of the ring-theme with the court proceedings and its play of roles and confusion, we are shown the double-sided risk emerging when a bond of pledged faithfulness, hence credit for an amorous partnership, is connected by chance with a commercial partnership. (In order that Portia forgives Bassanio's heavily symbolic faithlessness, the ring's return, Antonio offers a second, even more boundless pledge—that of his soul. [79]) That, to be sure, the risk-management typical of commercial partnership, although long-established by 1600,[80] plays no role in "Merchant" is among the play's aspects needing explanation: why has Antonio not simply insured his ships? It is the case that he intuitively spreads out the risk by distributing his ware and fortune among sev-

78 Luhmann 1986, p. 67.

79 See 5, 1, 250 f.

80 On the oldest Venetian document treating the foundation of a proper "partnerly" insurance enterprise in 1567 see Stefani 1958, I., pp. 101 f.

eral ships. But finally, he has no real concept of risk. In his melancholia—characteristic of *fortuna*-allegories[81]—he simply acquiesces in a danger threatening his ships and water, his treasure, in the end himself, instead of providing for it. He is sad from the start—but not because he failed to take out insurance. Rather, on account of his sadness he fails to insure himself.[82] Melancholia renders him a "want-wit" (1, 1, 7) who can see no calculated striving for fortune in the venture but can only fear misfortune, going under; in his laconic words, "all my fortunes are at sea" (1, 1, 3). Shylock, by contrast, looks clearly at *fortuna di mare:* "ships are but boards, sailors but men, there be land-rats, and water-rats, water-thieves, and land-thieves, I mean pirates, and then there is the peril of waters, winds, and rocks" (1, 3, 344 ff.).

Otherwise than Shylock, who demands a price for the venture of his loan, Antonio cannot calculate danger, and thus cannot create "risk." Not the established Christian merchant but the rootless Jewish usurer is the pioneer of a new and risk-laden art of trade: a theme transmitted from treatises on insurance well into the nineteenth century.[83] In any event Shakespeare's drama reveals itself, despite all personal complications and strivings for fortune, as ultimately a sea-play: for just as the bond-plot only superficially centers on incorporation of economic value but actually on *fortuna di mare*, the casket-plot at first view only stages what in the form of social dance is a form of "ladies' choice." As psychoanalytic interpretations in particular have suggested, the play's casket-selection episode operates with symbolic substitutions for "the woman": as with a woman, a casket, when the same as a tomb or coffin, can hold an entire body within its confines, so that with the beloved the suitor chooses, perforce and fatefully, the likeness of his mother, in the end Mother Earth. Seen in this manner, Fortuna is, to speak with Freud, merely a death-goddess.[84] Significantly, Shakespeare chose the term "casket" where one of the episode's important literary models, the *Gesta Romanorum*, refers to a "vessel."[85] The casket that is here fortunately chosen under the sign of risk thus even corresponds in its source-history to the ship's hull, termed a *kasko* in shipbuilding.

Now in that *kasko* is derived from Spanish *cascara*, which in turn stems from Vulgar Latin *quassicare*, "shake" or "shatter," and in that, however, the terms *ris-*

---

81 On the similarity of nature between Saturn und Fortuna and the main features of those born under Saturn's sign, "counting and calculating, weighing and measuring," see Reichert 1985, p. 123.

82 See Wilson 2007, p. 129.

83 See Koch 1995, p. 151.

84 See Freud 1953 – 1974a, p. 295.

85 See Cerasano 2004, p. 88.

ico and *risco* stem from Greek *riza* (with the ancillary meaning of "cliff"), the casket-plot also works with a completely different series of substitutions. Within this series, what is being chosen is not so much Mother Earth as its resting place but rather the sea—traditionally precisely not the secure ground but the unsettled lack of grounding for all things: a groundlessness posing the challenge of venture, of an opening, against the peril of shipwreck, to a future that cannot be known. Bassanio is not off the mark in comparing his choice of casket to coastal navigation upon "a most dangerous sea." (3, 2, 1465) Understood as an existential choice, the casket episode is aimed at what since late antiquity was allegorized as both divinely pleasing and the life voyage, albeit seldom tied to the "harbor" of amorous partnership. On the other hand, understood in terms of the secular art of trade and business, the episode points, in a highly non-metaphoric manner, to mercantile maritime voyaging, to an adventure whose success or failure depends on *fortuna di mare*. In the *Gesta Romanorum*, although that failure—the shipwreck decisive for "Merchant"—is missing, all other elements of the play's bond-plot are present.[86] Early Modern Fortuna, precisely this play shows, is recognizable more as an ambivalent maritime divinity than as a goddess of death. But forestalling her dangers and even using them would only be possible with risky partnership par excellence: maritime "casco insurance."

The oldest documentation of this maritime insurance stems from 1347; ancient maritime lending, the *foenus nauticum*, offering merchants, ship owners, and shippers credit for major maritime enterprises, can be considered its preform. In turn for his capital investment, the creditor could set an extraordinarily high rate of interest (roughly 30 percent). For if the debtor's ship went down, the creditor also lost his claim to repayment. *Foenus nauticum* in any case cannot really be considered an insurance-like institution—for a start, unlike what would emerge as insurance, no specifically calculated premium payment could be levied in advance: Ancient merchants almost always accompanied their ware on the sea, in the case of shipwreck, their survival was highly unlikely, recompense hardly necessary. Maritime lending actually shared, Paul Millett indicates, "some of the characteristics of consumption credit."[87] Merchant could only really have an interest in insurance when, since the waning of the Middle Ages, they became settled rather than accompanying the ships, thus having to cope with the sea's perils not on the sea but on the surface of paper.

After Justinian's time already, but especially when drawing interest was subject to limits through Pope Gregory IX's ban on usury, the practice of maritime

86 See Stevenson 1913, p. 261.
87 See Millett 1983, p. 44.

lending—that branch of business indispensable to maritime trade—was increasingly less rewarding. What capitalist would wish to embark on risky business without corresponding compensation? Importantly, in the case of maritime loans the creditor not only took on the danger of not getting his money back; he also took over the bulk of the immense dangers posed to goods and business in crossing the watery abyss. In addition, while even contributing his funds before the ship set out to sea, he only received payment for his defrayal of danger—alongside his lent capital and customary interest—with safe return of the ship. For his part, the loan's recipient owned the lent capital without having to financially stand surety for the connected dangers.

Especially when Mediterranean trade flourished in the High Middle Ages, a solution had to be found allowing those assuming risk sufficient compensation but not running against the ban on interest. Consequently, what previously had been accomplished by *foenus nauticum* through an excessive interest rate—but never in the sense of an independent and partner-like incurring of danger expressed and regulated as such—was now seen to in two different ways: on the one hand, as a non-interest lending operation; on the other hand, by signing a contract that specified compensation for danger.[88] In the course of the theological discussions unfolding since Pope Gregory's "Naviganti" decree of 1234 (concerning usury and maritime lending), just this risk assumption and its inner causes would constitute the criterion distinguishing forbidden lending from admissible partnership. When the creditor lent money exclusively for the sake of profit, he was engaging in usury. But when he assumed risk not merely for the interest but also for the sake of trade and if possible also had charitable motives, he could also request appropriate payment for this. This interpretation of risk-assumption and its attachment to specific motives created the concept of risk at the same time that it freed commercial partnerships from strangulation through the ban on interest. The two differing conceptions of money and property at work here, viz. either that money was sterile and property was use of consumable goods or that money could produce added value and property was characterized by risk assumption, took account of the fact that in loan-based business, otherwise than in partnerships, interest was uncoupled from the earning power of the lent capital and its risk.[89]

In the longer term, this development led to a strengthening of the model of partnership, as a *societas* under the sign of just distribution, thus encouraging broadly shared risk assumption—a development that in both respects was to

88 See Blumhardt 1911, p. 70; Perdikas 1966, pp. 430 f., 502 f.
89 See Noonan 1957, pp. 137 f., 151 f.

the disadvantage of the small money-trade. Already on account of its communally supportive effect, the institution of insurance would now enjoy increasing legitimacy, including among theologians. In general, the canonistic controversies created an awareness of the new economy—indeed a first theory of its mechanisms and laws, especially its risk-orientation.[90] Eventually, through forming agreements about risk—or what Benvenuto Stracca would call the "sale of danger" in his 1569 treatise on insurance—the rigid ban on interest was overcome, a separate insurance contract formulated, and the way paved for the partnership between capital and labor decisive for Early Modern capitalism. Since then, interest was justified as a premium on risk, so that capitalists soon also discovered life-annuity purchase and life insurance as the pre-modern "speculation business par excellence."[91] It was also justified through the danger threatening economic life with, as Blumhardt described it more than a century ago, "purchase of credit, long-distance purchase, offering guarantees, and exchange"—indeed with all forms of money transactions.[92] Nevertheless, the true starting point of interest-approval and the modern art of trade was the "purchase" of the danger tied to *fortuna di mare* in upper Italian mercantile cities since the fourteenth century.[93] "Risk" was first discovered for its sake, fixed terminologically, understood as a trading good.[94] And for its sake, maritime insurance was created as the first insurance of any kind.[95] In this way, insurance was born from the sea.

---

90 See ibid., pp. 193 f., 202 f., 281, 407; Perdikas 1966, p. 509.

91 Braun 1963, p. 26.

92 Blumhardt 1911, p. 74. On trust in the credit-worthiness and commitment of underwriters or *bancherii* in the case of damage, something that first made it possible to do business with an insured party not on the basis of a later repayable sea-loan but on that of a lawful insurance benefit, see Perdikas 1966, pp. 505 f., 508.

93 It is the case that into the late medieval period, various guilds and mutual insurance firms are documented that tried to cover the dangers of maritime trade and commerce. But their financial possibilities and understanding of solidarity were quickly overwhelmed by maritime "adventures" with their extraordinary accumulation of danger. Hence after 1300 in upper Italy, a number of individual insurers emerged, probably together with some Flemish and Dutch maritime insurance chambers. For this—somewhat controversial—chronology see Trenerry 1926, pp. 39–42, 122, and 275; Van D'Elden 1987, pp. 196 f.

94 On the mediating function of the Catalan Dominican monk Raymond of Penyafort, through whom the mercantile concept of *resicum* moved—from its starting point in *foenus nauticum*—into scholastic discussion, see Le Goff 1988, pp. 71 f.

95 This sort of insurance, for the first time not operating on a mutual basis, is characterized by (1) determining a "risk" against which one can be insured; (2) the insured object having a determinable value, for which reason life insurance was initially omitted; (3) the insurance only covering damage and offers the insured party no possibility of making a profit;(4) an appropriately

Classifying insurance contracts as aleatory or chance-centered business, however, a position taken from the seventeenth century onward that probably reflected the emergence of probability calculus, only obscures the starting conditions for risky partnership, centered as they were around mercantile and money-transaction activities.[96] True, in both insurance and betting liability is activated through a previously undetermined event. But in the case of insurance, an interest in the capital value of the insured object—say ships, goods, the life of a person—has to exist, possible damage has to be covered, a premium calculated in advance paid. This notwithstanding, *assecuratio impropria*, for example betting on a ship's successful return or a person's death, was widespread, opening the floodgates to insurance fraud.[97] For this reason statutes such as the English *Act Concerning Matters of Assurance amongst Merchants* (1601) stipulated the conditions for a valid *police*, in the Italian sense of "promise" or "obligation."

Circumstances, the value of ship and goods, the "personal worth"—moral and professional[98]—of the vessel's master, the reputation of the party involved, the length and dangers of the route, determined what was then set as a risk in maritime insurance matters. In maritime insurance, statistically precise calculation of a particular risk has never been possible, because of the complexity and number of possible causes of damage. Even when detailed statistics about clients and dangers are available, these have to be first combined with ongoing shipping reports, then intuitively extrapolated for each occurrence of risk. Just this endows even commercially established and routinely executed calculations of risk with what Frank Knight termed "the characteristic form of 'enterprise,'" indeed the form of a straightforwardly speculative enterprise.[99] Put differently, measureless uncertainty dominates what can be measured, simple danger definable risk. Premiums are here thus formulated through experience and competition, less through rational calculation than through de facto calculative behavior resting on both observation of past and present circumstances and ongoing market trends. At the same time, insurers and the insured continuously observe each

calculated insurance premium being paid; and (4) the insured party not being able to insure his own insurer.

96 See Koch 1995, p. 154.

97 On the previously negatively charged concept of interest and the occasional ban on gambling-like maritime insurance, for instance in Holland in 1568 by Philipp II, see Bonß 1995, p. 169.

98 See ibid., p. 162.

99 Knight 1933, p. 232. On the variability of maritime-insurance risks and the formula of a corresponding "variation measure" see Liebig 1914, pp. 218–220; Röpling 1960, pp. 17–20.

other, for which reason determining risks is finally based on a complex market-development of "double contingency."

There were good reasons for insurance contracts being agreed on mainly in bourse-centers from the sixteenth century onward. It was not only that the joint-stock companies now being established—companies, with their overseas activities, that had developed out of the *commenda* und *societas*—had an intensified need of resources and were thus dependent on the mobile capital investment of various shareholders, as well as on comprehensive insurance protection. The regional centers for trading, finance, and insurance thus came together, the case with Antwerp, then Amsterdam, and then London.[100] Furthermore, the skills involved in both insurance and the stock exchange have always been rooted in the art of observation: a kind of reality-doubling in which the same risk and value is produced with which one intends at the same time to trade. Precisely through the self-reference of acting and trading, the activities of insurance and stock exchange involve observation of what lies outside their own view and scope, a capacity to operationalize their non-knowledge while at the same time feeding back to contingent future events (say on the bourse through certain purchase options, in insurance contracts through an altered adjustment of danger).[101] In the context of this double contingency of trade, with its sole orientation toward a contingent future, the insurance and financial markets became a space of action forcing continuous risk-generation; but in the form of non-specified risks, these markets continued to be subject to the vicissitudes of fortune, sometimes to an intensified degree.

This is the basis for the maritime imagery through which the first true treatises on the bourse try to describe the speculative process. In the 1668 *Confusión de confusiones* of Joseph de la Vega, himself an often-failed shareholder, the Amsterdam trading floor is presented not only as a "labyrinth" and stomping ground for "sharpers," a "touchstone for the rational and gravestone for the foolhardy." Rather, with the bourse tied metonymically to the sea since the founding of the *Vereenigde Oostindischen Compagnie,* we here read of the exchange's "Neptunian" invention, of its flooding by "frothy waves of speculation," indeed its rendering into a "dangerous and deep sea" upon which "every gust of wind signifies a storm and every wave a shipwreck," and in which "many sirens sit whose song kills."[102] In a deceptive and seductive manner, we read, "things that only possibly occur" are treated as facts. And if here in the end "the expectation of a fact"

100 See Mokyr 2003, pp. 208f., und Barbour 1950, pp. 23, 33, 85.
101 On this typically Early Modern "reality doubling" see Esposito 2007, pp. 8, 18–21, 55.
102 Vega 1919, pp. XXXI, 1, 8, 18, 22, 34.

makes "more of an impression than the fact itself," then the effects have evidently freed broken free of their causes.[103] In this drama of contingent futures, the principle of (sufficient) reason appears fully overcome, life-world horizons fully fictionalized—at least until the moment of a stock market crash or shipwreck. Arguably, risky trading allows a clear drawing——and lasting transgression—of the borderline between the trading and financial systems (the stock market and the insurance industry) and their elementary environment (the sea and its perils): the environment is first and foremost observed and tapped by the economy. In any case both risk and value-production depend on a market system and a stage of mutual observation that the Early Modern theater for its part observed and evaluated.

In Elizabethan England, most theater was organized as a kind of joint-stock company, taking business risks with its productions in a growing mass-entertainment market and being forced, in addition, to create value through virtual goods. Beyond this, linguistically, figurally, and thematically the produced plays reflected a generally establishing culture of credit.[104] They demonstrated the value-creating function of imagination, trust, and even simply time and treated the adventure of riskily questing for fortune on various levels. The theatrical stage was an observational medium for the Early Modern mercantile art of mutual observation. It was, in addition, a medium that in a metapoetic manner registered the observation of that observational process's limits. Finally, in Shakespeare in particular it addressed the manner in which individuals situated themselves vis-à-vis their own elementary environment (on insular England, always "the sea") —and thus in face of what, as danger from the outside, threatened one's own values, behavioral forms, even existence. With the capitalist economy's expansion, such environmental observation became a crucial aspect of the art of trade. The maritime insurance industry was especially dependent on what the seafaring environment heard as rumor, assertion, or news.

For this reason, already in Venice insurers had their own spying rings and intelligence services to control mariners, ship-owners, and merchants—but also to observe all possible ship movements and report on maritime threats (for example from the weather, from pirates). This was a technically equipped communicative association activating—then as now—all available personal and medial resources to the insurers' end, and for its part giving impetus to de-

103 Ibid., pp. 65, 215. See also ibid., pp. 78 f.
104 On "verbal usury" as already inveighed against by the Church Fathers and particularly marking Elizabethan drama, with its numerous word-plays, see Shell 1982, p. 49.

veloping novel intelligence techniques.[105] This observational art was complicated by "retroactive" insurance allowing a ship already at seat to be covered through back-dating of the contract, although the danger existed that the danger being insured against had already occurred. In Venice, such contracts only were valid when neither the insurer nor the insured party could have received information about the unfortunate voyage or fortunate return of the ship involved, because—measured by the sailing and postal speed of the time—it had to simply be too far away.[106] Nevertheless, these "lost-or-not-lost-contracts," as they were called in England around 1600, were basically only a form of risk-centered business, a bet whether a rumor, assertion, or piece of news turned out true. They promoted deceptive schemes of all sorts. But otherwise than in Spain and Antwerp, they were allowed in London's insurance market.[107]

Against this backdrop, the rumors that Antonio's ships have sunk and the news Portia receives per post of their fortunate return sets Fortuna's influence in the play in a "strange" light. "You shall not know by what strange accident / I chanced on this letter" (5, 1, 2748 f.) Portia declares, a statement, as nonchalant as it is unchallenged, suggesting she has directed mutual observations presented on stage to suit her wishes—perhaps not first during Shylock's trial but already from the start. Where earlier she has interceded as judge over Antonio's and Shylock's fate with the help of a purported letter of recommendation from a putatively eminent jurist (4, 1, 2078.ff), she now disposes over the plot's future course with the help of the strangely "chanced" letter. Thus chance receipt of maritime news, however orchestrated, determines the play's *fortuna di mare*, meaning its entire outcome. Fortuna thus seizes the "chance" to re-appear as a sovereign goddess in a largely secular, calculating age less through unfathomable influence (or simple "mechanical sloppiness" of Shakespeare the dramatist[108]) than through an intervention in the news system of maritime trade as effective as it is hidden. In the end, around 1600 not only the size of insurance premiums but also individual insurability and credit-worthiness, a person's busi-

---

105 On the production and communication of "maritime intelligence," especially by Lloyds' use of the most recent medial techniques, and connection of the information service back with the stock exchange, see Liebig 1914, pp. 185 f.; Gibb 1957, pp. 153–156; Lehmann-Brune 1999, pp. 147–154.

106 See Nehlsen-von Stryk 1989, pp. 203, 207.

107 See Wilson 2007, p. 135. On the "operative potential" of the—barely organized and regulated, hence creative but also liable to gambling—London insurance market around 1600 in comparison to other important maritime and insurance locations, see Kepler 1975, pp. 44, 55 f.

108 Reichert 1985, p. 121.

ness and personal reputation, depended on rumor and, even more so, "authentic" maritime news.[109]

This may suggest, on the one hand, that the Early Modern drama of the search for fortune centers, just like insurance, on what Alexander F. Bene, in the latter context, termed "incidence of danger," the possible "realization of danger," the "uncertainty" about whether something has or has not been lost.[110] On the other hand we might conclude that the Early Modern drama of partnership, its promises and duties, is always exposed to the perils of communication and its shrouded channels,[111] and that as soon as this fact is made known, or as with Shakespeare is relentlessly explored, what is being undertaken between partners is always "risk communication." If commercial and amorous partnerships appear constantly imperiled by deceptive witnesses and falsified information, then in the heart of all insurance, personal and commercial, we find an *assecuratio impropria*, a bet on the uncertain. Evidently melancholic, careless, or greedy characters such as Antonio, Bassanio, and Shylock push these dangers and uncertainties toward a specific enjoyment of the other that transgresses the boundaries of partnership's desire, symbolized as natural love or legitimate business.

And nevertheless, the most authentic capital of every partnership is trust—"as tradesmen buy upon credit, so they must sell upon trust" was a seventeenth-century expression, followed immediately by the concession that in business as in love, a particular mix of certainty and uncertainty, trust and mistrust was in play.[112] Security depends on trustworthy partnership, however improbably that may be under a sign of absolute striving for fortune. The gamble of love understands devotion as daring; in word and deed it enjoys the improbability of its success. But "risky partnership" understood in the literal which is to say legal sense not only calculates the insured dangers but in a purely calculating manner also enters the improbability of real partner-like behavior into the books, the relevant terms being "moral risk" and "technical risk." Calculation and enjoyment are, all said and done, two sides of Early Modern partnership, with happiness, as Shakespeare observes, being located more in the striving than in the possession (cf. 2, 6, 922). Whether on the field of love or business —the modern art of trade has to consider the danger of being betrayed. The ad-

109 See DeRoover 1945, p. 191.

110 Bene 1928, pp. 8, 16.

111 On (maritime trade-resembling) amorous communication see Schneider 1994, p. 31: "The transmission speed of doctrines of love and messages between spatially separated lovers (mail) are elementary data for the age of their validity. Without press and without postal service modern amorous language would not exist."

112 Cited from: Waswo 2004, p. 57.

venture consists of this danger, from which the enjoyment is derived. Trust here emerges as something also based on fictions, which are not so much the alternative to solid, credit-worthy reality as rather a mirror in which instability and contingency are reflected, together with likelihood (or probability): the likelihood made of the reality that one's own trade, actions, and observations brings into being.

Defoe's Fortunate Island

Starting in the nineteenth century, "The Tempest," Shakespeare's last play (1611), has repeatedly been tied to the fate of a ship named the *Sea Adventure.*[113] Belonging to the Virginia Company of London, this vessel, with 150 emigrants, seamen, and high officials on board, became stranded in the Bermudas in 1609. Named after Juan de Bermúdez, that group of islands had been discovered by the Spanish in 1503 but had been neglected by them since then; the survivors built two new ships from the *Sea Adventure*'s wreck in order to reach their voyage's original goal, Virginia. But a number of seamen and passengers flouted the captain's and admiral's commands, viewing their authority as ended with the shipwreck, and instead ran into the wilderness or announced the intention to settle on these imagined "fortunate islands."[114] The Virginia Company had been granted a charter as a joint-stock company in 1609, upon which they canvassed for a new financier. However, already that autumn news reached London of the likely loss of the *Sea Adventure* in a storm; and a year later, a letter began circulating from William Strachey, both the company's secretary and a witness to the "tragical Comoedie" of the shipwreck and the landing on "those infortunate (yet fortunate) Ilands," which the letter described.[115] Although the epistle was not published rumors now spread in London that in the Bermudas and evidently Virginia as well, massive indiscipline must have become widespread—a state of affairs threatening future investment by the city's merchants and thus leading to introduction of emergency legislation.[116] Through Silvester Jourdan's *Discovery of the Barmudas* (1610), Shakespeare learned of that "richest, healthfullest, and pleasing land."[117] He probably also had been able to read Strachey's letter, because his theatrical joint-stock company, The King's Men, maintained an informal ex-

113 See Chalmers 1973, pp. 48–51.
114 See Linebaugh and Rediker 2008, pp. 21, 35.
115 Cited from Salingar 1996, p. 210.
116 See Sokol 2003, pp. 80–83.
117 Jourdan 1971, B5.

change with Strachey's own company.[118] After all, both companies had an interest in gaining partners through the circulation of previously undiscovered but probable stories.

In both cases, possible worlds were announced, their reality or realization no longer guaranteed by the sovereign, who nevertheless contributed licenses and charters. On the non-loci of faraway islands and on Early Modern stages, then, old practices of power no longer prevailed—these had been replaced by new approaches to business and representation. In "The Tempest," with Prospero a destitute sovereign thus rules. Or better: on the island to which he has landed by "Providence divine" (1, 2, 271), he now holds sway over elementary forces. His rule no longer rests on the legal grounding of terra firma, but feed on fictions and phantasms, on a fantastic redistribution of the real and the possible, of first and second nature as tested out at the court of James I as esoteric "daemonology." This "brave new world" (5, 1, 2235) of the island is the laboratory for a new form of exercising power. It is not only that by contrast with "The Merchant of Venice" a ship (that of Prospero's intriguing brother) here appears on stage. With the deployment of the image of the ship of state, the play is supplied with the emblematic framework of *gubernare*, in order to simultaneously dismantle this old topos of the art of governance.[119] The shipwreck on the shores of the magic isle opens the possibility of a "sea-change" (2, 2, 564): a fundamental shift of form and essence, become idiomatic since Shakespeare, whose course and effect is comparable to a rite of passage like a baptism. Reflected here is the sea's transformative power, not only expressing itself in fortunate passages —already global in 1600—but also in being swallowed up by the deep. Just as with their rescue the shipwrecked experience a new birth, the sea tears the drowned out of historical time, subjugating them to its force. As we will see in this book's conclusion, this can consist of decline and forgetting but also in profound change. And just such change is here tied to both generic typology and *homo politicus*.

A sea change does not only transform an existent entity into another. Rather, the sea shifts everything in its mercy into the existential modality of the possible: into what can be either being or non-being, what is contingent but not impossible.[120] In Shakespeare's work the sea is thus not only the "instrument" of iron "Destiny" (3, 3, 1629 f.). If in the later plays we find a movement from the fatalistic-genealogical model manifest in the historical plays to anthropological and

118 See Greenblatt 1988, p. 148.

119 See also Fowler 2000, pp. 37–40.

120 On the sea as a sphere of modallogical transformation see chapter 8.

social concepts tied to the problem of deterritorialization, in "The Tempest," quite possibly Shakespeare's final play, the sea becomes a purifying element that will make possible "heart-sorrow / And a clear life ensuing" for the erstwhile conspirators (3, 3, 1659 f.). On the island, that land swum away from a continent, the sea change allows a new beginning by breaking the past's determining force, its omissions and transgressions and opening the possibility of new options for action and being. The spell of pre-history is only broken on the island, the senses healed. Prospero ends up renouncing power and its delusions —and Shakespeare, some five years, it seems, before his death, the theater and its semblance of real presence. In this way the island arguably marks, in early seventeenth-century England, a poetic sea change from drama to narrative, from historical plays to the novel of adventure.

Since 1600, the wonderful and fantastic, the possibilities of a new world, become emblematic in the magic of its uninhabited and haphazardly encountered islands, were no longer merely the theme of fathomless and utopian imagination. From now on, they would be the object of speculative process: investment in the future, in possible worlds and the realization of poetic and commercial projects. Various maritime travelers would now quite literally invest their hopes in *fortuna di mare*, betting on their own survival and homecoming, a widespread practice around 1600 that with the fortunate return would bring in around five times the wagered sum.[121] Imaginative works such as Shakespeare's but also Early Modern adventure novels placed the forlorn island's possibilities in a thematic spectrum extending from its colonization and thus rigid attachment to the putative motherland to complete sundering from terra firma and its legal, power-political, and socio-anthropological basis.[122] Lonely and unexplored islands are located at a juncture between territorialization and deterritorialization, striation and flattening, reality and possibility—precisely the quality that circa 1600 rendered them, as outposts for a global investment network, no longer simply interesting for utopian fantasies but also for the logic of capital and capitalism.

The prototype of all future "island fictions" was, of course, Daniel Defoe's *Robinson Crusoe* (written 1719 – 1720). In this novel, the question of an island's possibilities is initially posed as a question about primal human nature. In this respect, it is here useful to recall the three forms and stages of *status natu-*

121 In "The Tempest," the practice is thus referred to with the expression "putter-out of five" (3, 3, 1620).

122 On the corresponding distinction between "oceanic" and "continently" islands see Burnet 1719, I., pp. 187 ff.; in the sense of an aesthetic-political program see Deleuze 2002, pp. 11 f.

ralis within the natural-law doctrine of the time. First, the "state of nature in itself" (*status naturalis in se*) describes an originary state of "thrownness" in which human beings are limited to their innate drives and capacities; but as Samuel Pufendorf observed in 1678, even the most miserable castaway will have preserved a certain civilizational capacity, namely "a memory of the cultivated life and a knowledge of its arts."[123] Next, Pufendorf emphasizes that not only the presence of others but also the absence of any efficient government marks "the state of nature in relation to other men" (i. e., considered as a relation between free and independent men: *status naturalis in ordine ad alios homines*). Finally, "the state of nature in relation to God the Creator" (*status naturalis in ordine ad Deum Creatorem*) points to the individual human being as situated under divine rule[124] while still being forced to postulate an authority guaranteeing the flourishing and proper orientation of the common weal. Where the state of nature as such only concerns the isolated individual, and sociability comprises an order that while tailored to the presence of others is simply arranged by human beings and is thus by no means stable, the work of God is the necessary condition for positive laws to consistently have a general benefit.[125]

That Crusoe immediately christens his shipwreck's locus the "Island of Despair" thus fencing in his future premises or promptly claims the entire island for himself suggests that we may here speak of the primal act of an initial settlement, and of a *nomos* from which not only his future sovereignty but also his *oiko-nomia* is derived, as is finally the narrative economy of the text itself.[126] However, the adventurous quality of this novel is not only owed to the haphazardly prompted exploration and conquest of a New World, but also to an exploring and conquest of chance itself. After he has taken and named the island he deems his own, Crusoe extracts some tools and weapons form the wrecked ship, in the process nearly suffering "a second shipwreck, which...verily would have broken my heart."[127] This almost-shipwreck appears to be a providential warning sign, his exertions now being increasingly aimed at safety and prevention, under the maxim "provide for the accidents that might happen, and for the time that was to come."[128] Seen in this manner, the fences he will set up in the future will serve less to mark territory than to protect against unspecified dangers.

---

123 Pufendorf 1990, § 5, p. 114. See also Schuster 2000, pp. 121–123.
124 See Pufendorf 1990, § 8, p. 119.
125 See Behme 1995, pp. 55, 73.
126 See Schmitt 1973, p. 493; Schmitt 1995d, p. 584; Schmitt 2003, pp. 45 ff.
127 Defoe 1994, p. 55.
128 Ibid., pp. 66, 103.

Following Max Weber, Crusoe has often been described as presenting the prototypical mentality of *homo religious*, and following Karl Marx as a prototypical *homo oeconomicus*. But Defoe's hero is not only a model Puritan in his unshakable faith in God and not only a model capitalist in his untiring industry, through which he already constructs the infrastructure of his future economic management in a state of nature. His "governmentality" is characterized precisely by a bringing together of the dynamic of an all-comprising *theo*nomy and that of a boundless *eco*nomy in one and the same idea of providence, hence of governance-focused fiction. This *homo gubernator*[129] will allegorize his adventurous development, which at first view is simply unfortunate, as the history of a rescue on a fortunate island—and in this connection obtain what the record of his life unmistakably demonstrates: the goodwill of Fortuna and fortune in the sense of wealth. Crusoe swims out to the wreaked ship no less than 42 times, for the sake of supplying himself with necessities for his "present subsistence." Only this starting capital furnishes him with a horizon of both possibilities and a future, for "the hope of furnishing myself with necessaries encouraged me to go beyond what I should have been able to have done upon another occasion." We need to understand such hope both theologically and probabilistically, Robinson himself reckoning the probability of the ship remaining in his reach and thus serving as a king of Noah's ark for him as "an hundred thousand to one."[130]

A painstaking bookkeeping of all events and developments, expectations and fears will allow him to grasp his development as an adventure in the aforesaid sense of *ad-venire:* something that heads his way and that befalls him. Behind the contingency of his insular existence and with—emphatically so—the help of his writing practice, Crusoe here registers the working of providence. What he must believe in, already because it represents the condition of possibility for his self-preservation and survival, is "that I was brought to this miserable circumstance by His direction."[131] He sees this belief, on which all the speculation about his rescue depends, confirmed in the audit of his affairs—and this from the start. As soon as it is in any way possible to establish a "non-stable ensemble of language, instrumentality, and gesture," what might be described as a "writing scene,"[132] Robinson proceeds to the balancing of his fortune. He thus perceives the "direction," as an abiding credo based on "the experience of the most miserable of all conditions in this world, that we may always find in it

129 For a more detailed discussion, see Wolf 2012.
130 Defoe 1994, pp. 52f., 66. – See also Schmidgen 2001, p. 26; Campe 2012, p. 173.
131 Ibid., p. 94.
132 Campe 1991, p. 760.

something to comfort ourselves from, and to set in the description of good and evil, on the credit side of the account."[133]

It is the fiction of providence thus allows Crusoe to draw a three-time surplus from a disastrous balance-sheet composed of shipwreck and stranding on an uninhabited island: to possess the material of autarky in salvaged tools; to have for the first time found a language for contemplating his existence, a language owed the literary added value of his life story; and finally, to be empowered as *homo gubernator*, not only capable—again for the first time—of independently seeing to his own affairs but also of presiding over a fortunate, fortune-bestowing island, and being able to count on further governance by a supreme authority. Accounting under the sign of providence allows an identity between theodicy and "ecodicy." As its governor, Crusoe not only cultivates his island but at the same time registers it in his books. Or better: he creates it first and foremost as an administrative order of things as soon as he replaces the basal paradisiacal distinction between good and evil with a mercantile distinction between profit and loss.[134] His writing is always simultaneously a kind of bookkeeping, and the island becomes a continuously updated deployment of all useful objects and creatures, developments and activities. As we see it in Defoe's novel, the world has shifted—and this is attested to precisely in the dynamic double-entry recording system, with its steady shifting between inventory, memorial, journal, and quaderno[135]—from a fixed and substantial inventory of things, each in its place, to a scene of transfer presenting things as only consistently mobile, sometimes present sometimes not, always part of a transactional event.

Bookkeeping is a form of self-care and at the same time a school of experience. Just as the *raggione* (in English the "invoice," "the bookkeeping," simply "the business) brought rationality and calculability to fourteenth-century Italian merchants, with his double-entry diary Crusoe gains all the qualities distinguishing the modern capitalist's conscientious self-care: he is restless, awake, prescient, always careful to exercise self-control and self-stimulation, without things managing to evade him in the process, despite their unceasing vanishing and becoming. For its part, attentiveness to first and second nature, that school of experience accompanying steady bookkeeping, was tied in England to Francis Bacon's empiricistic program. (At the Dissenter's Academy, with its reformist didactic program, Defoe was introduced both to practical knowledge in Bacon's

133 Defoe 1994, p. 69.
134 See Vogl 2015, p. 28.
135 On bookkeeping as a means for administrating cameralistic domains in the seventeenth century's second half, see Jäger 1876, pp. VII, 110.

spirit and to the Royal Academy.[136]) Defoe's narrative procedure is thus "realistic" in the sense that it aligns the regulative fictions of faith in providence with the practices of methodical observation. In this respect, it is no accident that Robinson salvages the essentials of Bacon's *Great Instauration* from the stranded ship: compass, powder, writing instrument. On the one hand, the painstaking diary entries, focused on the manifestation of earthly things, serve as a coordinating net for the unexpected,[137] rendering both singular and contingent phenomena decipherable (the former as following rules, the latter as part of a providential structure). On the other hand, the stranding itself contributes to the cognitive process of experimental science—as Robert Boyle observed in 1669, failed experiments can open up "new regions" of knowledge just as shipwrecks "do sometimes cast us upon new discoveries."[138]

As Luca Pacioli's pioneering treatise on bookkeeping already called for,[139] Defoe's narrative economy provides a continuous conversion of events into narrations and narrations into business transactions, which are then registered in the books as debt or credit entries. Crusoe explicitly hands readers a copy of his journal so that they may control the narrative's course for rationality and factuality, like a kind of audit.[140] In the framework of double-entry bookkeeping, interchange between individuals—including that between authors and readers—is always structured in terms of credit and debit. But at the same time, every human interchange is oriented toward the Great Other: versions of "in the name of God and good profit" are standard introductory formulae in the oldest extant account books.[141] Seen in this light, for Crusoe even chance, announcing merely contingent incidence and actuality within the "state of nature in itself," becomes an advent of divine grace in the "state of nature in relation to God the Creator." To be sure, this advent itself adheres to the stipulations of an act of exchange. Or put more precisely, we here have an act of exchange signed and sealed by Crusoe with God in the style of a conditional or insurance contract.[142] For as a premium prepayment, belief—understood as such on the condition of a calculability of both theonomy and economy—obliges God to offer compensation for losses in the transfer of goods always prevailing in the order of things. And

136 See Vickers 1996, p. 125.
137 In respect to Bacon's compass-conception, see Blumenberg 1985a, pp. 359 f.
138 Boyle 1669, p. 115.
139 See Pacioli 1997, p. 16.
140 Defoe 1994, p. 72.
141 Pacioli 1997, p. 14. See also ibid., p. 91.
142 On the conditional and insurance contract from the perspective of natural law, see Pufendorf 1991, p. 103.

such compensation can be offered through haphazardly compensatory justice in terrestrial matters (for example in line with the probability theorem later known as the "law of large numbers" originally discovered by Jakob Bernoulli) or also through the currency of salvation in an afterlife. The precondition is simply that the human being believes in God and God does not damn the human being to sheer contingency but rather admits him to the providential economy.

Seen in terms of theology of creation, God here is always pursuing a kind of policy of risk.[143] Even if Himself a fiction, as that authority managing contingent because terrestrial affairs in general, he is logically a downright necessary fiction: "it cannot be conceived, without great inconsistency of thought, that this world is left entirely to man's conduct, without the supervising influence and the secret direction of the Creator," we read in the last part of the *Robinson Crusoe* trilogy. "This I call Providence, to which I give the whole power of guiding and directing of the creation, and managing of it, by man who is His deputy or substitute, and even the guiding, influencing, and overruling man himself also."[144] Tied to the mere imaginability of such a final authority is a salvational promise or at least an existential guarantee: one less concerned with individual management or global preservation than preserving individual futures and a world of possibility. As already for scholasticism, the impact of such "governing providence"[145] is described to the unseen hand of God, and this already in the novel's first volume, with increasing clarity: during Crusoe's voyage "a visible hand of Heaven against me" becomes discernable—a sign of a storm's menace rendering him an *asebēs* and Jonah-figure. Later, this instance of divine governance appears as the hidden guide and governor of all things and human states of mind. With His discovery, Crusoe consigns himself to the "dispositions of Providence"; and finally a "secret hand of Providence" surfaces, a hand that first authorizes him to be Friday's teacher as "governor" of the island, then to engage in colonial trade.[146]

As if retrospectively spelling put the birth of governmentality from the miraculous effects of divine *gubernatio*, Crusoe's career recapitulates the rise of Europe's Early Modern elect nation, from the navigational arts to governmental technique and onward to global capitalism. True, for Defoe, the economic theorist, this salvational *and* mercantile economy requires royal re-adjustment and still follows various rules of mercantilist economic policy.[147] Considered "cyber-

143 See Gregersen 2003, pp. 373 f.
144 Defoe 1899, p. 179.
145 Defoe 2004, p. 85.
146 Defoe 1994, pp. 19, 94, 109, 173, 267.
147 See Novak 1962, pp. 29–35, 140 f.

netically," Defoe describes open causal sequences, not closed regulatory processes. These would first be devised by Adam Smith: arguing that although operating unintentionally when it came to the public good, the profit-oriented, economizing individual already promoted that good by being active in a free-trade system, Smith saw this as attesting to the presence of what he famously termed an "invisible hand."[148] This hand was a providential authority without which the consistently auto-regulative trading cycles, as conceived within liberal rationality, would be inexplicable. But that precisely in the economic realm a regulative providence-fiction such as the invisible hand might not necessarily be guaranteed, that in its place disguised manipulations might be at work that in turn gave impetus to self-reinforcing *de*regulation—that was a fear of author Defoe from the start.

As a merchant and project-manager he had already experienced ruin in the 1690s, for example through amateurish activities as a marine underwriter, when he lost his own—uninsured—ships. He wrote his *Essay Upon Projects* (1697) after having to declare bankruptcy and beginning to seriously grapple with questions of economic policy.[149] Tellingly, the book starts with Noah's ark as the first project in human history, one that in addition owed itself to a "peculiar Direction from Heaven."[150] A more concrete basis for all such reflections was the self-regulation of mercantile administration and global-commerce initiated around 1700, with booking routines and shipping routes, commercial projects and global discoveries, now being conjoined. The *Essay* ends up offering guidelines for socially precautionary governmental policies, for this as well "seems to be a Project that we are led to by the Divine Rule."[151] In order that the economy can—autoregulatively—manage its disturbances and contingencies, Defoe demands precautionary institutions, especially forms of insurance, as "they offer to bear part of the Hazard for part of the Profit." But he speaks out against a deregulated

---

148 Its metamorphosis in Smith's writing corresponds to that in *Robinson Crusoe*. In Smith's *History of Astronomy* (c. 1759), we read of "the invisible hand of Jupiter," a transcendental authority catalyzing "irregular events of nature" such as "lightning" and "storms." In the *Theory of Moral Sentiments* (1759), it represents personal intuition serving as an ethical capacity in the sense of divine governance and coordinating sympathetic intersubjective communication. Finally, in Smith's *Wealth of Nations* (1776) it allegorizes the self-regulative processes at work in the free-trade system, processes becoming recognizable in the gravitation of factual around a natural price. See Smith 1967, p. 49; Smith 2002, pp. 87 f; Smith 1976, vol. 1, p. 477. See also Macfie 1971, pp. 595–599; Andriopoulos 1999, pp. 739–745; Grampp 2000, pp. 442f., 462f.

149 See Gibb 1957, p. 20; Koch 1968, pp. 81f.; Schuster 2000, pp. 63–65; Riehle 2002, pp. 19f.

150 Defoe 1999, p. 13.

151 Ibid., pp. 8f., 49.

insurance market and against pension funds and life insurance, considering them to be masked betting contracts.[152]

For Defoe, insurance's *raison d'être* was communal precaution, for which reason he favored reciprocal insurance, so-called friendly societies. "All the Contingences of Life might be fenc'd against by this Method," he insisted, while also indicating that such monetary solidarity could only be maintained with a recognition that "mankind must be *sorted* into Classes; and as their Contingences differ, every different Sort may be a Society upon even Terms." He saw maritime insurance as a special case, not only standing at the beginning of insurance-history and risk-generation, but also confronting a broad range of perils. Mariners are, Defoe indicates, "*Les Enfans Perdue*, the *Forlorn hope of the World*," always in "constant War with the elements."[153] His suggested response is to manage English seafaring as a well-organized state enterprise, with admiralty offices and mariner lists, with the royal sovereign serving as general insurer for freight, hull, and also mariners.[154]

Defoe's project was never realized. Instead, the marine insurance market developed into what in 1776 Adam Smith called a "lottery of the sea."[155] At the seventeenth century's end Lloyd's Underwriters established itself as the most important maritime insurer, first pursuing a form of betting-based insurance, then, when that business model was banned, as a kind of speculative risk-investment.[156] Insurance contracts of this sort spelled the ruin of individual underwriters, because they were liable with all their assets. Beyond that, what was intended as a project of solidarity found itself consigned entirely to the free market of speculation, which, as a result of the amalgamation of capital investment and insurance, long-distance trade and overseas transport, increasingly intertwined with and became absorbed into the stock exchange business.[157] In this context, around 1700, the deceptive and fraudulent maneuvering took especially perfidious forms in that type of trade promising security and protection.[158] "Contingencies and accidents" thus shifted from being an external threat, a danger to the economy, to "venture capital," hence in the end to an internal danger for the

152 Ibid., p. 46. See also Schmitt-Lermann 1954, p. 59; Braun 1963, p. 99.
153 Defoe 1999, pp. 48 f.
154 See ibid., pp. 49 – 52, 118 – 125.
155 Smith 1976, I., p. 122.
156 See Lehmann-Brune 1999, p. 44.
157 On the (contemporary) comparison between the maritime insurance market and trading in derivatives in the first-, sixteenth- and seventeenth-century bourses, see Moss 2002, pp. 32 – 34.
158 On this and the abuse of reinsurance, see Vance 1908, p. 16; Hagen 1920, pp. 137 f.; Barbour 1929, p. 586; Röpling 1960, p. 59; Sinz 1979, pp. 58 – 61.

business—for both its course and reciprocity. Defoe understands the inventive fraudsters who continuously think up new "Methods of Trick and Cheat" as project-makers in the worst sense: in systematically making "a Real and a Suppositious Value" indistinguishable, they pose a menace for both private and, finally, state assets at their most sensitive spot, that of their fictionality.[159]

As Defoe suggests, a highly developed capitalist economy like that in England circa 1700 had long since detached itself from terra firma—what he describes as the "best bottom" for all assigned value—for the sake of grounding its value-creation in credit, "virtual" products, and promises for the future, indeed even in risks.[160] The economy had, as it were, liquidated its territorial foundations and abandoned itself to the groundlessness of endless streams of human beings, goods, capital. And no overarching fiction such as the "invisible hand" could now vouch for its regulation. The promise of reality or realization represented by credit "acts on all substance, yet it is itself immaterial; it gives motion, yet itself cannot be said to exist; it creates forms, yet has itself no form," because it consisted of nothing other than the purposeful managing of, first, the imagination, then trade and action. Defoe here compares credit with a clock-hand whose regular and precise movement points to the harmonious movement of the clockwork: the entire economy; people, he observes, allow their own actions to be steered and timed by this "hand."[161] In the credit-based economy, an invisible hand no longer regulates visible and real action. Rather, it is now a visible hand, activating—or rather deregulating—the imaginative faculty and thus merely virtual action.

All too often, the erstwhile *manus gubernatoris* had now become an archetype of the manipulations making use of incessant circulation of chimera. Precisely the bourse, having gradually advanced at London's Royal Exchange to a vanguard and laboratory of this officious pseudo- and shadow economy, was thus for Defoe the staging ground for always menacing deceit and ruin, worked at systematically with the help of false information and rumors by the stock-jobbers and brokers, refined tricksters all. As Defoe puts it, with their "system of cheat and delusion" such disguised "usurers" manipulate a trading process based not on exchange-operations but on mere payment promises.[162] Hovering behind his argument is a sense that a possession whose referentiality has been lost upon the horizon of a contingent future in no way enthrones *homo oe-*

159 Defoe 1999, pp. 17, 28.
160 Ibid., p. 28.
161 Defoe 1710, pp. 4, 6.
162 Defoe 1850, pp. 135, 137.

conomicus as an autonomous, decision- and action-capable subject; rather, it delivers that subject to a deregulated, irrational market process, one permeated by all kinds of fiction. Market players only themselves grant the credit economy the credit of faith because they have a need for substance, existence, and form, however fictive the traded goods may be.[163] Around 1700, precisely this willing suspension of disbelief coupled fictional texts with the market.

As a literary author, Defoe certainly had to begin by acquiescing with market rules and fictional concepts of a deregulated bottom-up economy. But he opposed it at a decisive point: while the market treated literary authorship as the effect of a mere masquerade and fictional texts as fully fictive products (fictive because not endowable with substance, hence not capable of being possessed and thus freely available), he insisted, against all pirating of editions and products, on what later would be termed "intellectual property." Already in the *Essay upon Projects*, we find him calling for something like copyright and patent protection for project-makers.[164] In the meantime, in 1710, the first copyright law had been legislated, the Statute of Anne, granting "right to copy" not to publishing houses but to authors: a right initially limited to fourteen years, but still exclusive and freely transferable. It would only be around 1800 that authors gained full authority over their works[165]—a concept with aesthetic and legal dimensions derived from ideas of personhood tied to natural rights.

The concept possibly had another source as well: an economy of self-referentialized paper currency, a development, unsuccessful around 1700 but functional by 1800, that inherently produced both unlimited transferability and, nevertheless, credible promise. Starting in 1797, the Bank of England vouched for the sustainability of the new economy, as it was freed from the duty to substantialize bank notes with hard coinage so that it could now "pull solvency and insolvency together...by a thoroughgoing 'temporalization' of the system."[166]

For Defoe, the distinction between material and immaterial goods, and between real and fictive value, had clearly already lost its clear-cut rational basis. But he presents the non-distinction at work here as the fatal tendency of a "potentializing" of all things, in order to arrest the tendency through an idiosyncratic concept of authorship. In the preface to *Robinson*, we read that its private adventure is already "worth making public" because it justifies "the wisdom of Providence in all the variety of our circumstances," and that the editor be-

163 See Sherman 1996, p. 70.
164 See Defoe 1999, pp. 15–18.
165 See Bosse 1981, pp. 25f. On the expansion of the concept of "work authority" to that of "work policy," including its economic and legal aspects, see Martus 2007, pp. 13–22, passim.
166 Vogl 2015, pp. 52f.

lieves that Crusoe's narrative is "a just history of fact" without any "appearance of fiction" whatsoever.[167] Here Defoe is coupling, on the one hand, the economy of regulated providence with the economy of deregulated fictions, and, on the other hand, the editor's commercial faith in facts with the reader's aesthetic faith in fiction. He leaves the two contradictions' resolution to the deregulated market, which trades with what the editor says is missing: providential fiction. In the third and final volume of Defoe's Crusoe series, *Serious Reflections during the Life and Surprising Adventures of Robinson* (a volume consisting of essays in Crusoe's voice), we then read, in the aftermath of the first two volumes' commercial success, that these are less the precondition than the product of the providential program, which now is explicitly treated.

Crusoe asserts that his story is "allegorical" and "historical" at once, since there is a "man alive" whose fate is both the basis for and allegorizes Crusoe's own adventures.[168] After all, no one is more duty-bound to the hidden signs of destiny *and* their unrelenting production than an author who feels called to success—for which reason the "historical" life of an author of providence-fiction can for its part become an allegory. Even if Crusoe's career is meant to allegorize that of the aforesaid man, we cannot simply say that that man is the creator of Crusoe, the novelistic figure. Rather, his life is the "reality" of Crusoe's adventure, which, to be sure, only the novelistic character can address in language: the "man alive" experiences "innumerable ups and downs in matters of fortune," but maintains himself above water through "exquisite management," fighting "with the worst kind of savages and man-eaters," suffering repeated wrecks, "though more by land than by sea," finally finding himself in "a state of forced confinement, which in my real history is represented by a confined retreat in an island."[169] It is not just that here the insular sojourn allegorizes the "state of forced confinement" called writing, hence that behind Crusoe's actions stands author Defoe's writing hand. Inversely, the bankrupt Defoe's temporary imprisonment, rendering the previously prominent man into a phantom, is here certified through the novel's protagonist. Defoe's career as *homo oeconomicus* with its "ups and downs in matters of fortune," with struggles for survival in a context of predatory capitalism, with "management" and terrestrial shipwreck, in fact appears far less real and more fictive than the "real history" of the insular adventurer.

167 Defoe 1994, p. 7.
168 Defoe 1899, pp. IXf.
169 Ibid., pp. X–XII.

No longer, then, was a simple balance possible between fictive and real-life accounts. A new regulative authority was needed, above all pointed to in Defoe's later writing. It is his last novels, the paratexts are all composed under the sign of a fully *deregulated* economy of the fictive, within which the assertion of factuality and truth can be considered the surest signal of fiction. Nevertheless, these novels no longer stake the claim, in style of the older fiction, that here a manuscript that was leaked or fell to an editor is simply being delivered to the reading public, as the authentic account of a vouched for biographical reality. In *Moll Flanders* (1722) and *Roxana* (1723), "serious inferences" of a "third hand" are conceded, these meant to be understood as "a Pledge for the Credit of the rest."[170] Crusoe already offers this sort of pledge in *Serious Reflections*, when we read, with a view to the "man alive," that "to this I set my name."[171] But finally, Defoe's fiction awaits a signature belonging, not to a protagonist, but to the author. Namely, with such a signature, a bottomless economy that has always already delivered its credibility to a dynamic of credit, could reterritorialize itself in a name vouching for authorship, hence for *accountability*. We may view the fate of Lady Roxana (as laid out in *Roxana*'s "Continuation") as the allegory of a fatal business dynamic that fiction had prescribed itself as both merchandise and market parasite: After the "fortunate mistress" invests her fortune in a treading ship's freight and at the same time in the gains of a privateer, the two ships sink each other. What is left to Roxanna is "fatal loss."[172]

To delay precisely such fatal, all-devouring loss, in 1776, with official registration of Lloyd's as a society, the practice of underwriting—until then extremely liable to fraud—was made a legally regulated insurance practice. Now with their signatures individual insurers, so-called "names," pledged both their responsibility to the insured and the "interest" they took in this.[173] In the literary market, just this underwriting procedure would insure against the fatal loss threatened by fully deregulated trade with immaterial goods and the liquidated distinction between fact and fiction. If Defoe, in consigning regulative fictions to this trade and this liquidation, had himself spelled out this marketplace logic to the last detail, his poetics nevertheless offers a different prospect: checking the deterritorializing dynamic of promise and credit through an economy of pledging and insurance. The masquerade of the "fictor" in the name of the author only comes to rest under the sign of "intellectual property." For only the author's name, as

---

170 Defoe 1989, p. VIII.; Defoe 1982, p. 35.

171 Defoe 1899, p. X.

172 Defoe 1840, p. 130. From the perspective of a history of "probability," see Campe 2012, pp. 172ff. On the editorial status of the "Continuation," see Furbank 1997, pp. 299–308.

173 On this and underwriting in an American framework see Wertheimer 2006, pp. 7–11, 92f.

Thomas Wegmann observes, “authorizes, covers, and vouches for the price and value of printed material.”[174]

174 Wegmann 2002, p. 154.

Chapter 3
Disorientation: Poetic Experiments with Hydrography

"Writing in Water"

If since the beginning of Western seafaring the sea has been demonized, then not only because of what Hans Blumenberg points to as having lawless, unreckonable, and disorienting status. Evidently where, to cite Schmitt again, no "unity of order and orientation" is present, no *nomos*, only *tychē* or *fortuna* prevail. If as something like a primal division defining rational judgment, first and foremost incision or setting of orientation markers is involved here, then the sea necessarily emerges as the "other" of that *nomos:* "the sea knows no traces of what once was."[1] But the sea, it would seem, is not only the other of the *nomos* but of writing as well. It seems to be the indescribable par excellence. As such, it is the unmarked space of a culture that precisely does not exhaust its possibilities in terrestrial pursuits but rather (at least since Greek maritime voyages) in ventures upon the dangerous element, comprising failure. Just as the secured stock of writing is liquidated by oral discourse and in the stream of time, in order to then flow back into writing, into the Early Modern period and beyond culture was grasped as an oscillating between the firm and the fluid.[2]

As the notorious hostility to the sea unfolds in Plato's writing, we are confronted not only with a topos of placelessness but—in a peculiar turn—with one of both writing and its critique: as argued in *Phaidros*, living speech alone contains knowledge "of what is just, honorable and good," hence of the *nomos*. This knowledge, we read, is analogous to the care taken by a judicious farmer who "sows his seeds in suitable soil" and is "well content if they came to maturity within eight months."[3] The "gardens" of writing can, by contrast, only amount to a "pastime," including in the sense of "collecting a store of re-

1 Blumenberg 2010, p. 21. On this genealogy for *nomos* see Schmitt 1973, pp. 490 f., Schmitt 1995d, pp. 576 f., and Schmitt 2003, pp. 42, 67 ff. For a critique of this conception see Balke 1996, pp. 320–328.

2 See Assmann 1991, pp. 182, 184, 191.

3 Plato, *Phaedrus*, 275 f. Cited from Plato 1961, p. 521 (transl. R. Hackforth). Plato distinguishes between the *nomos* as general law and the law specified by legislation, for whose effectiveness the statesman is responsible, as *nomos empsychos*, "living law." See *Laws*, 721 f.; *Statesman*, 293 f.

https://doi.org/10.1515/9783110610734-007

freshment" for one's "memory, against the day 'when age oblivious comes.'" But writing, for Plato, can never be truly productive, suitable for a productive matter or idea; rather, to the contrary, "once a thing is out in writing, the composition, whatever it may be, drifts all over the place."[4] Most likely because writing, particularly in the Greek variant of a vocalized alphabet—allows recording, analysis, and manipulation of spoken speech, for Plato it is the very medium of wily artfulness.[5] Rational, fruitful activities such as oral dialectic and cultivation generate Being and preserve it in its true form; writing, on the other hand, moves just as much as seafaring along a path of lawless, non-rational, ultimately unfruitful becoming. Because writing and the sea thus represent both *logos* and the other of *nomos*, Plato's "writing in water"[6] designates an operation that is not so much contradictory as doubly deterritorialized.

"Hydrography," a discipline arriving on the scene and institutionalized in the Early Modern period, appears all the more fathomless. Because "writing in water" has to contend with a permanent "distance between sign and signified" while also involving knowledge "that cannot be represented directly,"[7] as a putative discipline it is initially the object of allegorical depiction. Already Cesare Ripa's *Iconologia* of 1593, most likely the most well-known of the many Early Modern iconological reference works, contains a graphic allegory of hydrography. (Fig. 4) It exhibits the reference work's media itself—writing and the image, *scriptura* and *picture*. Beyond that, it exhibits this sea-science's three most important media: the ship, the chart, and the compass, an instrument emblematically separating later seafaring from antiquity. As Ripa suggests in the *subscription*,[8] hydrography's first concern is surveying the seas, and imagining them in their evidently unimaginable dimensions—a double operation from

4 Plato, *Phaedrus*, 275 f. Cited from Plato 1961, p. 521.
5 Siegert 2003, pp. 34 f.
6 Plato, *Phaedrus*, 276. See Plato 1961, p. 522.
7 Mödersheim 1998, p. 216, and Kleinschmidt 1979, p. 393.
8 From the work's French version: "Hydrographie. Sa Figure est celle d'vne vieille femme, vestue d'vne Robe de gaze d'argent faite en ondes, ayant par dessus sa teste quantité d'Estoilles; en sa main droite vne Carte de nauigation auec vn Compas; en la gauche vn Nauire, & à ses pieds vne Boussole. / On la peint vieille, pour la raison que nous auons dite en la Figure de la Geographie. Quant à sa Robe de gaze, elle est vn symbole de l'Eau, & de soin mouuement; le principal objet de cét Art consistant en la description des Mers, dont elle prend les dimensions auec la Boussole entierement propre à la Nauigation, qui est le sujet pour lequel on luy met vn Compas dans vne main, & vn Nauive en l'autre." Ripa 1999, p. 191.

time immemorial always undertaken in both a technical and figurative sense.[9] Second, the sea is meant to be described as such, demanding creation of a new symbolic or graphic system, as an artless "writing in water" would be both inconsequential and senseless. Third, Ripa indicates that the compass is suitable for seafaring in a very particular way—at sea, we may extrapolate, problems are present far more acutely than on land, problems to which the compass offers a solution: to both be located and arrive at the right place.[10]

Arguably, Western culture constituted itself topically from the start. Reflecting this topical focus, works emerging in that cultural framework have been stamped by a regular obsession with "on-site" location, and a consistent treatment of religious, existential, and political questions as spatial in nature. The sea may have been predestined to serve in a metaphorics of the life journey because of one fact alone: movement through its smooth, non-solid space drastically presents a need for self-localization. Not only ancient *periploi* and poetry but also the first Early Modern navigation manuals emphasize the aggravated circumstances under which both sign-marking and establishing life-enabling *nomoi* took place at sea. Sea voyages "doe differ from viages by lande," we read in the first English translation, published in 1561, of the *Arte de Navegar*, a manual written by the Spanish cosmograph Martín Cortés de Albacar appearing in 1551. "For the lande is fyrme and stedfast," the passage continues. "But this is fluxible, wauering, and moueable. That of the lande, is knowen and termined by markes, signes and limittes. But this of the Sea, is uncerten and unknowen."[11] Even today movement in this moveable, uncertain, unknown realm generates scientific desire to come up with ground-producing media as soon as fundamental questions are raised. This is with good cause, for at sea the unity of order and orientation is never given but has to be made; and what makes the unity possible are tools—charts and, first and foremost, different instruments.

On the high seas, far from the coast and its secure orientating markers, the compass is an indispensable navigation instrument for Early Modern mariners.

9 On the distinction between hydrography, concerned with the sea in a geographic and topographic way, and later oceanography, which also takes in description of the maritime milieu and organisms, see Carrer 2009, p. 128.

10 On Early Modern hydrography in general see Apelt 2003, pp. 95–104. On the field's institutionalization in Colbert's France see Carrer 2009, pp. 108–118, and Geneste 2003, pp. 34 f., 43 f. On the hydrographic service of the British admiralty see Day 1967, p. 343, passim. On Western hydrography and oceanography from the Early Modern period into the twentieth century, see Deacon 1971, esp. pp. xii, 41 ff., and Schlee 1975, esp. pp. 16–37, 80–98, 244 ff., 285 ff.

11 Cortés 1561, part III., chapter 2, folio lvl.

Fig. 4: Cesare Ripa, *Iconologia* (1593)

In traceless waters, the compass creates or reveals the reference points serving as a basis for the ship's self-location, hence its capacity to impose some form of order, whether religious, legal, or political, on what lacks any locus. The compass is an instrument of or at least a catalyst for knowledge-generation. Its deployment is equally connected to knowledge of navigation, astronomy, and cosmology, on the one hand, forms of writing that can be termed either scientific or poetic on the other hand. Without the compass, Pedro de Medina writes in 1545, all other navigational instruments and procedures would only be worth half as much.[12] This measuring instrument makes self-localization possible by opening a view into the earth's magnetic field. It was thus the first such instrument that

12 See Pedro de Medina, *Arte de navegar*, Valladolid 1545, cited from Schück 1911–1918, I., p. 3.

used a needle not only to visualize elementary forces but also their long-distance influence. This innovation was arrived at through the following four steps: first, discovery of a stone with polar-magnetic characteristics, the magnetite; second, the discovery (already post-antiquity) that steel or hardened iron could be magnetized with such a stone; third, discovery that as soon as it was freely movable, a magnet oriented itself toward the north with one pole; and fourth, the technical developments, starting in the High Middle Ages, allowing the magnetized iron needle to be used as a compass.[13]

From the sixteenth century onward, innumerable scholars as well as poets have speculated on the origins of compass-related discoveries and developments in, for example, ancient China and Greece, in Arabic seafaring and Southern Italian Amalfi.[14] The question of when the compass became a regular navigational instrument has thus been the subject of both debate and verse—and, from the beginning, the forces it renders visual have been a special focus of discussion and allegorization. In light of the fact that before the nineteenth century there was no understanding of magnetism (let alone a theory of the phenomenon), it is not surprising that it repeatedly prompted all sorts of cosmological invention. Pliny the Elder thus informs us of the shepherd Magnes, who felt a mysterious force pulling at his nailed shoes; Ptolemy refers to magnetic islands in the Indian Ocean onto which iron-nailed ships are firmly attached—probably an inverse conclusion from legendary reports that above all wooden nails were used for ships in the Indian Ocean region.[15]

When Ptolemy was translated into Arabic in the ninth century CE, magnetic mountains appeared in fabulous texts, for example the Third Kalander's Tale of what has come down to us as the *Arabian Nights*, while the crusaders were setting out for an eastern edge of a world whose secrets and treasures must have been perceived as a dangerous distraction from the proper path of salvational theology and trade. The magnetic mountain appears in *Herzog Ernst*, a Middle High German epic written around 1180,[16] as it does in the early thirteenth-centu-

13 See ibid., II., p. 4.

14 See May 1949, pp. 259–263, Hourani 1975, pp. 106–108, Breusing 1982, pp. 79–90, Hertel 1990, pp. 50–55, Aczel 2001, pp. 53–75, Meyer-Haßfurther 2005, pp. 10–14. On the "scientific" dispute around invention of the compass that already emerged in the sixteenth century, see Schück 1911–1918, II., p. 13. – See also the reference to a (putative) invention of the instrument in Amalfi in the didactic encomium "La Navigation" by Jean Alphonse Esménard: Esménard 1806, p. 112.

15 See Balmer 1956, pp. 43, 526–530.

16 On both the magnetic mountain and coagulated sea in *Herzog Ernst* see Sowinski 1979, verses 3886–3935.

ry *Tristan* of Gottfried von Straßburg, where Isolde is compared "to the Sirens, / who with their lodestone / draw the ships toward them":[17] Like the sirens, her magical power opens barred hearts, but also always threatens, in her beauty and love, distraction from the straight and narrow path. Finally, with invention of the magnetic needle, the magnetic mountain shifts from being an erotic and poetic to a geographical myth. In the mid-thirteenth-century Middle High German *Kudran* (or the *Gudrunlied*) it is displaced to northern waters, in line with the needle's northern directive. For thirteenth-century Italian poet Guido Guinizelli, the mountain functions as a kind of mediator between polar star and compass needle. And finally, it will be ascribed with the dangerous misdirection already drawing the attention of medieval mariners.[18]

Early Modern emblematics could draw on this stock of partly legendary, partly cosmological images in allegorizing—through an overlayering of ships of state, church, and life—the cleverness and faith of both regents and nameless Christians. In the contemporary translation of a book's *subscription* by Diego de Saavedra Fajardo we thus can read that

> no heart exists / that has not found itself touched by some divine magnetic stone / and like a compass's needle is in ever-unfolding movement / until it has found the glowing northern star / around which the star runs its course; hence it behooves us not to live in silence/ until we recognize / and honor that uncreated star / in which true ease is found / and from which the movement of all things stems.[19]

With seafaring as its unmistakable model, the compass of faith is thus a registering implement—or better: a sensorium—for an unfathomable attraction in any case calling for comparison with the mystery of true belief. That compass is a visualized expression of the directive to the correct locus that only divine mercy can offer the human heart. And by way of the needle, it produces natural commonality and sympathy with the extra-earthly force standing in the heavens as the northern star and in neo-Aristotelian reference works as the prime mover.

But emblematics has interest in the compass for reasons beyond its deictic, orienting function. "When the compass's needle is magnetized with godlessness,

---

17 Straßburg 1993, I., verses 8085–8131.

18 See Balmer 1956, pp. 533–535.

19 ist kein hertz nit / welches sich nit von einigem Göttlichen Magnetstein gerührt befinde / vnd wie ein Compas ziegelein auß einer sonderbaren natürlichen gemeinschaft [...] in einer immerwerdenden bewegung ist / bis sie den gläntzenden Nordtstern gefunden / vmb welchen das gestirn jhren lauf hat; Also geziemet es vns nit in der stille zu leben / biß wir erkennen / vnd ehren jenen vnerschaffenen stern / in welchem die wahre ruhe zu finden / vnd von welchem die bewegung aller dingen herrühret. Cited from: Peil 1983, p. 721.

fraud, and wickedness, it may not show the way without error," we further read in Saavedra.[20] Hence alongside the orientation toward God, granted for the sake of grace, the compass also displays a false or simply magical belief leading those fools into existential error, hence constituting a betrayal of creation. Just as to do justice to its task the needle has to be properly magnetized, belief, love, hope need to be planted in the sluggish human heart to get it moving. Both a state and individual human lives are only properly managed and oriented when in such divine harmony. The attraction between magnet and polar star sets ships on their proper course; similarly, Jacob Cats observes, divine love "drives us everywhere, so that we are never separated from it."[21]

One of the oldest significant documentations of the compass's needle stems from Alexander Neckam, who in his *De naturis rerum* of 1187 referred to the object in a water-bowl, hence a liquid compass.[22] For Neckam already, the instrument gains a sense of direction in both a seafaring and salvational context: mariners plying the water in a period of poor visibility or at night, readers are informed, can gain orientation through the magnetic needle, which miraculously is always directed toward their guiding, polar star, *stella maris*. Here practical self-localization thus replaces magical or—as often suspected—devilish usage of magnetism's mysterious force.

In order to transform a simple needle into a compass-instrument, Neckam tells us, mariners touch it to a magnet, so that the needle "initially turns in a circle until it stands still, finally pointing northward."[23] However unimportant the distinction between celestial, magnetic, and geographic poles was in this framework, revaluation of the concept of "orientation" would turn out decisive:

---

20 Wan das zungelein im Compas mit Gottlosigkeit / betrug / vnnd boßheit bestrichen ist / so mögen sie den weg ohne jrthümb nicht weisen. Cited from ibid., p. 720.

21 *uns an allen Oertern treibt / Also dass man von ihm auch ungeschieden bleibt.* Jacob Cats, *Silenius Alcibiadis sive Proteus*, Amsterdam 1622, 49, 3, cited from Henkel 1967, col. 1774. On emblematically shaped use of images, see also Lohenstein's *Arminius* (1689/90): "and if we unfurl all sails of our sagacity / all sweat at the oars of our strain; we don't arrive anywhere else / than where the compass of eternal providence leads in that it either lets us choose its intent without forcing our free will; or else tosses us on its unfathomable way through storm / to a place we would have never imagined in dreams. Nevertheless he cannot go under / nor fail to reach harbor; who has GOD as his star on this world's sea / and his conscience as a magnet." Lohenstein 1973, I., p. 1105b.

22 See Röller 2010, p. 20. On the spread of knowledge about magnetic directional force in medieval Europe and its usage for a compass needle (esp. by monks, seafarers, and poets), and on the connected competition between ecclesiastical and imperial claims to knowledge and power, see ibid., pp. 51–57.

23 Neckham 1863, chap. XCVIII: *De vi attractive*, p. 183.

by contrast with those moving on land, corresponding itineraries, and world maps, mariners no longer looked eastward, hence in the direction of Jerusalem, but toward the northern star. In Portuguese, becoming "dis-orientated" was thus now *desnortear*, in French *perdre le nord*.

In 1269, Petrus Peregrinus could already describe a dry compass with a compass dial, circular graduation, and sight, so that now only a suitable mounting, the linkage of needle and dial together with a closed case, was needed to have the compass or *bussola* in its classic form.[24] The expression *bussola* is first documented in the 1380 Dante commentary of Francesco da Buti, as a corrupt variant of Medieval Latin *buxida*, "wooden receptacle."[25] For its part, *compasso* was formed from Latin *cum passare*, "to go with," "accompany," itself derived from *cumpassum*, "in equal step," "in equal measure" or else from *compassare*, "to gauge." "to measure"—all derivations that can be related to a permanent course correction.[26] Probably, starting from the meaning "circle," "decoration in circular form," the nautical division of the horizontal circle as *compasso* was also at work here. Earlier, the order of the different wind-gods—those elementary guides leading ancient mariners as soon as they left the coast and had to sail under cloudy skies—had been followed for that segmentation. With the increasing differentiation of winds according to strength, humidity, temperature, and saltiness, the compass dial was divided into increasing segments, corresponding to the now identifiable wind directions.[27] Wind diagrams had always had a mnemo-technical and toponymic function; even today, being able to recite wind directions by heart is captured in the expression "to box the compass."[28]

Although the old seafarers had no seaworthy measuring and positioning instruments, nor any maritime maps, they nevertheless developed other strategies for overcoming their helplessness, alongside these mnemotechnics and gazing at the stars at night: poetry and writing. Namely, when they returned form the placelessness of the open sea, they promptly transmitted their practical knowledge as sailing directions. This could have been done in the form of memorized,

---

24 On the European marine compass's last decisive renewal, attaching the dial to the magnetic needle itself (otherwise than was the case with the land compass long known in China), in order to make possible the steady targeting of a specific course, see Körber 1965, p. 53 and Breusing 1982, pp. 90 – 94. On the connection between the gimbaled and compass mounting see Kay 1995, p. 129.

25 See Aczel 2001, p. 36.

26 See Schück 1911–1918, II., p. 45 and Metzeltin 1970, p. 143. For the following see ibid., p. 144.

27 See for example Blake 2004, p. 10. On the ancient systems of 8, 12, and 16 winds and their translation through various cults and cultic objects, see Aczel 2001, pp. 39 – 52.

28 See Obrist 1997, p. 84 and Taylor 1962, p. 19.

metrically and metaphorically increasingly shaped oral reports, or else through the *periploi*, containing travel distances and coastal descriptions, themselves capable of expansion into treatises nothing short of epic in scope. As we have seen, the *Odyssey* may have been a prototype for a symbolical maritime orientation system in which narrative and navigation, tropes and topoi came together. And perhaps elementary guideposts like the wind prompted the poetry of the first maritime epics—not only because of their relation to a range of mythological ideas and cultic practices, but even more so because from that period's perspective, in order to themselves be perceived, the winds forced precise and vivid description of what they effected. In the end elementary forces like the wind or magnetism, impossible to address in themselves, lead our attention to "the universe we can see, hear, and feel distilled to its essential oils."[29]

In face of the poetic *periploi*, the medieval portolans with their harbor-descriptions seem prosaically compiled, strictly practically oriented sailing handbooks. But they contain a decisive new element—course directions.[30] Portolans furnish contextual information for directives meanwhile offered by the compass; beyond that, together with the *bussola*, they made possible the graphic system that, along with the portolan *charts* for the first time render visual the sea itself, that unimaginable space of potentiality. The fourteenth-century portolan charts are the first of all marine charts. They shift the list of manuals into a second dimension. Emerging from the demands of nautical practice, they are deployed as information for a range of situations—dynamic, changeable, simply possible. These charts are thus the visual counterpart of discourse, as what Isabel Capeloa Gil has termed "getting-there narrative."[31] They are not projections, but correspond to empirical navigational criteria from point to point. They and not the portolan *manuals* are what, as Konrad Kretschmer put it in 1909, "allow a direct display and precise fixing of one site's position to the next."[32] (Fig. 5)

Although the portolan charts only covered the sea's unmarked space in small vectors, they furnished the first comprehensive view of what later was called "Europe"—which from this reflection on itself would undertake a steady exceeding and expansion of its "world picture."[33] Flanked by data from the portolan manuals, oriented northward, supplied with standards and networks de-

---

29 Huler 2004, p. 202. On the "Beaufort scale" with its standardized but versified wind determinations, see ibid., pp. 69 ff.

30 See Kretschmer 1909, pp. 31, 180.

31 Gil 2008, p. 109 (with reference to Michel de Certeau's concept of discourse in *The Practice of Everyday Life*).

32 Kretschmer 1909, p. 100.

33 On the portolan charts as first representations of Europe, see Jourdin 1993, pp. 36 ff.

Fig. 5: *Pisan chart* (1290), oldest known portolan chart, whose accompanying manual was entitled *Il compasso da navegare*

duced from the compass dial, the charts allowed fixing of the proper rhumb line (a radial directional line that determines a located point, or a point meant to be located, through angles to and distance from a given origin) on the chart's surface, for the sake of then choosing the proper course on the sea's surface with use of the compass. Around 1300, this simultaneously symbolic, graphic, and operative system was encapsulated in the polyvalence of the term *compasso:* a manual and chart, a pointer and an instrument for drawing circles to measure distance. Through use of compasses, the winter high sea's dark emptiness could now be measured, the indescribable par excellence described. It was now possible to leave the visible markers of solid land behind and nevertheless not simply voyage into uncertainty, hence, to again cite Kretschmer, "to even find non-visible objects by maintaining the proper course."[34] And this even if there was uncertainty about one's own locus: in delimited Mediterranean space, once the proper course had been embarked on, the ship was bound to reach a coast before too many days had passed, the last stage then being straightforward and predetermined. The Mediterranean is thus both itself an *ordo* and a *compas-*

34 Kretschmer 1909, p. 66.

so, securing empirical and vectoral navigation, promising fortunate voyage and a specific form of the old topos of "being in."[35]

Dante's Poetic Compass

Against this hydrographic background, the navigational-instrument compass was taken up by poetry in the early fourteenth century. Before then, in 1258, in an epistolary report on his stay in England, Brunetto Latini, a Florentine statesman and scholar banned by the Ghibelline party, spoke of an amazing instrument known to English mariners:

> This discovery, which seems so very useful for everyone traveling by sea, has to remain hidden for later times, since no ship's captain trusts himself to use it and thus falling into suspicion of being a magician. Even the sailors would not dare to go to sea under his command it he took a tool with him that so very much seems to have been made under the influence of a devilish spirit. The time may come when these prejudices, constituting such a great impediment to studying nature's secrets, are no longer present. Then humanity will harvest the labor of men as learned as monk *Bacon*, offering them justice for their energetic work and insight; presently he and the others receive no other payment but disgrace and reproach.[36]

It would be up to Latini's pupil Dante Alighieri to show through poetry that the magnetic stone could help reveal creation's secrets without being a devil's tool. The compass's first poetic mention is thus in the *Divine Comedy*, in the twelfth canto of the *Paradiso* where Dante refers in three verses to the song of Bonaventura, the pious Franciscan author of a *Journey of the Mind to God:* "from the core of one of these new lights, / as the north star makes a compass needle veer, / rose a voice that made me turn to where it came from."[37] The *Divine Comedy* spells out a poetic topography that, as suggested earlier, can be described as a scholastic-theological variant of Aristotelian cosmology, with its concentric

---

35 On the secondary importance of maritime charts in North America, which had no compass for the Atlantic, see Schnall 1991, pp. 274–276. On systematizing the portolans in a clockwise direction and to the designation of "the whole process of navigating from place to place" as "caping," see Parry 1974, pp. 40–42. In that the compass-system only really functioned in the Mediterranean (see Randles 2000, text VI., p. 2), its assessment by the Hegelian geographer and philosopher of technology Ernst Kapp that the "sea charts and compass are a triumph of spirit over the water-masses." Kapp 1868, p. 648.

36 Letter of the poet Guido Cavalcanti, cited from Schück 1911–1918, II., p. 57.

37 "del cor dell'una delle luci nove / si mosse voce, che l'ago alla stella / parer mi fece in volgermi al suo dove." Dante, *Paradiso*, 12, 28–30.

spheres. (Fig. 6) Hell, where no star gleams and within which heavenly attraction is no longer palpable, so that everything has to remain caught in its place, extends down into the earth's center, the point of greatest distance from God.[38] The Mount of Purgatory rises on the southern hemisphere, covered by the sea hence empty of human beings. This is the starting point for ascent to earthly paradise, before entry into the spheres of planets and constellations, the crystal sphere, and Empyreum finally dissolves time, space, and all human striving into light and bliss. Within the crystalline sphere, there is nothing that would not be transcended within "being-in" and encompassed by the *primum mobile*—that highest unity *in* the world, bordering on the One and the Empyreum's highest being. This does not exert influence on the earthly as a cause or corresponding to natural laws, but rather directly, in the manner of light or magnetism.[39] And the border between this outermost sphere of the *maggior corpo* and the prime mover is not spatial but ontological, located between the world's finitude and God's infinity.[40]

In the *Divine Comedy*'s process of initiation, on the way to the beyond, where technical, theological, and philosophical knowledge translate into each other, the compass appears in person, its *cum passare* or accompaniment leading to the true goal here seen to first by Virgil, then Beatrice, and finally Bernard of Clairvaux.[41] On one side of this topography we find the terrestrial, spherically organized, which is to say spread out on a sphere's surface. The "orientation" here is not eastward to Jerusalem but actually—and in accordance with the age's navigation, which, to be sure, deemed it only possible under open skies—toward the celestial pole, the northern star.[42] Precisely upon the "vastness of the sea of being"[43] directives are needed in the normally darkened situation. For although every thing is assigned its place—and only that—its essence nevertheless ends up lying in something like finding a harbor: just as after contact with the magnetic stone the needle shows sympathy with the northern star, the prime mover has planted faith, love, hope in sluggish hearts, hence brushed them "magneti-

38 On this cosmological topography of Dante see Maurer 2008, pp. 14–17, 21.

39 See Binggeli 2006, pp. 63, 236, 239.

40 See Edson, Savage-Smith, and von den Brincken 2005, pp. 7, 9; Koschorke 1990, p. 29.

41 On the theological and narrative function of these companions in Dante see Wittschier 2004, pp. 36, 39; on transitional space in Brunetto Latini and Francesco Colonna, see Rimpau 2008, pp. 27–30.

42 Cf. Brunetto Latini speaking to Dante (*Inferno*, 15, 55 f.): "By following your star / you cannot fail to reach a glorious port."

43 Dante, *Paradiso*, 1, l, 113.

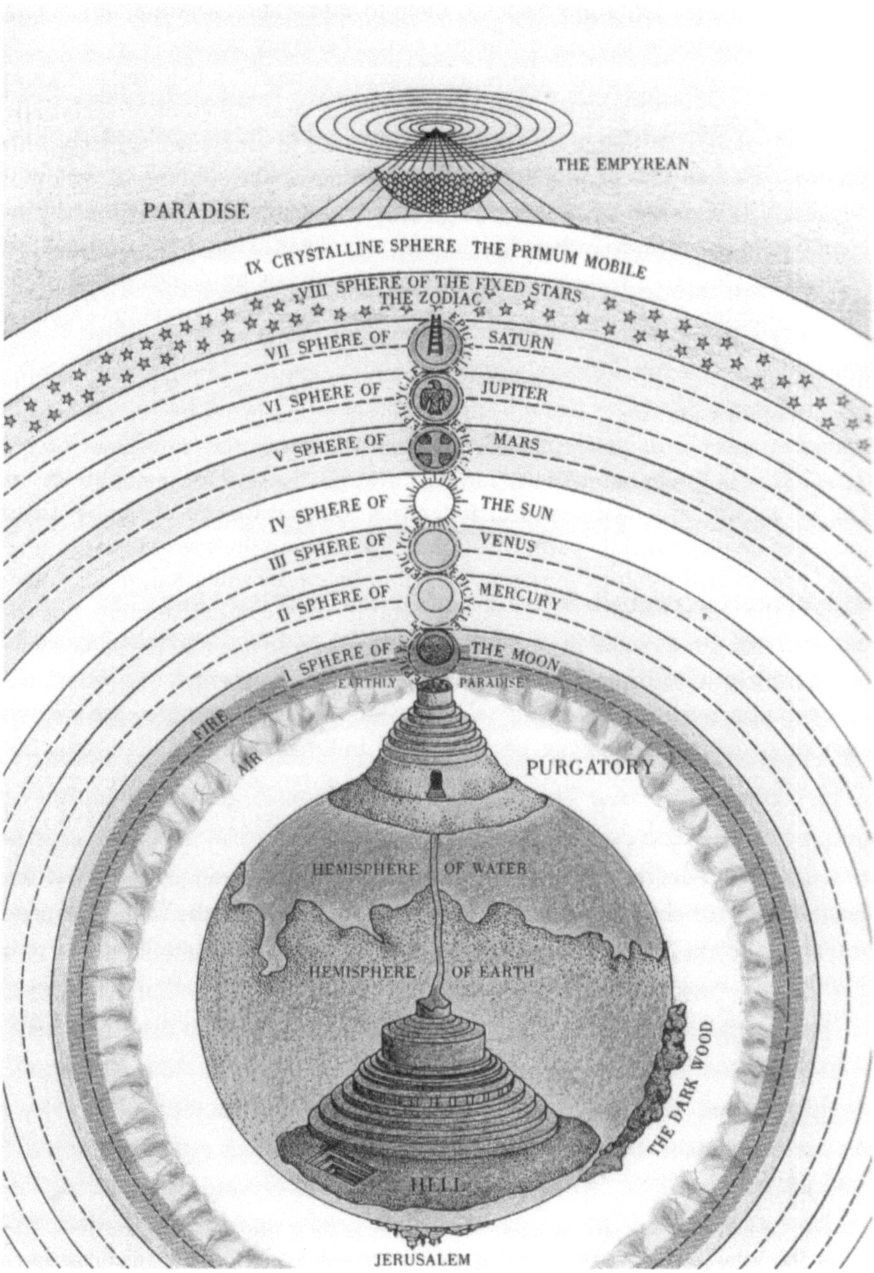

Fig. 6: Dante's poetical topography

cally" with divine mercy and disposed them toward what Katharina Münchberg terms, in reference to Dante, "mediation of the heavenly powers."[44]

As Dante observes in his "Questio de acqua et terra," a lecture he gave in Verona in 1320 on the relation between water and land, inherent in everything terrestrial, alongside simple nature expressing itself in a downward movement, is nature as ascension, expressing itself in the force of elevation or attraction of the stars.[45] The ground of all things, and of living beings together with their intersubjective interaction, is their final cause, we read in Dante's political treatise "Monarchia" (c. 1317). What is especially the case for living beings, Dante indicates, is that the simple fact of their having been created cannot be the creator's final goal; rather, it is the activity proper to their essence alone. In this way, human beings are duty-bound to a form of being that through the intellect fulfils its possibilities and its perception of the possible, as in its perception of the unique movement of the prime mover when observing the movement of the stars.[46] This is the fathomless grounding for Dante's seafaring metaphorics: as long as we are still in this world, indeed as long as we are on any path to God, even an otherworldly path, we all find ourselves on life's sea-voyage, a *navigatio vitae*.[47] Hydrography thus precedes annulment, or sublation, of all topography in the world beyond,[48] and our this-worldly instruction determines our situation in that other realm: at worst permanently fixed to one and the same locus of suffering, at best on a vector of infinite and blessed ascent.

Seen in this way, the outward voyage Dante's Ulisse recounts in the twenty-sixth canto of the *Inferno* inevitably had an infernal ending: Ulisse does not know the mystery of true faith. He sails into the Atlantic without a "compass," opening up the circuit, the round voyage, of the old *periploi* and crossing past the Pillars of Heracles, for Dante still the borderposts of Christian "being-in." He voyages westward toward the sun, stops then toward the south, where instead of the northern star only a four-star constellation leads the way; in Dante's work, it is not yet termed the "southern cross" as it is a body possessing no salvational meaning.[49] Then, instead of the hoped-for earthly paradise, Ulisse sights the highest of all mountains, within whose pull he finally goes under.

44 Münchberg 2005, p. 116.
45 Dante, "Questio de acqua et terra" §§ 47 ff., § 73.
46 Dante, "Monarchia," 1, 2, 7 f.; 1, 3 3 f., 6; 1, 9, 2.
47 Cf. the debarcation and star-search at the very start of *Paradise:* 2, 1–6, 29 f.
48 On the distinction between topographic space in the *Inferno*, in *Purgatory*, and in the space opened to transcendence in *Paradise*, see Münchberg 2005, p. 118.
49 See Hennig 1930, p. 550.

What Ulisse can only see as a magnetic mountain is thus what for Dante will already have been the Mount of Purgatory.[50] In firmly aimed steering toward his "guiding star" and the spiral ascent to the "One," Dante finally surmounts anything merely conventional and sign-based—his entry into the divine is "symbolic" not only in the poetological but also, even more so, in the existential sense. Ulisse himself moves on a curved path, but one fully attached to the curved earth, hence describing a two-dimensional movement: no ascent but a kind of determined aberrance in the this-worldly realm. The failure of spatial ascent corresponds to a failure on the level of signs. Because instead of following directions toward the One, he centers his efforts on a "polytropism" of conventional signs, of words and rhetoric, he lacks the ontological attractive force of what is figurally and existentially "symbolic." For this reason, he ends up in a *turbo* that swallows up his ship in the sea of confused sense and being: a *turbo* that is unmistakably both an act of God and a vortex of signs. The prime mover has, then, punished Ulisse for one specific transgression: use of conventional signs for the sake of lying, deception, false advice and bad faith.[51]

For this reason, the Ulisse who has failed at sea and thus, once and for all, can at least end his terrestrial wandering is punished in the beyond with infinite arrest: doing time without end in Malebolge, the circle of hell consigned to "simple" fraudsters (more malicious, "compound" ones are consigned to the ninth circle), and in the pit of the evil counselors. Even if we wish to see this version of Odysseus as a pioneer of the Early Modern world picture, for Dante he is still an exponent of a threefold transgression: legally, because he has abandoned the discretionary power of earthly rulers, which, as we read in the *Monarchia*, only extends to the ocean;[52] cartographically, because he had overextended the range of the (by definition Mediterranean) portolan; and metaphysically, because with the Mediterranean as "our sea," *mare nostrum*, he has also abandoned the "great sea of being." Ulisse sins through curiosity without *cura*, care. Precisely on account of purposive drive without mediation, he goes astray in a manner Aquinas terms *fuga finis*, flight ways from all purpose, all goals. And because his transgression is fully immanent, it has little salvational content. He does move past the sub-lunar order, but he fails to transcend it—the sole meaning of all exis-

---

50 On the role of Homer, the *Alexanderroman*, and the Sindbad *cycle* along with traditions such as that of the *Lebermeer* (coagulated sea) and magnetic mountain in the *Divine Comedy*, see Rüegg 1965, esp. pp. 183, 186 f.

51 On the different forms of being and movement and correspondingly different use of signs by Ulisse and Dante in the *Divine Comedy*, see Lotman 1990, pp. 179 ff.

52 See Dante, *Monarchia*, 1, 11, 12: "But there is nothing the monarch could covet, for his jurisdiction is bounded only by the ocean."

tence, and all writing as long as it treats Paradise, hence what Dante defines through the neologism *trasumanar*, "to transhumanize," a process of passing beyond the merely human that we have abandoned.

In order to approach this non-place named Paradise and even create linguistic loci for it, Dante uses hydrography as a medium.[53] He names his work a *legno*, a wooden ship, one that leaves the distinct sign, the *segno*, behind it. The *Divine Comedy* is for this reason not a topos-organized allegory. It is rather a dynamic relational system starting out from an uncertain locus of one's own (viz the dark woods and its paths in the first canto), in order to then empirically and vectorially wander—or better, navigate—from point to point, hence "through an eternal place" (*Inf.*, I, 114). *Cumpassare* is here at the same time a metaphor and medium for *transcendere*. Art is here no longer mere handwork, but steered by genius or intellectual intelligence, *ingenio*. It no longer follows a merely predetermined path, but continuously opens up its own coordinates: this in an unfathomable manner and in harmony with a still indefinable higher order.[54]

That this *ingenio* is perhaps merely, auto-affective (in Kant's sense) and thus in its own right arrogance and hubris; that the "fictitious words" Dante refers to in his treatise on philosophical wisdom, the *Convivio*,[55] and their fiction of not being such, possibly resemble Ulisse's deceitful address to his companion voyagers; and that in this way Dante's signs of being can no longer be read, to cite Barbara Vinken, as "signs of the Creator-Author" but only as offered for their own sake (and that of their earthly author):[56] all of this speaks for Dante as a Ulisse of poetry, and also for poetry and seafaring as two facets of one and the same process of transcending. Poetry and navigation would here no longer be read as subsumed to a Thomistic order, within which mere potentiality and contingency is accorded anything terrestrial, which only gains actuality and providential orientation through transcendent divine form.[57] Rather both activities now emerge as focused on a self-mobilization of loci oriented without divine governance, hence, again, empirically and vectorially, thus rendering what has no locus accessible in the first place.

Dante's hydrographic metaphors and his concept of sign-transgression are located in a horizon of nominalist representational doctrine, with its struggle

53 See Stierle 2007, esp. pp. 10–15, 37–39, 51f., 193–195, 225f.

54 On "intuitive knowledge" and on the relation of *sentire*, *imaginary*, and *memorari* in the *Divine Comedy*, see Hirdt 1989, pp. 56–66.

55 Dante, "Convivio," 2, 1, 3.

56 Vinken 1991, p. 16.

57 On the Aristotelianism of Aquinas see Bloch 1963, pp. 33, 50f.

against a feared vanishing of fundamental order.[58] Namely, as William of Ockham observed, signs now no longer generate direct access to what they signify. They no longer serve as direct (thought, spoken, written) representation of what is in any event present, but first show in representation what they are actually pointing to.[59] Put otherwise: the representational force of signs has to first and foremost be acquired. We always find ourselves within a "semiotics of experience," as rather than being firmly anchored in things and their places, signs—precisely because they become recognized as at our disposal—can become lost in what Patrick Hochart, referring to Ockham, has termed a "yawning maw" of pure operativity.[60] Propositions do not simply articulate an existing circumstance but rest on a mental, vocal, or written syntax that repeatedly first connected words with things. Like the instruction offered by *bussola*- and portolan-supported navigation, signs of this syntax are neither simply natural nor artificial, primal or derived. They result from an increasingly reflective use of media, as manifest in Dante's time with development of the compass and navigational charts. And it sets free the "virtue of fiction" that the *Divine Comedy* tries to legitimate, in order to assume it at the same time.[61]

Even if nominally propagating the fixed ontological order of Thomistic topography, Dante's epic, didactic poem shows how static division of space into separate loci can be transgressed and opened up dynamically. It shows how sign and syntax can articulate the "sea of being" in all its potency (how so to speak they can be used hydrographically), hence what "writing in water" would be like.[62] In this way, the poem offers a lesson in poetry, in what it can mean in an Early Modern sense. Namely, the narrated path is also the path of narration. Instead of fulfilling an abstract model of order, narration here becomes a practice involving the opening up of space. At the same time, in an exemplar manner Dante translates topography into narration and navigation into experience. Not only do Ulisse's and Dante's boundary transgressions spell out an initiation process—once successful, once not—including phases of separation, existential change, and rebirth. We here likewise find a transforming of

58 On Siger von Brabant's significance for the *Divine Comedy*, see Baur 2007, pp. 262f.

59 See Libera 1996, pp. 352ff.

60 Hochart 1973, p. 181. On the following, see ibid., pp. 183–192.

61 On the shift of *phantasia* from a simple retentive faculty to active *vertù* and on the corresponding separation between theology and philosophy see Hirdt 1989, pp. 60f., 70, 74f., 91f.

62 On the abstract medieval conception of space and the striving of all bodies to fund the space appropriate for them (the *auton topon*) in view of the *Divine Comedy*, see Münchberg 2005, pp. 112–114.

the combination of position and route (for the first time cartographically encodable around 1300) into a combination (fictional in the Early Modern sense) of *situation* and route.

For this reason, structuralistic theories of narrative such as Yuri Lotman's have seen in Ulisse's and Dante's movement past the old world picture, including its ordering of what can ever be said or experienced, as encapsulating the contingently eventful, subject-oriented, and thus quintessentially narrative quality of imaginative literature.[63] In this framework, we can understand the *Divine Comedy*'s significance as not being limited to Dante's variation on topoi and narratives of discovery, adventure, the underworld, the otherworldly voyage inherited from antiquity and the Middle Ages. Rather, Dante's poetics emerges as drawing on knowledge of theology, metaphysics, and cosmology to present a form of narrative practical knowledge for the first time making it possible to distinguish undirected curiosity about the world such as Ulisse's from purposive and self-limited studiousness such as Dante's, mere desire for experience from experience steered by knowledge and rules. In this respect, it is important to note that the author Dante does not stand in a relationship of personal expression or embodiment to his first-person narrator. Rather, a functional connection is at work here, one opening up the loci of a voyage into what has no place at the moment of its being narrated—this despite or actually because of the poem's epic preterit.[64]

We can here speak of a mediality of the artwork, in that fictional discourse becomes the means for its self-exceeding and annulment. By positing a distinction between reality and fiction while breaking loose from previously certified topoi, the text on the one hand, furnishes itself with an enabling structure for transgressing known worlds, on the other hand opens the prospect of a higher authority that (at the latest in the *Paradiso*) again annuls such distinctions, consequently both guaranteeing the truth of the narrated anagogic events and legitimizing poetic fiction as the means toward that end.[65] Furthermore, we can speak of autonomy of the artwork, to the extent that it grasps its own uncertain locus

---

63 See Lotman 1977, pp. 231 ff. See also Krah 1999, pp. 4–6 and Kinzel 2002, p. 29. On the convergence between Lotman's cultural semiotics and the "liminal theory" of van Gennep and Turner, see, together with this book's earlier discussion, Koschorke 2018, pp. 90 ff.

64 See Hamburger 1993, pp. 113 f., 337 f. On the complex question of identity between author (or "I-origo") and narrator (an identity that in Dante is broken in an exemplarily way) see Genette 1992, pp. 74 f.

65 Warning 2009, p. 34. See also Regn 2009, pp. 366–368, 384 f.

through auto-reflection and articulates it precisely in its contingency.[66] Finally, we can speak of the compass as possessing a threefold function: as a *bussola* for seafaring, it is an instrument of terrestrial world-production that visualizes invisible things through a specific dynamic and sympathy; as a compass of faith it is disposed toward theological transcendence in that it renders opaque terrestrial signs readable in a revelatory direction; and as a compass of literature it guides its transgression of the sign, in that it brings together image and language, intuition and concept, ideality and materiality.[67]

In understanding the compass as the skill of world-generation, Dante renders it into an allegory of both faith and writing—and poetic writing in particular. By for the first time linking invisibly acting forces with the "full play" of the "imagination,"[68] the *Divine Comedy*'s poetics look forward to the representational procedure of early modern scientific experiments, with their frequent focus on magnetism and electricity. Seen from this perspective, it is not surprising that Ossip Mandelstam, in his remarkable "Conversation about Dante," would name the medieval poet "a master of the instruments of poetry." "[H]ere the trembling hand of the compass not only indulges the magnetic storm but makes it itself," Mandelstam continues, then commenting later in his essay that "[p]oetic speech creates its own instruments on the move and cancels them out without halting."[69] Or, to return to the most important field for *compassare:* such a poetics creates its instruments in experience (for Dante, most prominently as *navigatio vitae*), and it leaves them behind in experience as well (for Dante, finally in visionary contemplation of the divine). The plausible imparting of such experience provides the first attestation that something significant, an improbable event or fundamental transgression, has ever taken place. Consequently, this poetics of experientiality[70] will be foundational not only for Early Modern narration but, just as much, for practical knowledge.

Experimental Writing

Emerging from Aristotle, the concept of "experientiality" will prepare the path, extending through scholastics into the Early Modern epoch, for the idea and

66 On the "terror of contingency" in Dante see Stierle 2000, pp. 29 – 31. On both self-generated totality and the metaphysics of art since Dante, see Lukács 1971, p. 37.

67 See Münchberg 2005, pp. 73, 80.

68 Priestley 1966, vol. 2, p. 16.

69 Mandelstam 2001, pp. 40, 47, 82.

70 Fludernik 2008, p. 73.

practice of experimentation—Greek *empeiria*.[71] Where for Plato experience signifies simple knowledge of isolated facts without attention to causes or sources, in his *Metaphysics* and the *Organon*'s "Posterior Analytics" Aristotle treats it as a foundational familiarity with individual cases taking in both schemata of action and elementary predication. For Aristotle experience thus establishes a first transition from the singularity to the universality of knowledge and art. To be sure, this transition is not identical with what later will be described as induction. Rather it unfolds in a non-methodic way, after the humanly specific mnemonic capacity has accumulated the scattered sensory perceptions.

The scholastic concepts of *experientia* and *experimentum* in any case disqualify the habitual and contingent dimensions of Aristotelian *empeiria*.[72] Both those concepts emerged from the Latin term *experiri*, "venture," "test," "find by experience," an etymology running against the stockade of the scholastic concept of experience. For in the medieval period truth was meant to be looked for in a *turn* from experience's singular moments. No knowledge was possible on the basis of an *experiential* centered on the particular and the contingent. And when Thomas Aquinas spoke of a *scientia experimentalis*, he emphatically did not mean discovery or invention of new things.

Until the Renaissance, *experientia* would thus largely coincide with *experimentum*. For in the field of natural philosophy, as long as commentary on authoritative arguments comprised the main activity, later distinctions such as that between representation and construction would hardly play a role—likewise with problems such as the instrumentality and mediality of experience and the intervention of cognitive passions like curiosity. In order to gain knowledge of experimental procedure, free textual genres were needed, in the sense that objects and representational forms had to first be determined. Literary art was one such genre: on the one hand during the Middle Ages it was not recognized as a separate artistic form; on the other hand, through the organization of the universities and the emergence of new areas of knowledge the old connection between *artes* and *auctores* was dissolved, which in turn furnished poets with new freedom.

Perhaps Dante stands at the end of this development. What is certain is that he contributed just as decisively to the revival of concepts such as *poetari* as he set the stage for the early Modern concept of autonomous literary creation.[73] When, as I have suggested, Dante is aligned with the transition from the

71 See Hager 1972, pp. 453f.

72 See Libera 1996, pp. 92f.

73 See Curtius 1953, pp. 153f., 483.

epic's closed narrative form to the open, world-encompassing form of the Early Modern novel, then one reason for this is the moment of contingency penetration on different levels into the space of what is narrated and what is narratable, and into that of narration itself. Already the world image formulated here stands under the proviso of simple contingency—hence divine mercy or will, letting the world exist (at least for the time) in its "suchness."[74] In His unlimited potential, God is by no means tied to the order of things catalyzed through his creation. Within the world and from the perspective of its exploration, the effects of His unfathomable will may appear arbitrary. But they manifest nothing less than providence itself—which is, however, tied to no fatalism: Freedom of action remains, so that for instance Ulisse's decision to turn to the contingency of mere experience, at the expense of exploring providence, is an unnecessary decision. Still, from the terrestrial look-out it must be considered contingent.

Contingency's capacity for non-being thus involves, on the one hand, freedom of action and thus one's own positioning vis-à-vis God (as contingency-fixed as with Ulisse, as providence-aware as with Dante). And it involves, on the other hand, natural processes and events in their capacity for non-being and unfathomable futurity—a duplicity of world-immanent occurrences amounting to secondary causes. These may be allegorized in the form of *fortuna*; but they cannot allow their subsumption into a providential plan and their dependency on the primary cause named God to be forgotten.[75] Poetry's task is consequently to render contingency observable on its various levels, while also conveying the providence—whether unmarked or marked allegorically or symbolically—essentially different from it. To be sure, through its experimental grasp of this-worldly phenomena the modern period will increasingly explicate this providential dimension under the sign of natural laws. Against that backdrop, the initial task of an "experimental" poetry is not only to diminish contingency's status, as something merely illusory, deceptive, fictive. Beyond that, starting with this capacity for non-being, it must conceive a space of possibility for various modalities of interpretation, representation, and action—modalities allowing poetic qualities to emerge precisely in their distinction from what is not poetic.

Even if Dante's poetry still articulates less empirical than theological and cosmological knowledge of the world, it nevertheless ventures outward to the limits of its own semantic nature, materiality, and processuality. As pointed to by the twinned figures of Ulisse and Dante as it were allegorically, the *Divine*

---

74 See Blumenberg 1987, pp. 139 f.; Waldenfels 2000, p. 4.

75 On contingency and providence in Dante, see Münchberg 2005, pp. 33 f., 155–160.

Comedy marks a particular moment: the moment when a *will* to knowledge encounters an *order* of knowledge whose paramount purpose is to discover revelation-laden signs of divine effect in the world. This will to knowledge grasps experience not only as the occasion of pious studiousness but as a curiosity about the world open to results. And already because we are in the sublunary realm of confusion, irregularity, and distance from God, this process is initiated not in the stars but on earth—and here, especially, the sea. For Early Modern researchers into nature such as William Gilbert, a phenomenon such as the magnetic needle's deviation will thus point to the deformation of great magnet Earth, to the uneven distribution of its bodies of land and water, consequently to the contingency within the earthly order of things.[76]

Dante's poetics takes up such questions and problems at work in natural-philosophical experimental science in that, as Mandelstam puts it, "every element of the modern experimental method may be found in Dante's approach to biblical tradition. These include the creation of specially contrived conditions for the experiment, the use of instruments of such precision that there is no reason to doubt their validity, and clear verification of the results."[77] For example, in the otherworldly sphere of paradise, the question of the double nature of light (located between material and spirit) and that of the eye's receptivity (hence of the autonomy of a visual-psychological realm) are raised at the same time as the question of the transcending of poetic semiosis and the capacity of perception in general, which reflects both the metaphysical problem of potentiality and Alhazen's and Roger Bacon's optical experiments.[78] And it was the introduction of one particular orientation technique, the compass, into the *Divine Comedy's* thematic texture, a technique making possible practical scientific exploration not only of familiar parts of the world but also hidden causes such as magnetism, that shows the work to be a dynamic referential system: one understanding its "instruments" as a metaphor and medium of its experimental *transcendere* and transgression.

That the conditional potentiality at work in the autonomous albeit factually-centered artwork emulates its absolute counterpart in the work of the divine creator instead of simply being its existential analogy (put otherwise: that it actually anticipates the experience of completeness meant to be granted by mercy alone) has implications for Dante's own poetic language: his own "fictitious words" here fall under suspicion of sinful *curiositas*—but also anticipate nature's

---

76 See Blumenberg 1987, pp. 191f., 514f.

77 Mandelstam 2001, p. 70.

78 Ibid., pp. 71ff., and Binggeli 2006, pp. 234, 236.

experimental exploration. Dante's discovery of fiction is the symptom of an epochal threshold. Previously, every new experiment either had to align itself with a valid topical corpus of certainties or simply abandon claims to truth-value. Now natural-philosophical knowledge, having moved past that hierarchical system to differentiate itself into a range of disciplines, will need to nourish itself from methodical experience and heuristic discovery.

Starting with the Renaissance, when it comes to elements of nature previously closed to practical experience, artistic and scientific discovery arrives on the scene. Up through the time of Francis Bacon, experience will be the object of a doctrine of perception and representation grounded in a deployment of medial apparatuses, as well as in novel techniques of writing for both the organizing and documenting of experiments and processing the resulting data.[79] As Joseph Glanvill would indicate in 1668 in a programmatic interim appraisal, decisive "Advantages for Knowledge" were now owed to various "Arts, Instruments, Observations, Experiments, Inventions and Improvements."[80] Replacing experience based on authority or chance, both designated as "haphazard experience," *experientia vaga*, was purposeful experience or experimentation, *experientia ordinate*, which Bacon in turn separated into what he termed "literate experience," *experientia literata*, referring to the planned organization of specific experiments based on previous research results, from "interpretation of nature," *interpretatio naturae*, systematic ascent from individual experiences to the natural laws.

The path-centered metaphorics Bacon uses to describe his experimental method is precisely aligned with the narrative of adventure, discovery, and compass guidance structuring Dante's voyage to the beyond.[81] The only difference is that what is even poetically precarious for Dante, because suffered by the creator wit reservations, has become normal scientific activity for Bacon: the confluence of theory with methodical practice. For now, instead of nature's contemplation in the evidence of its mere givenness, we find nature objectified, rendered accessible, indeed conquered. (As Michael Gamper has indicated, it is the case that although the main role was played by systematic observation of nature (and by contrast with the situation in respect to electricity), the "hidden forces and

79 See Bacon 2000, pp. 109 f.

80 Glanvill 1668, p. 6. On Bacon's introduction of "technological history" as a possibility to even start to break up the "canon with unalterable holdings," see Blumenberg 2009, pp. 59, 91.

81 See Bacon 2000, p. 10: "But just as in previous centuries when men set their course in sailing simply by observations of the stars, they were certainly able to follow the shores of the old continent and cross some relatively small inland seas, but before the ocean could be crossed and the territories of the new world revealed, it was necessary to have a knowledge of the nautical compass as a more reliable and certain guide."

causes of magnetism" were partly also uncovered through experimentation.[82] It no longer contains regions preferred by inherited topographies and hierarchies. Rather, what renders the research process possible itself becomes an object. Bacon's method follows the narrative of discovery for just that reason: it defines itself precisely through the non-specific potentiality of its transgression.[83]

Bacon, co-founder of the Newfoundland Colonization Company and one of the first stockholders in the Virginia Company,[84] thus formulated his *Great Instauration* in a manner fully contradicting the previously propagated form of "being-in": voyaging through the Pillars of Hercules was no longer hubris but rather a duty; shunning it an error and sin of omission. As already demonstrated in the Iberian enterprises of discovery, the "instauration" was only possible through a concentrated acquisition of practical knowledge—this compelling, for instance, voyaging beyond the bounds set by round trips in "our" known sea (whether the *mare nostrum* or *mare academicum*). If such major voyages also led into uncertainty and darkness, they were nonetheless methodically guided and could thus count on the New World's gradual and continuous discovery. Gunpowder, tried and tested not least in the New World's conquest, now became the insignia of military Europe. And because with the New World the compass opened up the opportunity for globalization of trade, that instrument would end up becoming the emblem for entrepreneurial Europe. But beforehand it emerged as the guiding instrument in experimental science: within the fathomless ocean, Glanvill explained, it had carved out first paths, thus enormously expanding natural history. Whoever its inventor may have been—it had contributed more to the world's advancement than a thousand Alexanders or Caesars, and ten times more to the growth of knowledge than Aristotle or all the disputations taking place since the dawn of philosophy.[85] In his *Of the Advancement and Proficience of Learning*, Bacon observed that because printing continued the "mutuall Trafique" of maritime commerce and its long-distance "participation" through both "the vast sea of time" and *mare academicum*, the printing press itself had emerged as the symbol of learned Europe.[86]

"We are certain of our path, not our place"—"place" in the sense of our own position, which always must be determined anew—was Bacons new guiding

82 See Gamper 2009. p. 29.

83 See Blumenberg 1985a, pp. 389 f.

84 On Bacon's biography and role in the history of science, see Krohn 1987, p. 45, passim.

85 Glanvill 1668, pp. 79–81. See also Bacon 2011, p. 412.

86 Bacon 1640, book I, 6, 64. On this emblematic triad see Glanvill 1668, pp. 75 f.; on its origins with Giovanni Tortelli (1450) and Polydorus Vergil (1499) see Witthaus 2009, p. 79, n. 8.

maxim, expressed in his work on "theory of the heaven."[87] Spatial limitation to the Old World's order, the outer limits here being marked for Bacon, as for Dante, by the Pillars of Hercules, is akin to retreating from that bright future the New Science is meant to usher in. Hence Bacon's systematic reassessment of *curiositas* itself reflects, in part, the theological foundations of his concept of science. As Bacon explains it in *Valerius Terminus*, the expulsion from paradise resulted from human arrogation of "moral knowledge," not of knowledge of nature. Advancing the latter is nothing short of a divinely endowed task. For paradise and human sovereignty, a sovereignty revealing itself in—both signifying and creative, discovering and inventing—prelapsarian language, can only be regained when creation has been made open.[88]

Against this backdrop, Bacon's concept of experiment is based less on mathematics[89] than on the power of the word. He sees science as rooted in the aforecited *experientia literata*, which along with natural-philosophical observations in the more narrow sense also comprise experiences recounted through religious transmission, myth, and poetic parable. In that creation is structured in a more verbal than geometric or logical way, such "literary experiences" explicate the "Book of Nature." In a narrative manner they present, for Bacon, what in the framework of traditional rhetoric might be termed the *inventio* of reference. And finally, this side of all causal chains, he describes these "experiences" as directed back onto the language of paradise, the common basis for the creation of both concepts and nature.[90] As in *On the Wisdom of the Ancients* (1609), Bacon thus collects and deciphers religious myths and poetic parables as hypotheses or models for knowledge of nature. And in works including his posthumous *Nova Atlantis* (1627), he in turn uses them to propose the scenario of a future renewal of science.

Here a narrative of discovery and a research program once again coincide: at the non-place of a haphazard ideal state, removed from the borders of both Old and New Worlds, where knowledge and power, research and law, are brought together.[91] All state resources are mobilized in order to maximize experimental knowledge-production, the thus acquired natural resources then utilized for further accumulation of power. This scenario of naked experimental politics is fic-

87 *Certi viæ nostræ sumus, certi sedis nostræ non sumus*; Bacon 1826, vol. 9, p. 187. See also Konersmann 1994, p. 73.

88 See Blumenberg 1985a, pp. 385 ff.; and Blumenberg 1987, pp. 635 f.

89 On Bacon's odd rejection of all mathematizing see Gaukroger 2001, p. 21–25.

90 See Daniel 1982, pp. 227–231.

91 Bacon 1965, esp. pp. 172 ff. On Bacon's utopism as a cipher of his methodological failure see Waage 1978, p. 216.

tional, as already made clear by its past-tense presentation and reference to facts that (at least until the point of the telling) can at the most be recounted. The fiction can best be termed descriptive and prescriptive rather than narrative, as neither conflicts nor transgression have a place in the world here recounted in the utopian past. Nevertheless, the text constitutes a narrative experiment, its premise being that every experiment certifies what Klaus Meyer-Abich has referred to as "the transition from theory formulated in mind of practice to theoretically founded practice"[92]—thus implying the narrative of its social realization, which can in turn be spelled out experimentally. When in *Nova Atlantis* Bacon's narration unfolds from the perspective of a realized future, then its utopism renders plausible the success of a scientific program comprising a bet on the future from the start. In *The New Organon*, Bacon declares, already in the style of Pascal's wager, that opting for the experiment of an experimental society is absolutely called for as "the danger of not trying and the danger of not succeeding are not equal, since the former risks the loss of a great good, the latter of a little human effort."[93]

For Glanvill, especially the "House of Solomon," the research society that secretly develops into the political control center on the science-conforming island of New Atlantis, was a "Romantick Model" retrospectively needing to be viewed as a "Prophetick Scheam of the Royal Society."[94] When Glanvill looked back at the accomplished Great Instauration of knowledge, it was clear that Bacon's imaginary House had in any case paved the way for the Europe-wide emergence of numerous societies and academies. England here took something of a special path in that the main focus was not on mathematical work as in France but on empirical research. In addition, with the English Restoration, the Royal Society and the East India Company operated as twin enterprises in the realm of practical science.[95] Knowledge and power, in particular natural-philosophical and technical knowledge on the one hand, marine and trade dominion on the other hand, were to render England, in the words of Thomas Sprat in 1667, "not only *Mistress* of the *Ocean*, but the most proper *Seat* for the Advancement of *Knowledge*." Above all the British island's geographical position, Sprat observed, predestined it to "Discoveries above others in the *intellectual* Globe, as they have done in the *Material*."[96] Already because no benefactors were available yet to support research voyages in the modern sense, adventurers, merchants,

92 Meyer-Abich 2006, p. 171.
93 See Bacon 2000, p. 88.
94 Glanvill 1668, p. 88; Glanvill 2005, p. lxv. See also Dear 1985, p. 147.
95 See Kanthak 1987, p. 63.
96 Sprat 1734, p. 86.

and seamen were integrated into experimental science. Or inversely, research programs could be arranged according to their willingness to cooperate and timetables. There was of course study of extant notes and reports. But for the most part mariners, merchants, and colonists were handed "queries" and diaries to document "inquiries" they were to themselves make concerning natural phenomena and distant land formations, together with what was offered by their nautical and navigational instruments.[97]

Before his death in 1662, Lawrence Rooke, who had been professor of astronomy at Gresham College, formulated the first guidelines in this respect. His "Directions for Seamen Bound for Voyages," appearing in 1665 in the first number of the Royal Society's *Philosophical Transactions*, presented a catalog of what was to be essayed, observed, and recorded for the sake of both hydrography and natural history in general: longitude and latitude determined through the compass needle's declination and inclination; the tides including their "accidents"; coastal views (recorded in sketches); soundings; terrestrial and marine samples; wind-related phenomena; all the celestial phenomena that could be seen at sea.[98] In the next number of the *Transactions*, Robert Hooke offered an appendix to these guidelines, focused on "Nice observations of the Variations and Dippings of the Needle, in different places, and in the same place, at different times," but also pointing readers to the documentation of all the dangerous places and passages at sea, furnishing the construction plans for corresponding hydrographic instruments, and indicating "The Form of a Scheme" (a reference to research plans or guidelines).[99] Through such directives and observational media, mere experience was thus to be placed on experimental foundations, hidden forces of nature—above all and in exemplary fashion magnetism—brought "within the Power of the Sense."[100]

In any case, decisive for the entire enterprise was heightened attention to the phenomenon of deviation and irregularity within nature, consequently the steady openness to new experiences; and equally the minute bookkeeping in respect to observations, experiments, discoveries, hence the schematizing and standardizing of all empirical data—this to be delivered on return to the appropriate authorities, themselves familiar with the data-processing involved. Here the Royal Society emulated the organization of knowledge already pursued by

97 See Carey 1997, pp. 275f., 282.

98 See Rooke 1665/1666, pp. 140–142.

99 Hooke 1666, pp. 433f., 439, 445, 449ff. On the context and the convoluted reception of these *Directions* see McConnell 1982, pp. 5–11.

100 Hooke 1705, p. 38. See also ibid., pp. 34, 46, 53f.

the Iberian *casas*, then starting in the 1630s by for instance the *Collegio Romano* (directed by Athanasius Kircher and commissioning both Catholic mariners and Jesuit missionaries).[101] At the same time, science and seafaring were now defined as two sides of the same quest for practical knowledge, a process leading to a scientizing of navigation—and, as it happens, to a standardization of the travel report as a genre. The new standards concerned not only the worldly material informing narrative but also, tied to it, its "experientiality": the introduction of a difference between the fictional and factual and the constitution of a reliable narrator as witness.[102]

Where the experimental science pursued at sea relied more on observations of nature than on experiments, the experimental science undertaken in the laboratory in many respects actually adhered to the nautical model. It was Robert Boyle who both offered virtuoso demonstrations of experiments in the narrow sense and worked at establishing common standards for the experimental report as a textual genre. By precisely furnishing all "circumstantial details," this kind of report was meant to give a voice to the investigated material itself (as opposed to, say, the eminence of those participating). For various failed maritime voyages had led through haphazard fortune to the discovery of "new Regions much more advantagious"—and this should be no less the case with the failure of "Philosophical Trials": "those unexpected accidents that defeat our endeavours do sometimes cast us upon new discoveries, of much greater advantage than the wonted and expected success of the attempted Experiment would have proved to us."[103]

Erring and failing were, then, integral to the experimental process, offering a chance to lead it onto previously undreamt of paths. Importantly, this contingency is not merely a regrettable accompaniment of experiments that as such serve to confirm already probable hypotheses alone. Rather, it imparts the event with its decisive moment, often representing its real reporting and narrating value. The knowledge of—reducible or irreducible—contingencies and—putative or actual—disturbing factors determining an experiment's success or failure promotes

101 On the *Collegio Romano* and Kircher's effort to solve the longitude-problem at work in his "magnetic geography" by organizing empirical knowledge and corresponding cosmological speculation, see Gorman 2004, pp. 240–248.

102 On the dependence of the travel report of for instance Dampier on Rooke's *Directions* see Thompson 2007, p. 83. On the—narratively resolved—paradox that the new experimental science both had "nullius in verba" as its motto and pursued its natural research on the basis of books, see Carey 1997, p. 283.

103 Boyle 1669, pp. 34, 115.

discovery, brings new knowledge to light.[104] Precisely the loadstone, together with its occult qualities, its countless possible malfunctions and deviations, serves as a model for this experimental science, for which reason Boyle placed himself among the "Magnetical Writers."[105] And further: Boyle was the Royal Society member who would promote systematic maritime research most strongly. His interest in the physical properties of water and air, his measuring instruments, his cooperation with a salvage-diver, his official research commission for a sea-voyage to Tangiers—all this rendered him, alongside his laboratory assistant Robert Hooke and before Edmond Halley with his standard-setting voyages and research, into one of the most important pioneers in the realm of hydrography.[106]

Romantic Disorientation in the Work of Poe

Since the High Middle Ages and the dawn of Early Modernism, the compass served as a hydrographical emblem—a tool encapsulating that idiosyncratic confluence (cf. *cumpassare*) of technique, knowledge, and writing. However, from the Early Modern period onward, the tool's disorders repeatedly made new aporias manifest within the enterprise of "writing in water." From the beginning, the great excursions—with Dante's Ulisse perhaps representing their literary harbinger—would give a hard time to the old *compasso.* In longer ocean voyages, the portolan charts proved inadequate because the now noticeable curvature of the earth meant that the rhumb lines could not be precisely fixed, so that alongside the courses also distances were erroneous.[107] Hence with advent of the new world picture, the old means of representation were overtaxed.[108] Developed since the rediscovery of Ptolemy's *Geography,* the world maps with grid could not simply rely on the old imaging procedures—especially since localization according to longitude and latitude was based on a totally different principle than the earlier vectoral technique. It is true that Pedro Nunes's introduction of the "Loxodrome"-curves that describe straight distances on a sphere's surface,

104 See ibid., pp. 75, 88 f., 92 f.
105 Boyle 1676, p. 20. – See also ibid., pp. 1–3, Boyle 1669, pp. 97 f., and Boyle 1664, p. 15.
106 See in detail Deacon 1971, pp. 117–129, 154, 166 f.
107 See—reflecting the Portuguese viewpoint—texts included in Randles 2000: text 6, pp. 1 f.; text 14, pp. 86 f.
108 I am here using the term "world picture" in Heidegger's sense of new procedures in gaining practical knowledge. In this regard and with a view to more recent theory of space see Dünne 2011, pp. 23–30.

always intersecting the meridian at the same angle and approaching the poles in a helical manner—along with Gerhard Mercator's development of conformal cylinder-projection opened up a new representational space useful for sailing.[109] But this representational capacity was limited to a small section of the world (or of its maps) and increasingly failed with proximity to the upper latitudes, where the standards were evidently increasingly distorted.

In addition, by Columbus's time it had been recognized that declination, the magnetic needle's deviation from true north, increased in a westward direction or simply—for instance on the Azores' level—reversed itself.[110] An effort was made to make just these needle aberrations useful for navigation by assigning every location a calculable declination-value—this in the firm belief in a stable distribution of terrestrial magnetism—while as it were incidentally moving to solve the notorious longitude-problem.[111] Systematic observations of the compass and its malfunctions were now commissioned, with the nautical instrument controlled by its azimuth counterpart (which aligns the determined position with astronomical observations). In other words, second-order instrumental observation now emerged, in order to indirectly track down magnetism's hidden sources.[112] With increasing instrument-based maritime exploration, the sea increasingly appeared, in Edward Wright's words written in 1599, as "an inextricable labyrinth of error, out of which it will be very hard for a man easily to unwinde himselfe." For this reason, Wright rendered "Certaine Errors in Navigation" into a detailed study of its own.[113] Here especially the complex interplay between maritime charts and use of the compass was meant to be placed on firm footing: through geometric and astronomical methods for systematically correcting deviations, and through the first direct mathematical description of the principle of projection that Mercator had implicitly formulated.[114]

---

109 On Mercator see Günther 1982, pp. 274–274. On the rediscovery of Ptolemy through Manuel Chrysoloras in Constantinople see Carrer 2009, p. 68. On the difference between perspective and projection and the different types of projection see Stockhammer 2007, pp. 19–29.

110 On the legend of Columbus's discovery of declination during his first voyage, so that he accounted for it in his second voyage, see May 1949, p. 262.

111 This belief was cultivated by Mercator, as well as by Nautonnier and young Kepler. Acosta, Stevin, and Halley would introduce certain meridians for the deviations, while Gilbert initially referred to the irregularity of the earth's surface as the reason, before then conceding it was the magnetic field itself and abandoning this kind of longitudinal determination. See Balmer 1956, pp. 128–140.

112 See Taylor and Richey 1962, pp. 31–33, und Meyer-Haßfurther 2005, p. 24.

113 Wright 1657, p. B2 r.

114 See ibid., pp. B3 r, C1 r ff.

Against this backdrop, maritime excursions led to a double—empirical and mathematical—"experimentalizing" of hydrography. For example Wright and others considered logarithms to be a mathematical "technique" meant to above all benefit seafaring and navigation tables, for which reason Wright quickly translated John Napier's *Mirifici logarithmorum canonis descriptio* (1614) into English, dedicating the volume to the East India Company. It was nautical demands that gave mathematics impetus to become an applied science, a science no longer focused strictly on working out general truths but turning to discovery of new objects and procedures of knowledge—hence leaving the safe harbor of Euclidean geometry for an exploration of new descriptive and calculative means. In order to convert, for instance, rhumb lines from spiral shapes on a globe to navigation-suitable lines on a rectangular projection, two things were required: a cartographic correction of the displacement in the map-grid possible with every projection; and, in respect to navigation technology, a fixing of longitude after covering each particular distance on the line. Thomas Hariot's solution to this problem was insertion of isosceles triangles, thus putting together a smooth and steady curve from an infinite number of discrete sections, allowing investigation and description of the continuum as with an unfamiliar area or unexplored coastline. From Hariot's procedure to Bonaventura Cavalieri's "method of the indivisible" and John Wallis's "arithmetic of infinitesimals" and onward to the infinitesimal calculus of Leibniz' and Newton, mathematics was practiced as a discovery-centered and experimental procedure that would develop novel methods moving forward from the hydrographic paradox of "writing in water."[115]

On the basis of this new formal means and with help of increasingly refined measuring techniques, empirical hydrography would in turn undertake its first pure marine research expeditions, so that at the beginning of the eighteenth century Edmond Halley was investigating the trade winds, currents, and—above all—compass-declination on a global scale. The result was the first "magnetic map" in a Mercator projection including isolines through the points of equal declination.[116] When Halley's findings were tied to further revision of geomagnetic hypotheses—from the assumption of a static axial or tilted dipole to that of a rotating or even non-connected dipole[117]—this simply pointed to a further destabilizing of the ordering and location system that the compass, and the projection of the "world picture," were meant to represent. For alongside the nee-

115 On Hariot's procedure and his analysis of coastal formations and their mathematical-historical results, see Alexander 2002, pp. 143–148, 167–170, 179–185, 194–197.

116 See Balmer 1956, pp. 180–183, Taylor 1971, pp. 240 f., and Reinke-Kunze 1994, pp. 9 f.

117 On this series see Jonkers 2003, p. 36.

dle's declination, its variable "inclination" to the earth's midpoint, had also been discovered, together with the needle's "variation," a declination that was not only changeable in location but temporally as well.[118] Finally, around 1800, "deviation," a deflection of the compass through nautical iron, began to be observed —and also, to crown it all, a magnetizing of iron ships, and even ships with iron parts, in the shipyard already, so that they gained a specific "magnetic signature" changing according to the angle at which the ship intersected en route with the force-line of the earth's magnetic field.

In Greenwich in 1838, a Magnetic Observatory was opened whose director, the royal astronomer George Biddell Airy, was tasked with arriving at correct compass compensation. Airy recommended placing magnets and quantities of iron around the steering compass, in a manner suitable to each ship while still on the dock. In longer voyages on the high seas and during storms, this corrective often turned out inadequate. By contrast with the trading fleet which followed Airy's method from the beginning, the admiralty decided on introducing a dry "standard" azimuth compass, and on regular measurements of deviation together with continuous recording of their values. Although one ship per day had already been lost because of misleading compass instructions, it was not until the sinking of the *Tayleur*, an iron emigrant ship of the White Star Line, in 1854—it ran into the Lambay Island rocks on its maiden voyage and 362 persons drowned—that a public controversy developed. That same year, William Scoresby, an amateur of magnetism who had previously been a whale hunter and later would become a Bradford pastor, declared war on the Airy method in a speech before the British Association for the Advancement of Science in Liverpool. The method, he declared, was dangerous, having led to most of the losses at sea: a ship's magnetic signature was suddenly subjected to unpredictable changes, facing which the compensation procedure could only offer deceptive safety. From that point onward, various English shipowners and underwriters avoided deploying and insuring iron ships. Liverpool shipowners founded a Compass Committee, and numerous voyages were now undertaken with an uncorrected "pole-compass" located on the mast, outside the nautical magnetic field.[119]

At the controversy's core was the question of whether the shift between "magnetic latitudes" during extended voyages and the tremors a ship experienced in storms negated any compensation measures, indeed became uncontrol-

---

118 On the geometric hypotheses and concrete verification methods tied to inclination and variation see Balmer 1956, pp. 141–147 and Jonkers 2003, pp. 63–75.

119 On the context of and course taken by the Airy-Scoresby controversy see Cotter 1977, esp. pp. 591f., 595f.; and Gurney 2004, esp. pp. 189–212, 218–222.

lable confounding factors—a theory that Scoresby tried to demonstrate in 1856 through an expedition to Australia. Perhaps the practical outcome of this controversy (retention of the Airy-method, but optimized through introduction of Scoresby's pole-compass) was less decisive than the recognition that once extended to a global scale, the location-ordering system tied to that instrument was bound to fail. In other words: On the one hand, it was the case that controlling and offsetting all of compass declination's variables seemed by no means inconceivable, since the various malfunctions possibly expressed a law calling for identification. But on the other hand, this seemed nothing less than an infinite task, as precisely taking account of the variations and deviations within the dynamic system comprised by ship, compass, and terrestrial magnetism would have required endless equations. What contemporaries observed in respect to global interference by wind also applies to magnetism: "investigation / when each particular factor is pondered / would be neither an end nor goal," but lack of both was, as he put it, "beyond human capacity.[120]

At least since Coleridge's "Rime of the Ancient Mariner" and Edgar Allan Poe's sea stories, Romantic literature is marked by innumerable, obsessively laid out wanderings. As we have seen, their source has been understood as a global situation of what Lukács termed "transcendental homelessness." With the opening to an "infinite universe" of a previously "closed world"—the terms are Alexandre Koyré's—the human being appeared not only to have lost his place in that world but (as was said of the cosmologist's particular vantage-point) any idea of being part of a finite, closed, and hierarchically ordered whole.[121] Since then, also from the limited perspective of maritime practical knowledge, "world" can no longer be allegorized, topographied, written, with the implication of specific markings, thresholds, and borders, and with them a set of concepts, values, and goals in harmony with a proper order of being. Rather, what is global now both reveals itself and withdraws as a space of force fields whose gathering and exploring requires experimental procedures and operationalization of knowledge. It also requires a type of orientation steadily compelled to reflect on and scrutinize the conditions for its own possibility—if not to fully derive itself from experiences of disorientation. This explains, the characteristic "transcendalistic" way of writing repeatedly attributed to Roman-

---

120 Halley 1723, pp. 75, 106. The citation could not be localized in the original English-language source.

121 See Koyré 1957, pp. 5 ff., passim.

tic sea-narrative.[122] And perhaps we can say that the two guiding variants of Early Modern narrative of wandering gravitate around two sorts of nearly emblematic topoi, both vouching for wordliness and simultaneous withdrawal of a world picture: on the one hand the cape, which gradually orientates a voyage that has no orientation in its transcendental homelessness, this in order to have it begin at the same point repeatedly anew;[123] on the other hand the pole, which, as the configuration-point for all self-localization approaches, in the end fully disorientates the systematic orientation of hydrographically supported seafaring, driving the ship into shipwreck.

Starting with Pedro Nunes's loxodrome and Mercator's cartographic projection, the terrestrial poles were defined as the vanishing point of all topographic description and thus of any construction of a world picture. Already Boyle, with his hydrographic experimental program, Halley with his research expeditions, and Siméon Denis Poisson, who between 1821 and 1838 produced a series of memoirs on compass deviation on the occasion of British surveys of the Arctic, shared the same conclusion: that from a hydrographic perspective, the abyss of all representation had to lie at the poles, also in respect to the earth's magnetic field and the global system of seas and wind.[124] In 1861, Matthew Fontaine Maury, founder of modern oceanography, named the Polar Regions a "circle of mysteries" whose exploration he referred to as follows: "It is no feverish excitement nor vain ambition that leads man there. It is a higher feeling, a holier motive—a desire to look into the works of creation, to comprehend the economy of our planet."[125] In this way the poles took their place in the list of topoi of primacy and research on origins, including river sources (especially of the Nile) and writing systems (especially the hieroglyphs deciphered by Jean-François Champollion). Geographically as well, around 1830 the Polar Regions were the last *terrae incognitae*—it was even unclear whether they were best considered land or sea.

The power of fascination exerted by the poles not only drew in hydrographs such as Maury but countless literati and adventurers. Striking enthusiasm thus informs Edgar Allan Poe's comment on an 1836 session of the congressional

---

122 See for example Auden 1967, pp. 6 f., 12f., 25, 38, and Frank 1995, pp. 202–207. For a focus on German Romanticism, esp. Kleist, Eichendorff, and Brentano, see Blume 1958, esp. pp. 146–152.

123 On the themes of cape and wandering, esp. with a view to the Flying Dutchman, see Wolf 2010. On the oldest known source for a locus-related compass-corrective, the anonymous Portuguese work *Das Agulhas* ("Of the Needle," 1502), which was meant to remedy needle-declination at the Cape of Good Hope, see Jonkers 2003, p. 150.

124 See Boyle 1725, pp. 7, 217 f., 224 f.; Blake 2004, p. 20; Cotter 1977, p. 590.

125 Maury 1861, p. 199.

Committee on Naval Affairs, where plans were discussed for a state-commissioned South Pole expedition:

> yea, to cast anchor on that point where all the meridians terminate, where our eagle and star-spangled banner may be unfurled and planted, and left to wave on the axis of the world itself!—where, amid the novelty, grandeur and sublimity of the scene, the vessels, instead of sweeping a vast circuit by the diurnal movements of the earth, would simply turn round once in 24 hours![126]

Coming together at the poles are stasis and self-circling, territoriality (which can be marked with flags and reclaimed) and deterritorialization (a non-place where ships begin to gyrate). What can be determined and calculated is here transformed, at the longitudinal convergence-point, to something fundamentally new and refined. If the pole is part or rather the vanishing point of a poetic topography, then we need to view it as what Bettine Menke has termed a "metatextual metaphor" and "mythopoetic self-thematizing of poetry," in whose wake countless intertextual traces—starting with Dante's *Divine Comedy* if not already with the *Periplus* of Pytheas—are gathered around the topos of an imagined absence of writing, signs, and traces, hence in a putatively pure "white."[127] That drawing near a pole is tied to hesitation, interruption, and finally an extinction of writing would be powerfully documented in later expeditions and notes by, for instance, Franklin, Shakleton, and Scott.[128] And that nothing is to be found at the real pole, or at least nothing significant, was revealed in discoveries such as that of the true North Magnetic Pole near Cape Adelaide in 1831 by John Ross: "The bank where the greatest magnetic descent was located," Heinz Balmer informs us, "was a plain, flat shore unanimated by any elevation."[129]

Poe's only novel, *The Narrative of Arthur Gordon Pym of Nantucket*, appeared in 1838, when three expeditions were simultaneously en route to the South Pole. The novel tells of Pym's nautical departure from the whaling town of Nantucket, a brutal mutiny at sea, a long sojourn on an island (where river-heads, hieroglyphs, and "savages" attest something mysteriously primal), and devious Pym's unavoidably steady casting up toward the South Pole—where he is finally swallowed up in a cataract in the presence of a gigantic white figure. Poe's adventure story has sometimes been interpreted as recounting an ecstatic voyage into the unconscious, sometimes as a neo-Platonic allegory;[130] Jules Verne

126 Poe 1984a, pp. 1247 f. On the context, see Brunotte 1993, pp. 158 f.
127 Menke 2001, p. 145.
128 See Ortlieb 2010, esp. pp. 125, 130, 137, and 139.
129 Balmer 1956, p. 547.
130 See for example Harvey 1998, p. 190; Leer 2003, pp. 201 f.

would still be drawing on it nearly six decades later, in *Le Sphinx des glaces* (1897). But the mystery of Pym, which Verne promises with the gesture of an—also in the philological sense—highly fussy realist is only illuminated through its mythopoetic recoding: now located at the pole is an actual magnetic mountain (since Ptolemy, as noted, the classical representation of mysterious attraction), loaded through the trade winds' influence; coincidentally possessing the shape of that archetypical bearer of secrets, a sphinx; and white as a result of countless polar winters.[131] Verne's text, including its putative "reality effects," explicitly lives on Poe's text, which in turn unrestrainedly and often feeds intertextually on numerous other texts: various contemporary nautical manuals, expedition reports, stories about mutinies, William Scoresby's descriptions of his Arctic voyages.[132]

And further: the editorial authority who comments on the papers of newly deceased Pym, in the process finding fault with the skepticism and lack of acuity of his ghostwriter, Mr. Poe, moves the hope disappointed together with the breaking off of Pym's report, hope of receiving a description of the pole itself, to the expected report on the ongoing report on ongoing expeditions. That authority reveals his own corrections and conjectures as involving fastidious but finally secondary work, in that it only operates upon a textual landscape, not in the slipstream of "real" information. Closed into self-referentiality, writing over writing here is recognizable as fiction—albeit philologically exact fiction—that faces Pym's in itself hardly credible because "curious" report, document of delirious pole-experience and thus a warrant of factuality.[133] Poe's aesthetic is in debt to German and English Romantic poetics less in the sense of relentless self-referential writing processes than in that of experimental procedures that reflect on a planned and systematic observation of "real" forces, natural phenomena such as magnetism and electricity; this reflective process both comprises epistemological critique (as a transgression of pure reason moving toward a sphere of the "absolute") and traces out a logic of representation (as a problem of "analogical" or "negative" representation). As Poe himself programmatically declares, his writing gains its aesthetic impact from just this exploration of mysterious effects.[134] In the style of transcendentalist critique, his poetics measures cognitive capacity by deriving measurement of the world from its cognitive aporias.

131 Verne 2010 (e-book; n.p.).

132 On the many sources see Huntress 1979, pp. xxviii–xxx; Kopley 1992, pp. 18–22, 35–42.

133 See Poe 1984b, pp. 1007 f., 1180–1182; Greenfield 1992, pp. 174, 178; Harvey 1998, pp. 184, 188.

134 See for example Poe's "Philosophy of Composition" in Poe 1984a, esp. p. 13. On Poe's aesthetic in general see Polonsky 2002, pp. 43, 52. On Romantic experimental poetics, focusing on

Poe's "Ms. Found in a Bottle" (1833) was already meant as a preliminary study for *Pym*, exemplifying its poetics in a double sense. Against a global deductive-scientific appropriation of natural phenomena, against "the habit of referring occurrences, even the least susceptible of such reference, to the principles of that science,"[135] this report on a pole-experience offers science turned contingent and an act of perception become event. And against the merely fictive imagining of an absolute or transcendent state, the report offers a narration treating its own emergence as text and book, the procuring of the "materials with which I write, and have written"—along with the contingency of being read (and as literature at that) in the first place.[136] For as a message in a bottle, its surviving the deep sea is inevitably uncertain, just as any comprehensibility in the midst of the waters' roar is very unlikely. It is jetsam, something that needs to arrive in the right reader's hands after being washed up on the coast. In any event, Poe's description of a cosmographic and hydrographic procedure in "Ms. Found in a Bottle" is not only metaphoric but concrete as well. Namely, having ran into dangerous heavy seas shortly before his return from his first voyage of discovery, Columbus had recourse to bottled mail as a last means to send out the cosmographically momentous news of the New World;[137] and centuries later, in the 1830s, maritime researchers for the first time used this medial technique in a planned and systematic way: to measure the world's oceans and register the direction and speed of their currents.[138]

Poetically (but with media-aesthetic precision) Poe, the man of literature, anticipates what in 1836 Poe the journalist will demand from American hydrography, which he describes as "a dangerous and arduous occupation, amid the perils and casualties of an intricate navigation, in seas imperfectly known."[139] According to the cosmologist Poe, undertaking or even gaining collateral for world-exploration through assumption of a First Mover would be a foolish decision. Rather, those forces, like magnetism, pushing research on nature to its limits should be conceived as mysterious news, a "letter corked in a bottle" (the term Poe uses in "Heureka. A Prose Poem" [1848]) emerging from the dark

Novalis, but also discussing Hans-Christian Ørsted's electromagnetic experiments and (for the first time so-called) "thought experiments," see Daiber 2001, esp. pp. 13–22, 66, 86, 257–259.

135 Poe 1984b, p. 189.

136 Ibid., p. 195. See also Hauss 2001, pp. 139, 144, 156f.

137 For the composition and preparation of Columbus's bottled epistle and its providentially mediated claim to truth, see Columbus 2006, pp. 5f., 34f.

138 On Admiral Beechey's measuring technique see Orsenna 2008, pp. 96–101. On hydrography as message in a bottle and as flotsam, see Schlee 1975, p. 43; Reinke-Kunze 1994, pp. 34f.

139 Poe 1984a, p. 1227.

seas—but without any belief that with these forces, we have already found nature's transcendental or unifying principle. As Poe put it in a letter of 1848, mere natural scientists are least of all in a position to interpret the phenomena they experimentally bring to light. Something is necessary moving past the reductive procedures of an Aristotle or Bacon, intuition and not least imagination, the poet's creative powers.[140] Poe's writing intends to capture just those forces evading the conceptual systems of pure natural science but also breaking through the bounds limiting texts of pure poetic imagination. This is the reason that various correspondences have been found between "Ms. Found in a Bottle" and John F. W. Herschel's *Treatise on Astronomy* (1835), where readers are promised a pursuit of natural research to the edge of the other-worldly. And this is the reason that the sense of danger in Poe's narrative—a sense manifest initially in fastidious nautical and astronomical observations, finally in an intuition of being imminently swallowed up—is tied not only to feared perils of the sea, but to the general threat posed by the pure poetic imagination.[141]

"Ms. Found in a Bottle" depicts the collapse of the nautical orientation system—indeed of orientation and observation in themselves. For here, all natural reference-points for the self-positioning of ship and passengers vanish. In the end, even the standpoint of the observer, the story, and its author are swallowed up in the depths. In this process, the report is increasingly carried forward by intuition, and even more by the certainty of drawing closer to the point of one's own extinction and impossibility. All of this takes place in the course of a movement not so much describing curiosity-based seeking as beings steadily and inevitably engulfed. Poe's text leads, as it were, Early Modern practical knowledge, represented emblematically by the age of discovery, up to the conditions of its own (im)possibility. It forces an exploration of the premises for a world-knowledge become purely immanent, its transcendence or—in the framework of Kantian philosophy—transcendentals, and in the process to ultimately lose oneself. Poe's version of a "narrative of wandering" pushes that genre to the point of disorientation—as if the Flying Dutchman, wandering around in "bad infinity," repeatedly had to step into the *Divine Comedy*'s voyage into the beyond. It is only that now neither Dante's ascent, nor his God's-eye perspective seems possible.

For just this reason, the Flying Dutchman in fact appears at the turning point of Poe's story: a super-sized old East Indiaman at full sail in a heavy storm, with now useless navigation instruments and both a long-since outdated sea-chart

140 See Poe 1984b, pp. 1261–1263, 1273 f. 1289 f.

141 See Friedl 1990, pp. 42 f., and Poe 1984b, pp. 190 f.

and a royally signed commission on board.[142] And for this reason the ship's crew is, as in Dante, of unfathomably old age, the captain seeming a doppelganger of the narrator—except that his words, by contrast with Ulisse's flaming discourse, are only audible as if from afar.[143] With a great hunger for discovery, Ulisse and his crew seek the abyss of their worldly knowledge, while at the same time fearing this abysmal turbulence. In Poe's "Ms.," the ship's crew is only driven forward and sucked in by the pole. They anticipate their whirlpool-drowning as a revelation—or as release from their wandering at sea. The age's former hunger for discovery fulfils itself in destruction. Nevertheless, after absentmindedly dabbing some sails stored on deck with a tar brush, the narrator realizes that the ghost ship's wandering will lead to a fundamental discovery. Once the sails are hoisted and stretched out in the wind, the unplanned traces he has left will spell out the word DISCOVERY.[144]

Hence future discoveries will be expected not within a topographic grid or perspectivized space, not within the geometry of a subjective show-room with its metrics and stable, clearly defined distances—but within an initially non-intuitive topology of wrinkled, inwardly folded manifold phenomena and their surprising vicinities. That with the signifier DISCOVERY, hence with discovery per se, no signified, no future, hidden, or deeper meaning is being announced, but rather a (literally "senseless") de-anchoring of sense and signified, is an insight the narrator gains as a final note. He is both swallowed and tempted into the cataract's immemorial chaos by those nameless and indefinable elementary forces mingling oceans of heaven and sea: "It is evident that we are hurrying onwards to some exciting knowledge—some never-to-be-imparted secret, whose attainment is destruction."[145]

Poe's polar voyages straight into abysses of perception and cognition: abysses we might connect with the "sirenic," tempting and bewitching primal ground of the poetic itself. It is the case that concretely, Poe's narratives draw on those yawning gaps in knowledge encountered by contemporary hydrography, with its research on currents, map projections, and hypotheses regarding terrestrial magnetism. In a note on "Ms. Found in a Bottle," Poe represents or rather supplements one such yawning knowledge-gap with another: "the Pole itself" is here

142 See Poe 1984b, pp. 193 f., 196 f.

143 See ibid., pp. 197 f.

144 Ibid., p. 195. For a wrinkled handkerchief as an example of such a topological constellation see Serres 1995, p. 60.

145 Poe 1984b, p. 198. On the non-denotable elementary forces, see ibid.: "If I trembled at the blast which has hitherto attended us, shall I not stand aghast at a warring of wind and ocean, to convey any idea of which, the words tornado and simoon are trivial and ineffective?"

"represented by a black rock, towering to a prodigious height," with the Indescribable in the mythic magnetic mountain having its analogical counterpart in the "bowels of the earth" and thus of worldly knowledge in general, albeit its negative representation.[146] The absolute center of knowledge is itself decentered, displaced, and only recursively describable. And as in an endless equation, the ocean and its powerful pull, the pole and its mysterious magnetism, map projection and its distortion, all stand for one another.

We already find the idea of a hollow earthly sphere that swallows up what is terrestrial and containing its own ocean in its interior postulated in *Mundus Subterraneus* (1664), a work by the earlier-mentioned Jesuit cosmograph, hydrograph, and universal savant Athanasius Kircher. Poe probably chanced on the book though Gerhard Mercator's map of the North Pole, that, according to a note allegedly added later, he received following his text's composition, like an unexpected message in a bottle. Hence the vanishing point of his riddling narrative is meant to become visible through an equally riddling sea-chart whose legibility is in turn to be guaranteed through Poe's text. As long as they are deployed for the unknown or variable value of their complement, text and map, narration and projection form an equivalence as infinite or indissoluble as are—according to the information in Poe's text and Mercator's map—cartography, magnetism, and currents at the North Pole. (Fig. 7)

Mercator is said to have grasped the complex connections between cartography and magnetism for the first time when he drafted a map of the Holy Land and made use of Jakob Ziegler's *Tabula Universalis Palestinae* (1536) to that end: a map that, despite the standard illustration of Jerusalem, was oriented toward the north and also for the first time referred to the compass needle's misdirection. (This involved depicting a compass rose with tilted needle on the margin of the cartographic representation.) Furthermore, in 1539, following a number of voyages to the upper northern regions, the Swedish cartographer and cleric Olaus Magnus published his *Carta Marina*, a large-format woodcut map of Scandinavia with an inserted magnetic island. (Fig. 8) This symbolic representation of non-representational magnetic forces, which in any case only appeared explorable through deviation from the true geographical pole, was taken over by Mercator.

His work, meant for instance to render measurable the difference between geographical and magnetic poles, formed a cartographic-historical caesura vis-à-vis such "mythic" depictions. This notwithstanding, Mercator adopted such depictions, just as, not wishing to withhold any cosmographical information, he

146 Ibid., p. 199.

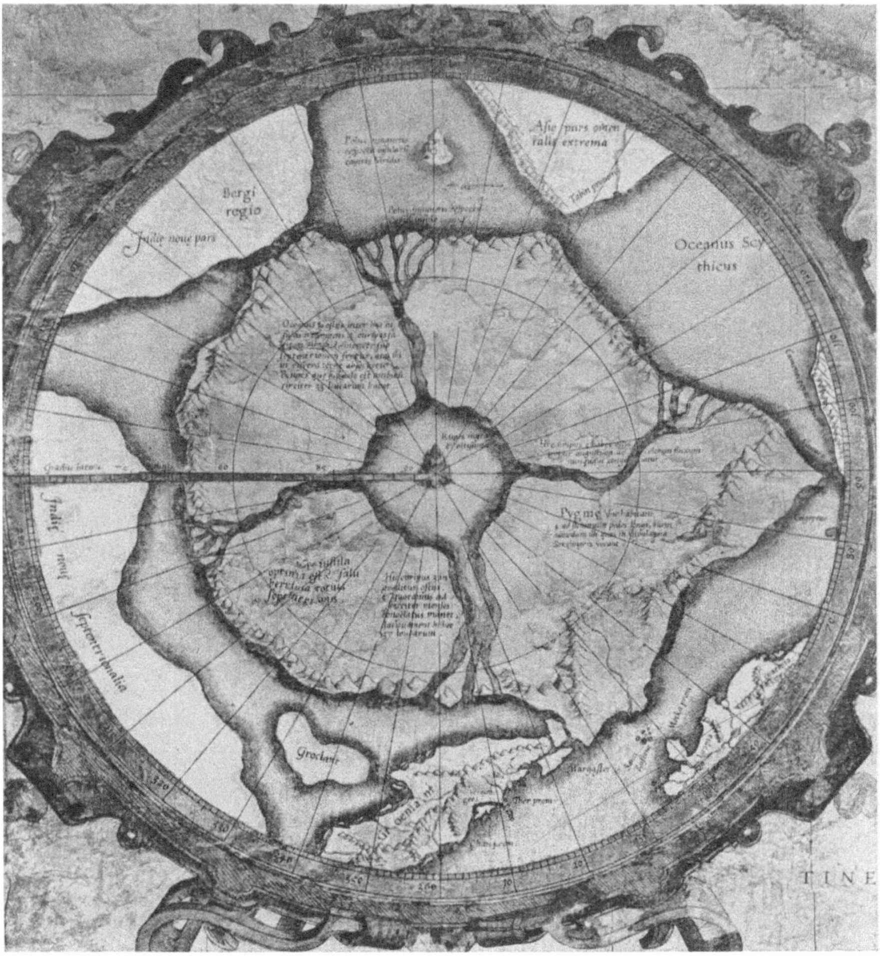

Fig. 7: Excerpt from Gerhard Mercator's world map of 1569

processed all manner of representations of the North Pole and reports by previous or present northern voyages. These included the notions of the pole presented by Gemma Frisius, Martin Behaim, and Johannes Ruyschs; the latter figure drew a world map for the Roman edition of Ptolemy's *Geography*, here grouping four islands around the (cartographically elided) North Pole. This archipelago was first discussed in a travel book of Jacob Cnoyen, who purportedly traveled across the Polar Sea in 1364 at behest of the Norwegian king. This was before an English mathematician is said to have set out to measure the pole, where be-

Fig. 8: Excerpt from Olaus Magnus's *Carta Marina* (1539)

tween the four islands he discovered four raging streams leading to the polar maw and the earth's interior, an enormous magnetic rock being located at the crossing-point of the four streams.[147] (Fig. 9)

Mercator brought all these conceptions and fables, reports and data about contemporary northern voyages, together, their contradictions convincing him all the more that (to make use of Kant's later terminolgy) an "analogical" and "negative" representation of the pole was needed. As visible on his map, following the circulated range of inclinations, he assumed two magnetic mountains, one on the endpoint of the Cape Verde meridian, the other at that of the Corvo Meridian, between which he assumed the true magnetic North Pole. In addition, in a note to his world map he pointed to the main map having to break off at the pole, where, after all, latitudinal lines run into infinity, for which reason he was adding a separate polar map.[148] Liberated of representation logic, the North Pole zone would later be depicted in yet another polar map, based on his world map; the polar region was here expanded southward, for the sake of registering future new discoveries and their toponyms. (Fig. 10) Although orientation around the pole is impossible because here, on the one hand, Jerusalem lies outside the topographic (and also toponymic) horizon, and, on the other hand, the compass needles begin to spin and maps to "de-marcate" their position-fixing, this map is nonetheless compass-like in its circular form. To be sure, the *compasso* has become a region of the world and of knowledge, one that although appropriating the will to know in a downright magical way, never-

147 See Balmer 1956, p. 534; Jonkers 2003, pp. 45–47. On the genesis of Mercator's cartographic projection and representation of the poles, see Snyder 1993, p. 45; Crane 2002, pp. 115, 147, 243–247.

148 See Balmer 1956, p. 540: "For Mercator the magnetic mountain was connected on a profound persona level with objective considerations; here it was part of science."

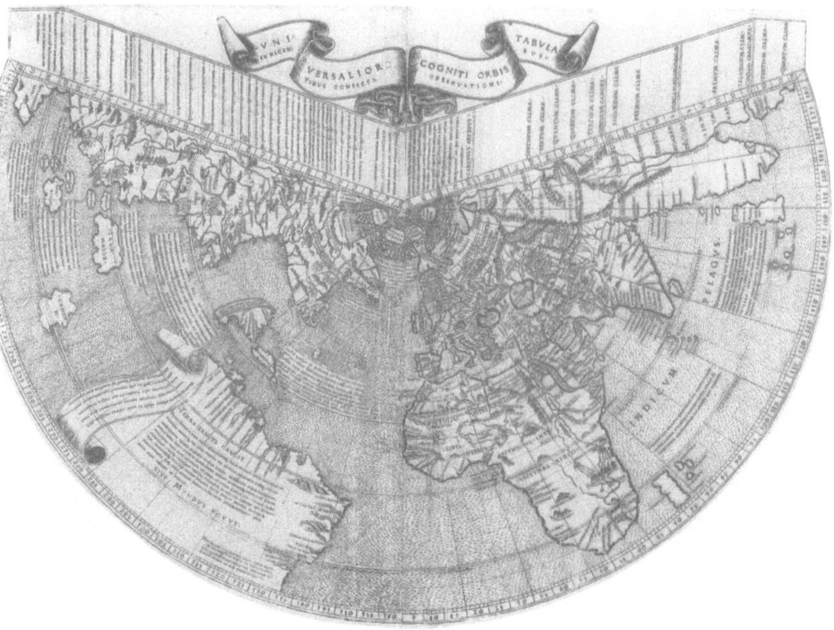

Fig. 9: Excerpt from Johann Ruysch's world map (1507)

theless appears to swallow up all of its own efforts at approximation and representation. A look at the compass here does not produce much orientation. Rather, the compass—as Nicolas Crane observes[149]—appears to look out like an eye at the necessarily disoriented observer. In this respect Poe's literary and Mercator's cartographic "poetics" converge. Against this backdrop, Poe's figurations of the indescribable are meant to be read not only in terms of their literary sources, not only in the intertextual sea. Just as decisive are the hydrographic source-codes of his imagination—their emergence from what Poe (rather than Mercator) conceived as an abyss of knowledge that in the end represents itself as having been swallowed up by a self-swallowing sea.

149 See Crane 2002, pp. 312f.

Fig. 10: Gerhard Mercator's polar map (first published 1589)

From Topography to the Poetic Function

This figuration or rather de-figuration by the forces at work in a maelstrom and the event of an engulfing stand at the center of all Poe's maritime stories. That moment of "being-in" that previously took effect on ships and in their stowage spaces, in crates, caves, and canoes either as longed for safety or horrific enclosure, is radicalized in *Arthur Gordon Pym* into being swallowed by a polar cataract. For its part, "Ms. Found in a Bottle" moves forward from the moment when the narrator senses the first signs of advancing danger on the gradually opening groundlessness of the sea.[150] Finally, "Descent into the Maelström" (1841) centers around nothing other than the narratability of maelstrom and engulfment themselves. In this story, in the Norwegian North Sea island-area of Lofoten the narrator climbs a mountain from where a local fisherman shows him the legendary maelstrom: through the dizzying view down over the edge of the rocks but also through the hypotypose or hyperbolically vivid descriptive qualities of his report of the experience. Three years before, we read, together with his brothers he went out on his fishing boat (his smack) to fish between the islands behind Moskoe—hence where the maelstrom forms each day—for the sake of richer catch than what was found in safer fishing grounds. As the fisherman puts it: "In fact, we made it a matter of desperate speculation—the risk of life standing instead of labor, and courage answering for capital."[151] But this calculation—speculation on *fortuna di mare*—does not pay off. The men's timepiece fails, so that they overrun the period of slack water, find themselves in a violent whirlwind while the maelstrom forms, and then, together with its *fortuna*, as in Revelation 21,1, there is "no more sea."

Tidings of a hellish maelstrom forming the sea's source and abyss and swallowing up smaller and larger ships and enormous whales begin with Adam of Bremen's eleventh-century ecclesiastical history.[152] "Descent into the Maelström" begins with what is presented as a motto of Joseph Glanvill about God's unfathomably deep impact on various levels of both human affairs ("special providence") and natural phenomena ("general providence"): an impact marked by such "vastness, profundity, and unsearchableness" that it renders all our theoretical and linguistic models useless.[153] But Poe not only presumes the cognitive caveat inscribed in Glanvill's motto "Plus ultra", rather scrutinizing it as well. In this way the story renders the doubling of the maelstrom's hyper-vivid description—its twofold hypotyposis—into a poetological experiment, a test of the durability of different poetics of knowledge under the sign of an inescapable loss of representation.

As the frame-narrator indicates, the extant descriptions of the maelstrom's suction (for instance the description of Jonas Ramus) can do as little justice to the violent "spectacle" of being swallowed up in the maw as the phenomenon's

150 On rescue and enclosure in *Pym*, see Peck 2001, pp. 98 f. On the process described in "Ms. Found in a Bottle" of the sea first becoming transparent, then exerting suction, then engulfing, see Poe 1984b, pp. 190, 198.

151 Poe 1984b, p. 438.

152 See for instance Stammler 1967, col. 2942 f.; Simek 1996, p. 108.

153 Poe 1984b, p. 432. Seger 2001, p. 236, has pointed to the motto as a fabricated citation borrowing from Glanvill's *Essays on Several Important Subjects in Philosophy and Religion* (1676).

hydrodynamic, experimentally grounded explanation (as a collision between tide-determined currents and the cliffs and reefs on the seafloor). If, Poe suggests, the suction's vortex is actually experienced, even simply by an observer from a cliff, then Kircher's cosmographic myth of the chasm can claim to possess the strongest supporting evidence.[154] This is especially the case in respect to the intradiegetic narrator and his report on his own experience of being sucked into the maw: at the moment where the two seas of heaven and earth seem to plunge into each other, and the earth evidently swallows everything earthly into itself, the *turbo* is no longer, as it still is for Dante, an event narrated and observed from God's perspective. It has become an event that takes in narration and observation themselves and in the very absence of relation characterizing both founds an ultimate relation, the relation to God. Which is to say: the non-ground of knowledge only announces itself in the abysmal groundlessness of existence itself, and just this is "so wonderful a manifestation of God's power."[155]

Poe's figurations of engulfment and having being engulfed have been read as a revival of apocalyptic scenarios, but also as the modeling of a psychotic process; after all, the observation that "crisis sucks us in, it has the character of a vortex" applies well to these scenarios.[156] In the terminological framework of a psychoanalysis of elementary phenomena, they have been described as a topology of impossible enjoyment, with the striving for unity at work here not only aimed at a non-pictured and indescribable object but at the matrix of all "thingness," "a kind of substance of substance, a mother- substance."[157] What is designated as a structure of the transcendental and a precondition for all experience can only refer to empirical objects and the world in general through mediation by the space-time schematism or the synthesis of a phantasm;[158] but here the external and internal sense of space-time experience collapse into one another, just as all phantasmatic objects are themselves swallowed up: the phallic object (the flagpole rammed into the North Pole) loses itself in a maternal matrix; a maw gapes open at the place from where one is aware of being looked at (the eye of the polar region); and also the voice that topologically could still establish a relation to the great Other (for Dante still to another named God) goes under in a disarticulated raging.

---

154 See Poe 1984b, pp. 435–437.

155 Ibid., p. 443.

156 Caspar Kulenkampff (1959), cited from Metzner 1976, p. 223. See also ibid., pp. 226–231.

157 Bachelard 1983, p. 46 (translation modified).

158 See Baas 1995, pp. 62, 67.

Precisely by pushing up to—indeed through—the screen of what is transcendental and phantasmatic, as topographically laid out after 1800 in the Polar Regions, Poe's poetics is equally precise and creative.[159] On the one hand, with this forward, penetrative movement, faith in a first cause or mover is eviscerated. What Poe postulates is a displacement of experience and what makes it possible furnishing at least one thing in place of this loss of orientation-founding final things: Lacan's "certainty of anxiety."[160] On the other hand, not only the disorientation of a constant inversion of figure and background, signifier and signified is tied to the plunge of the symbolic into an abysmal reality, but just as much what Sam Weber, himself commenting on Lacan, calls a "dizzying shapeshifting" between topographical and unmarked space—between figure and backdrop.[161] In this way Poe's narration culminates in an immersive poetics. And this narrative poetics of a collapse—of a self-engulfment of schematized (re)presentation—points back to that procedure of defiguration that Early Modern painting and hydrography developed to the same degree. "Where the eddies emerging from the seething mass mix with the current, a deeper base is created," writes Leonardo. "Water creates its base by leaps."[162]

The mutual engulfing of figure and background in Leonardo's studies of water already announce a drowning of perspective—hence of that symbolic form[163] simultaneously guaranteeing and concealing the framed "presentedness" of the world from a point of fixed vision and configuration, the construction and fiction of a consistent sense of space and world.[164] Early Modern hydrography is itself characterized by space not being defined as an immemorial precondition for everything seen and experienced, but rather as something that emerges in the course of an immersion, a plunging into the sea, a forward movement and its successive notation. For hydrography was undertaken on board ships, without disposal over exact chronometers and thus a referential system for global time: a lack of temporal-spatial self-localization, which can hardly be compensated for by the only approximate and occasional orientation on the (itself moving) heav-

---

159 See Full 2007, p. 219: "The loss of the form of security conveyed by transcendent structures is the process that has to be paid for creative freedom." On Poe's transcendentalistic "fiction of fiction," see also Schwarztrauber 2000, pp. 41–44, 400f., 579f.

160 Gondek 1990, p. 231.

161 Tuschling 2009, p. 126.

162 Vinci 1996, p. 40.

163 See Cassirer 1957, pp. 159ff.

164 On immersion as a collapse of perspectivistic construction see Neitzel and Nohr 2006, p. 16; Schwingeler 2008, pp. 42–54. For a transfer to narratology, see Ryan 2001, pp. 121–157, 348f.

enly vault. Here objectless and defigured images do not emerge from a fixed topographical order that offers a medial (re)presentation of its spatiality and places. Rather, they emerge from self-recording—from a continuous graphic movement. They invent their "spaces" themselves, for which reason they can only vouch for their representationality and realism through themselves.[165]

In his descent into the maelstrom, Poe's inner narrator finds himself in just this immersive situation. His chronometer has stopped working, and the heavens have shifted from serving as a system of astronomical navigation to being an upward-gaping maw, from whose innermost interior only the moon seems to now shine.[166] The story's space is thus no longer covered by any realistic topography but stretches out between coordinates only capable of defining the narrator's experience. Reporting on both the object (the maelstrom) and situation (engulfing) cannot be direct but only through the narrative's idiosyncratic self-referentiality. For one thing, fictionality—the manifestly evident "as-reality" presented in the fisherman's narration (contrasting with an evidently merely fantastic "as-if reality"—is what vouches for the "writing in water" of Poe's text; at the same time, it is the doubled narrative authority that does so: the text of the cultivated but also strongly imaginative frame narrator; the voice of the inner narrator, someone both familiar with place and skilled, but also someone visibly careworn, with the stamp of danger (or death).

With his descent into the maelstrom, the fisherman has endured a straightforwardly traumatic experience of engulfment: an experience destroying the ideality of intuitive forms, hence the premise of all subjectively attributable experience, which as such thus cannot be symbolized and can only be attested to belatedly, by means of peculiar "reality signs." For this reason, credibility and truth, story and history, cannot be aligned in his report. That this man, allegedly a Norwegian highly familiar with the surroundings, is a simple liar or perhaps a wily stranger trying to appropriate the identity of a fisherman who has met with an accident, this semblance of truth, is opposed by the narration's scarcely probable assertion that he only appears like "a *very* old man—but I'm not."[167] The very fact that his "old mates and daily companions" do not recognize him any more than they would deem a "traveler from the spirit-land" one of their own is meant to vouch for the truthfulness of his report and the authenticity of his sea change.[168] Poe here is writing in the wake of the *Odyssey*'s anagnorisis

---

165 On this and with reference to correspondingly disfigured painting see Schäffner 1996, pp. 9–15.

166 See Poe 1984b, p. 441.

167 Ibid., p. 432.

168 Ibid., p. 448.

scene in which notorious liar Odysseus is only identified as truly Odysseus the homecomer through his scar, but also in that of the first Early Modern collection of "authentic" shipwreck reports: the *História trágico-marítima*, in which East India voyager Jorge de Albuquerque, after enduring a shipwreck and immeasurably long odyssey, can only be recognized by his "astouned" friends through "certain marks on his body."[169]

If we wish to believe him, Poe's fisherman has been abruptly aged in the indeterminate time-space of his engulfing. The event which he reports on has inscribed itself in his body. For this reason, the trauma's traces, for instance his snow-white hair, will serve as control data for his hydrographic experiment and as certification of his experimental narrative report. Certified identity and socially sanctioned authorship, from which he could recruit credibility, are not at his disposal. He is thus dependent on "willing suspension of disbelief"[170] on the side of the frame narrator—and with him the reader as well. The fisherman only gains proper credit when his report's readers do not limit their concept of truth to a simple conjunction of the visible and sayable, thus adhering to Aquinus's famous definition of truth as the similarity between things and intellect: a similarity already tested at the start through the maelstrom's doubled, visual and narrative hypotyposis. He will only gain credit when readers show trust just at the locus where the vortex of the real has led to what can neither be said nor imagined.

What seems to suggest itself here is understanding Poe's text as a testing of the doctrine of the sublime. That doctrine's exemplary scenarios are in fact largely aligned with the story's starting situation: the gaze from the cliff onto the sublime natural scene, facing which one does feel safe, but which deep inside prompts agitation, for the sake of moving from it, from trepidation and malaise, toward a higher, ideal freedom. In the shape of the ocean—according to Edmund Burke, in *On the Sublime and Beautiful*, "an object of no small terror"[171]—the sublime reveals itself to both narrators at a safe distance, thus standing under the rubrics of "self-preservation" and "society."[172] Already for this reason, nature here appears to be sublime; in Kant's formulation, "it raises the imagination to the point of presenting those cases in which the mind can make palpable to itself the sublimity of its own vocation even over nature." To be sure, Kant re-

---

169 "Misadventures of the *Santo António* and Jorge d'Albuquerque," in Boxer 1968, p. 157.
170 This is Coleridge's famous formulation in the fourteenth chapter of the *Biographia Literaria* (1817). On Poe's construction of poetic credibility in *A Descent Into the Maelstrom*, see Person 1993, pp. 47–49, 51, 54.
171 Burke 2008, p. 131 (from *On the Sublime*, part II, Section II "On Terror").
172 Ibid. p. 110 f.

minds us, we should not only contemplate the ocean "in periods of calm, as a clear watery mirror bounded only by the heavens." Rather, we must "be able to find it sublime" "when it is turbulent" as well, "an abyss threatening to devour everything."[173]

What is truly sublime is moving from a safe standpoint and point of observation to as it were fully enter into the presented image, like Poe's *interior* narrator. Having plunged into "the very jaws of the gulf," the fisherman's experience of immersion and sea change is also spiritual, in a baptismal sense: He recognizes "how foolish it was in me to think of so paltry a consideration as my own individual life, in view of so wonderful a manifestation of God's power."[174] Hence having initially placed himself before a nature totally overwhelming his sensory capacity and sensibility, the insight gained into his own existence's contingency and nullity has led him to a rational idea of God. And starting with that idea, having recognized his capacity to attribute higher intellectual purposiveness to the natural event, he finally discovers his freedom to attribute laws to manifest phenomena—first moral then natural laws. He is driven by "unnatural curiosity," a kind of curiosity of the second order, no longer comprising visual pleasure (as *curiositas* was understood by Augustine) but rather now angst-based pleasure, facing the unfathomable.[175] In this half providence-endowned, half indifferent condition, with his engulfment he gains the distance of reflective curiosity: he considers the walls of the maelstrom-funnel with its "perfectly smooth sides" as a kind of picture frame, within which what is formless and shapeless can gain perspective.[176] On this basis he can grasp his experience as an experiment in whose course the Archimedian laws can come to fruition, however inchoate the process. Hence hydrodynamics here gains its substantive weight not within the external perspective of strictly theoretical analysis (as engaged in by the frame narrator) but only from the interior of an immersion.

Among the objects circling in the maelstrom but not yet sucked to the ocean's floor, the fisherman grasps onto a water-drum, according to his observations the object with the strongest buoyancy. He binds himself firmly to the drum, thus rendering himself, in purposively grounded self-abandonment, into a kind of bottled message from the sea of darkness, before he is at last spat out on a Lofoten bank. The intradiegetic narrator has here released himself to the elements like a fish to water. At the same time, in place of having the

173 Kant 2000, pp. 145, 153.

174 Poe 1984b, pp. 442f. On the sublime in *A Descent into the Maelstrom* see Seger 2001, pp. 229f., 235f., 238–240.

175 Ibid., p. 445.

176 Ibid., p. 444.

fixed or territorial standpoint of deductive knowledge as his starting point, he has conquered a provisional observational level and viewpoint in the gulp and pull of this elementary experience: a perspective, in addition, that takes in precisely those laws governing irregular and turbulent phenomena. In a paradigmatic manner this combination of immersion and reflection describes, as Marshall McLuhan has observed, the epistemological position of modern media theory.[177] What may distinguish Poe's story from the media-theoretical approach may be a moment of aesthetic experience: a moment that as a result of its fictional and negative presentation cannot be annulled. But what distinguishes the fisherman's experience significantly from that theory's normal operations is the very thing that makes his sea change clear.

The profound shift in perspective and nature involved here can itself be described in concepts framed by the transcendental. For where the fisherman's brother goes under as "a raving maniac through sheer fright,"[178] hence loses himself in imaginative capacity gone astray, the fisherman soars into the sublime—a doppelganger-relation between drowning and rising that repeats the countervailing movement between Ulisse and Dante. But what is outstanding in the fisherman's sea change is less elevation to an ideal of reason as rather a descent into the "anesthetic," to the ground of schematism and into the "depths of the human soul."[179] The brothers do not differ as a result of the opposition between "sheer fright" and heroic courage but that—to apply Heidegger's categories in *Being and Time*—between paralyzing "fear" of something inimical to existence on the one hand, "anxiety" that, on the other hand, instead of being fixated on objective or localizable threats, concerns the non-place of one's own "Being-in-the-world," tearing existence out of its "Being-in" and placing it before itself.[180]

Freedom as ideally promised by the sublime is not the other of anxiety. That before which on feels angst is, in the words of Poe's contemporary Kierkegaard, "freedom's reality as possibility for possibility"[181]—the real freedom to transgress the transcendental schemata of merely ideal freedom, in the end swallowing up oneself. Angst is thus freedom's dread before its own possibility. From Kierkegaard to Sartre, the dizzying abyss threatening engulfment is presented as a phantasmatic topology of existential anxiety. Gazing into the abyss inevitably displaces the subject into the "dizziness of freedom." And once this has been ex-

177 See McLuhan 1964, pp. 97 ff.
178 See Poe 1984b, p. 444.
179 Kant 1998, p. 273.
180 Heidegger 1962, pp. 169 ff., 225 ff., passim.
181 Kierkegaard 1957, p. 38.

perienced—once one's very self has been recognized as the ground and groundlessness of, on the one hand, assumption and fear of danger and, on the other hand, the possibility and freedom to plunge into the abyss at any time—then there is an awareness of not only feeling dread but, so to speak, of "being" it.[182]

The aesthetics of sublimity was earlier defined as a school for overcoming fear, especially fear of nature; through a movement into the sphere of the transcendental, this shifted into an anxiety at work in inner nature, and into the releasing of an endogenous, self-revolving, and "fantastic" angst or dread.[183] While in the early eighteenth century fear was still being expressed at sight of the "bottomless sea" as this evoked—even for physico-theologians such as Barthold Heinrich Brockes—the "space of the abyss,"[184] by the century's end the ocean and its engulfing power already served as an image for the abyss of human nature. "Just as the storm churns up the bottom of the sea and lets us see what the waters otherwise have hidden, misfortune makes the true character of a person clear," we read in an anonymous text of 1791.[185] Poe converts the motto of his older "Ms. Found in a Bottle" in just this sense: "He who only has a moment to live / has nothing more to hide."[186]

But with "A Descent into the Maelstrom," Poe in the end transgresses subliminal aesthetics for the sake of an "anaesthetics" of angst, leaving behind both anthropological questions regarding true character and hydrographical questions regarding true places. For just as the North Pole here no longer appears as a piece of topographical data allowing purposive fear, but rather as an unmarked or arbitrary space opening a topology of anxiety, a distinct *character*, one marked by physiognomically inscribed fear, is no longer to be found in this text. It is here helpful to recall Gilles Deleuze's observation that the face is a kind of expressive material—on the one hand a complex interplay of material areas, on the other hand a virtual connection of singular features whose expressed contents alters with the slightest change in the whole. In that light, when he says that "the whole expression of my countenance had changed,"[187] then it would appear that the fisherman, whose identity is vouched for by nothing else than his expression, cannot be one and the same person who simply changed with descent into the maelstrom. The intradiegetic narrator, this witness of the maw's interior, is no individual who purportedly experienced dread.

182 Ibid, p. 54; Sartre 1966, pp. 67 f.
183 See Begemann 1987, pp. 11, 142 ff., 260 f., 311.
184 Brockes 1999, p. 22.
185 See Anonymous 1791, p. 185.
186 "Qui n'a plus qu'un moment à vivre / N'a plus rien à dissimuler." Poe 1984b, p. 189.
187 Ibid., p. 448.

Rather he is, entirely, anxiety or angst, in both appearance and essence he has entire become what is expressed by that affect. And just as such a pure, "de-individuated" affect-figure, he bears witness not to the identitarian continuation of his self but only and alone to the event of his self-engulfment.[188] What if not experience should amount to what we call a sea change?

The stories of Poe that most incisively present an imaginative moment where research on nature has reached its limits may be read as promoting speculation along the following lines: poetic expression generates space for itself only to the extent exact knowledge is abandoned. Nevertheless, Poe consistently aims at an interplay between understanding and the imaginative faculty.[189] His attentiveness as a narrator, poet, and thinker is steadily focused on the poetic function tied to knowledge from the start, not only when blind spots appear. Perhaps he emerged as an author of epoch-making sea-stories because hydrographic practice has always been closely tied to a specific "poetology of knowledge": in part reflecting its unrealistic effects, cartographic projection has always been associated with the imagination's phantasmatic activity,[190] while orientation has been a guiding question in both anthropology and theory of the subject at least since Kant.[191] Concepts, Kant argues, are "always...appended to *image* representations" that "make these concepts...serviceable for *experiential use.*" If the need for orientation involved not only my position vis-à-vis "objective data" of say a geographical or astronomical nature but already "a *subjective* ground of differentiation," consequently "thinking" and the range of rational operations,[192] then the directive-problem rendered emblematic with the compass has arrived at a transcendental-philosophical (non-)foundation. For this reason, continuous, isotropic and metrical space is merely the external topographical side of that form of intuition owing itself just as much to a non-researchable constructive achievement, one carried out in experience itself.

188 On the affect of angst as a "complex entity" expressed in "dividual" form, as opposed to an individual's self-expression, and on both the expression of vertigo before an abyss and the "affection-image" of a face and of "any-space-whatevers," see Deleuze 1996, pp. 102–107, 110 f., 120 f.

189 On the concept of ratiocination as "the combinatory capacity of the imagination qualified as fancy, steered by reason," see Smuda 1998, p. 123.

190 On the reciprocal metaphorizing of projection and discovery see Knox-Shaw 1987, p. 30.

191 Cf. Heidegger, 2002, p. 133: "Namely, the more extensively and the more effectually the world stands at man's disposal as conquered, and the more objectively the object appears, all the more subjectively, i.e., the more importunately, does the subiectum rise up, and all the more impetuously, too, do observation of and teaching about the world change into a doctrine of man, into anthropology."

192 Kant 1996b, pp. 7, 9.

A poetics of narration such as Poe's takes account of such a transcendental empiricism in that on the one hand, as already in Dante, it employs—to cite Hans Krah in reference to Lotman's topological model of narration—"spatially concrete facts by way of semiotic operations as vehicles for non-spatial facts,"[193] hence for example geographical border-transgressions as representations of epistemic and cultural constellations. On the other hand, as is especially clear with Poe's hydrographic aporias, to the extent that spatial facts increasingly lack concretion and for their part reveal qualities that are more disorientating than eidetic, topographic material no longer serves to illustrate a foundational topology. The abstraction of space now emerges, together with the vanishing of guaranteed references on the topological level: their absence on the level of epistemic and cultural conditions establishing the possibility (or indeed impossibility) of what previously had the status of naturally given space. Precisely in the course of such recursion, in a reentry of the difference between topography and topology within the topographic itself, previously hidden causes, forces, and references are newly discovered.

In this specific poetological turn, which understands space and its semantics, the world, neither as a given nor as merely generated, but rather grasps it through what Blumenberg terms a "syntax of elements," fiction and practical knowledge come together.[194] If—as is the case with magnetism and the Polar Regions—nature as such does not disclose itself to intuitive perception, thus also not being topographically assessable, poetic processes are no longer concerned with retrospective but rather with a kind of anticipatory imitation of nature.[195] As soon as the ideality of the forms of intuitive perception such as space reveals itself as failed, the "writing in water" that Plato declared as impossible as it was nonsensical becomes a domain of both the literary imagination and practical knowledge. Each forms a functional structural, whether verbal or mathematical. And the "immanent consistency" of each sign system is "the only—though adequate—approximation to the given reality."[196] The functional structures show this immanent consistency most clearly in an operative manner: in one case through practical testing (for instance, hydrographically) of actual capacity to move toward a blindly ascertained locus; in the other case through fictional effects such as the evidence informing merely narrated "as-reality."

193 Krah 1999, p. 4.

194 Blumenberg 1979, p. 42. On the basal distinction between theories of coherence and correspondence from the perspective of a "general theory of narrative," see Koschorke 2018, pp. 273 ff.

195 Blumenberg 2001, p. 45.

196 Ibid., p. 41.

Also common to both semiotic structures—literary texts such as Poe's and hydrographic procedures such as that of the compass—is their self-referentiality—the orientation of their signs toward themselves and their organization. In both cases, we can speak of "poetic functions," naturally dominant in the literary domain but also having an influence on areas of practical knowledge such as hydrography.[197] For example the reality of epistemic objects such as magnetism, electricity, and gravitation was initially justified through the hope that such conceptual entities could be imagined, experienced, even narrated, as soon as the proper visualization, operationalization, narrativization procedure were only found. Until that point, the "fiction of the reality of realities"[198]—a fiction that, similarly to the narrative procedure at work in fictional literature, promises to generate reference to the world through systematic self-reference.[199]

Stories such as Poe's, in which, unlike in scientific texts, the poetic function is dominant, explore the fictive contents of facts and the horizon of possibility of the inventive faculty. In their reflexive curiosity, they can observe observation of the world and thus render visible what it means, as Rüdiger Campe has put it, "to imagine the world with a certain intention and predisposition."[200] For this reason they have the status of "texts of performance," not speaking a "language of statement."[201] To be sure, in its research on natural phenomena it was precisely hydrography that increasingly drew attention to the performative and poetic functions at work in its knowledge-oriented processes: a development in science history touching on several of Poe's narrative motifs. For example, in 1861 Jules Michelet explained the fascination exerted by the Polar Regions in terms of bewitchment by both what is useless and impossible and the enticement of "sublime horror," hence an attraction highly advantageous to scientific research in various realms including geography, meteorology, and terrestrial magnetism. For his part Maury understood the distribution pattern of shipwrecks and diagrams of the movement of drifted wrecks or ship-parts as a basis for exploring the sea's abyss and the Polar Regions.[202]

197 Jakobson 1960, pp. 6 ff., passim.

198 Blumenberg, 1979, p. 48.

199 On the heuristic and mediatory function of "magnetic symbolism" in the emergence of doctrine of electricity, and on Newton's first precise demarcation of gravity from magnetism see Humboldt 2004, pp. 650–657; Balmer 1956, pp. 160–163.

200 Campe 2006b, p. 40. See also Peters and Schäfer 2006, pp. 25–43. On observation of the *facultas fingendi* see Weigel 2004, pp. 187, 192 f.

201 On this distinction of John L. Austin from the viewpoint of reception aesthetics see Iser 1970, p. 10.

202 Michelet 1861, pp. 289, 292, passim. See also Maury 1861, pp. 14, 63, 303.

As a Dutch captain noted, mariners could thank mainly Maury's oceanographic labor of enlightenment for no longer suffering from a sense of "our own nothingness," of being overwhelmed by a "vast and fathomless abyss."[203] Maury's achievement had both empirical and speculative aspects. On the one hand there was feedback from the experience of mariners themselves: during their voyages they maintained logs for Maury, recording, at least once every four hours, the local compass deviation and current situation, as well as the direction and strength of the wind; once delivered on land, they would receive updated maps and tables from Maury.[204] On the other hand, Maury's global hydrography was based on a correlation of air and water currants, terrestrial magnetism, and atmospheric electricity—in other words on infinite equation systems which he allegedly arrived at by observing the Polar Regions.[205]

In this regard, we might say that in his pioneering oceanographic enterprise, Maury extended something like the phantasmatic umbrella of an electromagnetic sea over the waters' terror-inducing, engulfing abyss. Once the laws of this atmospheric ocean were grasped, even in the maritime ocean's placeless depths, orientation—whether through continuous compass correction or through future communication via a sea of electromagnetic waves—would be possible. Maury here already relied on the idea of a topological field inferred from both Hans-Christian Ørsted's demonstration of electromagnetic interactions and their global conception by André-Marie Ampère a conception also applied to terrestrial magnetism. In the end, in a fundamental "crisis in intuition" (the phrase is Hans Hahn's),[206] Michael Faraday's further developed theory of electromagnetism and that theory's mathematization by James Clerk Maxwell would topple the theory of terrestrial magnetism, already because now there was no idea of the earth as a huge bar-magnet.[207] But just as with increasing distance from any descriptive possibility, electromagnetism was put on the slate with all the greater zeal, the impossibility of describing terrestrial magnetism seemed to guarantee the hydrographic enterprise. There was now an effort, for instance, to determine the processes of oceanic movement by measuring its electrical potential—induced by the earth's magnetic field—with a "geomagnetic electrocine-

203 Captain Jansen, cited in Michelet 1861, pp. 287, 91.

204 See Reinke-Kunze 1994, p. 13; Carrer 2009, p. 123.

205 See Maury 1855, p. 123; Maury 1861, pp. 168, 199.

206 See Hahn 1980.

207 On the transition from causal observation to field description in theory of terrestrial magnetism, see Jonkers 2003, pp. 118–128. On the "crisis of intuition" as characterizing an epoch marked by mathematical formalization, see Hahn, "Crisis"; Volkert 1986, pp. 89 f., 168–176, 310 f.

matograph." Or the previous form and strength of the terrestrial magnetic field was extrapolated from the declination-determined distortion of historical portolan charts, hence tied together hydrography and theory of magnetism within a framework of history of science.[208]

In the course of all these developments, as the nineteenth century proceeded the magnetic compass would increasingly lose its traditional hydrographic function. When in 1854 it became clear that a ship's magnetic signature could change in the course of the voyage itself, hence that the nautical interferences themselves oscillated as soon as they moved in space and time, that instrument became the sorry emblem of a future relativity theory that, wandering over the world's seas, appeared only to be waiting for its salvation. This he achieved in 1908 in the form of the gyrocompass, which with its free, direction-seeking axis used not only terrestrial magnetism but also the earth's rotation.[209] Nevertheless, the compass never quite shed the emblematic qualities ascribed to it by the founders of Early Modern experimental science. In 1910, we thus find Albert Einstein, having written an expert opinion in a patent dispute involving the gyrocompass (a dispute between Hermann Anschütz-Kaempfe and Elmer Sperry), singling out the compass as a model for his theory of atomic magnetism.[210] Likewise, in order to describe a science informed by a world picture stamped by quantum mechanics and relativity theory, concerned mainly with forces rather than material and thus largely intuition-free, in a lecture delivered in 1954 Werner Heisenberg compared the current epistemological situation to that of a captain whose ship is so much constructed of steal and iron that his compass's magnetic needle only reveals the ship's iron mass, no longer points northward. No goal can be reached with such a ship; it would simply move in circles, delivered to the wind and currents. But to further recall modern physics' situation; there is actually only danger if the captain is not aware that his compass no longer is reacting to the earth's magnetic forces.[211]

Heisenberg was speaking of a navigational instrument decoupled from modern science's graphic and symbolic systems; that for this reason had become functionally useless, even dangerous when used naively or all too trustfully; and that moved the world-exploring subject (the captain or helmsman) into a state of disorientation and with it possibly angst or dread. What Heisenberg dismissed was an obsolete world picture of topical orientation tied to this instrument. But what he simultaneously laid claim to for future research was just

208 See Munk 1992, p. 445; Harley and Woodward 1987, pp. 384f.; Jonkers 2003, pp. 5, 232f.
209 On the functional principle of the gyrocompass see Balmer 1956, p. 230; Körber 1965, p. 65.
210 See Galison 1987, pp. 34–47.
211 See Heisenberg 1958, pp. 30f.

that poetic function that was always inherent to the compass in literature: its world-generating productivity, made possible precisely by self-referentiality and non-arbitrary deviations (hence through what the formalism and structuralism of Shklovsky, Mukařovský, and Jakobson has described as a defining feature of poeticity); and its specific sort of world-relationship, grounded in the recursive distinction between *compasso* and its exterior, form and unmarked space, fiction and the real: in what Niklas Luhmann has described, in "Literature as Fictional Reality," as something specific to literary observation.[212] It is at least a fortunate coincidence that Heidegger delivered his lecture "The Question Concerning Technology" as the next in the series that had included Heisenberg's lecture. Heidegger here describes technology as something that "reveals" (ent*birgt*) and simultaneously "conceals" (ver*birgt*) the "unconcealment" (*Unverborgenheit*) of natural forces, finally also disguising their "bringing forth" (*Her-vor-bringen*) as such. He thus initially depicts technical instruments like the compass as they appear in strict histories of technology. But if, as Heidegger puts it, "the essence of technology is by no means anything technological," but is rather something poetic, then the "essence of the compass" reveals itself precisely in its poetic functions.[213] It is only through a *cumpassare* of medial, cognitive, and literary history that the compass, that "poietic" technical instrument par excellence, can be terminologically fixed.

212 See Luhmann 2008, p. 281–287. On Ernst Cassirer's analogy between concept-formation or mathematical symbolism on the one hand and instruments such as the compass on the other hand, see Röller 2010, pp. 176f.
213 Heidegger 1977a, pp. 4, 12, 13.

Chapter 4
The Ship of State: Cultural Techniques of Early Modern "Sea Appropriation"

The Allegorical Ship of State

From its beginning, the Early Modern period staged itself as a nautical epoch. In its political iconography, rulers are consistently navigation specialists. Educated Europe was familiar with the figure of the "political helmsman" from Plato. From Aristotle it knew that the state was a whole composed of things and human beings that could only get anywhere or even survive through appropriate functional distribution—the reason for political *technē* being so aptly comparable with management of a ship. And first Alcaeus, then Horace, and finally Jean Bodin had already shown that a body politic threatened at its core by elementary enemies or war, a disintegrating polis, could be described as a ship in maritime and organizational distress.[1]

The same epoch that propagated faith in blessed governance formulated the following rulers' motto, reinforced by providence: "As long as my helm remains straight and steers the ship, I can leave everything else to the one God."[2] (Fig. 11) But as contemporary emblematics articulated in countless variations, holding this course had to be more than rigmarole, *schwetzen*. It needed a certain art and scheming, *Künst vnd Ränck:*[3] first and foremost the art and strategy of self-governance, so that temptations could be steadily avoided—on his life's journey, the ruler had, as it were, to keep enemy fleets everywhere in sight, maintain himself in constant readiness for battle.[4] (Fig. 12) Tied to this, he had to protect

1 For an overview of the history of the metaphor of the ship of state, see Schäfer 1972, pp. 260 f., 274 f., 284, and Quaritsch 1979, pp. 251–286. On the problem of every reading of Alcaeus as the source of the ship of state topos (a topos only emerging later on), see Rösler 1980, pp. 118–124.

2 Subscriptio to Gabriel Rollenhagen's emblem (fig. 13), freely based on *The Aeneid*, 10, 218, in: Henkel 1967, col. 1454.

3 Subscriptio to Zincgreff's emblem with the motto *Gubernando non loquendo* (fig. 14), in: ibid., col. 1455.

4 See the Lipsian motto of sovereign self-mastery or the self mastery of the sovereign: "Just as a lonely ship is tossed to and fro in the sea by all kinds of wind: that's what our inconstant feelings suffer / when not strengthened and shored up by the ballast of reason" (*Gleich wie ein ledig Schiff durch allerley wind hin und wider im Meer umbogeworffen wird: so gehets auch unserm unbestendigen Gemüte / wann es durch den Ballast der Vernunfft nicht gestercket und befestigt ist*). Goldast 1965, p. 15.

https://doi.org/10.1515/9783110610734-008

his subjects from themselves, from the self-destructive fighting and quarrelling of mutinying sailors. (Fig. 13) Inwardly oriented, he had to rule them in the sense of *protectio*; outwardly oriented he had to shield them in the sense of *defensio*. And he had to protect his own ship of state from the battle through clever retreat. (Fig. 14) And despite all that, he had to move his own state forward like a galley against the wind: magnanimously, determinedly, and regardless of opposing elementary forces. (Fig. 15) Finally, he was meant to govern his ship of state "artfully" and with "experience / not violence."[5] (Fig. 16)

Fig. 11.: Proper Government

Fig. 12: *Gubernando non loquendo*

5 Subscriptio to Zincgreff's *Consilio non robore*, in Henkel 1967, col. 1474.

Fig. 13: Baleful Discord

Fig. 14: Clever Retreat

As extended metaphor that furthermore metaphorizes a real technology of transmission, the allegory of the ship of state appears in the epoch's emblem books as if through its own volition. What suggests itself here is connecting the ship-trope at work within politics from the sixteenth century onward to the emergence of the Early Modern "state." And against this backdrop, it would appear only logical to see the art of ruling a state as having its perhaps

Fig. 15: Determination

Fig. 16: Wise Government

most evident metaphor in *gubernare*. But the period here in question was one in which the rulers had to find a resilient and flexible form for giving their territory a solid structure—their "reason of state" often being tied to maritime rule, so that they also had to supply their ships with an optimal form and constitution. In such a period, political rhetoric could not simply have recourse to an iconography preserved since antiquity. The ship as metaphoric vehicle was no longer a fixed topos attesting to the state's politically proper and regulated, good and appropriate forms.

For ships as for states, the Early Modern period first sought the best form. Both were "under construction," so that deciding whether the form of state had to correspond to that of the ship or, inversely, the ship's organizational form had to correspond to that of the state, was nearly impossible. For this reason, the art of proper governance was not already prescribed by a Platonic ideal of the ship of state.[6] Rather, it was initiated beneath its organizational level. As a specifically Early Modern form of "governmentality," this art was involved in the state's development to the same extent that once the state was established, it helped keep it on course, tapping and securing its real territories and resources beyond rhetorical and legal claims. Consequently, *gubernare* was more than an allegorical detail in the imagistic field of the ship of state trope. Already in ancient Rome, the rudder served as the insignia for governmental authority in general, not only authority of the laws.[7] In the Early Modern period, it then came to all the more clearly symbolize the statesman's regulation of self and subjects, as well as providence, state-steered cleverness, a *technē* such as diplomacy. If reason of state "is what allows the state to be maintained in good order," then governance "is the right disposition of things arranged so as to lead to a suitable end."[8]

In the sense of "simple technique" on the part of the state,[9] *raison d'état* (often referred to at the time as *ratio status*) concerns the construction, maintenance, and preservation of the ship of state. But steering ships and men requires governance. In mid-sixteenth-century literature of rule and related emblematics, *raison d'état* was still described as a necessarily sovereign authority in a state of emergency—but the regime as an "art" based on simple "experience."[10] Not the form of the ship of state (and the techniques of leadership adapted to it) but its sheer existence (and simple maintenance of course) were subject to debate. As sketched out by Bodin,[11] sovereignty involved either rescue of a ship of state in distress or its tried and tested steering, grounded in natural authority. By the mid-seventeenth century, however, the form of the ship of state itself had become a test of sovereignty, the art of its steering a touchstone for the sovereign will to knowledge. Not only governance was transformed in this way into a reg-

---

6 See Paul Valérys pseudo-Platonic dialog *Eupalinos* on architecture and the art of shipbuilding: "[I no longer separate the idea of a temple from that of its edification....[Architects and shipbuilders] have no reason to be modest. They have found the means of inextricably mixing up necessity and artifice.... And it sometimes happens that the extremes of speculation put weapons into the hand of practice." Valéry 1956, pp. 70, 131f.

7 See Peil 1983, pp. 713f.

8 Foucault 2007, p. 377; La Perrière, *Le miroir politique.*, cited in ibid., p. 134.

9 Meinecke 1976, I., p. 9.

10 Starting with Plato, nautical imagery was evoked for matters of state, especially under the sign of danger. More sweeping catalogs of the good ruler's attributes tended to draw on pastoral and weaving imagery, the ship in distress at sea mainly referring to steadfastness in times of political danger. See Plato, *Statesman*, 279–283, 298d–e, 302a–b, and *Republic*, 488.

11 See Bodin 1992, p. 23 (book 1, chap. 8); Bodin 1955, p. 209 (book 6, chapter 5).

ular technology. Now rule in general turned to what had been situated outside sovereignty's borders: the sea. Henceforth the ship of state would not simply cruise in the *mare academicum* of learned treatises and emblematic library holdings. As an allegory and instrument of sovereign rule, it would also actually be built.

Sovereign of the Seas

There is an emblematic *architectura navalis*—and this not only in the sense that Joseph Furttenbach's like-named shipbuilding manual (1629) could also be read allegorically. Rather, the emblematic pictorial program of sovereign representation would now be actually implemented on the docks, the flagship of such an *architectura navalis* naturally being named the "Sovereign of the Seas." The sovereign Charles Stuart was very personally the driving force behind the construction of this royal warship, overruling reservations by the marine authorities: possibly, the ship would require deeper water than that present in the English harbors, or else it could at the first opportunity capsize on account of its heavy guns, as had once the *Mary Rose*. Built between 1635 and 1638, the *Sovereign* distinguished itself from previous royal ships not only through the extra-personal claims made by Charles I for its use and performance.[12] It was also one of the largest and heaviest armed sailing vessels to have ever seen service. Ten times as expensive as a traditional marine ship, it established new standards in the areas of figural decoration, rigging, and manner of construction. Above all, this *Ship Royal*, as it was named in the fleet lists, was a medium of sovereign representation. It made manifest the monarch's supreme position, his *sublimitas* between divine grace and hereditary charisma. And in connection with his demonstration of military potency, he created what Louis Marin has termed "power of institution, authorization, and legitimation.[13] (Fig. 17)

That on account of political developments that escalated soon after its launching, the ship hardly was deployed is thus almost besides the point.[14] Rather, what is decisive is that with its three continuous, heavily mounted battery decks, it represented a transitional ship-type between the Elizabethan galleon and the British late seventeenth-century battleship. In addition, in 1665, in both the Anglo-Dutch sea war and the battle of Lowestaft, deployment of

---

12 See Friel 1995, pp. 14–16, 29.

13 Marin 1988, p. 6.

14 See Lambert 2000, p. 33.

Fig. 17: The *Sovereign of the Seas*, full view

the *Sovereign of the Seas* established the keel-line battle order. The ship was likewise connected with the policies of maritime sovereignty already expressed in its iconographic program. (Fig. 18) The figural decoration on the stern, as well as the beasts and badges on the galleon referred to Charles Stuart's alleged lineage and thus to his claimed inherited sovereignty over the four English seas. Through this genealogical-juridical argument, the ship was tied to the commissioned work of court lawyers and historians, who since the time of James I were trying to back up the authority of the British rulers over adjacent waters by means of historical documents. For example, various scholars worked in the archives to demonstrate claims to sovereignty in the context of the *Reglement for Preventing Abuses in and about the Narrow Seas and Ports* (1633). That same year, Sir John Boroughs, the Tower of London's keeper of records, wrote a treatise on *The Sovereignty of the British Seas*—this possibly marking inauguration of the ship's planning and construction.[15]

15 See Busmann 2002, pp. 179, 187.

Fig. 18: The *Sovereign of the Seas*, stern view

In any case under the Stuarts there was no hesitation to belatedly prepare missing documents and to instantaneously furnish other documents of a dubious nature (or at least needing interpretation) with a convenient reading. The so-called *Charta Aedgari*, presumably a twelfth-century falsification, was given the suitable wording through a small but decisive change: where the original text of the introductory address had the–in their own time faked—words *regum insularum oceani que Brytanniam circumjacent*, the text now read *Regum Insularum, OCEANIQVE BRITANNIAM CIRCVMIACENTIS:* from the ruler of the surrounding islands, a sovereign of the entire "British ocean" had now emerged. As something like a corporeal certification, the *Sovereign*'s bulkhead was furnished with an effigy of the selfsame charter's alleged author, Edgar (who ruled between 959 and 975), depicting the moment when he rides over seven subjugated kings. And as something like official certification, each of the ships 102 bronze guns has a heraldic shield with the inscription "Charles has firmly grasped Edgar's scepter of the sea" (*Carolus Edgari sceptrum stabilivit aquarum*).[16] Maritime sovereignty was thus declared an ancient and now renewed grounding for the British royalty, with even the judge and jurist Edward Coke, ordinarily a critic of the monarchy, duly declaring in 1631 that the "greatnesse and glorie of this Kingdom of great Britaine consisteth not so much in the extent of his Majesty's territories by land, as in the souerantie and command of the seas."[17] Coke here aligned his views with the Stuarts' singular policies, reflecting intent on absolute extension—or rather overextension—abroad of their rule over people and territory.

These policies were unprecedented in that not even the Tudors, to whom English maritime power owed recuperation from a long agony, went so far as to stake such claims. For Elizabeth, the late medieval British guiding maxim "To keep the sea environ"[18] was compatible with that freedom of the seas she felt called on, like the Dutch, to defend against the Iberian world power. It would only be James I who in 1604 declared the so-called King's Chambers off limits to foreign ships, and a zone in which a royal license for fishing was also needed. For his part Charles I would then lay claim to the entirety of the four seas around Britain: the Channel, the North Sea, the Irish Sea, and the North Atlantic. Warships of foreign nations were here not allowed to exercise any defensive functions, capture other ships, or use force of any sort. All shipping commerce here stood in *pace Domini Regis*, under Charles' protection and rule—a claim that would be ceremoniously confirmed by a flag salute and,

16 See ibid., pp. 190 – 193.

17 Cited from Fulton 1911, p. 212.

18 Hertzberg 1878, p. 25.

once thus confirmed, recorded for the sake of future claims.[19] For Elizabeth, this greeting ceremony had been merely a matter of honor; for Lord Protector Cromwell the contractual obligation of other nations to undergo the ceremony was something to celebrate as a power-political success. And while the 1201 ordinance of King John constitutes the first extant evidence for the practice of lowering the top-sails and striking the flag, its source was less a complex of legal questions and more the practical problem of one's own battle ships not being able to keep pace with presumed pirate vessels. Hence at that earlier time the coastal waters were deemed less the king's property as a zone calmed for and of traffic.[20]

Under the premise of maritime sovereignty, traffic on the waters around the British island would become tied equally to archival practice, ceremonies of rule, and legal arguments. With his *Mare Liberum* (1609)—emerging from his earlier *De Jure Praedae* (1604/1605), a commissioned text justifying a Dutch capture of a ship in the Malacca Straits—Hugo Grotius mobilized the idea of the "community of the human race" and its legitimate "benefits" against the Portuguese and their Far Eastern trading monopoly.[21] In turn, James and Charles would now commission proper counter-opinions to limit Dutch fishing and trading power. In his *Abridgement of all Sea-Lawes* (1613), for example, William Welwood invoked the bible and James' divine mission as authorizing, against core principle of human community, the doctrine of *Mare Clausum*. Welwood countered the old argument that the sea's liquidity prevented ownership by pointing out that with navigational methods of location, justified claims to territory could indeed be marked here as well.[22] "Visible marks" such as rocks, coasts, and islands merely presented what a divinely commissioned human being could execute, designation of the "diuisible parts" of the sea for the sake of appropriating them. Human beings possessed, he insisted, "the helps of the compasse, counting of courses, sounding, and other waies, to find forth, and to designe *finitum in infinito*, so farre as is expedient, for the certain reach and bounds of seas, properly pertaining to any Prince or people."[23] Likewise, in his *Mare Clausum* of 1618 John Selden asserted that despite the lack of firmness and the constant movement, "Instruments" could inscribe "a distinction of Bounds" into the sea. Sel-

---

19 On the significance of legal ceremonies in the Early Modern connection between sovereign claims and spaces outside ancestral territory (alongside colonies especially the sea), see Benton 2010, pp. 23–25.
20 See Fulton 1911, pp. 6f.
21 Grotius 1919, pp. 25, 83. See also Muldoon 2002, pp. 16f.
22 See Fulton 1911, pp. 12, 353.
23 Cited from Cormack 2001, pp. 162f.

den here referred to "that useful invention of the sea-man's Compass, and the help of Celestial degrees either of Longitude or Latitude, together with the doctrine of Triangles arising therefrom." And further: "The Sea and Land mutually imbrace one another with crooked windings and turnings...the Seas are sufficiently distinguished by their Names and Bounds."[24]

The "radical title—primary ownership, and *nomos* of "sea appropriation"—that Carl Schmitt would treat in his *Nomos of the Earth*[25] thus constituted not only a question of representation but of navigation and naming as well. In actuality, still in the time of James' regime all three dimensions appeared together in John Speed's *Theatre of the Empire of Great Britain* (1611). (Fig. 19) Precisely Speed's cartographic theater,[26] tied as it was to England's early choreographic tradition, aimed to show how even the non-territory of the sea could be represented as a sovereign region by means of navigation and the name's genealogical performance. Through their mere existence, the argument went, England's inhabitants had natural property on the British island. But their interconnection with the sea authorized those among the inhabitants who for their part safeguarded primary ownership to also apply their means of terrestrial representation of rule, namely border demarcation and emblems, to the adjacent uninhabited coastal zone, unclaimed by anyone. When the compass rose was tied together with the "badge" of a ruling house, navigation became an emblem of maritime rule.

As shown by this theater of sovereignty, for James the empire was, perforce, a virtual phenomenon. Welwood's dispute with Hugo Grotius centered on just this virtuality. In his initially unpublished rejoinder to Welwood written in 1614, Grotius argued as follows: ownership could not result from naming and border-demarcation alone, as then any astronomer or surveyor could easily claim the moon and stars to be his own; rather a "physical act" of appropriation was required.[27] Concretely he here meant the dominance of a fleet or, near the coast, the reach of guns. But on a systematic level he was here pointing to the distinction between *imperium* (or legal sovereignty) and *dominium* (or factual use).[28] In his chamber proclamation of 1605, James had still spoken simply of *imperium* and thus of a sovereign legal authority that could only refer to its subjects and fixed territory. But in 1609 he proclaimed imperial authority over foreign

24 Selden 1972, Book I., Chap. XXI., p. 127, Chap. XXII., pp. 136, 138.
25 Schmitt 2003, pp. 44–46.
26 On Speed's cartography and its context see Barber 1992, p. 84; Klein 2001, pp. 106 f; Sanford 2002, pp. 15–19.
27 See Cormack 2001, pp. 165 f.; Fulton 1911, pp. 349, 356.
28 On this distinction, labile in the Early Modern period, see Benton 2010, pp. 4, 122 f.

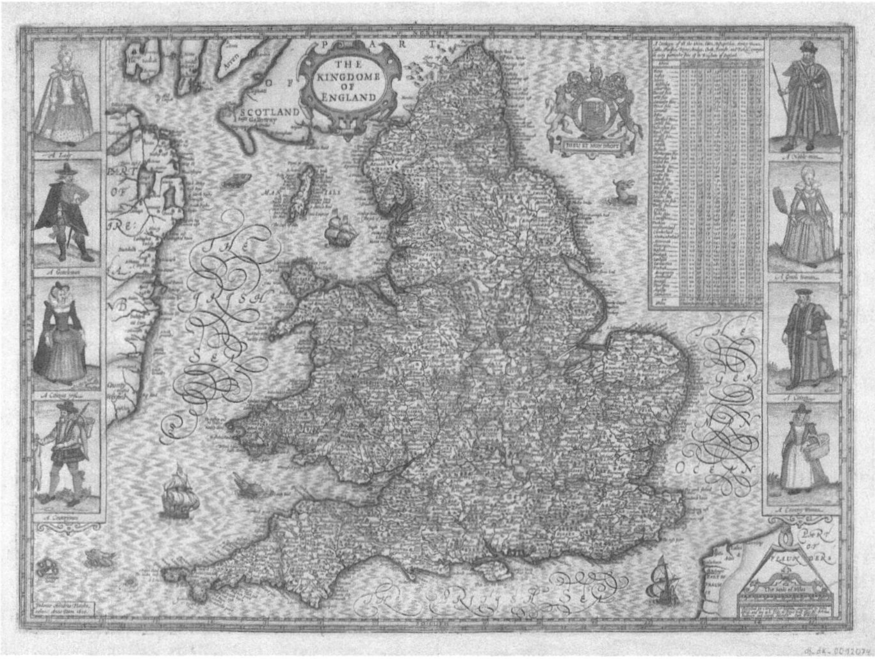

Fig. 19: John Speed, *Theatre of the Empire* (1614 (second edition))

fishermen as well, not his subjects and only associated with him through the dominion's adjacent zone. His sovereignty was thus to be freed from an ancestral, land-based locus, what *has* no place now being "taken" by means of names such as "our Coasts and Seas."

Selden carried this strategy forward on the level of legal doctrine, gradually dissolving the definitional border between *imperium* and *dominium*, in order to transfer the dominion's legal title to sovereignty and in the end define that sovereignty as unlimited.[29] The Stuarts' strategy might thus be described as proceeding with what Carl Schmitt described as "sea appropriation"—taking hold of the naturally uninhabited sea as putatively ownerless property, Roman law's *res nullius*—under the slogan of sovereign rule, but in the mode of effective occupation, continuous exploitation. This strategy corresponded to the transfer of terrestrial legal concepts to the sea, which in the process could appear as former no man's

29 On Alberico Gentili, who expanded the concept of territory to the sea in his *Hispania advocatio* (1613) and in his commissioned legal work for the English, see ibid., pp. 121, 124–129.

territory now entered into the Stuarts' sphere of rule and possession. However, various jurists not at the Stuarts' mercy would now look for a position mediating between the genuine freedom of the sea as open to all—Roman law's *res omnium*—and a territorial concept of power. These would be the theorists stamping the form of modern international maritime law. In the late seventeenth and early eighteenth centuries, Cornelis van Bynkershoek, for example, again proposed a distinction between the usage claims of *dominium* and the legal titles of *imperium*, here understanding sovereignty as what it had been since Bodin: discretionary power that could only be unlimited because it remained limited in the interior. In any event, as a concept, the so-called three-mile zone, interpreted as legal recognition of factual territorial rule, did not originate with Bynkershoek. And the concept did not mark a first placing of technical over legal power of disposition.[30] Rather, what now became clear was that the sea could not be an originary area of application for any legal concepts, and that here representation always depended on transfer from the land—on terrestrial arguments and figures.

The Construction of the Ship of State

Strangeness to law and a corresponding need to mitigate that reality by transferal of solid terrestrial grounding not only characterizes the sea but also the maritime transmission medium itself: the ship. To seize the sea, the ship must first be mastered. Already ancient Greece and Rome recognized that ships housed danger-bearing communities, for that reason alone having to become "states" in the midst of stateless space. In the Early Modern period, however, viewing a ship as a body corporate in which the chief or captain assumes unlimited command was the case since the Portuguese and Spanish voyages of discovery in practice, since Jean Bodin in legal theory. For this reason, as we will see in greater detail in chapter 5, it is starting in Early Modernity that the ship, with its balusters and public staging, became the scene of a regular sovereignty drama.[31]

Martial law was first established by Richard I, for the royal fleet. In the end, a lasting state of emergency prevailed there. It was laid claim to by Charles I as a badge of his sovereignty, and not only for his commanding representation on English ships. When in 1634 Charles wished to make his fleet ready to attack the Netherlands, but parliament had already been dissolved and thus could

30 See Akashi 1998, pp. 69 f., 126; Schmitt 2003, pp. 182 f.

31 See for example *Dictionnaire de marine* 1702, p. 196.

not authorize any special levies, he thought of an old Plantagenet law, according to which in situations of acute danger the king could commandeer so-called ship-money from the coastal cities. The flagship of the then-developed "ship-money fleet" would be none other than the *Sovereign of the Seas*. But for the fleet's further maintenance, Charles soon elevated the emergency law into a special tax, thus rendering a state of emergency into a rule and provoking a protest by both parliament and the cities.

To a significant degree, this arrogated sovereignty is what led in 1642 to civil war, the fleet now almost entirely standing against the king. Although the temporary and then final end of the Stuart monarchy did not lead to abandoning the principle of sea power, it did prompt a turn from practices of power and representation displaying a logic of sovereignty. In order to govern the ship of state, henceforth not so much conceptions of imperial rule would be decisive—these had themselves had very concrete consequences, leading to a marked upswing in ship construction[32]—but rather media of sea appropriation. The fleet would thus no longer only be a symbol of power and a compliant instrument of the sovereign's rule over territory and subject. Rather, it would now be understood as a complex structure made up of human beings—and things.

What now drew attention was not the pomp and representational value of royal ships, but rather what the fleet-reformer—and famous diarist—Samuel Pepys referred to as the "extraordinary Decays under which the Body of your whole Fleet in Harbour now lies." The countless memoirs written by Pepys reflected the excessive labor of writing undertaken for the sake of "Re-establishment of good Governance upon his Majesties Ships."[33] A check was placed on corruption among the fleet's personnel; officers now had to take a written examination, and both nautical and tactical manuals such as Admiral Russell's *Sailing and Fighting Instructions* (1691) were commissioned and printed. The fleet's deficit was balanced through an imperative of frugality and careful bookkeeping; at the same time regular payment was implemented, together with disciplining through muster rolls and journals, which is to say the mariners' professionalization. In turn, this development served as a basis for transforming the previously royal—and correspondingly easily instrumentalized—fleet into a national structure comprised of real ships of state—a better organization for the new battle situations, one which could also build upon reliable logistics.[34]

---

32 See Harding 1999, p. 60.

33 Pepys 1906, pp. 20, 69.

34 See Hill 1995, pp. 55, 67; Kennedy 1983, pp. 65 f.

Part of this general overhaul involved initiation of regular, carefully documented maintenance work, so that the dockyards and the construction activity taking place there also began receiving close attention. In the dockyards, silent, opaque practices were still being cultivated behind which Pepys presumed "a mere betraying [of] the King's service";[35] inspectors were now to objectively scrutinize and document these practices. True, Pepys' ordered reforms were by no means as singular as they made themselves out to be. They only continued a confrontation that Early Modern maritime sovereigns had pursued with their shipbuilders for a long time. Henry VIII, for example, had demanded a look into their activities for the sake of surveiling and improving them. After his first plans failed, the second plans did better, bringing in Italian shipbuilders from the docks—Italy, and Venice in particular, was then still the world's foremost shipbuilding center. That city-state's status was evident in the magnificent and speedy galleys produced there, and in a technique of skeleton-construction soon replacing the shell construction used in North Europe; it is also manifest in the manuscripts treating ship building that emerged there starting in the fifteenth century.[36]

Described cursorily, here it was still official orders that produced new shipbuilding practices and knowledge. For as state employees, the master shipwrights of the Venetian Arsenal had to follow the decrees of the Senate, which determined the main measurements for individual ship-types, in order to develop a standardized fleet. To adhere to these decrees when for instance hull-width or even the ship's volume was concerned, the master builder used patterns quickly replaced by the main frame and other frames at the end of the hull. He documented his work for the authorities and colleagues first in mnemonic rhymes, then with cutaways of different parts, and finally with detailed description of the construction. Once the ship was represented in central perspective, the required construction documentation was issued.

Through this chain of descriptions and representations, the ships were ready for serial production. The construction work, differentiated between engineering and carpentering activity, was now controllable, systematic construction-improvement feasible, together with a comparison or even explanation of hydrodynamic, hydrostatic, and stability characteristics. All of this in turn served as a basis for development of Northern European construction rules, in whose course refined drawing methods quickly emerged—a process whose success was displayed in printed manuals on naval architecture. Such books required less an

35 Pepys 1995, p. 132.
36 See Blume 2010, pp. 296 f., 302.

emblematic than a technical reading. Nevertheless, the models and plans for ships never embodied construction technique alone, but also always techniques for exercising power. Anthony Deane's *Doctrine of Naval Architecture* 1670) was published, at Samuel Pepys' urging, within this knowledge-political context, making possible the modern hull-frame.[37] Following Italian dominance, for a time the Dutch would dominate shipbuilding activities: this in part because of highly standardized and mechanized production procedures, connections with Baltic timber resources, and synergy with Dutch manufacturers of nautical instruments, partly because of a feedback loop between suppliers, shipowners, insurers, and merchants.[38] Nevertheless, England would soon supplant the Netherlands, ship manufacturing then remaining an English domain—although one that meanwhile was subject to challenge from across the canal.

Politics of Nautical Science

Especially following the Anglo-Dutch alliance of 1678, France in turn aspired to a dominant maritime position—a development already made clear, superficially, in Louis XIV's 1689 ban on presenting compliments to English ships with dipped flags. But in France as well, sovereignty at sea had to be proceeded by thorough reform of the fleet. And here as well, this meant renewing the fleet's personnel, disciplining the ordinary seamen and officers through the mentioned administrative measures. The French *gens de mer* were consistently skeptical, even recalcitrant, although the reforms promised liberation from the brutal practice of impressments, the miserable provisions, the frequent lack of wages. But the ship officers, with their handed down privileges of estate, offered straightforward resistance, which the French monarch's sovereignty had to first break through. Louis and his marine ministers threatened draconian penalties with abuse of office and refusal to obey orders; but the old cliques, indispensable in the enduring maritime conflicts, retained their audacity. In the long term they could only be disciplined through a system of training and examination ideally rendering marine officers into hydrographers.[39]

In comparison to post-revolutionary England, the French reform of the fleet unfolded in a strictly juridical, but above all pronouncedly scientific or scientific-political way. On the one hand maritime trade and the manpower on all French

37 See Alertz 1991, pp. 104–106, 267–269.
38 See Wallerstein 1998, pp. 45f., 59.
39 See Asher 1960, pp. 31–36, 92; Lutun 2003, pp. 244f.

ships was regulated by bringing together scattered common-law rules, now codified in the *Ordonnance de la Marine* (1681).[40] On the other hand there was an effort to collect and systematize all available nautical knowledge, a process embodied in the *Neptune françois* (appearing from 1693 onward), a general atlas of all rectified maritime charts. Beforehand, Georges Fournier's *Hydrographie* (1643) had accounted for a broad range of nautical knowledge in a kind of repertory of known chronicles, facts, and narratives. However, even if this work seemed to exhaustively cover all systemic and historical facets of such knowledge, from shipbuilding to a ship's command structure, from the Deluge to the French navy, it did not show any analytic or scientific perspective. Rather, this form of hydrography was symptomatic of an economic, fiscal, social, and religious regime under the sign of handed down policing, disciplining, and norming.

A new approach was rather marked by the Academy of Sciences that Colbert brought into being—an institution devoted to basic research in the fields of navigation, cartography, and hydrography. To be sure, the real innovative accomplishments took place outside the academy In 1697, the mathematician Paul Hoste, for example, together with Admiral Tourville, produced an *Art des armées navales* geometrically treating marine battle in columns, thus inaugurating an analytic perspective on naval warfare and its tactics. In any appendix to this book, Hoste also took up his theory of ship construction, here sketching his conception, based on both Aristotelian and Archimedean mechanics, of an optimized design and maneuvering of ships. This discussion marked a first synthesis of theory and practice in shipbuilding; it included a presentation of mathematical equations for empirical scrutiny. The ideas at work here were meant to be developed in a work on naval architecture, but in 1700 Hoste died, before its completion. What his pioneering work in any case shows is that the basic arts of maneuvering and shipbuilding were for an overly long time left to practitioners' silent, implicit knowledge. In order to make the knowledge palpable and capable of optimizing, and to move the shipbuilding program into a framework of controlled practice, Colbert formulated a range of regulations. But the instructions for constructing models, for taking over British and Dutch building methods, were simply disregarded by most of the French shipbuilders. For the king's scrutiny, Colbert now staged various exemplary maneuvers and even a speed-re-

---

40 See Asher 1960, p. 45 for Colbert's complaint about the lack of French codification of maritime law directly before the formulation of the *Ordonnance*. On the general program of the *Ordonnance*, hydrographical service, the duties and rights of the captain and crew, and the tasks of the ship's clerk, see *Ordonnance de Louis XIV* 1687, pp. 2f., 34f., 63–75, 79.

cord for shipbuilding: representative spectacles to which Louis did not even appear.

Colbert thus decided to win the monarch over at the location where he first saw any ships of state—at the Grand Canal of the Versailles palace complex. But despite all his diplomatic arts, he could scarcely manage to stir even slight interest on the king's part. It took the aforementioned Anthony Deane to move a step forward. In 1675, during the brief English-French alliance, Deane traveled to Versailles to deliver two yachts. On this occasion, Louis asked him, cluelessly and directly, why a ship could actually sail against the wind. The renowned Deane evidently had no good answer, so he handed the question on to a young man from Colbert's court. This courtier was Bernard Renau, a gifted autodidact and admirer of Descartes. He took the sovereign's question as seriously as possible, recalling a work by the mathematician Ignace Pardies that had appeared two years before treating the movement of objects and resistance of various media: one of the many examples in his vector-based analysis of movement involved the force effect of wind on a ship and the resistance at work in different forms of the bow.

Pardies could not yet know anything about the connection of force and speed, so that the entire geometric description was erroneous.[41] Nevertheless, Renau was inspired by the work to write a *Mémoire* that has been described as the first effort to offer a theoretical basis for ship design. The text was meant to allow builders to calculate the characteristics and performance of a ship in advance—to determine the hull-line with the least drift. From Descartes, Renau took over the idea of a "geometric bow" that, by contrast with the "mechanical bow" used until then in shipbuilding, was meant to be describable through an algebraic equation. But as an autodidact he did not execute the analysis properly, so that the entire derivation went astray. For this reason, in 1680 he and Colbert had a try on a half exemplary, half experimental comparative maneuver, meant to be executed with a conventional model, named *frégate*, and an ellipsoid prototype, named *modèle*. Despite hopes that the maneuver would point the path ahead, it also failed—and in a double respect. Although staged on the Grand Canal, it again received no royal attention. And because the *modèle*'s displacement was ten times that of the *frégate*, no significant comparison was actually possible. With this failed representative maneuver, Renau abandoned his ellipsoid theory of the hull, turning instead to the maneuvering problem on a purely theoretical level, since the sovereign spectator was in any case absent. (Fig. 20)

---

41 See Ferreiro 2007, pp. 68 f.

Fig. 20: Fleet on the Versailles Grand Canal

In his book *De la Théorie de la Manoeuvre des Vaisseaux* (1689), Renau first considered the forces exerted on sails, hull, and rudder, then derived the drift geometrically, analyzed the maneuver in various wind conditions, described the maximal rudder rotation in turning maneuvers, and finally presented the best sail-positions for different wind directions in various charts. Again, however, his inadequate mathematical knowledge was a fatal flaw: although unlike Pardies he correctly postulated that resistance was proportional to the speed squared, he did not know how to convert this trigonometrically, thus arriving at absurd calculations of the drift. Renau made no use of the new calculations and did not build upon any more general theory of movement. In not being able to integrate the variable influences of rudder, wind, and water-resistance, what he formulated was more a geometric than a physical problem. His theory was nevertheless admired by Pepys, at least respectfully acknowledged by Hoste. Furthermore, with his book Renau sparked a debate about drift in which initially Christiaan Huygens, then also Johann Bernoulli participated. Bernoulli did not develop his critique in geometric terms but by using diagrams of vectors and parallelograms showing forces—an approach that his son Daniel then developed further into a vectoral analysis of forces. In the end an analytic solution was arrived at through a sort of mathematics of flowing quantities allowing consideration of boundary values, minimums, and maximums.[42] (Fig. 21)

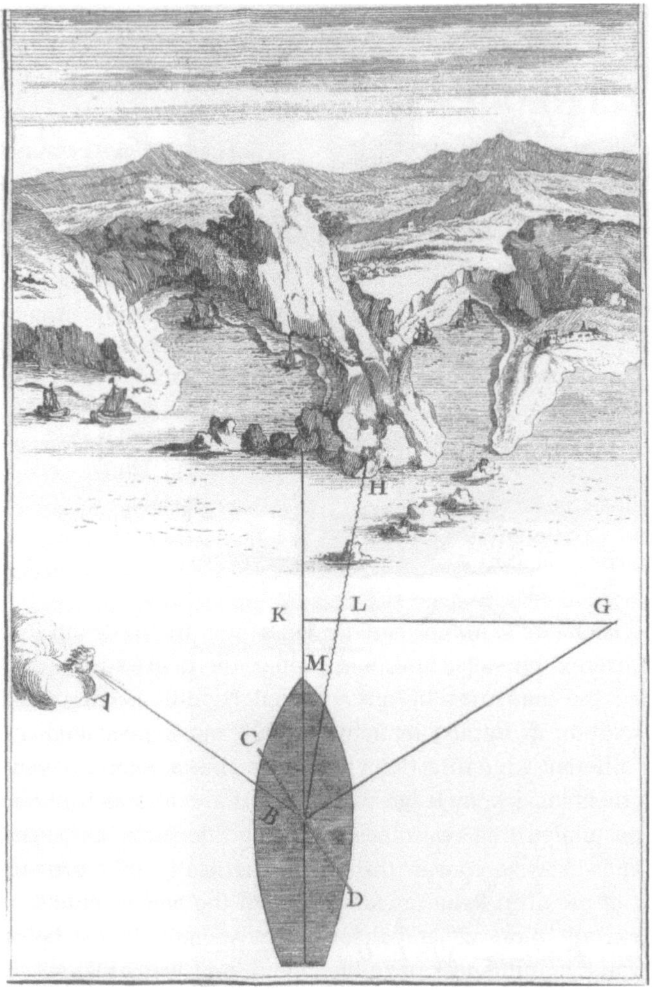

Fig. 21: Renau, *Théorie de la maneouvre* (1689)

42 Ibid., pp. 92–94; Séris 1987, pp. 112, 125.

In the long term, this confrontation between a practice needing little correction and an endlessly problematizing theory was centered not only on maneuvering but on, for instance, the question of the optimal hull-form: a form that could either be simply produced through skill and rich experience or, to the contrary, made use of as the goal and outcome of hydrodynamic reflection that would simultaneously demonstrate the possibilities of infinitesimal-calculus. There was also the question of a ship's stability: in this case as well, theoretical penetration resulted less from practical troubles than from the opportunity being offered to freshly explore the mechanics of solid bodies in liquid. The prelude for this was the determination of the center of gravity and propulsion, sketched out by Simon Stevin in 1608 in passing. Stevin offered an Archimedean foundation for hydrostatics, assigning every floating body two gravity-centers lying perpendicularly over each other. It is highly probable that by observing the way ships were loaded, he reached the mistaken intuitive conclusion that the center of propulsion had to always lie over the center of gravity.

In any event, since then the Archimedean principle, the lever rule tied to it, and the resulting insight into ship stability were rediscovered after nearly two millennia: that to hold itself upright in a floating position, a body needs a balance between force and momentum; that when a body loses that balance, the resulting forces of propulsion and displacement produce a "righting moment"; and that the propulsion's center of gravity corresponds to that of the volume of the immersed partial body.[43] (Fig. 22)

The Archimedean method of calculations was insufficient to guarantee stability for ships of every form. But with the totaling for finite geometric series, more complex volumes and their gravity centers could now indeed be calculated. It also offered a means to calculate curvilinearly circumscribed surfaces. For by increasingly replacing these surfaces with rewritten and inscribed staircase-figures composed of increasingly smaller right angles, a successively lower level of errors was made possible. This procedure may have been the starting point for Bonaventura Cavalieri's "method of indivisibles" of 1635, which itself would be applied cartographically, for the approximation of loxodromes in Mercator projections. When John Wallis then brought this geometric method into an arithmetic of "infinity," the path was free for the infinitesimal calculation of Newton and Leibniz.[44] Welwood's and Speed's legal and representative method of "divisibility" could only describe the sea's infinity in imaginary fashion, in the finitude of genealogical segments and cartographic excerpts. By contrast, the sci-

43 See Nowacki 2001, pp. 24, 26 f.

44 See Timmermann 1977, p. 50; Alexander 2002, pp. 194–196; ch. 3 of this book.

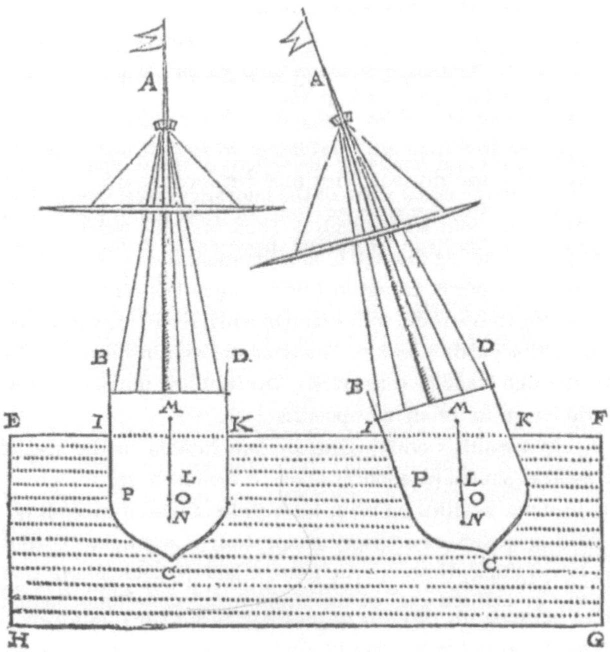

Fig. 22: Stevin, *L'Art pondéraire* (1634)

entific treatment of the form of the hull, of nautical maneuvering, and of ship stability led directly into the infinitesimal. That here, as well, a nautical sovereign was at work is made clear by the subsequent growth of knowledge about nautical stability. In his *Scientia Navalis*, published in book form in 1749, the prolific and pathbreaking mathematician Leonhard Euler newly defined hydrostatic stability by confining himself to infinitely small heeling angles. He here subdivided the ship's body (as in principle bodies with other forms) into cross-sections, integrating the hydrostatic pressure into every infinitely small surface element. For his part, Pierre Bouguer, in his *Traité du navire* of 1746, analyzed the stability of real ships as well, the book thus serving as the most exhaustive study of ship design. Bouguer's criterion of stability not only opened up new research questions such as that of the oscillation of swimming bodies and the application to longitudinal stability.[45] (Fig. 23) Described as a "metacenter," that

45 On Boguer's calculation of the metacenter and its disciplinary context, see in detail Ferreiro 2007, pp. 187–257.

criterion also stabilized practices tied to ship constructions at the end of the Baroque period's nautical research. It was no accident that Bouguer discovered the "metacenter" at exactly the same time as Duhamel du Monceau opened his "small marine school," meant to train former ship builders as engineers.[46] That since then ship commanders could be instructed by construction experts —something that previously would have been deemed outrageous—and in fact were obliged to take courses in engineering school is a clear sign of radical social change. Preceding this ascent of individuals formerly designated as "carpenters" to "constructors," now drawn into all sorts of nautical matters and considered equal to marine officers, scientists, and administrators, was the disciplining and disempowering of an entire profession.

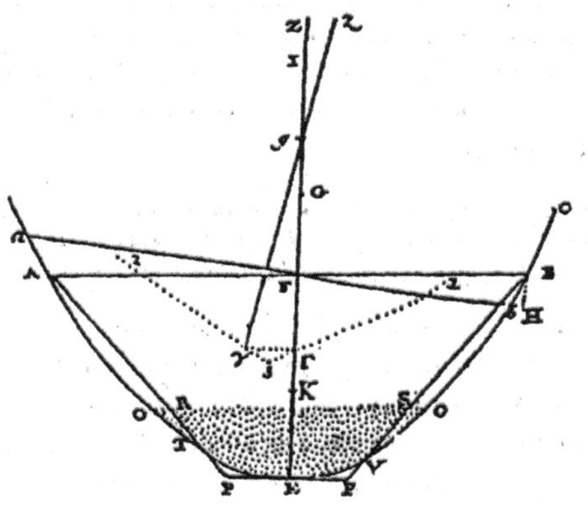

Fig. 23: Bouguer, *Traité du navire* (1746)

Such scientization was no neutral development, purely technically oriented. To the contrary, it was the systematic continuation of the previously recalcitrant shipbuilders' disciplining and the standardization of their work. The application of geometry and analysis to ship design and maneuvering, the development of mechanical, hydrodynamic, and hydrostatic theories and their linkage with ship architecture may have eventually made the sea's conquest possible, on both real and infinite (not only imaginary and finite) levels. At the same time, in the context of science politics, the development forced an explanation of what can be designated as implicit or even arcane knowledge, as yet uncodified,

46 See Monceau's conversion of Boguer's calculations and theorems into rules of hand in Monceau 1973, Vorrede, p. 511.

in the future hidden "practical reason"[47] of shipbuilders. In the end the cultural techniques of Baroque sea appropriation had become enlightened technologies of governance.

The Deconstruction of the Ship of State

The new and foundational approach evident in the Baroque period in respect to guiding state and governmental concepts was reflected in, among other things, the transformation of political and iconographic concepts. Initially, in using the metaphor of the ship of state arguing, in line with Bodin, Baroque theory of sovereignty followed corporative guidelines comparing state formation and shipbuilding not so much as forms of architecture but more as paradigms of Aristotelian hylomorphism. For just as " a ship is no more than a load of timber unless there is a keel to hold together the ribs, the prow, the poop and the tiller... a commonwealth without sovereign power to unite all its several members, whether families, colleges, or corporate bodies, is not a true commonwealth."[48] In Bodin's comparison, the shipbuilder thus offers a guarantee for the formal and functional synthesis that the Early Modern state would represent as the transcendent body of all natural bodies. The shipbuilder disposes over the sovereign authority that in political reality is expressed less as the state's *technē* than as representative action and the head of state's unlimited authority. For this reason, the analogy between the nautical and the state-political was from now on reduced to the sphere of governance: a *caput* called "sovereign" on land and "captain" at sea was to guarantee proper leadership of human beings and things. Even, still, in Fournier's *Hydrographie*, the ship's "captain" serves as the "telos" of this state-resembling nautical association, "like an epitome and summary of all desirable good qualities in all the other officers, resembling the end and perfection of all."[49]

The mid-seventeenth century saw a mechanization of the political body, perhaps given most impetus in the writing of Thomas Hobbes. In ascending order,

---

47 For a definition see Cohen 2010, p. 2: "intelligence distinguishing people who excel in the arts of action. Practical reason is embodied intelligence, drawing on the diverse aspects of our humanity. Rationality is one of its tools, but so are the senses, intuitions, feelings, and the body."

48 Bodin 1955, p. 7 (book 1, chap. 2). On the absence of architectonic thinking in Bodin and its ascent in seventeenth- and eighteenth-century political theory, see Foucault 1984, pp. 40 f.

49 "comme vn Epitome & Abregé de toutes les bonnes qualitez qu'on desire en tous les autres Officiers, estant comme la fin & la perfection de tous"; Fournier 1667, book 3, ch. 3, p. 89.

he deduced the state body's "artificial human being" from the elements of his *philosophia prima*, his physics, and—only lastly—his anthropology. Hobbes exemplified his resolutive-compositive method by means of the clock—its mechanical gears offered the most vivid possible analogy to the state as a machine that was constructed and set into play, and that had to be regularly maintained. The concept of the "state-machine" took up the rationalization process initiated in the Baroque age at the point where theoretical rationality and political practice came together. To this extent, this trope might be considered an "absolute metaphor" designating the political sphere not merely through figurative discourse but as that sphere as such. In any case a mechanical conceptualization of the state freed its affairs from the influence of corporeal ideas and both ethical and prudence-centered behavioral doctrines. What was political could now be conceived not only on the level of practice but that of *poiēsis* as well.[50]

Hence on the one hand, starting in the mid-seventeenth century and in the course of a scientization of things nautical, ships were increasingly understood as machines, their functioning as something to be thoroughly studied and constantly optimized.[51] On the other hand, this theoretizing and technologizing of what had been a purely artisanal domain appears not to have been directly reflected in the ship of state trope's image-field. Namely, this trope has always been marked by a remarkable constancy and classicism and—here marking a difference with for example the political clock-metaphor—has never been subjected to probing critique.[52] For this reason the ship of state appears to have been downright predestined for politically oriented conceptual schemas tied to a logic of sovereignty, as well as those that are corporatist and praxis-oriented. But even such conceptualizations could only hide their constructed nature to a decreasing degree. For instance, emblematics did not, in fact, produce a simple connection between concept and intuition. Rather, here the aesthetic-conceptual analogism of inscription and image was meant first and foremost to transfer a contingent political situation—for which distress at sea stood—into an obligatory ethical order: an order for which a ruler-author expressly had to vouch. In this manner emblematic allegory increasingly revealed the analogically produced order to be, for its part, staged; it thus no longer reduced an object like a ship to an ethical or moral idea, but expanded it "according to a whole network of natural relations."[53]

---

50 See Stollberg-Rilinger 1986, pp. 19 f., 32, 52 f.

51 See Séris 1987, pp. 120 f., passim.

52 See Peil 1983, p. 893.

53 Deleuze 2006, p. 144. See also Moog-Grünewald 2000, p. 206.

Through this process, the at first sight thoroughly conventional allegory of the ship of state is opened to technological knowledge—something we need to understand as the sign of a gradually altered mentality of governance. Concrete pictorial details already point to this, for example precise representations of ships that, as in Zincgreff's *Gubernando non loquendo*, were unceremoniously copied into one or another text in the context of allegorical iconography.[54] The concept of the ship of state was thus finally rendered as technological as was its pictorial representation. And the concept was no longer projected on the basis of an idea of the political sphere, for the sake of then declaring political realities in terms of the concept. The old topos was deconstructed, ships and states being now examined like machines, with a view to their construction, constitution, performance: a development that could be derived from the fundamental conceptual shift from an absolute to an intensive space, mathematical-mechanical analysis of forces here replacing classical Aristotelian topology.[55] In sum, both ships and states developed into the object of a specific technological knowledge, whether of nautical or political matters.

In the seventeenth century, a science emerged of circumstances of state—a science that increasingly shifted from describing the state's extant constitution, resources, and assets to probabilistic calculations. Knowledge concerning affairs of state was expanded to take in dynamics, contingencies and probabilities; at the same time new representational procedures (lists and tables, diagrams and curves) were steadily developed. Not only were maritime trade and commerce, as important and at times even decisive state assets, connected with this new statistical reality. A completely new and highly complex relationship between nautical and political spheres was at work in two different ways: in complex procedures through which the state controlled shipbuilding and the operation of nautical machines; and in the numerous methods and forms of knowledge now accompanying shipbuilding.

---

54 The ship represented with technical expertise in Zincgreff's emblem had already surfaced in map-books such as the *Hispaniae Novae Descriptio* (1579) and in a series of engravings by Pieter Brueghel the Elder, and then later in John Speed's *Theatre of the Empire of Great Britaine* (see for example fig. 4.9.); it would then free itself from cartographic and artistic contexts.

55 See Leonhard 2006, p. 31: "The relationship between element and system thus concerns not only physical theories. It also concerns theories of society once a society is understood as a system of animated bodies....But it equally concerns a theory of the image, once it leaves behind a static, resting ground and wanders into the processual and dramatic....For this reason Aristotelian topology entered into an insurmountable crisis, the traditional model of change of locus being replaced by the Baroque model of transmission of forces."

In seventeenth-century shipbuilding, it was the differentiation of plans and construction that above all led to improved methods of drawing and representation, alongside professional specialization and production of ships in series. This was the basis for freeing shipbuilding from the shipbuilder's implicit and arcane knowledge, from their traditional art and experience. Late Medieval shipbuilding had still been oriented around successful models, rules of thumb for the hull's measurement and ship's form, and for characteristic parts of the ship such as the keel and main bulkhead—a procedure that theorists like Bodin also took account of metaphorically in their doctrine of the ship of state. But with production of models accompanying shipyard work, including drawn geometric-digital planning diagrams, it was possible to systematically develop specific types of ship and construction characteristics, and to study hydrodynamic and hydrostatic dimensions of shipbuilding.[56]

From that period onward, the ship would no longer be the best-guarded secret of an autonomous trade or a "diagram" of the safeguarded procedures"[57] of a machine. Rather, it became the medium for an experimental system that—more than being a mere testing procedure—promised still unknown answers to questions that, in Hans-Jörg Rheinberger's words, "the experimenter likewise is not yet in a position to clearly raise." Shipbuilding's experimental parameters allowed a "materialization of questions" that could not have been posed without the emergence of *scientia navalis*.[58] And in the social, technological, institutional, and scientific configurations accompanying shipbuilding, not only hidden facts were systematically disclosed, but "epistemic things" were experimentally created.

Such things do not have to be material. They can consist of "structures, relations, functions. When it comes to Baroque thinking—a thinking characterized by an interacting development of mathematics and mechanics—the following, general point seems to hold: in its predilection for variable quantities, it above all profiled the concept of function, while at the same time rendering functional the concept of the object itself. The object, Deleuze argues, "is no longer defined by an essential form, but reaches a pure functionality...inseparable from a series of possible declensions or from a surface of variable curvature that it is itself describing."[59] Approached in this way, immersed in the water's surface the ship appears as nothing less than the ideal-typical object of Baroque functional thinking, linked not so much to Platonic-Euclidean geometry as to Archimedes'

56 See Alertz 1991, pp. 43 f., 110.
57 Baruzzi 1973, p. 31.
58 Rheinberger 2001, p. 22.
59 Deleuze 2006, p. 20. On the following see also ibid., pp. 39, 153.

dynamic concept of space. Instead of consisting of ideally formed material, it folds specific material in a manner rendering its forms into vectors within an infinitely variable play of forces.

The ship might also be understood as an exemplary realization of Baroque architecture: on the one hand a completely external exteriority limited to pure representation without "essence" or interior; on the other hand an infinitely folded interiority with no direct relation to exteriority, while entirely enclosing it like a monad. We can understand this exteriority without interiority—the "infinite receptivity" of the Baroque façade—as a quality defining, precisely, the sumptuous display ship, now appearing as a "room…where all activity takes place on the inside." More concretely, we can understand it as for example a steering cabin in which "linear and numerical tables" infinitesimally comprise what is offered in external perspective and arrange it in an "information table."[60]

This sort of "osmotic interrelation between exterior and interior space" that Erwin Panofsky described as a feature of Baroque art[61] can also be described as a characteristic fold between the aesthetic and epistemic sides of the Baroque ship: both sides, but also the series of their development, are folded into one another in such a way that their difference unceasingly newly unfolds on both sides. For this reason, in the Baroque nautical representational art and *scientia navalis* belong together in their very differentiation. We see this in, among other things, in the double exploration of the infinity represented by the sea in Baroque seafaring. Where the new practices of knowledge and representation in nautical stability management, in hydrostatics and hydrodynamics, opened up the infinitively small in infinitesimal calculation, for their domains of rule the maritime powers claimed, to speak with Welwood, "diuisible parts" of the sea and thus the *finitum in infinito.*

The ship in this way emerges as a double medium of sea appropriation: power-politically but also at the same time hydrographically, for only the most seaworthy and maneuverable ships allow a—per se imaginary—legal demarcation through occupation of the "smooth" oceanic space. In this sense, the "nomos of the sea" is not based on any originary allocation, however energetically especially archivists and historians at Stuart's mercy tried to show this. For an enlightened international lawyer such as Emer de Vattel, writing in 1758, it was evident that the "right of navigating and fishing in the open sea" was "a right common to all men," for which reason "a nation, which, without a legitimate claim, would arrogate to itself an exclusive right to the sea, and sup-

60 Ibid., pp. 30 f., 39.
61 Panofsky 1995, p. 45.

port its pretensions by force, does an injury to all nations; it infringes their common right."[62] This legal freedom of the sea was not seen as conflicting with the ability of individual powers to establish a maritime domain through superior nautical capacity. Hence in 1750 Johann Gottlob Heinrich Justi argued that as soon as "a general belief that not all states can be maritime powers becomes evident in Europe,"[63] then a general European balance of power would emerge, with on the one side sea powers (especially England), on the other side land powers (especially France).

The *nomos* of the sea cannot be based, again, on any primal allocation because the sea, this outside to legal and political thinking, is "from the very start a differential quantity: it is what evades taking and without whose existence every act of taking would grasp at nothing."[64] Sovereignty at sea is always a function of nautical knowledge and technology, just as inversely there could be no technologized and scientific nautics without a sovereign will to knowledge. From the navigator's perspective we might say more pointedly: the actual sovereign is the sea. As Hoste programmatically indicated, the only decisive thing is, in the end, "what the sea desires" and what orders it gives to navigators.[65] From a political perspective, to the contrary, the state sovereign's will was inscribed into the project of the maritime voyage's scientization from its onset.

Coinciding with the end of the Baroque period, the project's completion not only offered answers in the framework of a sovereign will to knowledge in nautical matters. The stability calculation that since the 1730s Bouguer aimed at the epistemic object of the "metacenter" continues to form a basis of a ship's architecture. But it did not emerge from a pressing need within nautical practice, for example because of certain catastrophes at sea. Likewise, the calculative expenditure was not justified in terms of the ship's better maneuverability.[66] Rather, the problem's scientific introduction allowed a stabilizing of the complex practices and social relations that had developed around ships. As if concern about a "social metacenter" was the thing at stake, the now-constituted *scientia navalis* allowed better control while at the same time creating a heightened sensibility for disturbances.

---

62 Vattel 2008, p. 251.

63 Justi 1759, p. 7.

64 Balke 1996, p. 328.

65 See Hoste 1697, p. iii v: Voilà ce qui me fait suivre une route nouvelle, dans le dessein que je me suis proposé de donner des régles, pour la construction des Vaisseaux: j'ai résolu de commencer par la théorie, en apprenant aux Constructeurs, ce que veut la mer, avant que de leur donner des régles sures & aisées de leur art.

66 See Ferreiro 2007, pp. 25, 187.

Hence instead of directing the social, economic, and technological field of navigation simply as a matter of law, what emerged here was nothing short of a technology of governance: it pervaded the "actor-networks" of objects, persons, and institutions, actions and ways of behaving, forms of knowledge and variations of medial architecture, that together constituted a new social machine. This continuity between knowledge and power was made possible by the ship as instrument of control and command, as epistemic object, economic vehicle, material artifact. The ship became a medium of power as soon as power became tied to all those social practices and types of knowledge, forms of production, communication, and transport, tied to the sea and to maritime commerce and, particularly in the Baroque period, having a capacity to define the constitution of an entire state.

Although in France the architecture of ships had been well developed for a long time and the theory of shipbuilding could quickly gain a foothold, even in the age of Louis XIV the predominant focus was on terrestrial existence. By contrast, England, where seafaring was bureaucratically and commercially much more developed albeit far less "scientized," would end up opting for a maritime existence—in other words, for political and economic dominance over the sea. Consequently, the decisive factor was here neither legal claims not scientific achievements but rather cultural techniques such as those of navigation and the extent to which they were absorbed into the constitution and productivity of the machine of state.

David Hume's designation, at the end of the Baroque age, of *scientia navalis* as a paradigm of the political sphere attests to the henceforth non-ideal, non-allegorical relationship between ship and state:

> The mathematicians in Europe have been much divided concerning that figure of a ship, which is the most commodious for sailing; and Huygens, who at last determined the controversy, is justly thought to have obliged the learned, as well as commercial world; though Columbus had sailed to America, and Sir Francis Drake made the tour of the world, without any such discovery. As one form of government must be allowed more perfect than another, independent of the manners and humours of particular men; why may we not enquire what is the most perfect of all, though the common botched and inaccurate governments seem to serve the purposes of society, and though it be not so easy to establish a new system of government, as to build a vessel upon a new construction?[67]

Although Hume defined the state's construction as a more complex project than building a ship, he nonetheless saw the emergence of mathematical nautical science as attesting to the need for something like state-technological knowledge.

67 Hume 1875, p. 480.

For even if Columbus's and Drake's ships could survive their voyages without being constructed in a standardized scientific way, nothing spoke against optimizing the form of the ship. Likewise, the survival of various forms of state developed out of experience alone said nothing about the political sphere's potential for perfection. With Hume it became clear that the political imaginary had been freed from the sign system of inherited commonplaces—that, as seen in the metaphor of the ship of state, it had to nurture itself on more complex forms of knowledge and representation.

If that metaphor is also a symbol in Kant's sense, creating a "rule of reflection" over the state (which as such is removed from direct intuition), its construction and constitution, then the nature of this reflection alters once some sort of ideal-typical and allegorical ships no longer represent the "entirely other object" of the metaphor's vehicle but rather the complex epistemic substance involved in a scientific approach to construction.[68] The political sphere is thus no longer something given through an idea (however mediated its illustration). Rather it always has to be first, and repeatedly, constructed—and optimized. That it was the Englishman Hume who rehabilitated the ship as an object of reflection on future state forms is no coincidence. With Hume, the political was for the first time—and long before observers of the French Revolution—grasped as an experimental system. Without a doubt this meant that theorists of the state might have to strike their sails before its constructors. As Hume put it, "in some future age, an opportunity might be afforded of reducing the theory to practice."[69]

68 Kant 2000, p. 226.
69 Hume 1875, p. 481.

Chapter 5
The Figure of the Captain: Doctrine of Maritime State of Emergency

Fictions of the *Caput*

"The captain, in the first place, is lord paramount," declared Richard Henry Dana in 1839.[1] On board ships on the high seas, those peculiar floating places within the placeless, bodies freed from terra firma and its order for the sake of forming small autonomous state structures, the commander serves as an emergency figure par excellence. In the end he has "no companion but his own dignity, and few pleasures, unless he differs from most of his kind, beyond the consciousness of possessing supreme power, and, occasionally, the exercise of it."[2] As described in a mid-nineteenth-century report passed down from the American brig *Pilgrim,* the figure of the captain comes into its own as a limitless sovereignty of word and deed. "'I'm no negro slave,' said Sam." "'Then I'll make you one'" replies a certain Captain Thompson. And when the shackled mariner, tormented by vicious whiplashes, calls out the Lord's name, the sovereign commander roars "'Don't call on Jesus Christ...*He can't help you. Call on Frank Thompson!* He's the man!'"[3]

For Americans in the 1840s, this scene was doubly shocking and scandalous. It was not only that here a captain or, as he was sometimes called, a "master next God," had placed himself blasphemously *above* the Lord—but also, that like a slaveholder the captain had cast scorn on rights to freedom and human dignity guaranteed in the homeland. When in 1840 Dana, who meanwhile had completed his legal studies and would emerge as an expert in maritime law and an active abolitionist, published the diary entries from his time as a voluntary sailor between 1834 and 1836 under the title *Two Years Before the Mast*, he lay the groundwork for the North American tradition of the realist sea-novel. Here, as Dana himself asserted, the existential circumstances on seagoing vessels were not depicted from either the lofty bird's-eye perspective of the officers or the clueless outer view of passengers. For the first time, these circumstances were subject to "accurate and authentic narrative," offered by a "*voice from the fore-*

1 Dana 1996, p. 10.

2 Ibid.

3 Ibid., pp. 80, 82 (italics in original).

https://doi.org/10.1515/9783110610734-009

castle" belonging to a reporter who was both legally schooled and experienced on the high seas.[4]

What Dana had accomplished as an observer of civil transport and trading ships was repeated a decade later (1850) by his admirer Herman Melville for the American navy in *White-Jacket*. Here Melville focused in detail on the U.S. fleet's articles of war, which following a British model granted sovereign authority to individual commanders, while at the same time mingling the legislative sphere with that of institutional discipline. With the articles of war, flogging had become a legal matter, and for that reason alone—not on account of its abnormal frequency—was in basic opposition to the free, democratic order. "By this article the Captain is made a legislator, as well as a judge and an executive," we read in *White-Jacket*. According to Melville's "genealogy" of war articles, these are rooted "in a period of the history of Britain when the Puritan Republic had yielded to a monarchy restored." Already for this reason, he argues, they are against the constitution's spirit—but especially because they make wartime state of emergency into a rule for peacetime as well. In the naked violence of flogging, the navy reveals "the professional and intellectual incapacity of her officers to command," also and precisely when they legitimate themselves through the despotic fiction of infallible command.[5]

The birth of the realistic sea-novel took place out of the spirit of critique—and this was not a critique exhausting itself in reporting on floggers and flayers to which "masters next God" have degenerated. Both Dana and Melville initially understood the fact that in the nineteenth century abuse of sailors had become a rule on American ships as constituting a legal problem. Already in 1839, in a legal case study of two abused sailors, Dana demonstrated the inordinate sense of power and inadequate culpability of American sea-captains; two years later he published a legal manual for simple sailors.[6] For its part the legal acuity of Melville's *White-Jacket* appears to have contributed to the ban on flogging legislated by the U.S. Congress in 1850.[7]

Both Dana and Melville sought a form of writing capable of probing the figure of the captain and its fictive substrates.[8] The sorts of literary battle-declarations at work here already point to the historical and legal mythology surrounding the sea-captain and his status. Within the Western seafaring history, the figure arrives late on the scene. The crews on ancient ships did have clearly de-

4 Ibid., pp. 1 f. (italics in original).
5 Melville 2002, pp. 143, 297 f., 149, 217.
6 See Dana 1839, esp. pp. 97, 104 f.; Dana 1856, esp. pp. 99–105, 208–230.
7 See Matthews 2014, p. 401, n. 77.
8 See Dana 1996, p. 83 for the putative call of sailor Dana to future authorship.

marcated hierarchies, taking in, for example, the Greek *triearchos* and *nauarchos*, *kybernētēs* and *proreus*, *keleustēs* and *pentekontarchos*. But either these functions reflected complex religious, navigational, and economic categories distinct from any sense of a political and religious monopolization of power; or as in the case of Roman seafaring they expressed sovereign power, *dominica potestas*, over mere slaves, so that monopolization was simply unnecessary.[9] We have documentation from the Middle Ages of official interventions in Northern Italian maritime locations and Northern European Hanseatic cities meant to furnish crews—especially on long voyages and military expeditions—with an autocratic authority. Still, as a rule commercial maritime travel was seen to by a group of free and equal seamen, undisturbed by terrestrial officials. Even following the thirteenth-century nautical and bureaucratic revolution, which led to a differentiation between shipowners, merchants, and seamen, the public legal powers granted commissioned ship-masters were mainly for the sake of guaranteeing the business transaction and safety of goods—there is no question here of any absolute authority, a mariner continued to be first among equals. Final discretionary competence lay with the ship-council.[10]

Likewise, before the epoch of European state formation the ship's organization cannot really be thought of as having the form of a state—just as the figure of the captain could not emerge before the epoch of sovereignty theory and corresponding practices of rule. The captain is the chief agent of Early Modern efforts to transfer terrestrial concepts of rule to the state-free and traditionally law-free space of the sea—and thus to exercise "command of the sea" in the sense of a sovereign territorial state. He is a figure and fiction of law and politics aimed at compensating for the incompetence of law and politics in maritime matters. For gaining or retaining maritime command, fictions and figures were naturally less decisive than nautical, administrative, technical, and economic needs. But nowhere did the fictions and figures produce more practical consequences than in Early Modern life on board ships. The history of seafaring thus contains fictional and figural lessons both anticipating and grounding the way of writing, scenes, and plots of seafaring literature.

In William Bourne's *A Regiment of the Sea* (1574), probably the first English-language nautical handbook, we read the following about the shipmaster:

> he ought to be sober and wise, and not to be light or rash headed, nor to be too furnish or hasty, but such a one as can wel gouern himselfe, for else it is not possible for him to gouerne his cõpany well....And the principall point in gouernment is, to cause himself

9 See Hanses 1983, p. 22.
10 See Friedland 1995, esp. pp. 259 f.

> both to be feared & loued, & that groweth principally by this meanes, to cerishe men in well doing...and to punishe those that be malefactors and disturbers of their company, and for small faults, to giue them gentle admonition to amende them: and principally these two pointes arte to be foreseene by the maisters.[11]

Good ship management is good human management. Where Bourne still sees cause to describe this double art of governance in detail (the virtue of self-governance, the political doctrine of prudence, a system of reward and punishment), later manuals such as John Smith's *Sea Grammar* (1627) make it formally short: "The Captaines charge is to command all."[12] On the basis of this commanding authority, portraits of an ideal captain emerge, portraits presenting the exemplary qualities of a political leader as with a mirror for princes. In Georges Fournier's *Hydrographie* (1643), for example, the "Capitaine du Nauire" is described as a "Telos" of this state-resembling nautical enterprise, "like an epitome and compendium of all the good qualities that one wants in all the other officers, being like the goal and perfection of all."[13] Finally, what clearly surfaced here, in view of an enlightened understanding of governance, is not only this on-board regime's despotic constitution. Beyond that, its grounding or rather groundlessness can be found in the state of emergency that in voyages on the high seas putatively began at departure. In a travel journal Herder kept in 1769, he thus observed that the ship

> is the prototype [*Urbild*] of a very special and strict form of government. Because it is a small state that sees enemies everywhere around it, the heavens, storms, wind, sea, currents, cliffs, night, other ships, shore, a government belongs here resembling the despotism of an initial hostile period. There is a monarch and his prime minister, the helmsman: everyone behind him has his assigned place and office, with neglect or rebellion from these ranks being especially punished.[14]

Herder's approach here was the same as that of shipowners who, as in a legal dispute over a case of disobeying a command that unfolded in Lübeck in 1804, still maintained as almost self-evident, that "every skipper is the ship's monarch during the voyage."[15] Nevertheless this form and constitution of on-board authority was anything but self-evident. It was based on a specific constel-

---

11 Bourne 1963, pp. 170 f.

12 Smith 1968, p. 34.

13 "comme vn Epitome & Abregé de toutes les bonnes qualitez qu'on desire en tous les autres Officiers, estant comme la fin & la perfection de tous." Fournier 1667, book III, ch. 3, p. 89.

14 Herder 2002, p. 19.

15 Material in Lübeck city archives, cited from Witt 2001b, p. 249.

lation of nautical, legal, and power-political factors whose development is attested to eloquently in the derivation of "captain" from Latin *caput*, "head, summit; beginning, end, main thing." At first a military title granted heads of British admiralty ships, the term was gradually transferred to seafaring in general. In the context of Early Modernity's corporative and hylomorphic thinking, a ship consists of "members" under the command of a "head";[16] for this reason, for Bodin the captain's status is aligned, as we have seen, with the formal principle that renders a mere pile of wood and human beings into the purposive association of a ship (and with it a state). Some other aspects of this guiding nautical figure reveal themselves along an etymological path: contained in "capitulate" is not only its present meaning but—as in the famous *capitulación* between Columbus and the Spanish king—"negotiate over a contract." Entirely in line with the age of global discovery, *capere* takes in "seize" and "grasp" or "understand," "appropriate," "chain," and "master." "Chapter," directly stemming from Old French *chapitre*, itself originates in Latin *capitulum*, a diminutive of *caput*. There are examples of the German term *Kapital* starting in the sixteenth century; the term was derived from Italian *capitale*, denoting "main sum" and "wealth," in actuality the headcount of a cattle-herd. The geographical term "cape" emerged in Late Middle English from Old French *cap*, stemming from a Provençal term itself based on *caput*—that word in turn eventually taking on the meaning of "legal clause" or, once again, simply "chapter."

The figure of the captain appears to have itself emerged from a chapter of world history. We read in Gaspar Correa's *Three Voyages of Vasco da Gama and His Viceroyalty*, that in November 1497, commissioned by the Portuguese crown, a convoy commanded by da Gama sought the route to India. Underway since July, the convoy was now blocked at Africa's southernmost end by a violent storm. The crews, we are informed, prayed for survival and pled to reverse course, Da Gama, however, swearing, to remain at sea until he had rounded the cape. With land long-since out of sight and the winds finally abating, the mariners mistakenly concluded they had indeed doubled the cape. Encountering the mouth of a river, they steered into Africa's interior, then hesitating for weeks to voyage onward, fearful of enduring more heavy storms on their fragile ships. But Da Gama insisted on staying true to the promise he made his sovereign and not deviate an inch more from the planned course—the king, he reminded his men, had forbidden any interruptions of the voyage by going ashore. While most of the despondent mariners now "abandoned themselves to the chances

16 Camões 1880, vol. 1, canto II, v. 85, p. 73.

of the sea,"[17] some plotted to overpower da Gama and sail back to Portugal in the hope of royal mercy. When the "captain-major" grasped what was transpiring, he expressed assent, but demanded the signatures of three outstanding seamen for the sake of good order and justification before the king. But when these men appeared, he had them shackled, together with the shipmaster and pilot, forcibly collected all the charts and navigational instruments of the latter two men, and threw this material overboard.

What took place under the order of Captain Major da Gama was a capital event. Namely, ships were the advance guard of the Iberian territorial authorities. In 1494, two years after Columbus's voyage of discovery, with papal blessing and through the Treaty of Tordesillas, the world's seas were divided into Spanish and Portuguese areas. Ships were also the media of a new "world picture"—in Ptolemy's just rediscovered geography, Africa was still described as grown together with an unknown southern continent. But starting at the time that, as the story long had it, Prince Henry the Navigator established a maritime academy —a center for avant garde European "practical knowledge"—in 1415 at Cape St. Vincent at Portugal's southwestern end (according to Camões, the "coronet" to Europe's Iberian "head"),[18] Portuguese captains discovered and conquered one African cape after another, decade after decade.

That these capes could only be apprehended through concerted navigational and political action becomes clear in the significant shifting of the definitive cape. Already unseen by Bartolomeu Dias in a heavy storm in 1488 and unknowingly doubled, it appears to have only become accessible to da Gama's methodical expedition after the fact and through astronomical self-localization; where initially this *cabo encuberto* was termed "Cape of Storms," then "King's Cape" and "Cape Brandanus," as a turning point in Western history it was finally christened the "Cape of Good Hope"; and even when Cape Agulhas, the "needle cape," was clearly determined to be the southernmost point of the African mainland, the political significance continued to take precedence over the geographic data.[19]

---

17 Correa 1869, p. 56.

18 Camões 1880, canto III, v. 20, p. 93. For a critique of this traditional account see Domingues 2007, p. 63; Randles 1993, pp. 20–22. Randles shows that what is at work here is a gradually emerging historiographical legend that, starting in the epochs of Damião de Goes and João de Barros' and mediated by the English travel-author Samuel Purchas, remained widespread in Europe into the nineteenth century. Both the decreed keeping secret of valuable practical knowledge and the systematic control Henry exerted over his chroniclers may have contributed to the legend's formation. See also Kohler 2006, p. 139; Meyn 1984, p. 52.

19 See Hamann 1968, p. 328.

Cape or *caput*, Derrida observes, "refers...to the head or the extremity of the extreme, the aim and the end, the ultimate, the last, the final moment or last legs, the eschaton in general."[20] Henry the Navigator, Grand Master of the Military Order of Christ, placed both geostrategic and salvational-theological hopes in mastering the cape and the connected possibility of enveloping unbelievers. In doing so, he disregarded the fate of Guido and Ugolino Vivaldi, brothers who went missing on Africa's western coast in 1291[21]—and whose sensational odyssey had been incorporated by Dante into the figure of Ulisse in the *Divine Comedy,* the hero exhorting his reluctant crew to the African passage to "[c]onsider how your souls were sown: / you were not made to live like brutes or beasts (*Inferno*, 26, 118 f.). Da Gama's men peremptorily invert this rhetoric, lamenting "that they did not choose to die like stupid people who sought death with their own hands" and exclaiming to their captain that he will have to answer to God for their deaths.[22] Whereas Dante's Ulisse suffers actual shipwreck on the Mount of Purgatory, finally landing, because of his sly discourse, in the eight circle of hell, reserved for "simple" fraud, da Gama's oath at Table Mountain is conveyed as divinely inspired anger. In place of pagan *kybernēsis*, attested to by Dante's wily, wandering Ulisse, the captains of the age of discovery possess the divine gift of *gubernatio:* what from a nautical perspective is owed, as during da Gama's heroic voyage of world-discovery, to violently forced cooperation by local Moorish guides[23] appears from a world-historical perspective to be God's guiding intervention. In 1500, Cabral thus writes of the voyage that "we ourselves would not have been able to do [it], had we not had it from His Hand and consequent on His Will."[24] It is just this hand that the major-captain relied on in having his ship's navigational instruments tossed overboard.

The man who recorded the episode at the cape, Gaspar Correa, was Albuquerque's scrivener. He has often been described as Portugal's first real historian, and he was indeed the first Portuguese historian to travel personally to India. But his account of da Gama's voyage is based on notes of the priest João Fig-

20 Derrida 1992b, p. 14.

21 Otherwise than was the case with the Portuguese expeditionary entrepreneurs, the Vivaldis undertook their voyage without the support, resources, and "geopolitical" visions of a sovereign. Although motivated by the obstacles to Mediterranean trade with the Far East (dying out of the crusades, papal ban on trade with the Saracens), their enterprise was not purely commercial in nature: two clerics prepared for missionary work were on board. See Rogers 1955, p. 44; Verlinden 1986, p. 40.

22 Correa 1869, pp. 51, 58.

23 See the first-hand account of da Gama's voyage to India in Giertz 1986, pp. 58–70.

24 Letter of Cabral to the Samudri Raja of Calicut, cited from Subrahmanyam 1997, p. 162.

ueira, who was probably no regular crew-member but merely an accompanying party, consequently excluded form the ship's operations, whose manuscript may in any case only have reached Correa in a fragmentary state.[25] Da Gama's description, as taken up in the chronicle-like *Lendas da India* (probably 1561), correspondingly reveals various inconsistencies, for instance between the time of year, latitude, and day-length of da Gama's putative odyssey toward the South Pole. Likewise, the protocols of fellow-traveler Girolamo Sernigi do not mention any violent storm or suppressed mutiny,[26] nor do the accounts of either "Portuguese Livy" João de Barros or national poet Camões: all praise the brave, obedient crew under their "valiant Capitayne."[27] And last but not least, for Correa it was not so much da Gama's actual bravura feat, his direct voyage from Malindi to Calicut, that was legendary; rather, in his text the legend of indomitable command in the name of God and king appear as a fact of national history. In the process the figure of da Gama blends with that of the Portuguese explorer Bartolomeu Dias—who in early 1488, following circumnavigation of the southern cape, had to turn around at his crew's wish—at that point, after all, the captain, the officers, the pilots, and the crew together still formed a ship's decision-making body.

Although Columbus was the captain of an entire fleet, he was also simply first among equals on board his Late Medieval carrack, possessing no forcing lever to use against insubordination and thus fearing nothing more than a joint decision to turn back.[28] It appears that at one point there was a plan for mutiny that Columbus only foiled by referring to increasing providential signs.[29] It was only after his successful voyage that Columbus gained the status of a blessed and sovereign ruler's doppelganger: in 1493 he was granted the discretionary power of viceroy and governor of Castille and Léon. To be sure, these offices had not existed—they rather designated a special legal status better tried out on sea than on land, perhaps institutionalized in the New World but certainly not in the Old.[30] And just as in the New World Columbus, now "Viceroy and perpetual Governor of the mainland," would punish any statutory violation with all severity as an attack on his sovereignty, da Gama, following his successful testing as a discoverer-captain, led a reign of terror on the Indian Ocean and in India itself—a reign marked by draconian punishment and retaliatory mea-

25 See Huemmerich 1977, pp. 119 f.
26 See Ravenstein 1963; Hamann 1968, p. 365.
27 Camões, *Os Lusíadas*, canto I, § 44, p. 19.
28 On "small-scale democracy" on Columbus's ship see Gay 1996, pp. 226–228.
29 See Columbus, 1992, p. 122; Luzzana 2004, pp. 66 f.
30 See Siegert 1994, p. 311.

sures. After being named, following Columbus's model, a viceroy and Admiral of the Indian Ocean, he would dispose over a similar ceremonial apparatus and initiated open competition with his counterpart in discovery.[31] With the successful expeditions of Columbus and da Gama, the age of discovery had begun in Spain and Portugal. This meant that on all the world's seas and oceans, the fiction of a sovereign captain now had the normative power of factuality.

A good decade after Correa's report appeared, Jean Bodin's *Six livres de la République* was published. Although this text is anything but a treatise on maritime law, here the ancient allegory of the ship of state was for the first time taken at its word in a legal context: on ships, which are exposed to elementary forces and the constant danger of mutiny, a state of emergency, we read, is always in force. And the only thing that can keep control over such an on-board emergency is unlimited power over life and death. Importantly, for Bodin the ship is less meant to be interpreted as a state than the state is meant to be understood in terms of a ship. Without a sovereign, it is menaced with going under, "just as a ship without a master is liable to be wrecked by the first wind that blows." For this reason the sovereign is the figure who "ought to have the rudder in his hand to move at his discretion if the ship is not to go down while waiting on the opinion of its passengers."[32] Chronicles such as Correa's thus appear to be nothing short of models of what Bodin, the theorist of state, conceived as sovereignty in the context of terrestrial civil war and states of emergency on the mainland. But before such chronicles, it was the expedition reports, written immediately after the voyages for delivery to the authorities, that were directly aimed at the sovereign patron and written to glorify his regime—already Henry the Navigator had personally supervised the work of his reporter and chronicler.[33] Even log books (with all their recorded mutinies, punishments, and acts of nautical heroism) are always shaped by the fiction of maritime sovereignty. The conception of the sovereign state is thus supported by the despotic on-board regime as prototype—a regime first set up or at least described as such at the instructions of a domestic sovereign. The autocratic ship of state is, then, a circular construct, the sovereign's authority having always had only the sovereign as author.

Against this backdrop, Correa's "Legends of India" need to be understood, on one level, as legendary scenarios of a sovereignty then conceived from Bodin to Hobbes in terms of theory of state. At the beginning, da Gama very much heeds the advice of his crew and sees to each individual's welfare. Like

31 See Giertz 1986, pp. 23 ff., 163 ff.; Subrahmanyam 1997, pp. 241, 304.
32 Bodin 1955, p. 209 (book 6, chap. 5); Bodin 1992 p. 24 (book 1, chap. 8).
33 See Meyn 1984, p. 52.

a shepherd, he grants his flock the "target" of "good grazing ground," or else it is "brought back to the fold."[34] But suddenly, in divine rage he swears himself and his men to the inexorable. They then make a simple calculation: On the one hand, unlike themselves, the captain has no fear of death; on the other hand, he is one and they are many. However, da Gama promptly demonstrates prudence, seizing the moment. Naturally "he was not so valiant as not to have the fear of death like themselves, neither was he so cruel as not to feel grieved at heart at seeing their tears and lamentations."[35] After he has the three representative seamen sign the declaration, contract-negotiation, as etymologically contained in "capitulation," turns into straight-out submission, they and the rest of the crew now seized with deathly fear not only of elementary forces but of the captain-major, before whom they will bow as something self-evident, those in chains now pleading more than ever for mercy. For his part, the absolute ruler now displays *clementia*, granting limited, purely functional participation in the ship's guidance. How could they only hope for mercy from their mainland ruler? Evidently it was no self-evident thing that mutineers were doomed to die.

If we wish, then, to invert Bodin's tropic procedure and read his provisions of state theory as an explanation of on-board authority, then it appears that the crew has no right to resist as long as the investiture of the ship's head was a legitimate process. At most, it is "certainly permissible not to obey him in anything that is against the law of God or nature—to flee, to hide, to evade his blows, to suffer death rather than make any attempt upon his life or honor."[36] Sovereign rule is thus limited by divine or natural law. But rebellion in thought or deed merits nothing less than execution. For the head of that nautical organism meant to have the form of a state is inviolable. In actuality, as a legal concept, factual finding of mutiny developed at the same time as the concept of sovereignty—until then the reference was at most to "mutinous behavior" and variants.[37] The term "mutiny" per se referred to the open refusal, verbal or physical, of legal authority. But there was disagreement concerning the question of whether a plot without concrete implementation, simple refusal to obey a command, or a single person's silent intentions constituted elements of such an offense. Be-

---

34 Foucault 1981, p. 229. On the distinction between the Greek governmental art of *kybernēsis* and the Christian concept of the pastorate, which alongside the purposeful management of the polis promises the flock's salvation, see ibid., pp. 234 ff., passim. On this and the decline of the pastoral concept of governance in the sixteenth century, see Foucault 2007, pp. 303 ff.

35 Correa 1869, p. 61.

36 Bodin 1992, p. 120 (book 2, chap. 5, on tyrannicide).

37 For this and the following, see Guttridge 1992.

yond the question of the mutiny-concept's limits, even the term's meaning was itself controversial. For after all, until well into the sixteenth century various conflicts in this area were resolved amicably: although for example Elizabethan mariners knew themselves to be under martial law, they nevertheless at times rebelled on account of poor provisions, pay, or leadership. But such revolts were by no means put down through immediate, bloody implementation of such law. Rather, they were ended through what Gervase Phillips, referring to a report of 1523, describes as "a mixture of threats, entreaties, and compromise."[38]

Commanders in Danger

At Elizabeth's time, the English fleet consisted mainly of private and privateering vessels. Then, starting in the seventeenth century and in the context of the Stuarts' quest for sovereignty, the fleet was strictly reorganized. Now, civil rights protected on the mainland (at least in times of peace) were cancelled for ships sailing for both the royalty and the East India Company. Ships thus came to be understood as always endangered miniature states; martial law was freed from its actual area of usage, war, and set against maritime common law. The state of emergency, that "temporary excrescence, bred out of the distemper of the state,"[39] thus emerged as a rule and institution on ships. Or to speak with Carl Schmitt, subjected to martial law, the ship became a space "set free for the detailed technical implementation of a military operation."[40] In this place within what has no place, law of war was sundered from both law and war. Upon the paradoxical because extraterritorial terrain of the state's ships, the state of emergency, both a test case for and source of sovereign rule, located neither inside nor outside the legal order but furnishing it with legitimacy through its ever-threatening indeterminacy, had found a fitting stage. Here namely—and most clearly in the conflict-situation of mutiny—the two, heterogeneous yet connected elements of every legal order meet: as Agamben describes the encounter, the normative-legal dimension of *potestas* and the anomic-metalegal dimension of *auctoritas*; the limiting and ordering authority that is force of law together with its authoritative counterpart, the unlimited force that is command over

38 Phillips 2001, p. 320.
39 Sulivan 1784, p. 2.
40 Schmitt 2014, p. 150.

life and death.[41] The former type of authority gains vigor from the latter, the latter legitimation from the former.

Just as on ships the captain's presumed authority only gains an "appearance of originality" through an always possible suspension of the legal order,[42] the seamen embody a rebellious nature, standing as they do for a threat to just that order. The legal elements constituting mutiny thus are grounded in the drama of a core conflict in which defiant mariners face the figure of the captain, personified authority. (Historically mutinies have almost never been directed at the officer class as a whole but rather at individual officers. A compelling argument has been made that, against the presumption of various historians, what usually has been in play here is not so much a class conflict as one over fundamental authority.[43]) The captain's authority stems from the fiction that he possesses force of law that, in disposing over naked life, can generate law. Corresponding to this, on the flip side, is the fiction that mariners are—more than any terrestrial people—mutinous, hostile to authority. On account of its "danger,"[44] this fiction is not only implicit in every seaman's civilian work-contract but in fact constitutes "a generally applicable but informal codex": one stipulating submission to a captain's coercive authority.[45] "There is no justice or injustice on board ship," a mariner observed in 1785 with striking clarity. "There are only two things: Duty and Mutiny."[46]

If there is a historical model for this on-board regime, it is a disciplinary institution of the Roman Empire: the *imperium militiae*, in force in military camps and everywhere "outside the walls" of Rome. But around the turn of the nineteenth century, already on account of the increased discussion of human rights, placing ships in a zone external to the mainland's law and politics was increasingly subject to critique. The nautical reformers considered it a pedagogic challenge to lead mariners in a way ending the conflict between official-technical disciplinary needs and claims to human freedom. Rather than a strict on-board regime forcing seamen together into a nautical machine by way of a rigid chain of command, spontaneous and decentralized cooperation was now a desideratum. For the on-board masquerade, this meant that a "master" in gentleman's costume had to replace the godlike "commander." For paramount focus was now to be not on regimenting the stubbornness of mariners, but rather on

41 Agamben 2005a, pp. 39; 86 ff., passim.
42 Ibid., p. 85.
43 See Cabantous 1984, p. 141.
44 Maier 1932, p. 3.
45 Witt 2001a, p. 268.
46 Cited from Weibust 1969, p. 372.

finding the ideal regulatory principle of on-board life between their practical reason and its disciplining.

The ship's master now had to be concerned not only with supervising subjects in need but also with maintaining proper communal life on board. He was meant to regulate both the ship's human and technical operations and find a compromise between the two sides. Already in 1731, the Admiralty Regulations had prescribed conscientious and regular delivery of an on-board journal, its purpose being "to note therein all occurrences...mentioning therein the condition of the ship, men, stores and provisions."[47] By the century's end captains had become supervisory authorities who increasingly operated through writing. They no longer mainly registered nautical data but above all produced minutes of behavior then centrally assessed as conduct-books to be used for future nautical activity. If the mariners were subsequently instructed in appropriate concern for themselves or even in keeping their own diaries, which could then be also assessed as needed, then observation tripled—observation of the observation of self-observation—was possible: an absurd idea under previous conditions. Health care at sea developed similarly. Long before the implementation of epochal medical reforms, for example introduction of James Lind's anti-scurvy menu, mariners underwent professional medical observation, so that regulated "co-operation between the surgeon and other officers" rendered the ship into an experimental laboratory for nautical medicine.[48]

By mid-century, with the start in reforming the on-board regime, corresponding efforts began to govern the crew "as regular as any private family." Expectations developed that a captain would show "a parental feeling for those whom he commands," since he was "the guardian of their lives, their happiness and their comfort," as Commander John Davie wrote in his *Observations and Instructions* of 1804. "His seamen become fearless of danger; they dread nought but the anger of their commander." Admittedly, the commander, as a manager of danger who was himself feared, had to also see to the crew's "internal management," as "necessary as self-government to sea officers." Previously personified irrationality, mariners now appeared to be "a thinking set of people" whose stubborn but not unjustified "pursuit of happiness" would have to be "beyond the reach of... authority" for a commander, but whose common sense and "natural character" could be felt out with sufficient observation and then, through indirect steering, made useful for the ship's leadership.[49]

---

47 Admiralty Regulations of 1731, cited from Lavery 1998, p. 14.

48 Ibid., p. 481. On the development of *medicina nautica* on the ships see for example Matschke 1975, pp. 85 f., passim; Blasius 2001, p. 141, passim.

49 Citations from Lavery 1989, pp. 26, 45, 269, 263, 624, 355, 359.

Until the end of the eighteenth century, Lord Nelson served as a figurehead for powerful yet humane command. As was long after praised by maritime historians including Alfred Thayer Mahan, Nelson reshaped a simple "rule of obedience" into a "spirit of obedience," developing a decentralized but efficient system of "participation" accompanied by "liberty of judgment." Nelson solved the basic cybernetic problem of variable size of command—with the very limited communication possibilities around 1800, a particularly acute problem in the battle situation—through an "impetuous spirit for co-operation, addressed to men over whom he had not immediate control," an approach rewarded by an "intelligent and honest will to forward his purposes."[50] This small miracle of a self-regulative, flexible, and nevertheless purposively functioning system is still considered exemplary in our "era of electronic micro-management."[51] In the course of this efficient reform of the on-board regime, the symbolic structure of authority had been fundamentally if quietly altered. No longer were complaints about on-board command to be incriminated as mutiny, but initially to be paid heed to as the symptom of a systemic disturbance or imbalance. And the captain was no longer meant to stand on the quarterdeck, as in an allegory of the sovereign in a state of emergency. Rather, what was now called for was him remaining "as it were behind the curtain, to move the machine in secret and be reserved to be brought forward on great occasions."[52]

This shift in role and function on the part of the figure of the captain corresponded to a changed conception of mariners and the means used to lead them. The former absolute ruler was now warned that "all attempts to drill the seaman into unnecessary military parade will tend to breakdown the whole of his natural character." For the sake of free on-board self-disciplining, protection of this "natural character" sometimes went so far that there was an effort to understand the sense of seamen's rituals that had been considered excess or wildly irrational behavior: "There are a good many arguments for and against the tolerating this custom of shaving and ducking the Johnny Newcombes on their crossing the line," noted a captain when his ship crossed over the Tropic of Cancer. This custom, we read, corresponded to "the Roman Saturnalia, when the slaves were al-

---

50 Mahan 1941, pp. 125 f.

51 Palmer 2005, pp. 316 – 318.

52 Captain Griffiths, *Observations on Some Points of Seamanship* (1824), in Lavery 1998, p. 364. On the problem of mutiny see ibid.: "In recurring to anonymous letters I confess the impression on my mind would be that where the crew, or part of them, come forward to avow these complaints when the inquiry is instituted, generally speaking something is wrong in that ship. I do not mean downright tyranny or oppression but some fundamental error in the management, something that makes the shoe pinch unnecessarily."

lowed to take all kinds of liberties with their masters," for the sake of, once the order of things had been turned on its head in carnival-like fashion, returning to the legitimate state of affairs, now reinforced *ex negativo.*[53] Even admirals recognized a need for reform in respect to the organization and exercise of sea power. Renewal was to begin with the seamen, for "power renders mankind so prone to error," and "the safety, nay the existence of the whole machine depends on sailors."[54]

Good seamen were a guarantee for maritime safety, but also for the continuation and expansion of British sea power. Especially in the period between the Seven Years' War and the American Revolutionary War, such power was lacking —something reformers such as James Hanway underscored starting in the 1750s. Hanway's willful mix of philanthropy and patriotic agitation would furnish the Royal Navy with fresh recruitment in a period of personnel shortages—and real recruitment, not only scattered reluctant individuals from England's underclass, previously forced onto ships through impressment. To work against that practice while also providing for British youth, Hanway founded the Marine Society in order to take up, outfit, and train volunteers, mainly from that underclass. "If the children of the poor are sent to sea before their *constitutions* and *turn of mind* are formed," promised Hanway in 1759, "they will be habituated to a sea-life, and the duties of a ship will become much less perilous as well as less toilsome to them, than it is to any others."[55]

But such measures were urgently called for not only from the perspective of navy specialists and pedagogues but also from that of geopolitics. "Increase alone can make our *natural* Strength in Men correspond with our *artificial* Power in Riches," wrote Hanway in 1762, "and both with the Grandeur and Extent of the *British Empire.*"[56] Hence were the country to be freed from the burden of its poor, impressment ended on the coasts, and heightened economic and safety standards to prevail on sea, this realization of reform plans for the navy would represent an optimization of the entire British life sphere between land and sea: "the tendency of this plan will maritimize the inhabitants of this island," is the way Hanway put it.[57] In order to undertake this "maritimization" in a fundamental way, his "Proposal for County Naval Free Schools to Be Built on Waste Lands" (1783) envisioned transforming older military seafaring aca-

53 Ibid., p. 366; diary of Robert Clarke, *HMS Swiftsure*, 15 Jan. 1815, in ibid., p. 371.
54 Admiral Philip Patton, Strictures on Naval Discipline and the Conduct of a Ship of War, Intended to Produce a Uniformity of Opinion among Sea Officers (1804), in ibid., pp. 624, 634.
55 Hanway 1759, p. 11.
56 Hanway 1762, p. 26.
57 Hanway 1783, p. 108.

demies into pedagogic institution, together with furnishing ordinary future seamen, when possible even children, with both practical training and character formation. Although this plan was not realized in all its details (for example with dummy ships on land and accompanying agricultural training), starting with purchase of a training ship by the Marine Society, it nevertheless led to a tradition of sailing-school ships—in the merchant area as well—that continues today.[58]

Under such a sign, ships' captains were the agents not only of the old institutional power-structures but also of a comprehensive project of reform. In that capacity they had to repeatedly struggle with problems of authority—in the end they were no longer peremptorily accorded putatively primal force of law on the basis of absolute authority. The probably most famous of all mutinies, that on the *Bounty* in 1789, had its source in just this difficult shift of roles—and not in horrific flogging by a sadistic captain. As a helmsman, William Bligh had contributed in an essential way to the success of Thomas Cook's third voyage around the world, upon which various organizational and pedagogic reforms were tried out. Likely because his services were underappreciated, so that he was overlooked in promotions, Bligh displayed zeal for reform when commissioned by the president of the Royal Society with the transport of breadfruit from Haiti. He wished, absolutely, to undertake a perfect voyage according to the latest standards, and this despite adverse circumstances: the *Bounty* was classified as a cutter, thus having neither a regular captain nor a complete hierarchy of responsibilities.

Where expedition ships normally possessed a separate captain's cabin, with the quarterdeck serving as the stage for nothing less than representational theater, this ceremonial typology was inoperative on the *Bounty:* Bligh's cabin and the deck were both under the control of the botanical enterprise and resembled a plant-nursery. The voyage itself had been declared a botanical expedition, for which reason there were no navy men on board. Bligh's status was half that of a governmental official, half tied to trading law. His commission thus stemmed from a gray zone between command and commerce.[59] The crew thus promptly interpreted Bligh's reform-laden commands, prescribing for instance a specific diet and forbidding a number of mariners' rituals such as the equatorial baptism, as an attempt at self-enrichment by adapting a steward's function and interfering in its legitimate interior hierarchy. For this reason it quickly formed a closed front against the reformist commander, the subsequent floggings appearing—even if relatively rarely carried out compared to other voyag-

58 See Taylor 1985, pp. 179, 185.

59 See Alexander 2004, pp. 69–72, 95; Dening 2002, pp. 19–28.

es—as naked violence, precisely in the light of the prescribed diminishing of Bligh's authority.

We could say that Bligh failed to recognize the symbolic premises for authority, in that he destroyed their fictive and ritual basis through his patriarchal and provident role, while at the next opportunity having mariners that had been degraded into children clearly feel the naked violence of command. He consistently formulated his commands in "bad language," namely with consciously ambiguity and spiked with a moralistic undertone and personal attacks. At the same time, he interpreted any smallest gesture by the crew that failed to fit into his choreography of perfect command as a deliberate attack on his position. As Bligh himself put on record in the subsequent investigation, his orders ended up not simply being defied but denied through assertions they had never been issued; he thus eventually had them vouched for in written form.[60] But for the reformist commander, the "secrecy of this mutiny" remained "beyond all conception." In his detailed and suggestive report, he acknowledged no "grievances, either real or imaginary."[61] But the moment in which his barge was launched likely made manifest a basic decision: those who came with him were let off, those who remained on ship consigned to a death sentence. While the events were examined in a court martial in Portsmouth, Bligh had already taken up a new command. Otherwise than was the case with posterity, which would derive the myth of the *Bounty*'s mutiny from writings by the boatswain's mate and the brother of master's mate Fletcher Christian, who had led the mutiny and seized control of the ship on 28 April 1789, the authorities relied on Bligh's report in passing judgment on all the mutineers, those brought back and those still absent. In putative atonement for unforgivable lèse majesté, those condemned ceremoniously confessed their great guilt immediately before being executed. But this staging seems not to have had the desired effect. For the audience included a majority of the seamen who would organize two great mutinies at home a few years later—the mutinies of Spithead and Nore.

These two events of 1797 led to a state of emergency in England itself, the country being in a state of maritime war with revolutionary France. Following nearly 80 percent of all English sailors being called up for service, and volunteers being furnished with inadequate provisions, their pay sometimes withheld, dissatisfaction and indiscipline spread among the seamen in Spithead, seat of the Royal Navy. At first, the admiralty opted for simply ignoring the complaints. But when various ships were detained in a coordinated way, specific commands

60 Ibid., pp. 59 – 61, 145; Alexander 2004, pp. 128 f., 138, 141 f., 172.
61 Bligh 2001, p. 12.

disregarded despite the threat of French invasion, and concrete demands made, sometimes with reference to Thomas Paine's *Rights of Man* (higher pay, better provisions and medical care, pensions, vacation, removal of disliked officers, royal amnesty), recognition of the situation's seriousness set in. Lord Howe, the fleet's former chief commander, succeeded in at least partly meeting the demands, in the spirit of both a provident patriarch and ambitious reformer. He returned the mariners to their patriotic duty, thus remobilizing the fleet.

Meanwhile, however, a new, better organized mutiny was brewing at the Thames estuary known as the Nore, publicly referred to as the "floating republic." Its "president," the carpenter Richard Parker, had established discipline among the mutineers. But he soon allowed them to terrorize those living on the coast, submitted exaggerated demands to the admiralty, shot at navy ships, and finally threatened to paralyze British maritime trade by blockading the estuary. In this case the officials excluded amnesty, increasingly isolated the mutinous ships, and finally arrested Parker and his four hundred-odd followers. To help prevent future episodes of this sort, the admiralty now undertook an exhaustive study of what had occurred—it would eventually take up thirty-two thick volumes.[62] The official interest manifest here in the sources and dynamic of nautical revolts would then foster nineteenth-century trials, including those where no mutiny may have been present but where its mere possibility was met with draconian measures.

One example is the trial in 1842 of the author and reformist captain Alexander Mackenzie, focused on a putative mutiny on board the *Somers*, an American training ship—he had ordered the execution of three sailors in training, citing mutiny as the grounds. During the trial, Mackenzie's lawyer argued that his client had observed "looks and gestures" empowering him to whatever measure he deemed suitable. The judge agreed—simple intent to disobey legitimate authority fulfilled the legal definition of mutiny. In another mutiny-focused trial, held in 1848, Captain Joseph Smooth pointed to the legal indeterminacy of the concept, in his view furnishing the ship's leader with the privilege of simply legally determining mutiny as he wished. Jurists such as William W. Winthrop insisted on demonstration of a clearly motivation to engage in mutiny, since protests, insults, and disobeying commands were separate offenses. But initially American legislation followed jurist Andrew Allan Harwood, who simply dispensed with the criterion of conspiracy—a mutiny, he maintained, could begin and even end with a single seaman.[63] This was doubted by British jurists—for them muti-

62 On the Spithead and Nore mutinies see Guttridge 1992, pp. 44 ff.
63 Ibid., pp. 176, 184 f.

nies remained a collective phenomenon. But they themselves disagreed when it came to clear signs for establishing mutiny as a factual finding. Fully in the style of the *Somers* trial, William Hickman, for example, referred to "a murmuring, or muttering even, against the exercise of authority, or about real or fancied grievances."[64]

Because addressing the legal problem of mutiny repeatedly involved interpreting signs and signals, conceptions and motivation, it would seem in a sense predestined to be fictionally dealt with. But in the case of prominent mutinies, the fictional substrate of authority and maritime rule is so exhaustively illuminated through official investigations, detailed personal reports and vivid memoirs, and historiographical studies that "fictional descriptions" are mostly superfluous. For this reason, in eighteenth and nineteenth century sea-novels mutiny usually remains a conventionally structured or quaint episode, with a scattering of sustained narrative treatments such as Melville's "Billy Budd" and Jack London's *Mutiny of the Elsinore* (both in fact first published in the early twentieth century) representing exceptions. But perhaps the texts of Melville and London attest to the importance precise causal analysis of mutinies had taken on for modernization of managing ships at sea. For simple ignorance about the private and working circumstances of mariners contributed to their generally being seen, well into the nineteenth century, as having self-evidently seditious natures. The phantasmatic identification of the maritime sphere with elementary danger as such was only overcome through the study of (in any case rare) legal cases of "mutiny," and of the nautical milieu.

Since that time it would become increasingly clear that mutinies are nearly always an expression of latent dissatisfaction, which could involve not only inadequate working and living conditions but also fundamental shortcomings in the ship's operation: a ship or navigational apparatus in a bad state of repair or maintenance; dangerous maritime routs and maneuvers; insufficient crew, with accompanying impact on safety; and finally deficits of knowledge or competency on the part of the captain, quickly leading to loss of authority.[65] As a group that has to come to terms with the "total social institution" of a ship and its "barracked socialization,"[66] the crew can indeed develop a dangerous self-intensifying dynamic that can reach the point of violent and thus "piratical" takeover of command. But seamen are only very rarely a source of danger. Rather, it is usually seamen who have a sense for a wide range of future prob-

---

64 Hickman 1851, p. 186.

65 See Cabantous 1984, pp. 57 f., 61, 119, 143.

66 See, with Erving Goffman as his starting point, Lisch 1976, esp. pp. 8–14, 51; Popitz 1992, p. 187.

lems representing a danger to the ship's operations—and who put aside their own needs and claims in the event of acute danger. Hence for mariners, one of the few things that can be counted on in states of emergency, say distress at sea or contact with an enemy, is that crews will display an extremely high level of cooperation and solidarity. The captain is thus not only entrusted with formal safeguarding of the system of command. His task is to simultaneously manage the ship and its crew and thus the complexity of dangers stemming from both the sea and among the seamen, threats of an operational nature and those stemming from his own (possibly neglected) observation of dangers. If it actually does come at some point to what might be considered "mutiny," the tried and tested captain is meant to follow three maxims: to immediately remedy justifiably voiced grievances; to uncompromisingly reject exaggerated demands; and above all to make decisions promptly.[67] For such discretionary competence remains the core of the captain's on-board authority, hence the basis for the possibility of successful management of danger.

Paragons and Delusions

As soon as Early Modern states began aspiring to maritime dominance, the figure of the captain took over their ships as in a coup de main—Correas da Gama offers vivid testimony to this development. It went more slowly, however, in the case of merchant vessels: following the specifications of scattered legal clauses and as agents of accumulating trading capital. Since the Middle Ages, seafaring had been regulated according to common law. Navigation was based on collective practical knowledge, maritime trading an activity in which seamen and merchants were often indistinguishable. It was not rare for all those involved to share "an interest" in the ship and cargo, and "good order" on board was guaranteed through a combination of established customs. The core division that would come to stamp commercial activity on sea only formed with the thirteenth-century nautical and bureaucratic revolution—with the introduction of the compass, more seaworthy types of ship, and double bookkeeping. On board the ships, now, were the skippers and seamen, who increasingly had to give up their cooperative participation in the ship's operation in favor of a pure wage-relationship; on land were the merchants and consigners, the only

67 See Guttridge 1992, pp. 73, 288, 295.

parties capable of raising the necessary capital for an undertaking that had become investment-intensive, together, increasingly, with lawgivers.[68]

Emerging now alongside common law were statutes and codifications of nautical law, these fixed by urban and territorial authorities who were mainly interested in protecting the interests of the major merchants and consigners. If the skippers had previously been co-partners, they were now increasingly contractual buyers or else jointly shipowning navigators. Their trading privileges and ship ownership were systematically limited—a process continuing into the eighteenth century. Eventually, they had nothing more than the status of employees—who were in fact largely liable for the company's success. The merchants, it was said, not only underwent great risks through the perils of maritime voyage; rather, the greatest threat came from the seamen, who, so the widespread belief, were not only undisciplined but tried to enrich themselves at every opportunity. In short, because the seamen no longer participated in profit and consequently possible loss, but were now simply wage-earners, they were now seen as themselves a danger—and a danger not normally covered by the now established institutions of risk-coverage and insurance.[69] For this reason, high officials and legislators worked at supplying shipmasters with far-reaching authority.[70]

These individuals now found themselves, however, in a paradoxical situation: degraded by merchants and shipowners into accountable and liable underlings, they at the same time had become the authorities' extended arm—bureaucratically empowered supervisory paragons for inherently mutinous and fraudulent seamen. As an employee "the captain," it is explained in an early nineteenth-century compendium of French maritime law, is liable

> for all, even light oversights that he is responsible for in his office...for example inserting false expenses, damages, etc. into his bill...if his oversight or offence gives cause for legal prosecution, forfeiture of rights, or other losses, as a consequence of the said principles, aside from the liable legal penalties, he must indemnify the parties involved.[71]

Not only was the captain legally obliged to record every last relevant detail of the voyage in his log. He also had to punish sailors against his own will to the extent it was stipulated by maritime and contractual law. Since other than was the case

68 On the transition, catalyzed by both capital and capitalism, from "profit-sailing" to "wage-sailing," see Jackson 1989, pp. 606f., 623f.

69 On the institutionalization of insurance, taking place simultaneously with the separation of work from capital and thus a new determination of role for crew and shipmasters, see Hanses 1983, p. 28.

70 See Welke 1997, esp. pp. 37–52.

71 Schiebe 1840, p. 56f.

with the shipowners and merchants, he could not insure himself against the dangers tied to his activities, the documentary evidence of his log was his only tool to parry claims of liability. Even simply formal accounting errors could ruin him.

Just this secretarial office rendered the captain into the on-board legal authority. On the one hand, this authority was limited through maritime law and the final authority of competent courts to "a kind of domestic authority" and to the sphere of "corrective and disciplinary law." On the other hand, the captain is granted "almost absolute authority on the ship."[72] On grounds of reason of state, on-board discipline on merchant vessels was declared a public matter and rendered both these vessels subject to the same "captain's authority" as warships. As the Prussian justice ministry indicated in 1841, a ship at sea was always in a situation of latent danger, the acuity of which only the captain could judge, and this in the same way as "a military commander in proximity to the enemy."[73] Hence when the territorial authorities were also now moving into the sea by way of commercial maritime activity, this was supported by the mercantile sector and shipowners less from concern for maintaining order than for possible damages and losses. In any case, even civil captains could eventually consider themselves to be, on one level, "masters next God." The paradox that juridifying nautical authority led to its nearly unlimited expansion in turn produced the no less paradoxical fact that now a mere hireling of merchants and shipowners had taken on the status of state representative with official competence and the military title of "captain." As holder of a public office, he did stand under state control, something that drew notice through the authorities' program to professionalize, impart scientific knowledge to, and examine the shipmasters as individuals engaged in administrative activities. But just this demarcation from the collective knowledge of seamen helped him acquire the necessary distinction on board—and, in general, his new social prestige as a guiding paragon of all the seaman's art.

Notwithstanding the continued legal controversy over the extent to which the authority of the captain occupied a realm of public or private law, tacitly he was acknowledged to be acting, as Eitel-Friedrich von Kaufmanns observed in 1931, "in a certain way in the name of the state" and as a "representative of the state's authority," for which reason, as indicated in Rudolf Wagner's manual of maritime law of 1906, "a limitation of the captain's command authority would simultaneously mean an undermining of the state's authority in the swimming

72 Ibid., p. 32, 40.

73 Cited from Welke 1997, p. 97, which furnishes archival information.

territorial areas [sic]."[74] As representatives of large-scale capital and state sovereignty, now to be granted not only civil but also penal competency, ship captains would now be protected by the officials and owners, so that even with clear offences they enjoyed far-reaching immunity before maritime courts. Their authority had emerged from a process of struggle to gain the most profitable organizational management; through opportunistic statutes, it had developed into an unintended and as such nowhere codified fiction of maritime legislation. The fiction is readily manifest with sober scrutiny: on what basis should employees of a private company be authorized to exercise genuine sovereignty of state over a specific area?[75] Nevertheless, various historians tried to "construct"[76] or "speculatively"[77] justify this "power of absolute command."[78] In this respect, the argument repeatedly used well into the twentieth century was nothing other than that present in the first doctrines of sovereignty: the captain's unlimited power was, one jurist indicated in 1926, "no artificial legal institution but a necessity of nature."[79] For ships were always in a state of "threatening danger" so that a permanent state of emergency existed on board. In this way the captain's authority rested on the fiction of not being based on a fiction.

No less than literature and popular mythology, law and politics have contributed their fictions to the captain-figure's appearance, first on the stage of world history, then on the stage of everyday operations at sea. This is very much the case with the seaman's saga: reports, for instance, about a Dutch merchant vessel that cannot move past Africa's southern cape on route to East Indies because of a ceaseless storm, so that the ship's commander threatens and curses its despondent crew, finally swearing that "despite storm and waves, thunder and lightening, devil and God he will travel around the cape, even if he has to sail to Judgment Day." Since then he wanders as a shadow or ghost over the seas, with no prospect of land or rest. Among eighteenth- and nineteenth-century seamen, the ship and captain were talked of as the Flying Dutchman, the appearance of this ghostly deathship always announcing storms and heavy seas, setting passing ships on a false course and "signifying danger and sinking."[80] In the archives of ethnographic saga-recording, what here can be brought togeth-

---

74 Kaufmanns 1931, pp. 63, 68; Wagner 1906, p. 317.

75 See Hanses 1983, p. 103. On the corresponding controversy in Germany, esp. between Rudolf Wagner and Philipp Zorn, see ibid., pp. 102–105.

76 Zorn 1897, p. IX

77 Weber 1907, p. 34.

78 Bernstein 1904, p. 96, , – See Welke 1997, pp. 239 f.

79 Schnobel 1926, p. 7.

80 Golther 1911, pp. 8 f.

er as a coherent primal scene of Early Modern globalization falls apart into endless, only loosely connected fragments. Starting around 1800, these were placed together in a range of literary (and soon also musical) adaptations, which thus became the core of different narrative renditions.[81]

We are offered some hope of illuminating the doctrine of shipmaster's authority by the Grimm-brothers' compendious historical dictionary, which informs us that the saga is "more historical" than the fairytale in that "it is attached to something known and familiar, to a place or name secured by history."[82] The saga, we read elsewhere, constitutes "a belated effort to explain the meanwhile dark meaning" of a name;[83] we might thus understand the saga of the Flying Duchman as inviting a typonymic analysis of what as the "Cape of Good Hope" offers a staging of the hopeless because eternal return of the Flying Dutchman. Already for the first Portuguese chroniclers of the India Run, this southern cape marked a turning point in world history, as the ultimate step in the journey taken by the Occident, itself only a cape of Asia, in its globalizing and universalizing mission.[84] This mission was proclaimed in an entire series of voyages around the cape. "Doing the cape," *Fazer cabo*, is the Portuguese term for the process engaged in by Henry the Navigator and his successors over decades: the carefully planned exploration of Africa's western coast meant "doing" one cape after the next until Bartolomeu Dias, in a secret mission for King John II carried out in 1486, succeeded in doubling the ultimate cape. When in 1500 on the Cabral expedition, Dias was killed in a hurricane before just that tip of land he had been the first to pass by, the duel between the

---

81 On the fragments and variations at work in the saga's transmission, see the anonymous 1821 *Blackwood's Magazine* article "Vanderdecken's Message Home," pp. 127–131; the anonymous 1841 article in *Das Ausland,* "Über den Ursprung der Sage," p. 945; Smidt 1825, p. 95 f.; Marryat 2000, pp. 14 f., Nork 1848, p. 942; Smidt 1849, pp. 5–7, 13, 15, 131; Bassett 1885, pp. 346 f., 362, 375–377; Engert 1927, pp. 27, 30; Wossidlo 1951, pp. 285–287; Woeller 1968, pp. 294 f., 300, 309, 311; Gerndt 1971, p. 15, 20 f., 111, 116, 128, 164, 211, Lamont-Brown 1972, p. 20, Eidelmann 1985, p. 253 ff., Beck 1985, pp. 391 f.; Barth 1994, pp. 311–313; Guillemot 1995, pp. 151–176; Frank 1995, pp. 78, 84, 105, 124; Flohr 2002, p. 105. For a more detailed discussion of this figure as the flip side of modern "practical knowledge" and capitalist globalization, hence as a demonstration of England's maritime, commercial, and publicist dominance over Portugal and the Netherlands by 1800, see Wolf 2010.

82 Grimm 1994, p. 11.

83 Fock 1914, p. 47.

84 Within this logic of universalization, the cape is understood as the projecting tip of *telos* and *archē* at once. It forms the starting point of a process of self-exceeding and embodies an "anticipation" that is simultaneously messianic, strategic, and geographic. See Derrida 1992b, p. 18. See also Hegel's description of the activities of spirit: "it...grasps itself in its self-dependence, submitting the world to its thought by creating it out of the Notion." Hegel 1978, p. 31.

cape and its conquerors became predestined to be the stuff of sagas. As tradition has it, a comet appeared at Dias's Atlantic crossing, together with "a huge black pillar of clouds" at the southern cape.[85] In *The Lusiads,* the titan Adamastor, emerging from a cloud as "the untamable" or "unconquerable," will personify watch over the overtaken world picture at the moment of its speculative—in the double sense of exploratively and commercially motivated—transgression.[86]

In Gaspar Correa's report on da Gama's heroic rounding of the cape, various details are simply taken over from Dias's travel report, while inversely on da Gama's real voyage neither the great storm nor the mutiny meant to have been sparked by it actually took place. Hence in respect to the cape's rounding, Correa, who probably made Camões' acquaintance in India,[87] always was engaged in saga-like historiography. But when his da Gama, during the suppression of the alleged mutiny (in Dias's text the reference is still to a legitimate majority vote), tosses all the charts and navigational instruments overboard, proclaiming "I do not require master nor pilot...because God alone is the master and pilot,"[88] then in the saga of the Flying Dutchman the double oath of this slave-driver becomes a curse, the dispensing with equipment a confused and disoriented voyage. (Serving as a possible technical backdrop to this transcendental homelessness may be the fact that—as Sophus Ruge observed in his book on da Gama in 1898—after passing the equator, the polar star, for a long time the most important orientation-point on high-seas voyages, itself "sinks away."[89]) To this we may add that on Portuguese voyages to the East Indies, by royal decree any excursion on land was strictly forbidden—a fact that in the saga may have become endless voyage.[90] And when, by contrast with the Portuguese, the Dutch—soon to be masters of the cape—decided on the organizational form of a trading company, the unrelenting nautical,[91] colonialist, and commercial competition[92] to

85 Lichtenstein 1810, p. 399.
86 See Pierce 1972, pp. 207–215.
87 See Hart 1962, p. 147; Freitas 1963, p. 140.
88 See Correa 1869, p. 62.
89 See Ruge 1898, p. 32.
90 See Flohr 2002, p. 54. Likewise, the fact that the Dutch denied access to their port of call at the cape to even leaking ships that did not belong to the East India Company (see Billecocq 1996, p. 19) corresponds to the merciless ban on landing imposed on the Flying Dutchman.
91 Especially the *Itinerario* of Jan Huygen van Linschotens (1596) shows the extent to which the ascent of Dutch seafaring profited from the Portuguese nautical knowledge. For Linschoten's illuminating observations concerning Portuguese deficits in for example navigation see Linschoten 1974, pp. 147 f., 169 f.
92 The guiding maxims of competition, displacement, and acceleration forming the basis for the rapid rise of this first global capitalism—and more concretely, the then popular Dutch

which the seamen of the Dutch East India Company were subject was manifest as—to use Hegel's term—the "bad infinity" of modern global capitalism.[93] This then ended up as the endless wandering of the Flying Dutchman within the propagandistic perspective of the most fierce competitor with the Low Countries, England. Until 1800 that country would rule the cape and, soon, the world; the saga's initial and most important documentation is consequently English. We thus see that until at least the early nineteenth century, maritime rule rested upon not only legal and substantial arguments but on elements of saga and other fiction—on, as English seamen showed in a framework of practical knowledge, regular phantasms.[94] The figure of the Flying Dutchman in fact only vanished from the world's seas and oceans with the "decapitation" of the cape from the geopolitical terrestrial and maritime map accompanying the opening of the Suez Canal; and with the triumph of the steamship, rendering unnatural movement in the midst of elementary forces, storm and slack, swell and flux, into the unremarkable operation of maritime commerce.

To the repeated justification of the captain's office in terms of on-ship authority's "natural necessity," the evident response is that while the captain's power indeed attests to a certain necessity, it is rather the necessity of fiction. Regardless of whether the ship represents a miniature state or a private enterprise, it is consistently a test case for the fictions required by the leadership of human beings and things. If Latin *fingere* generally means creation on the basis of an image, the captain's figure is based first and foremost on the vision or "world picture" of ruling the state-free space of the sea by means of territorially established legal concepts. As the Roman rhetorician Quintilian observed, the rhetorical tool of *fictio* allows a semblance of truth where (as is vividly the case at sea) insufficient legal or political grounds are present. And against that backdrop, as "shape or form" a *figura* attests to the realization or becoming

---

motto "trading must be free, everywhere, all the way to hell" (see Braun 1963, p. 81)—are unmistakably embedded in the Flying Dutchman saga. See in this respect Gerndt 1971, p. 25 for the saga's variants offered by Baron de Raigersfeld in 1830. On the "homelessness" and "rootlessness" of none other than the seventeenth-century Dutch capitalists and speculators, whom enemies traded with and even hostile privateers financed, see Barbour 1950, p. 130 f.

93 Chaunu 1959, p. 34 describes the *Carreira* as a *première économie-monde*, one that would only become organized along strictly capitalist lines with the Dutch and English maritime enterprises. On the Hegelian figure of "bad infinity" and its interpretation after 1800, see Gutzkow 1842, p. 155; Körte 1995, pp. 180 – 188.

94 For an explanation of the sailors' saga as a "phantasm" in the optical sense as well—an optical phantasm emerging repeatedly under the cape region's thermic and stream conditions—see the observations offered by the English captain William Owen on his exploratory voyage of 1824; Owen 1833, I., p. 242.

factual of *fictio*.[95] In addition, as was the case in Columbus's time concrete prophecy can be tied to a *figura*, the eschatological expectation of a salvational event. It was just in this sense that Portugal understood the rounding of Africa's southern cape as the aforesaid turning point in Occidental history—or at least as a sign from a providence whose most important tool would be captains of da Gama's mettle.

The relation between a *figura* and its maker, its *factor*, becomes more complex as soon as it no longer derives from the Creator but presents itself as factual or real: on the one hand, a figure such as the captain already betrays its artificiality and simple "as-if" reality through the fact that it stages itself as an absent higher (or at least territorial) authority. On the other hand, the office becomes its second nature and an "as-reality" as soon as it shows real impact in leading a ship, for example when, as signaled by the officer class, "the internal mutual process of recognition becomes visible, observable to others" (this is Heinrich Popitz's way of putting it in his book on "phenomena of power").[96] In addition, the repeatedly circulated "primal scenes" of captain's rule not only offer reports on the evidence for this role-centered fiction, but also manufactures it. In this process, just those marvelous chronicles such as Correa's and legend-filled reports such as Bligh's, whose peculiar portraits of the captain's figure would have unpredictable results for practice of the office, are especially close to strictly fictional or literary narrative texts. In the end they construct a story (to return to classical rhetoric: the *fictio*), invent events (the *fictum*), and imagine certain situations (the *suppositio*).[97]

We could also say that the history of the captain's forceful authority is a history of specific acts of fiction: of declarations and speech acts through which the speaker influences reality by means of his power.[98] This concerns, in general, the "legal fiction" introduced to adjudication through counterfactual assumptions[99] —but especially political-legal doctrine such as Bodin's. Here the traditional allegory of the ship of state was interpreted realistically, so that the sovereignty-grounding scenario can appear plausible in the first place. And in order to summon up and preserve the power of the legal order, an authority appeared on the scene that not only corresponds to an absolute ruler in the framework of scenarios and arguments. The very fact that a conception like Bodin's or report like Correa's could even be perceived as an act of fiction guaranteed the absolute ruler's

95 On this conceptual history of "fiction." see Stierle 2010, pp. 381 f., 388.

96 Popitz 1992, p. 199.

97 On this narratological distinction see Fludernik 2008, p. 74.

98 See in this respect Genette 1993, pp. 40 – 42.

99 For a study of the legal fiction of the "pirate," see Heller-Roazen 2009, esp. pp. 165 f.

authorizing agency. Sovereignty is thus a fiction whose author is signed up for by the sovereign himself.

In this context, the fictive substrate of the captain-figure thus comprises numerous elements of corporative and hylomorphic metaphorology. That the crew-head is accorded an originary force of law is all the more evident the more the seamen personify a form of anarchic, mutinous, and dangerous nature: like the sea itself, they are held to be an elementary enemy of all political and legal order. The captain may sometimes appear on the stage as an exponent of patriarchal ministration, sometimes as a pedagogic reformer. But the core of his position and vocation remains an absolute authority creating socio-legal order in face of the threatening, shapeless sea and seamen. When he confronts mutiny, peril, misfortune as in a state of emergency par excellence, a situation in which reality and imagination, latency and actuality in a certain sense become indistinguishable, he preserves his sovereignty precisely in passing existential judgment on such shapeless phenomena. To this extent the captain is himself a maker or creator, a *fictor*, capable of transforming fiction into fact and vice versa. Finally, as civilian merchant-ship captain, he emerges as the agent, indeed a guiding figure of modernity's *fictor* par excellence: of capitalism.

Navigation and Charisma in *Moby-Dick*

The captain is one of the modern sea-novel's guiding figures. In countless eighteenth- and nineteenth-century literary sea-stories, his masquerade served as a set piece through which often scarcely knowledgeable authors tried to fill out their insight into a milieu presented as without a history. At the same time, however, authors who were both sea-tested and legally informed such as Dana went beyond the characters and scenarios modern on-board theater was itself staging. Instead, they explored the fictive substrate of what was representing itself as legitimate and a necessity of nature, something feeding itself, finally, on legal casuistry centered on danger-prevention. Herman Melville—who would enlist himself on an English packet boat in 1839, then voyaged on a whaling boat he deserted in the South Seas, participated in a mutiny on the *Lucy Ann*, and returned home on an American war ship—would repeatedly try out this sort of analytics of forceful on-board authority. As already *White-Jacket* shows, with its corporeal abuse and especially flogging, "sovereignty at sea" is only emphasizing its display side. But it simultaneously is operating by means of a rigid linguistic regime forbidding any "reproachful words" by the crew, whatever their sort, while also approving the commander's use of "bad language" as a disciplinary instrument.[100] For Melville's analysis of discipline, the commander's role initially amounts to theatrical arrangement but finally represents a problem of linguistic performance.

In the story "Benito Cereno" (1855; based on a real episode reported by captain Asamo Delano), we are presented with a Spanish merchant ship's phantom command, whose real nature is gradually discovered by an American sea-hunter. The ship, he is led to believe, suffered distress at sea, losing most of its Spanish crew in the process, while the black slaves continue to follow orders of the captain, Benito Cereno. But his disturbed appearance, impolite behavior, the disorder on board, and the manifestly undisciplined crew "chained to one dull round of command" prompt a suspicion confirmed by the captain's sudden flight: in a mutiny, the slaves seized control of the ship and, following the seal-catcher's appearance, forced Don Benito to simulate command under threat of immediate murder. The words "Follow your leaders," chalked onto the ship's bow are retroactively readable as the motto of a slave regime, while the reality of any command is revealed in the mutineers' figurehead: where previously Columbus stood, they have erected the prepared skeleton of their murdered slavemaster, covering it in makeshift fashion when the seal-catcher arrives on the scene.[101] Absolute authority, mortal threat, and discipline are only the items of role-play that slaves can engage in like masters—but whose termination is reserved for the intuition of an American seal-catcher.

In *Etchings of a Whaling Cruise* (1846), J. Ross Browne lambasted the sadistic tyranny of a whaling ship's captain and slave-like working conditions on board, something reviewer Melville denigrated as reflecting a personal grudge.[102] In *Moby-Dick* (1851), the whaling boat *Pequod* is allegorized, to the contrary, as a Noah's Ark for the New World's heterogeneous populace, a heterotope counterpart to the Promised Land, where the seamen, instead of being constantly degraded by a commercial authoritarian regime, can hope for a primal democratic sharing of slaughtered prey. Here, instead of being subjugated from the start to the fiction of an absolute ruler, fraternal unity built around a primal hunting instinct is possible. The big Other may here once again appear as Leviathan, and also again have an agent and deputy—in shipmaster Ahab, through his whalebone prosthesis metonymically joined with his quarry Moby Dick. But for one thing the whale explodes the choreography and symbolic constitution of regulated representation of rule, both through an appearance that is downright noble

100 Melville 2002, p. 300.
101 Melville 1997, pp. 58, 55.
102 See Pechmann 2003, pp. 145 f.

and his repeated deep diving. For another thing, the giant mammal gains his power through the very reality of being hunted: as the collectively and infinitely coveted "thing," he becomes the mobilizing agent of communal formation.

The vanishing point of all that transpires in the novel is "the whale," no reality in the sense of something locatable and namable object but rather a borderline entity rendering external objects and internal representations indistinguishable.[103] As a "dead, blind wall," mere "whiteness," the whale stands as a phantasmatic screen before the non-symbolizable real.[104] As a "thing," it does not have a locus where word and object are connected. It is, in Lacan's words, a "beyond-of-the-signified" that is not nothing on the level of perception but literally is not.[105] For this reason the *Pequod*'s voyage is not already mortally threatened at the start but only at its end—that threat not emanating from the land but from the dark ground of the *cosa nostra* of collective interest. The crew frees itself from state and sovereign, from the "Leviathan" of the land by becoming the "Fast-Fish" of the whale.[106] In this connection with an elementary sphere, however, it abandons itself to that person most closely interconnected with the whale, to Ahab (this already through his prosthesis). Only Ahab can bind human beings to himself while at the same time rendering the whale into prey—and rendering himself into prey of the whale.[107]

Not a regime of sovereign nautical authority commissioned from the land, but a deterritorialized hunting instinct thus embodies shipmaster Ahab. Although his orders suffer no contradiction, they are in fact only a means to hunt the whale. His workings are not based on any legal principles but rather exemplify charismatic rule.[108] The *Pequod* sails through regions "where God does not reign" and where every "Marchant service be damned." It is only in the course of a masquerade centered on the White Whale that Ahab lays claim to a kind of primal kingdom, declaring himself "Khan of the plank, and a king of the sea, and a great lord of Leviathans". And it is only as such a double of Moby Dick that he embodies, simply, all of humanity: "Ye are not other men,

103 See Williams 2006, p. 76.
104 Melville 1997, pp. 58, 55.
105 Lacan 1992, p. 54.
106 See Ishmael's definition of two legal principles, both stemming from whaling yet universal: "I. A Fast-Fish belongs to the party fast to it. II. A Loose-Fish is fair game for anybody who can soonest catch it." Whether dead or alive, a fish is "technically fast, when it is connected with an occupied ship or boat, by any medium at all controllable by the occupant or occupants." Hence each and every "recognized symbol of possession" is derived from this factual rather than sovereign control. Melville 2000, pp. 395f.
107 Ibid., p. 568.
108 See Vogl 2007, pp. 121–125; Werber 2011, pp. 56–60.

but my arms and my legs; and so obey me."[109] What in *White-Jacket* was still the grotesque pretension of a despotic captain—that of being the absolute master of even (technically measured) time and space—[110] has become a gesture of intuitive navigation for Ahab: "fully acquainted with the ways of the leviathans," he dispenses with logline and quadrant, and nevertheless, armed only with compass and pocket-watch, space and time are fully at hand.[111] Without a rudder but only with a tiller cut from a whale's jawbone, Ahab's ship plies the water thanks to dead reckoning. And just as the *Pequod* scorns the royal science of navigation, thus undermining the authoritarian maxim of being a clearly defined place within the placeless, here the masquerade of sovereign authority and its commanded language acts end up without function.

All told, *Moby-Dick* describes an endlessly smooth surface without corners or edges, which as such opens up an indefinite horizon of possibilities. Starting in the Baroque period, through the labyrinth, Early Modern narration was furnished with the topography of a systematic, increasingly multi-cursal confusion; although increasingly incapable of being traversed with one and the same sense of direction, this topography nonetheless brought to light ramified structures with bifurcations possessing a certain probability of passage.[112] *Moby-Dick*, however, presents us with another narrative model of life and its paths for going astray: that of *navigatio vitae*, the narrative of life's movement, of continuous directional decision-making, not by way of architectonic milestones, but through continuous steering. At stake here is not only a process of finding home and harbor defining existence, fully in the ancient classical sense, as resting within one's proper locus. Rather, here one or another form of navigation must first be chosen. Existential, political, and religious decisions are made less facing bifurcations and more through options for and against a medium.

Ahab's quadrant, with which he tries to take the sun's bearing near the equator for the sake of inscribing the position-data initially in his prosthesis and fi-

---

109 Melville 2000, pp. 43, 71, 129, 568.

110 "Only the moon and stars are beyond his jurisdiction. He is lord and master of the sun. It is not twelve o'clock till he says so. For when the sailing-master, whose duty it is to take the regular observation at noon, touches his hat, and reports twelve o'clock to the officer of the deck…'Twelve o'clock reported, sir,' says the middy. '*Make* it so,' replies the captain. And the bell is struck eight by the messenger-boy, and twelve o'clock it is." Melville 2002, p. 23.

111 Melville 2000, p. 199. See also Röller 2005, pp. 58 f.

112 See Pias 2002, pp. 164–183. For a reading of *Moby-Dick* as a labyrinthine narrative space see Mumford 1963, p. 107: "We must gather our own strength together if we are to penetrate *Moby-Dick:* no other fable, except perhaps Dante's, demands that we open so many doors and turn so many secret keys; for, finally, *Moby-Dick* is a labyrinth, and that labyrinth is a universe."

nally in his sea-chart, confirms nothing more than self-localization. Although the "eye of heaven" beholds Ahab and the White Whale at the same time, it does not allow the quadrant to track down the whale. Cursing, Ahab thus destroys the instrument. From now on he will not look upward into the heavens, turning instead toward the horizon, a horizon of futurity and possibility. As in Melville's novel *Mardi* (1849), which recounts a "chartless voyage tither" with compass and lead alone,[113] from now on the voyage is limited to dead reckoning. In this way Ahab steps out of a cartographic knowledge-structure in order to find Moby Dick where he always was, on the outside. Connected to this crossing, to Ahab's opening to the forces at work on the outside, is a decisive new beginning.[114] For the vectors through which the old topoi tied to the "life's journey" are transformed by an excursion into the "limitless, uncharted seas"[115] are the very vectors first leading experience and narration to what is eventful and thus worthy of both those things. In respect to the theological poetics of knowledge, this is the case for Dante and his Ulisse in the *Inferno*'s twenty-sixth canto. When it comes to experimental science, it is the case at latest for Francis Bacon, with his programmatic assertion that "we are certain of our road, but we are not certain of our standpoint";[116] and according to Lotman it is the case for any narration moving past the bounds of a "map," hence of a subjectless, merely classificatory and typographical text.

However, with the entry into a smooth, deterritorialized space a sea change of the social sphere is also present, a sphere that on ships, this place within what has no place, always has to be initially modeled. The quadrant, according to Ahab only a "plaything of haughty Admirals, and Commodores,"[117] represents the heart of a royal science of navigation, a science continuously attempting to integrate the sea and the ship into a clear and manageable order of things. That being the case, navigation through mere dead reckoning, dispensing as it does with transcendent forms of orientation (such as the angle between sun and horizon), can be understood as thwarting one basic possibility: that of making what is deemed objective knowledge into a central point of reference, for the sake of the on-board commander's purported authority and higher legitimacy. As soon as Ahab's maneuvering is no longer oriented toward the heavens but only relied on logline and compass (chapter 124), he is no longer a sovereign captain.

113 See Melville 1970, p. 556.
114 See Casarino 2002, pp. 14–18.
115 Melville 2000, p. 183.
116 "Certi viæ nostræ sumus, certi sedis nostræ non sumus." Bacon 1826, IX., p. 187.
117 Melville 2000, p. 501.

Rather, he is now a "charismatic" who derives the laws of navigation (and simultaneously of social order) from an implicit knowledge unfathomable for others. This is all the more the case since Ahab now guides himself not through the visible sun but through invisible forces of the sort the compass makes useful.[118]

That his instruments for route-measurement are themselves attacked by the route, that a measure can alter when advancing along the measured path,[119] is made clear in the rotten logline of chapter 125. The line tears asunder, but "Ahab can mend all," the shipmaster pronounces in charismatic tenor.[120] In the case of the compasses in the previous chapter, things were a little more complicated: a heavy night-storm caused the instruments' needles to "turn" or "transpoint"—thus endangering Ahab's authority. For until that point, it was even the case for Ahab's relationship with his helmsman that "Starbuck's body and Starbuck's coerced will were Ahab's, so long as Ahab kept his magnet at Starbuck's brain."[121] It is as if the authority is annulled with the storm and the occasion for tyrannicide arrived—an event Ahab appears to have long courted through his reckless endangerment of ship and crew. Accordingly, Starbuck slinks full of homicidal intent from Ahab's cabin, but only to sink into Hamlet-like hesitation. "The wind has gone down and shifted, sir," he finally reports, "she heads her course."[122] But as becomes clear in the morning, the ship is by no means on course, Ahab, the only *kybernētēs* bound to the elements, recognizing that while the compass declares eastern bound the sun stands aft.

The trepidation that the reversal of compass and ship threatens to prompt among the sailor-folk, the fear of being blinded and seen in this blinding by the high eye of heaven, contains Ahab to the extent that he explains to his near-murderer and helmsman that a lightening strike must have reversed the needles' poles during the night. Although novel-narrator Ishmael may explain the connection between magnetic energy and celestial electricity in an erudite comment, solely Ahab recognizes the "[i]n-flashin of world" into the "injurious neglect" of "enframing" that Heidegger views as furnishing "insight" into Being itself.[123] And entirely in Bacon's sense, he uses this exclusive knowledge for "restitution and the reinvesting (in great part) of man to the sovereignty and power"

118 Parker 1996, p. 698 has suggested William Scoresby's *Journal of a Voyage to the Northern Whale Fishery* as a source for chapter 124 of *Moby-Dick*.

119 See Röller 2005, p. 58, referring to Hermann Weyl and relativity theory.

120 Melville 2000, p. 521.

121 Ibid., p. 212.

122 Ibid., p. 515.

123 Heidegger 1977b, p. 45.

that had at times been lost.[124] The crew now has greater fear of Ahab than of their fate. Beyond that, since steering with the magnet needles' reversed polarity, awkward to start with, is unbearable for the superstitious seafarers, Ahab "revives their spirits"—by producing new needles for them. He beheads a lance and connects this now useless killing instrument with "the smallest of the sail-maker's needles" to form the medium of his rule: "going through some small strange motions with it – whether indispensable to the magnetizing of the steel, or merely intended to augment the awe of the crew, is uncertain," he hangs the needle over the compass's binnacle and then waits until it comes to rest.[125]

After this test, Ahab can declare himself ruler of the lodestone. Comparing the orientation of this magnetic needle with the sun's position has become unnecessary. Rather, the needle shows that the sun is at the right spot. Like Ishmael, we may perceive Ahab's self-proclamation as master of external powers to be "fatal pride."[126] But in rendering the compass whole again and again setting the *Pequod* on the track of the whale, he has restored his authority, vanished the night before. He has proven himself as a charismatic leader in the full Weberian sense: as a therapeutician, magician, hunting leader, and prophet who has renewed his gifts, vocation, and mission, his supernatural powers or, indistinguishable from this, his recognition by those he rules.[127] If the rule specified in *White-Jacket*, "[i]n time of peril, like the needle to the load-stone, obedience, irrespective of rank, generally flies to him who is best fitted to command,"[128] is valid for seafaring in general, then this renewal of charisma is directly manifest in magnetization.

Among seafarers, magnetization of iron with magnetite was considered a magical practice. To that end, the stone was placed upon the center of a compass needle, the needle's northern half then being slowly stroked up to its tip. The stone, the belief went, needed to then be led back to the midpoint in a large arc so as not to alter the magnetization.[129] At time, the instruction was given to proceed as in knife-sharpening and to repeat the process a number of times. From fear of raising suspicions of witchcraft and alliance with the devil, the compass pioneers concealed their magnetizing below deck. But as

---

124 Bacon 1826, III., p. 222.
125 Melville 2000, p. 518.
126 Ibid., p. 519.
127 See Max Weber's famous typology of charisma and "charismatic authority" in Weber 1978, pp. 241 ff.
128 Melville 2002, p. 110.
129 This and following summary based on Meyer-Haßfurther 2005, p. 12.

soon as the needle was granted natural communion and sympathy with the extra-earthly force standing in the heavens as the polar star, the ritual no longer had to be hidden. The occult practice thus eventually turned into a professional secret assuring specialists in navigation of a monopoly while also allowing those who had it to impress and even command the superstitious seamen in a regulated, professional manner. In *Moby-Dick*, the pagan harpooners, as prototypically superstitious individuals, are as easily subject to "a certain magnetism shot into their congenial hearts" as Ahab would have always wished of the entire crew: "It seemed as though, by some nameless, interior volition, he would fain have shocked into them the same fiery emotion accumulated within the Leyden jar of his own magnetic life."[130] As a magnetizer, helmsman Ahab has not only shifted from cartographic to vectoral navigation but from a sovereignty nourished by political institutions or their remainder to social practices that his charisma produces and reproduces in the first place.

Up until Melville's time, the most prominent translation between magnetism's elementary forces and charismatic, potentially state-threatening social practices was brought about by mesmerism: according to mesmerist doctrine the invisible principle or *fluidum* works through the nerves on both the individual's living organism and society. Interruption of this process can cause individual and social illnesses that can be remedied by a healer or magnetizer. By the time of *Moby-Dick*'s writing, the mesmerism vogue had naturally made its way to the United States. In a book by a certain Joseph Deleuze translated from the French and published in Providence in 1837, we read, for example, that magnetization was "an extension of the power which all living beings have, of acting upon those who are submitted to their will."[131] As Gilles Deleuze suggests, this conception of power emerges in Melville when Ahab suspends the disciplinary principles of ship-management, consistently aimed as they are at a "secondary, sensible Nature," instead having "innately depraved beings participate in a terrible supersensible Primary Nature, original and oceanic."[132] Ahab's will, which he forces on his crew, is not the will of a sovereign but a cosmic will, just as for mesmerism the cosmos forms the unity of the transcendental will. As Charles Caldwell wrote in 1842,[133] this can develop into *persona* through certain facial expressions and gestures, or, as is the case with Ahab's magnetizing, be planted into individual heads through a play of eyes and hands. In any case mesmerism

130 Melville 2000, pp. 165, 518.

131 Deleuze 1837, p. 11.

132 Deleuze 2018, p. 56.

133 Caldwell 1842, p. 117.

was seen as demonstrating the human being's essential openness to cosmic and "outside" forces—to, Louis Alphonse Cagagnet explained, "ecstasy, promoted by magnetism."[134] All human beings, indeed all animated and non-animated natures, found themselves, as Melville's admired oceanographer Maury wrote, in a magnetic ocean revealing itself just at the intersection to the maritime ocean, itself the greatest of all magnetizers.[135]

To be sure knowledge concerning this magnetic-oceanic ecstasis likely arrived with Melville and his Ahab not only by way of mesmerists and oceanographers but of philosophy of will as well. According to Schopenhauer, although from a scientific viewpoint magnetism may derive from electricity, "there will always remain over original forces; there will always remain, as an insoluble residuum, a content of the phenomenon which cannot be referred to its form, and which thus cannot be explained from something else in accordance with the principle of sufficient reason."[136] Even "the force that turns the magnet to the North Pole"[137] is no mere manifestation but a thing-in-itself, hence will. In the case of the magnetic needle, magnetism still makes itself manifest as the lowest level of the will's objectification, for instance "when a magnet forces magnetism on iron."[138] But the individual can "accomplish...things which cannot be explained according to the causal nexus, *i.e.*, in the regular course of nature; if we find it in a sense even annulling Nature's laws and actually performing *actio in distans*, consequently manifesting a supernatural, that is, metaphysical, mastery over nature."[139]

Starting in the 1820s in the USA, this magnetism of the will and force of will on the part of magnetism became the core of a reform—or hygienic-salvational—program aimed at corrupt political rhetoric. The program was at work in John Humphrey Noyes' Oneida Commune, for example, and it informed Albert Brisbane's conception, following Charles Fourier, to steer and elevate human passions by means of a "social compass."[140] In that, as Mesmer already presumed, magnetic properties could be concentrated, accumulated, and stored, every transmission thus being an act of will, such unforced magnetism, free of rhetoric and command, could be understood as the political communication technique of the future. "The *will*," wrote Caldwell, "is *alone* sufficient (the subject being at a

134 Cahagnet 1851, p. 3. See also Caldwell 1842, p. 117; Radloff 1996, p. 104.
135 See Maury 1861, p. 132.
136 Schopenhauer 1969, vol. 1, p. 124.
137 Ibid., p. 110.
138 Ibid., p. 148.
139 Schopenhauer 1891, p. 331. See also Schopenhauer, 1969, vol. 1, pp. 110, 118, 130.
140 Fourier 1996, p. 105.

distance and unconscious of the operator's extension)."[141] And in fact, not that much later, in Durkheim's sociology and Mauss's ethnology among other theoretical writing, the "field of the social" would be described on the model of electromagnetic field theory and pre-symbolic dynamics.

The compass, this first nautical medium for hidden elementary forces, consequently had for a long time had the makings of being more than a mere navigational instrument—and this not merely because of the loadstone's "occult qualities," preoccupying an entire school of "magnetical writers" extending from Bacon's first adepts to Melville.[142] It was, of course, on ships, those experimental laboratories for politics, that compasses would become a medium for social experiments. Namely, if according to Heidegger "To flash [*blitzen*], in terms both of its derivation and of what it designates, is 'to glance' [*blicken*]", then the insight gained by what he terms "in-flashing," *Einblitzen*, is not "any discerning examination [*Einsicht*] into what is in being that we conduct for ourselves"; rather, it is "the disclosing coming to pass of the constellation of the turning within the coming to presence of Being itself, and that within the epoch of Enframing [*das Gestell*]."[143] In this respect, Ahab's experiment demonstrates one thing in particular: atmospheric disturbances offer insight not only into terrestrial magnetism but also to the social sphere's existential forms, that sphere's attractions and repulsions. But that in order to be effective, the will does not necessarily need sovereigns and their progeny, and that in order to exist the social sphere does not necessarily need firm political institutions, hence that as a "thing" belonging to the social sphere the will can have many "enframings"—Ahab's mastery of the ship, making prey of human beings through charisma, shows nothing less than that.

Melville's *Billy Budd* and the Dangerous *Fictor*

Melville described figures such as Ahab as a "fiction based on fact," as original characters who, precisely because they are distinguished by "perplexity as to understanding them," hence by inconsistency of character and "obscurity," reveal "nature herself"—thus allowing all those schematisms and fictions to become manifest that are tied to evocation of "nature itself" and purportedly factual necessities tied to it.[144] In any case the elementary power of these figures serves as

141 Caldwell 1842, p. 43. On Mesmer see Radloff 1996, pp. 92f.
142 Boyle 1676, p. 1, 20.
143 Heidegger 1977a, pp. 45f.
144 Melville 1971, p. 58f.

irritation for both the command's disciplinary theater and the regime of pedagogy together with its psychological arsenal. Melville's last narrative text, "Billy Budd, Sailor," which he worked on until his death in 1891, but that remained unpublished until 1924, centers on an angelic character, that of a foundling "peacemaker," who reveals not so much specific features as a double incapacity: that to feel real enmity and that—resulting from a stutter that worsens under stress—to transmit regulated speech-acts.[145] When in the captain's presence he is accused of planning a mutiny, words fail him; rather, with a guileless, instinctive hand movement he strikes dead the calumniator, the master of arms Claggart.

In doing so he commits a special sort of insubordination—one unfolding neither in thought nor as a real accountable deed, but only through his elementary nature, simultaneously undermining both the discourse and practice of command. His verbally incapacity—more specifically, his stuttering at the decisive moment—has rendered him ineligible for discursive legal practice: "Could I have used my tongue I would not have struck him."[146] And that up until his execution he reveals neither enmity nor mortal fear robs the command's linguistic order of its power, based as it is above all on the transcendental force of the threat of death. Elias Canetti has argued that "commands are older than speech," which along with Gilles Deleuze and Felix Guattari can be carried forward to the idea that speaking is simply an endless transmission of commands, and that "[t]he oldest command...is a death sentence."[147] In that light, we can understand speech's faltering at the moment of decision as shattering the chain of command. This faltering discloses an elementary nature that cannot be disciplined or pedagogically steered in the right direction, but from the perspective of sovereignty can only be killed.

The plot unfolds in the summer of 1797, hence against the backdrop of the Spithead and Nore mutinies. Nevertheless Captain Vere does not rely on enactment, through retributive theater, of a draconian command regime, rather shifting the proceedings to his quarters. But because he represents the contradictory type of a master with humane qualities, a commander as gentleman, despite being aware of Billy Budd's innocent nature, he has, it seems, no option but court martial and a death sentence: "Though the ocean, which is inviolate Nature primeval, though this be the element where we move and have our being as sailors, yet as the King's officers lies our duty in a sphere correspondingly natural?,"

---

145 Melville 1997, p. 408. On the incapacity of will and decision and a corresponding poetics, see Agamben 1999, pp. 253 f.

146 Melville 1997, p. 456.

147 Canetti, 1978, pp. 303 ff. See for example Deleuze and Guattari 1980, p. 111.

is Veres' rhetorical—in the word's double sense—question. For that an indissoluble contradiction exists between that elemental realm, treated in a tale situated in "the time before steamships," and the duties imposed on a territorial sovereign render his question both obvious and an anticipation of each and any response.[148] An irritating figure through linguistic-affective failure, Billy Budd forces the humane sovereign to avow his own failure.

This is because the execution is only self-evident on the level of appearance, signaling uprightness and decisionary emergency, and legal casuistry, leaving no doubt as to the execution's necessity. It is only self-evident in the sense of the "natural necessity" repeatedly conjured up at sea. But limiting oneself to such a theatrical-juridical perspective is the material of a naval chronicle, not an imaginative text announcing itself as "An Inside Narrative" in its subtitle. Melville's tale thus initially engages in an analysis of appearance and legal procedure, in order to eventually explore the doctrine of the maritime state of emergency. Billy Budd, we read, is not only externally a "Handsome Sailor," but also is one through (nearly) perfect harmony between "moral nature" and "physical make."[149] Even when impressed from the *Rights-of-Man* merchant vessel into the King's Service and the *Bellipotent* war-ship, he shows no rancor—indeed, to "deal in double meanings and insinuations of any kind was quite foreign to his nature." As a seaman Billy Budd has consistently proven himself "in the hour of elemental uproar or peril," only betraying his "occasional liability to a vocal defect" when "under sudden provocation of strong heart-feeling"[150] Incapable of simulation or dissimulation, indeed of any sort of wickedness, this as it were pre-Adamitic figure is anything but a mutineer. But he is also incapable of articulating himself as a legal subject. And in his guilelessness, also alien to him is the game of concealing and revealing, of calculated discretion and ostentatious uprightness, played by Claggart in his machinations and by Captain Vere, as well, in his conducting of the court martial—for instance when he finally encounters Billy Budd for an intimate statement or even confession.

Also on the level of legal procedure nothing is as clear on board the *Bellipotent* as it at first sight seems. For otherwise than still in the pioneering period of European maritime rule,[151] around 1800 even on war-ships captains were by no

---

148 Melville 1997, pp. 405, 459.

149 Ibid., p. 406.

150 Ibid., pp. 410, 413.

151 See for example the Portuguese *regimento* for the Count of Aveiras, 20 March 1640: "In your flagship during the voyage and in any port, save only that of Goa, you are empowered to try and sentence anybody committing a crime, including those deserving the death penalty, and there will be no appeal or stay of execution against your judgments." Cited from Boxer 1984, p. 43.

means allowed to organize their own courts martial and promptly carry out their own sentences. Against the appearance of scrupulousness, Vere commits an entire series of grave offences against the procedures specified in the *Articles of War* of 1749: he does not hand the serious case of Billy Budd to the Commander of the Fleet and likewise fails to leave the verdict concerning the putative capital crime to the admiral, both deviations from the usual legal path, at most legitimate in the case of a manifest mutiny. Again, contrary to what is specified as required except in cases of factual mutiny, his hastily convened "drumhead court" does not contain five or more officers. Furthermore, otherwise than stipulated in 1749, Vere both stands as commander of a court martial and takes over the role of witness, accuser, and judge. Despite what other jury members suggest, he excludes consideration of mitigating circumstances a priori together with possible appeal of the verdict to the next highest authority—a procedure specified even in cases of mutiny. In these kangaroo-court proceedings, Vere excludes the ship's public, Billy Budd's fellow seamen, not holding the proceedings in the morning on the quarterdeck in the usual way but secretly, unannounced, below deck. And last but not least, he refers repeatedly to the *Mutiny Acts*, although as he certainly knows these only apply to land forces.[152]

Despite his display of level-headedness and although he steadily emphasized concern for Billy Budd's fate, Vere is nothing short of obsessively fixated on a factual finding of mutiny. Excluding mitigating factors and alternative legal approaches, and interpreting the law's literal wording in a sense of equity, he instead invokes a kind of emergency law. In thus staging himself as a metalegal or law-creating authority, he repeatedly violates contemporary English legal custom, according to which declaration of martial law and application of measures in that framework is only the purview of parliament, the crown, and at sea the admiralty if need be. Vere justifies himself with the disciplinary and political argument that in the context of the recent revolts and the acute menace posed by the French, not convicting would threaten his ship: a crassly false assessment in that peacemaker Billy Budd's execution is far more likely to spark a mutiny. Vere is simply drawing on all possible legal arguments, ruses, and fictions to manufacture a state of emergency, a suspension of the legal order ostensibly to save it. To that end, he has recourse, to, on the one hand, a concept of mutiny that is indeterminate and incapable of any legal standardization, and, on the other hand, a claim to be facing an unavoidable choice between security and justice. For the sake of the higher because collectively vital good of security, he

152 For the precise legal-historical background, see Eder 2004, pp. 151–164; in relation to *Billy Budd*, see Weisberg 1984, pp. 149–153, 157.

thus configures obviously guiltless and guileless Billy Budd into a bearer of danger, sacrificing him like "young Isaac" at a higher command.[153] We might articulate this in the terms of Melville's contemporary Kierkegaard: in his individuation and in the course of a private decision, this Abraham, "by virtue of the absurd," executes an impetuously volatile yet "teleological suspension of the ethical."[154]

Lemuel Shaw has been identified as a model for the figure of Captain Vere.[155] As Chief Justice of the Massachusetts Supreme Court, Shaw had to apply the Fugitive Slave Act, passing judgment on various slaves escaped from the south in "fugitive rendition processes" of a summary nature and in which those accused had no right to speak. In addition, through his cousin Guert Gansevoort, the first officer on board the *Somers*, Melville was familiar with the background of the supposed mutiny; he knew about the behavior and arguments of Captain Alexander Mackenzie and the later strategies and decisions of the court. With the *Somers* scandal, the American public had become aware that with failure of their plans, ambitious reformist captains could simulate acute danger, taking drastic measures on the basis of the alleged emergency. In fact, with this case as a precedent, that process was now widely approved in American jurisprudence. This was the context for Melville's exploration, in an "Inside Narrative" presenting "fiction based on fact," of the obscure motivations for the actions not only of putative mutineers but also of the supposed mutiny's sovereign *fictor.* "True, the circumstances on board the *Somers* were different from those on board the *Bellipotent*," we read in "Billy Budd". "But the urgency felt, well-warranted or otherwise, was much the same."[156]

At the very time of the great mutinies and continuing maritime warfare, fewer commanders conjuring up states of emergency were in demand than those cooperatively managing the emergency at sea. "Nelson was the one, not indeed to terrorize the crew unto base subjection, but to win them, by force of his mere presence and heroic personality," observes the narrator in Melville's tale. While "Sir Horatio...is at bottom scarce a better seaman or fighter" than Vere, the latter man has "a queer streak of the pedantic running through him," is the way "some officers of his rank" would put it. "Ashore, in the garb of a civilian, scarce anyone would have taken him for a sailor."[157] Hence the narrator suggests that in the eyes of both seamen and officers, ambitious

153 Melville 1997, p. 463; see Solove 2005, pp. 2454 f.

154 Kierkegaard, 1994, pp. 42, 47.

155 Solove 2005, p. 2448.

156 Melville 1997, p. 462.

157 Ibid., pp. 418 f., 421.

Captain Vere presents a drastically inferior picture to "him who became Lord Nelson," despite showing no overt rancor toward the "Handsome Sailor," by contrast with Claggart, a man full of sinister envy and hatred. But from Vere's verdict in the Billy Budd affair, we gain an impression of someone ruled by a dark and unacknowledged urgency allying him secretly with Claggart—an urgency that led to measures of marital law aboard the *Somers* and that was felt by all the officers.[158] On the *Bellipotent* it is Vere alone whose default mode, following the death of Claggart (himself "a sort of chief of police"), is fear or rather confirmation of imminent danger.[159]

When the ship's physician notices Vere's violation of prescribed legal procedure and comments on his "excited exclamations," the narrator poses some questions in his stead: "Was he unhinged? But assuming that he is, it is not so susceptible of proof. What then can the surgeon do?...To argue his order to him would be insolence. To resist him would be mutiny."[160] The captain's possibly dangerous condition can only be articulated directly—through the "depravity" of his double, Claggart. Applying to both men is a psychological characteristic that the "an honest scholar" once divulged to the narrator:

> Toward the accomplishment of an aim which in wantonness of atrocity would seem to partake of the insane, he will direct a cool judgment sagacious and sound. These men are madmen, and of the most dangerous sort, for their lunacy is not continuous, but occasional, evoked by some special object; it is protectively secretive, which is as much as to say it is self-contained, so that when, moreover, most active it is to the average mind not distinguishable from sanity.[161]

In face of a "special object," a whale for monomanical Ahab, a "Handsome Sailor" for Claggart and Vere, Melville's demonic characters develop, as Gilles Deleuze describes it, a kind of "metaphysical perversion that consists in choosing one's prey, preferring a chosen victim with a kind of love rather than observing the maritime law that requires him to apply the same discipline to everyone"—a law grounded in the paternal law of the territorial sovereign. Within that dynamic, they can "participate in a terrible supersensible Primary Nature, original and oceanic, which, knowing no Law, pursues its own irrational aim."[162] Tied here to oceanic nature is no enduringly manifest madness, but the particular hidden madness "of the most dangerous sort" that Melville precisely describes. On a

158 Ibid., p. 454. See Weisberg 1984, p. 147.
159 Melville 1997, p. 422.
160 Ibid., p. 452.
161 Ibid., pp. 430–432.
162 Deleuze 2018, p. 56.

characterological level, Vere is the figuration of what the state of emergency represents for the legal order": "a temporary excrescence, bred out of the distemper of the state."[163] In this way Melville uncovers, behind the fiction of danger grounding the doctrine of the state of emergency, the fundamental danger of that fiction. Beyond that, he reveals the danger of that *fictor* who as captain can adjudge the existence or non-existence of danger, and with it the factual presence or absence of mutiny and finally of a state of emergency—sovereign, unchallenged, and at his whim.

In this way Melville spells out the lack of danger posed by Billy Budd –and that innocently and reliably he has stood the test of every nautical and social peril. With the functionary who possesses a free hand for forestalling danger revealed as a "dangerous subject," Melville's text reflects a broader development: the increasing transformation of moments of "danger" and "dangerousness" from a problem of juridical order to a key concept within new legal approaches to special and preventative law. Here for the first time the approach to danger at work in the humane sciences was being integrated into the new fields of psychiatry and criminology.[164] But what is especially clear with Melville's tale is that the doctrine of the maritime state of emergency has changed in a fundamental way: where previously officers, and especially captains, as the highest representatives of territorial state authority, appeared to guarantee lawful defense at sea, while sailors to personify the wild menace of the sea on board the ship itself, by around 1900 "dangerousness" will also be housed at the interior of state authority—and will now repeatedly manifest itself in its very agents.

Under this sign, Jack London's maritime novels, for example *The Sea-Wolf* (1904), are laid out as observations of anthropological experiments that spell out scenarios of a state of nature within the high sea's "elemental environment" and in the laboratory of a "miniature floating world." In that environment and laboratory, London wished to study the erosion of institutional order and territorial authority, together with the "atavism," dangerous but "magnificent," of the "struggle for existence."[165] Here life itself has become a perpetual mutiny initiating, at sea, a movement toward survival of the fittest and paving the way for the natural rule of charismatic leaders. Under these conditions, mutinies in the legal sense can only be caricatures of the earlier theater of sovereignty—only confrontations between half-witted criminals on the one side, intelligent functionaries

---

163 Sulivan 1784, p. 2.

164 On the psychiatric and criminological concept of "dangerousness" see for example Hücking 1935, pp. 31, 44 f., 49; Castel 1983, p. 53.

165 London 2000, pp. 21, 26, 64.

on the other: "It may be nineteen-thirteen mutiny on a coal carrier, with feeblings and imbeciles and criminals for mutineers; but at any rate mutiny it is, and at least in the number of deaths it is reminiscent of the old days," reads one passage in London's *The Mutiny of the Elsinore* (1914). "And still, the mutiny we are enduring is ridiculous and grotesque. There was never a mutiny like it. It violates all standards and precedents."[166]

But around 1900 not only mutiny as an on-board state of emergency no longer followed traditional standards and precedents—this was now the case with normal operations as well. For this reason, a sea-tested author such as Joseph Conrad will locate the ship's command on a level lying deeper than any legal or disciplinary authority. "There may be a rule of conduct," writes Conrad in *A Mirror of the Sea: A Personal Record*. "[T]here is no rule of human fellowship," he continues. "To deal with men is as fine an art as it is to deal with ships. Both men and ships live in an unstable element, are subject to subtle and powerful influences."[167] Enveloped in an unstable and indeed dangerous element, the ship's leadership offers a lesson in the matter of sea power—and power in general. Here, around 1900, when it comes to the ship as a complexly organized "total institution," such power can only be exercised through forms of cooperation and communication that, as Niklas Luhmann describes them, "are *not* based on coercion and *cannot* be implemented in cases of conflict. *That* is the domesticating of the Leviathan, real separation of powers, which as it were developed on the quiet."[168] Hence emerging behind the masquerade of modern rule at sea is the same practical knowledge of maritime leadership on account of which the ancient and medieval world held the figure of the *kybernētēs* or steersman in esteem. Tied to the modern captain's ascent was a degradation of those steering the ship; even in 1900, their earlier exclusive competency continued to be functionally transformed—already through official training courses at the navigation schools—into a means of distinction for captains.[169] Cybernetic knowledge was only rehabilitated, together with its practitioners, with the first crisis of representative on-board authority. But offering a portrait of those practitioners would mean writing a very different history of seafaring than the traditional one: not a history of floating states or mobile capital, but of real ships.

166 London 2009, pp. 228, 240.
167 Conrad 1995, IX., p. 27.
168 Luhmann 1969, p. 166.
169 See Welke 1997, pp. 172–176.

Chapter 6
The Ethos of Wreckage: Perils of Modernized Seafaring

Joseph Conrad and the Ethos of Command

Herman Melville's accounts of life at sea have been described as a last expression of the young American nation's spirit of discovery before the west's territorial frontier seized the collective imagination[1] Joseph Conrad has been approached as an author who spells out the doctrine of seafaring behavior prevailing in Britain's world empire, long-since anchored in control of the oceans but no longer expansive by 1900.[2] Both authors engage in an analysis of sea power moving from a critique of the sovereignty-centered logic and disciplinary organization of the on-board regime to models for managing seamen that have left behind juridical codes and institutional constraints: models that also understand voyage on sailing vessels as an elementary challenge. Particularly in view of increasingly industrialized maritime activity that was shifting to steamships, the approaches taken to the elementary socialization of on-board life by Melville and Conrad were naturally different. Conrad's main focus is not on charismatic or dangerous shipmasters, or on the phantasmatic "thing" on which maritime social existence nothing less than depends. Rather, he is concerned with the ethos of captains and their "craftsmanship," drawing parallels between ship and "human resource" management for the sake of presenting scenarios for the use of natural and social elementary forces.

From the beginning Conrad inquires into the elementary experience of command itself, aiming at the experience's encompassment in a half ethical, half aesthetic doctrine of what he refers to as "fidelity" and "faithfulness." On the parade ground of seafaring, he views these qualities as making their mark within four areas: tradition, competence, solidarity, and the ship itself. Because at sea the ship always contains a risk-bearing community, solidarity here has both a political and technical dimension. Mariners are "men whose material and moral existence is conditioned by their loyalty to each other and their faithful devotion to a ship."[3] In a seafaring nation such as England tradition is what furnishes the command with the correct means to coordinate communal faithfulness. The com-

1 See for example Peck 2001, p. 8.
2 Ibid., p. 166.
3 Conrad 1995, vol. IXX, p. 192.

https://doi.org/10.1515/9783110610734-010

mand thus maintains "fidelity to an exacting tradition" on the one hand, "fidelity to right practice which makes great craftsmen" on the other hand.[4] The four dimensions to fidelity—guidelines offered by tradition, professional competence, formation of a cooperative community, and technical skill—render plausible an analogy between seafaring and authorship. Conrad's own "command" and his existence as a writer are, as it were, emblematically connected. As in the patristic catalogs of symbols, the maritime voyage is here, once again, presented as a life voyage, the ship as a *principium individuationis.*[5] But in place of the theological figure of *homo viator,* human beings as terrestrial animals now occupy center stage, maintaining their existence *'Twixt Land and Sea*—the title of an anthology of stories Conrad published in 1912. Especially the on-board command is, in the words of Conrad's narrator and alter ego, "a symbol of existence"—"the endeavor, the test, the trial of life."[6]

The autobiographically based story "Youth" (1902) elaborates on the topos of a ship's first command as a heroic version of this existential experience. Here an elementary space is conquered at mortal risk—a space beyond all legal, religious, and commercial speculation. It is "the sea that gives nothing, except hard knocks—and sometimes a chance to feel your strength."[7] If in this context, the story's protagonist, for the first time commanding a ship, is subsumed to the ceremony of a conquering existential sovereignty, in another story, "The End of the Tether" (1902) an aged, nearly blind captain, for the last time undertaking a ship's command, is engaged in the masquerade of a deferred divestiture. Henry Whalley, once a legendary captain named "Dare-devil Harry," sells his own ship to offer his daughter a basis for existence. "[T]he hopes of his youth," we read, "the exercise of his abilities, every feeling and achievement of his manhood, had been indissolubly connected with ships." If a ship without crew is like a body without a soul, then a seaman without a ship is "an aimless log adrift upon the sea." For this reason, Whalley invests the revenue from his ship-sale in shares in a rundown coastal steamer, the *Sofala,* for the sake of once again taking over command of a ship. He thus takes "an interest in the ship" in a double sense.[8] However, the ship-owner Massy is also the machinist, which leads to an interest-conflict at the latest when the ambitious Maat Sterne draws his attention

4 Ibid., pp. 194, 197.

5 On Conrad's career as a seaman and the evident transfers of personal experience into his narratives, see Kramer 2002, pp. 157–163.

6 Conrad 1995, vol. VI, pp. 4, 12. On Conrad's sea stories as rites of passage and a "struggle for pure survival," see Nüstedt 1998, pp. 132, 151, 186, 201, 313.

7 Conrad 1995, vol. VI, pp. 30, 42.

8 Ibid., pp. 184f., 191, 208.

to Whalley's failing eyesight, recognizing the opportunity this offers for insurance fraud.

Natural and official person here turn discordant, it becoming evident that with the help of his "shadow," his Malaysian helmsman and confidant, the captain is hiding his growing blindness and actually leading a proxy existence. Consequently the *Sofala* becomes the stage for an archetypal dethroning, of the sort described by James Frazer in *The Golden Bough* (a book with massive cultural influence published in various editions and volumes between 1890 and 1915) as the custom, prevailing "two centuries ago in the caffre kingdom of Sofala," "of putting kings to death as soon as they suffered from any personal defect."[9] In his *Heart of Darkness* (1899), Conrad takes up the primeval natural-magic rituals Frazer believed were practiced in African sacral kingdoms as both a primitive origin of civilizational rule and its abyss.[10] Where in the evolutionary schema of British social anthropology, eventually science supplanted magic and religion, in "The End of the Tether" regicide is carried out in the form of a scientifically informed manipulation of a navigational instrument: Massy sees to magnetically induced declination of the compass.[11] In the course of the ensuing accident, Whalley does discover this fraud perpetrated against both the on-board regime and the insurance company; but his share of the ship would be irrevocably lost with a complaint. In order to at least retain command on a technical, administrative level, hence the all-important share he has reserved for his daughter, he remains alone on the sinking ship, the scrap-iron used for the declination transferred to his own coat-pockets. What will be an accident for the insurers is for him a tragic act of God.

Conrad's relatively neglected late story "The Partner" (1920) even more clearly juxtaposes the captain's ethos with simple speculation, trustworthy action with risky behavior liable to fraud. Here the first-person narrator, a writer searching for a theme, encounters a sullen ship-loader in a harbor hotel; the man recounts the background to a shipwreck whose remains have been a local tourist attracting for the past twenty years: a story that begins with a decision by the Dunbar brothers—Harry, a passionately committed sea captain, and George, formerly a bookkeeper—to invest their shared inheritance in a sailing vessel, the *Sa-*

---

9 Frazer 1944, p. 272.

10 Conrad 1995, 6, p. 216.

11 Ibid., p. 322: "Mr. Massy knew something of the scientific basis of his clever trick. If you want to deflect the magnetic needle of a ship's compass, soft iron is the best; likewise many small pieces in the pockets of a jacket would have more effect than a few large ones, because in that way you obtain a greater amount of surface for weight in your iron, and it's surface that tells."

gamore, thus resisting the trend of converting to steam. One day the Dutchman Cloete, a kind of modern project-manager, becomes George's partner and convinces him to enter into the lucrative trade in medicinal products. But to bring together the necessary starting capital, the Dunbars would have to pawn the *Sagamore*—for the seaman Harry, now the ship's captain, a simply unacceptable prospect. For this reason, Cloete presses the less principled of the brothers into a conspiracy: "wrecking" and "tomahawking," the intentional, unobserved sinking of the ship, would bring together enough insurance money.[12] The inventive Cloete thus engages Stafford, formally a seaman, now an impoverished petty thief, quickly a suitable wrecker.

Whether from nervousness or pangs of conscience, George soon resists the plans for fraud. But after repeatedly surfacing, Stafford is brought aboard the *Sagamore* as first officer—and then, one day, reports appear of the vessel's shipwreck near Westport Bay. George and Harry's wife follow the dramatic event from the coast, while Cloete has himself brought on board. What has occurred is in his eyes an actual accident and in addition "enormous luck," whether deemed "providential" or ascribed to the devil.[13] For this reason he declines to accede to Stafford's demands for payment, rather locking him in the captain's quarters while the ship gradually sinks. Harry, however, returns to the cabin to rescue the ship's papers (hence from a simple sense of duty) and is shot and killed with his own service pistol by the panicky Stafford. Later the official report will refer to the captain's suicide from despair, and Stafford will be obliged to confirm before Lloyd's that the shipwreck was due to pure chance. Fallen into melancholia, Henry's widow receives half the insurance money—the remainder being insufficient to get the medicinal-product business underway. George and Cloete have missed out on an "immense fortune" "by the price of a revolver-shot."[14]

Within Conrad's narrative arrangement this story is told to an anonymous writer by a nameless stevedore who for his part relies on the accounts of Cloete and Dunbar. While the stevedore already holds forth about the "silly yarn" told by tourists' boatmen, newspapers, and trade writers, he promises to himself furnish nothing but the "truth": one located beyond the picturesque "sea life" presumed to unfold on board ships, and also beyond all atmospheric maritime accounts and ennobled views of view-worthy wrecked vessels.[15] This truth only

12 Ibid., 16, p. 98.
13 Ibid., pp. 110 f.
14 Ibid., p. 126.
15 Ibid., pp. 89, 91, 109.

manifests itself in a "remote connection" reaching into George's office.[16] Consequently, on the one hand Conrad's story simply reverses the conversion of events into business matters that, as referred to earlier, Luca Pacioli already prescribed for proper bookkeeping in 1494. What is narratable is here deduced, as it were, from invoice items. On the other hand, with its precise causal and processual analysis, the story appears to probe the event of shipwreck in just the manner assigned to the Shipwreck Committee in England since the nineteenth century. But if any plot-core is present here, a core around which a "true" seaman's yarn is woven, then it is the "risk event" of an insurance case—and this in the light of its fraudulent staging or rather narrating.[17]

This story, laid out as a story in a story, plumbs the fictive and fraudulent aspects of risky partnership of the sort entered into by the Dunbar brothers but also by George with Cloote and the Dunbars with their insurers. Here conventional procedures—for example proof through witnesses and experts, documents and *prima facie* evidence, or the reconstruction of a closed chain of causality in general—necessarily has to fail, together with the principle of "consider the proximate cause, not the remote one."[18] When Stafford manipulates the anchor-chain in such a way that he does not even leave behind a "broken link," in the future reconstruction of an evidentiary chain just the missing link will be missed.[19] For this reason "The Partner" revolves around not only insurance fraud but, even more so, what always renders testing of insurance claims so difficult: distinguishing between events and their observation, together with the inseparable unity of both.

Maritime insurance, in particular, rarely involves clear facts and doubt-free statements. It is necessarily concerned with contingency, with events that have neither taken place without doubt nor were impossible; likewise with speech acts that are not only true or false but possibly subjectively in accordance with truth but are involuntarily fraudulent, or else are deceptive precisely because they are aimed at producing a certain truth. The security promised by insurance is fused with uncertainty. But if this uncertain grounding is taken into account, then the study of insurance cases requires second-order observation: not only observation of putative facts but also of their external and internal *circumstantiae*. The external circumstances here take in the questions of when, where, and how something took place, while their internal counterparts are con-

16 Ibid., p. 92.

17 For the "risk event," the insurance case, and the intertwining of the two in the context of insurance law around the story's time, see Bene 1928, pp. 10–16, 34.

18 See Mevert 1921, p. 47; Kreutziger 1966, pp. 176 f.; and Hauer 1990, pp. 28, 59.

19 Conrad 1995, 16, p. 115.

cerned with the witnesses and interested parties, their reports and arguments, intention and reliability.[20] In this way what is being observed is not so much a factual world as rather its observation. And this in turn means that the world's facts are not so much delivered to a predicate as to a perspectivist or narrative structure.[21]

But it also means that in the case of contingent event-structures, alongside analysis of their *veritas*, increasingly analysis of their mere appearance—their probability or verisimilitude—becomes dominant.[22] What in classical rhetoric is referred to as *narratio verisimilis*, "plausible narration," is nothing other than "circumstantial" narration that, in addition, accounts for the narrator as a second-order narrator.[23] If narratives, as opposed to drama, can convey inner perspective and experience, than fictionality is—as in Conrad's "Partner"—more than mere fiction, and not only a counterpart to truth.[24] On the one hand it can deconstruct any sort of certainty about certain putative facts and events. On the other hand, even if it offers no certainty, it can lead to what Coleridge famously termed a "willing suspension of disbelief." Everything depends on the narrators and their way of narrating. In the framework of Conrad's story, the dutiful Captain Harry, this prototypical adventurer at sea, would have to make a "reliable narrator." But he remains as mute as his antagonist Stafford, the fallen and now criminal seaman. George serves as Harry's counterpart, in that he is what in Italian is termed a *mercante residente*—someone engaged in a venture on terra firma. He is a pioneer not in overcoming dangers but in acting on and trading in risk. His role is less tied to his brother than it is, indissolubly, to his partner Cloete: "A man who's been in the patent-medicine trade," and who

---

20 From the perspective of maritime insurance see Elster 1931, pp. 20–23; from that of rhetoric and probability theory see Hauser 1997, p. 80. From a history of science perspective and with a view to narration, together with a focus on the reversed association of "inner" circumstances with the event, "external" with statements, see Campe 2012, pp. 97 f., 134 f., 167 f., 381.

21 The collating of different perspectives or forms of representation can be understood as a characteristic not only of Conrad's narration but, just as much so, of modern navigational technologies. In her book on the novel and the sea, Margaret Cohen thus speaks of "the *propagation of representational state* across a series of *representational media*. In other words, evidence in different kinds of media—compass bearings, depth, celestial calculations—need to be aligned so that they all give the same answer as to location. This process may sound straightforward, but calculations are made by people using sensitive technologies and, like any human process, subject to error." Cohen 2010, p. 209.

22 With a view to the Early Modern emergence of "probability" and a corresponding ennoblement of narration (alongside old, Aristotelian declarative truth) see Campe 2006a, p. 72.

23 See Campe 2012, pp. 130 f.

24 On this concept of fictionality see for example Hamburger 1993, pp. 55–59; Fludernik 2008, p. 73.

is clearly capable of anything "from pitch-and-toss to willful murder."[25] Conrad's narrator thus turns out to be informed by a narrator who for his part can only invoke two fraudulent and unreliable narrators.[26]

The interest that the first-person narrator of "The Partner" takes in the story does not concern its truth or originality. Instead, he uses the shipwreck and insurance case as grounds for practicing another kind of certainty or creativity, centered on concepts of fidelity and faithfulness. In Conrad's memoir *Mirror of the Sea* (1906) we read that "faithfulness is a great restraint, the strongest bond laid upon the self-will of men and ships on this globe of land and sea."[27] In the novelist's *Personal Record* (1912), we read that fidelity is one of the "few very simple ideas" upon which the present world can build, an idea "as old as the hills." Probity vis-à-vis other people and circumstances is aligned with rules of behavior not aimed at calculable gain or fast success but at non-pragmatic consistency and steadfastness.[28] With Conrad's principle of fidelity consistently designating both aesthetic and ethical qualities, that principle is also foregrounded in the context of the art of seafaring—for Conrad, the ship and its adequate command serves as "a moral symbol of our life,"[29] at the same time a moral symbol of writing and its fidelity. Hence as already was the case in the ancient world's catalogs of symbols, the art of writing and the art of steering are here in accord.

Under this sign, the narrator of "The Partner" serves as the eloquent spokesman and faithful partner of the mute captain, just as Conrad the writer is the double of Conrad the former seafarer. Both of Conrad's personae are allied with an elementary, nautical and finally ethical experience. And both demarcate themselves from a "culture of security" dominant around 1900, together with its bureaucratic and fraudulent dimensions. Conrad thus juxtaposes the sailing vessel's conscientious handwork with the speculation and manipulation of modern risk-grounded trade and action. In doing so, he reaffirms what speculative business had long-since seemed to obscure: the elementary distinction between land and sea. It is not the least of the paradoxes tied to Conrad's contradictory modernity that certainty and conscientiousness seem rooted nowhere else but in the sea—in the midst of the groundless, lawless domain of pure contingency.

25 Conrad 1995, 16, p. 94.

26 On the controversial relationship between modern courtroom testimony and unreliable narration, see Weitin 2009, pp. 257–265.

27 Conrad 1995, 9, p. 111.

28 See ibid. p. XXI; Najder 1997, p. 199.

29 Conrad 1995, 9, p. 188.

Gentlemen and Confidence Men

Conrad once described himself as a strange antediluvian animal that never could opt for voyage by steamship—and this not despite but on account of the evidently insecure and unreliable nature of voyage by sail. In a memorandum on equipping a British Navy sailing vessel used for training, he wrote that "[a] ship's safety, apart from the 'Act of God,' rests in the hands of the men who are aboard of her, from the highest to the lowest in their different degrees."[30] Seamen organize themselves into a danger-ready community spontaneously, in line with an inner capacity for bonding and non-codified rules. "No outward cohesive force of compulsion or discipline was holding them together or had ever shaped their unexpressed standards," we read in "Well Done" (from *Notes on Life and Letters*).[31] What furnishes them with form and function is not the binding force of a sovereign command or disciplinary institutions, and not even pressure such as that to maximize profits. To explain their cooperation, Conrad has recourse to a kind of regulative idea of fidelity, and with it to the ideal of a regulative *technē* allying the captain, as a manager of ships and men, with the writer as analyst of both.

Industrialized seafaring saw these "unexpressed standards" of the seamen's art replaced by standards of a purely technical nature. Especially passenger steamships, with their "sham shore conditions" basically spared voyagers' confrontation with the sea. That voyager did not live in the element like the seaman on a sailing vessel, but was mere freight. The passenger, not the captain on the bridge out of sight and hearing, personified the new form of maritime rule: a democracy that in the end is a technocracy, and for which the sea is, as Carl Schmitt puts it "no longer an element" but rather a "space of human activity and exercise of power."[32] For Conrad, this upheaval in the maritime travel industry represented a cultural break, a turn away from the old nautical specialists, their experience and craftsmanship, sealed by a fatal trust in "expertocrats" alien to the sea. Conrad saw the extent of this cultural break as manifest in what took on the reputation of the maritime catastrophe par excellence—the sinking of the *Titanic.*

That object of British maritime prestige sank at the same time as a "culture of experts" had first begun to develop; and perhaps a mass-media supported global public only first formed with the sinking. The shattered trust in technical experts sparked by the highly publicized catastrophe of 14 April 1912 itself was a basis

30 Conrad 1926, p. 114.
31 Conrad 1995, 9, p. 182. See also ibid., p. 117.
32 Conrad 1926, pp. 54f.; Schmitt 1995b, p. 398.

for innumerable cultural-critical reactions. Especially in Britain, Conrad possessed a certain authority in commenting on both the event and its aftermath on the official level. With a view to the American and British investigatory committee, but with pronounced skepticism regarding the new, developing global civilization, he used the occasion to attack the "Oracle which has failed, but still must remain the Oracle," those "high priests of the modern cult of perfected material and of mechanical appliances."[33] He thus publicly called for a vote of mistrust against a bureaucratically enthroned engineering caste that, with its concepts of normality and calculability, reliability and functionality, had miserably failed at sea. As is the case for a tradition of cultural criticism extant since Homer and Hesiod, for Conrad the sea is a sphere of elementary incalculability within which nothing either socially or technically self-evident can exist. But just for this reason he considered it a domain in which trust could form.

In the maritime horizon of steady challenge, the ship appears as a laboratory for technical and social cooperation. Nowhere else is society and its survival so based on trust. For this reason, nowhere else can the phenomenon of *confidence* be optimally studied: "ships are what men make them," writes Conrad in *The Mirror of the Sea*.[34] Handling a ship, he continues some pages later,

> requires not only the knowledge of the general principles of sailing, but a particular acquaintance with the character of the craft. All vessels are handled in the same way as far as theory goes, just as you may deal with all men on broad and rigid principles. But if you want that success in life which comes from the affection and confidence of your fellows, then with no two men, however similar they may appear in their nature, will you deal in the same way. There may be a rule of conduct; there is no rule of human fellowship. To deal with men is as fine an art as it is to deal with ships. Both men and ships live in an unstable element, are subject to subtle and powerful influences, and want to have their merits understood rather than their faults found out.[35]

There are thus solid reasons for specialists in ship-classification societies referring to the seaworthiness of a vessel in terms of reliability and trustworthiness.[36] But whether placed in ships or people, for Conrad trust is not nourished by an absence of error. It is, repeatedly, a function of Conrad's fidelity and faithfulness, only becoming evident in emergency situations and in face of unforeseen events. Put differently, in Conrad's work trust is a question of ethos and performance, not theory. And if trust does have a technical dimension, this is only in the

33 Conrad 1995, 19, p. 234.
34 Ibid., 9, p. 16.
35 Ibid., p. 27.
36 Manes 1918, p. 88.

old sense of *technē:* the sense of skill and artisanal competence. This is the basis for the analogy between management of ships and human beings, but also between Conrad's on-board command and his authorship—his double form of authority. Trust's source is finally not fulfillment of certain rules of conduct or mere procedural regulations. It can only emerge from a collective confrontation with elementary dangers (even those induced technically): a voluntary confrontation undertaken by individually situated, in exemplary fashion, in an insecure existence *'twixt land and sea.* Authority participates actively in this collective process, coordinates and takes responsibility for it. It embodies the circular relationship between trust and responsibility,[37] and this even in the case of its admitted non-knowledge or a failure arrived at through no fault of one's own. This is the reason Conrad deems worthy captains (and corresponding authors) to be "gentlemen." But he sees "experts" as merely potential confidence-men, as nineteenth-century stock-market speculators were also called.

Conrad's sentiments in this respect reflected, for a start, the interests shared by nautical specialists with the trusts of the large shipping companies, hence their alliance with major speculative enterprises. Beyond that, for Conrad an ocean liner like the *Titanic* was no ship in the traditional sense, that of a vessel exposed to elementary forces in order to also use them. Rather, it was "a sort of marine Ritz, proclaimed unsinkable and sent adrift with its casual population upon the sea."[38] What here ruled was not the ideal-typical *technē* of free, informal discipline embodied in every seaman but a kind of biopolitical technocracy. Such steam-operated passenger ships symbolized a regime of pseudo-security and a faith in progress that although guaranteed by technical expertise was finally simply speculative in nature. Conrad confronted the previous, mistaken expert opinion that the *Titanic* was "practically unsinkable" thanks to its watertight construction and sheer size[39] with the tradition-anchored practical experience of the seaman: in collisions gigantic mass and measurement led less to maintenance of solidity than to an inability to steer; and the self-closing compartments, only reaching to the water-line, thus not sealing off the underdeck to the top, were more a trap for boiler-room workers needing evacuation than safety in an emergency. Conrad saw the American senators on the investigative committee as lacking both legal authority and specialized nautical competence. But when it came to the technical specialists, he observed that "mere calcula-

---

37 See Engel 2002, p. 61.

38 Conrad 1995, 19, p. 227.

39 For the incorporation of this initially purely technical assessment into an advertising catalog of the White Star Line and from there into popular mythology, see Koldau 2012, pp. 20 f., 197.

tions, of which these men make so much, when unassisted by imagination and when they have gained mastery over common sense, are the most deceptive exercises of intellect."[40]

In view of the incalculable, those who count on calculation and—more than concrete craftsmanship—on the manipulative possibilities of algorithms evince an expertise that for Conrad is worth nothing and cannot be trusted. Trust, he underscores only forms on the basis of practical expertise (with seafaring here clearly simply representing one exemplary field among many), of testing in a context of socio-cultural complexity, limited technical possibilities, and elementary threats. Trust is not reduced by complexity; nor does it exclude asymmetric power relations. Rather, through repeatedly newly produced trust in authority complexity is developed and synthesized in an ongoing way.[41] Put otherwise: with social communication, it is not so much contents (for instance in the form of expertise) that are transmitted as roles (such as on board ships) created: roles between which authority repeatedly emerges and wanes. Trust has no reference and offers no certainty; rather, it is a medium of communication. Socially and economically, it is manifest as risky credit, better offered, Conrad suggests, to "gentlemen" than "confidence-men." But for Conrad it was clear that the nautical experts in the British-American investigative committee formed for the *Titanic* affair were in the end swindlers who dealt in trust. It was not only that during the hearing the various interested parties succeeded in having "their" experts testify. It was also that the expertise of the navy's shipbuilding engineers could have totally different import from that of civilian seafaring; that in the liner business there were no procedural rules once technical experts, captains, and owners came to contradictory assessments of dangers; and that such a case was itself treated as "a contingency which is remote."[42] All this would have enduringly undermined Conrad's trust in the nautical expertocracy.

Conrad's polemic was naturally not simply aimed at the sensational accident that befell the *Titanic*. It was tied to a broader perspective of cultural criticism. What since the Early Modern period, initially in seventeenth-century France, had been propagated as nautical Enlightenment now seemed questionable: at that time the inherited and implicit, arcane and future knowledge nurtured by shipbuilders and seamen had been collected and codified for the first time, then placed on "scientific," which is to say hydrographic and hydrodynamic, static and statistic foundations, and finally monopolized by the new experts

40 Conrad 1995, 9, p. 230.

41 See in general Hartmann 2011, pp. 9–13.

42 See "British Wreck Commissioner's Inquiry," day 24, 23545.

(by engineers and in the nautical academies). As we have seen, this scientization of nautical experience was a form of "disciplining" in the double sense: behavioral control and epistemic normalization of shipbuilders and seamen. Conrad viewed the industrialization and continued technization of seafaring, in whose course sailors were demoted to specialized laborers and marine officers to on-board operation managers, as simply the final outcome of this reform. The cultural break he lamented was thus most vividly manifest in the decline of a professional milieu and its "local knowledge," in the vanishing of a tradition of craftsmanship and the ethos associated with it. Authorities in the sense of "gentlemen" were slowly expiring. The rudder of the ship of state now seemed in the hands of mere confidence-men. Although the nautical knowledge emerging in the age of sailing ships may never have been a core element of the "intellectual culture of mankind," it was nonetheless indispensable for civilization, "which is founded on the protection of life and property."[43]

The Ship as Locus of Displacement

Until the late nineteenth century the ship was a key component in a globally interwoven economy, centered on large-scale feedback-links between commerce and speculation, industry and maritime traffic, in the process increasingly dependent on reliable and well-regulated transport media.[44] But because such ships had to serve as robust and yet flexible links in a world-spanning chain of economic interests and technical goals, on-board working conditions inevitably altered. With the stokers, machinists, and engineers, something like a new type of "worker on the sea" joined the on-board ranks, although initially the new workers had outside status, which is to say were not thought of as proper seamen.[45] On the official side, a strict separation was established between

43 Conrad 1926, p. 60 f.

44 It remains unclear if the discussed technology was a precondition for such economic activity or rather vice versa. Since the eighteenth century's mechanical *régulateurs* and *governors* could also easily have been developed with renaissance and Baroque technical means, but as the choice was made to use rigid steering processes akin to Colbert's mercantilism, a mentality shift toward liberalism and its "governmentality" could easily be postulated as a developmental precondition. Inversely it was technical regulations such as the steam machine of Newcomen and Watt that first allowed auto-regulative processes, unfolding as systemic relations between functions, to emerge as a guiding program for collective action. See Mayr 1969, pp. 122 f.

45 In the German language context, this status was confirmed in ambivalent honorifics such as *Funkenpuster, Füerpuster, Stoker, Ölprinz, Talgsmeerer, Kesselbums*, and *Ölkannenschwenker*. See Stammler 1967, p. 2936.

what the British navy termed deck and engine departments, with the latter lacking planned training programs, let alone guaranteed promotion chances, even in the early twentieth century. At sea, stokers, machinists, and engineers had no authority to issue directives, but simply the duty to maintain the machines at any cost and expose themselves to all types of danger in emergency situations.[46]

Because the procurement and maintenance of new types of ship was tied to considerable capital outlay, various shipping companies were transformed into corporations, further loosening the personal ties between crew and client. And because the machines on board meant high operational costs, the ship's command was subject to steadily increasing technical guidelines. More precisely, technical, operational, and communication processes were interpreted in advance in terms of strictly social hierarchization, so that the immense gap between on-deck officers and machine workers appeared all the more objectively justified.[47] Below deck, the maritime proletariat gathered: a motley international collection of anonymous industrial workers without training or vocational tradition, often without a proper contract. Scorned by the "real" seamen and underpaid by their employers, these men had no legal workplace protection and were thus exposed to steady stress and danger. Partly but not only because of their high suicide rate—a reality for coal trimmers in particular—they were considered something like the anti-type of traditional mariners.[48]

In this manner, the new class of proletarian seamen embodied a basic, in part cultural upheaval in maritime commerce. Where commanding a sailing ship had resembled both handwork and art, synthesizing activity and passivity in face of the elements of sea and wind and the danger of both, the steering and coal firing of a steamer exemplified a violent and antagonistic relationship with the elementary environment. This development was justified by prioritizing technical and economic factors over inherited claims and a long-practiced social dynamic. Seamen, whose first concern had been the ship and its crew, became appendages of machines—and also subsumed to a command that fully followed the guidelines of owners and entrepreneurs on land. "The Industrial Revolution had transformed the children of the sea into machine-builders and servants of machines," commented Schmitt in *Land and Sea* of 1942 with a view to steamships and radio communication. "Today...every shipowner can know daily and hourly at which point in the ocean his ship on the high sea is to be found. With this, in contrast to the age of sail ships, the world of the sea is elementarily

46 See Griffiths 1997, pp. 132–139.

47 See Gerstenberger 1996, pp. 239 f.

48 See Heimerdinger 2005, p. 68.

altered for humans. However, if this is so, then the division of sea and land, upon which the link between sea domination and world domination allowed itself to be erected, falls away."[49] Hence with seamen and ships no longer free of the grip of terra firma and its laws, certainties, and expectations of profit—with no further venturing to confront and cooperate with an element both imperiling the mariner and carrying him forward, this elementary difference rather being leveled on an everyday basis in favor of a territorial, economic, and technological regime—the demise of an entire way of life and vocational culture could only follow. Perhaps this also meant the demise of a unique narrative culture—the previously ceaselessly spun seamen's yarns now taken up in a global genre, connecting land and sea, of something like a "literature of the working world."

"He / Who enters here / Will have his name and being extinguished" is the inscription displayed over the crew-quarters on an ocean-steamer in the (German-language) version of B. Traven's novel *The Death Ship*, marking the locus as a non-place of infinite exploitation.[50] Speaking of a "non-place" is apt in part because around 1900 the existence of such "death ships" was disputed. Whether as the allegory of a murderous late capitalism or as a realistic report on the circumstances prevailing in the modern maritime trading business—in Germany of the interwar period, a new workers' literature was invested with the hope of addressing the non-places of the industrialized workers' world, bringing them into speech and thus public awareness. In 1926, with this first of his adventure novels stamped by "social criticism," B. Traven would become famous overnight. (The book would be translated into thirty languages and millions of copies would be printed.) The novelist would serve as a pioneer in the "Gutenberg Book Club" labor project, initiated in 1924, to establish a "culture institution of working people." *The Death Ship* became widespread reading in secondary schools and would be quickly understood as a key text in developing enlightened class consciousness. It quickly became a bone of contention within various social-revolutionary fractions but also an object of bourgeois professional literary criticism.

---

49 Schmitt 2015, p. 93.

50 *Wer hier eingeht, / Des Nam' und Sein ist ausgelöscht.* B. Traven, 1978, p. 131. This novel may have initially been written in English, then translated into German by B. Traven himself. The English-language text has appeared in several editions, the first being that published by Alfred A. Knopf, Inc. in 1934. Because of often very substantive differences between the text's English and German versions—differences that are themselves highly interesting but beyond the scope of this book—citations, with parenthetical page-references in the main text, will consistently be from the German version; all citations translated by Joel Golb.

"He / Who enters here / Will have his name and being extinguished": while this motto was greeted by some readers as an urgent call for worker's solidarity, by others rejected as a symptom of a false because defeatist assessment of the "social question," few readers would have not recognized its source in the inscription over the door to hell in the *Inferno:* "Abandon all hope, you who enter here."[51] For centuries the *Inferno* was understood in terms of a certain economy and topology of damnation. Hegel was still praising Dante for representing "damnation...not only absolutely in its universality," not only abstractly and as the mere negation of a Thomisticly constituted salvational economy, "but as a list of practically innumerable individuals brought forward in their particular characteristics."[52] Since the mid-nineteenth century, Dante's poetic topography also served as a model for a secular economy of perdition no longer damning namable individuals but a general class of nameless ones. This was the sense in which Marx described his critique of political economy in 1859, and again in 1867 in *Das Kapital*, as a daring scientific venture into the hellish region of unbounded exploitation for which certain lines from the *Inferno* (3, 14 f.) serve as a fitting motto: "Here must you banish all distrust, / here must all cowardice be slain."[53] B. Traven appears to have written his adventure novel along exactly the same lines: as a bold entry into the economy of infernal exploitation.

After World War II, *The Death Ship* was repeatedly read as a realistic workers' novel, its locus being the hellish working circumstances of the international industrial proletariat, illustrated through the boiler room of the *SS Yorikke.*[54] Through the figures of the American seaman Gale and his Polish companion Stanisław, the reading went, B. Traven's focus on an unlimited exploitation and alienation destroying both the individual and class consciousness had succeeded in presenting nothing less than anthropology of labor. Exercising free existential choice—with the ethos of a good maritime worker—Gale signs up with a dubious freighter, then keeping faith with the smuggling ship against all the maritime police's authority, to the point of a shipwreck aimed at insurance fraud. Even when Gale and Stanisław are "shanghaied" in the Dakar harbor on the *Empress of Madagascar,* in order to sink this prettified nine-thousand ton vessel suffering from

---

51 "Lasciate ogni speranza, voi ch'entrate." Dante, Inferno 3, 9.

52 Hegel 1975a, p. 589.

53 "Qui si convien lasciare ogne sospetto / Ogne viltà convien che qui sia morta." Dante, *Inferno* 3,14–15. Marx 1904, p. 15. See also Karl Marx, Capital, transl. Hans. G. Ehrbar (only available online), p. 650: "Branches of English Industry with No Legal Bounds to Exploitation."

54 For the topos of the hellish boiler room, see Egon Erwin Kisch's report "Bei den Heizern des Riesendampfers" (1914), in Kisch 1979, pp. 63 – 67. We here find steady references to the "realm of damnation," an "underworld," and a "hellish landscape."

construction flaws on the open sea, their labor on their own demise still preserved a sense of human self-realization.[55]

That, long before he hires himself to the *Yorikke,* after a shore leave Gale has failed to arrive back in time on his cotton freighter, the *SS Tuscaloosa*, then, lacking money and papers, has been deported from one country to the next, hence that he above all is forced to undergo an odyssey of purely administrative-technical nature, does not really harmonize with such an interpretation; this plot detail was thus read like a novel in a novel, a satire on bureaucracy as a proem to the workers' novel. But Gale's odysseys across Europe's borders and on the floating coffin are intertwined. *The Death Ship* needs to be read as a transcript of bureaucratic procedures, but also as a new sort of "sea novel." Gale and his fellow voyagers are no "sea laborers," needing to mediate, à la Victor Hugo, between a traditional seafaring ethos and increasingly technologized nautical knowledge. Rather, they are—bureaucratically—excluded from regulated work at sea, work that is exploitable in a regulated way. B. Traven's *Death Ship* is by no means the locus of an existential test vis-à-vis the watery element or the new challenges of henceforth industrially driven seafaring. Instead it is the ubiquitously denied locus of an administratively catalyzed nihilation of labor, a nihilation operating precisely *through* labor.

As opposed to Joseph Conrad, who only narrated life at sea from the commander's bridge, B. Traven has been credited with giving a voice to the oppressed crew, the actual maritime laborers. For Conrad, the disciplinary structure or "class society" on board amounted to a theme concretizing the epochal change from the sail to steam. As described above, for Conrad, in its topological forsakenness the sailing ship, freed from the authority of terra firma and always in a dual with elementary forces, had produced its own existential and communal form, under an imperative of unbreakable fidelity. "The ship, this ship, our ship, the ship we serve, is the moral symbol of our life," the novelist wrote in 1921. For its part, the steamship, with its radio technology this always locatable "swimming island," operated as a machine under "sham shore conditions."[56] The steamer's operation only needed the work done on terra firma: services on deck; stoking and regulation of the steam machine below deck.

On the one hand, as indicated Conrad here saw an entire form of existence and an entire vocational ethos as going under. And indeed, a 1903 inquiry by Berlin's Association for Social Policy, for example, confirmed that after the century's turn maritime crews were increasingly recruited from inlands, with "even

55 See for instance Kastely 1985, pp. 84, 90; see also Boehncke 1989, pp. 6–16.
56 Conrad 1995, 9, p. 188; 1926, p. 55.

untrained and declassed elements used as coal trimmers when they have the necessary strength."[57] On the other hand, in that the sailing ship was the archetypal vehicle for an "endless journey," the possibility of adventure on sea vanished as well. Since this change set in, steamships have either operated according to a fixed schedule or while leaving their routes open have been oriented toward precisely calculated loading availability. What can still be referred to as adventurous voyage is no longer a life journey into the blue but follows the exact directives of a flexible logistics.

Michel Foucault characterizes Early Modern seafaring as follows: "the ship has not only been for our civilization, from the sixteenth century until the present, the great instrument of economic development...but has been simultaneously the greatest reserve of the imagination."[58] The reason for this was a topological state of emergency: as a locus outside all loci, but that potentially can be located, it represents the "heterotopia" par excellence. For one thing, the ship is the extreme case of an excluded inclusion: as a place in what has no place, it is left aside while being used precisely in its placelessness. Just as on ships of fools exemplary passengers were locked up "in the interior of the exterior, and inversely,"[59] in the case of plague ships there was no interest in liberating their inmates from placelessness. The crew itself was meant to be eternally left in that space of contingency, that threshold between life and death, always represented by the sea from the perspective of terrestrial thinking, of a rational *nomos*.[60] For another thing, the ship is a locus within no locus that because of its very topographic lack of firmness can be used to open up what is placelessly elemental. In this way ships become the legal means and weapon for spatial and geopolitical ambition, in Lauren Benton's words, "vectors of law thrusting into ocean space."[61]

In this light we can understand the ship as a medium in a double sense. Firstly, it is a medium of existential transformation—albeit one that the sovereign tries to place at his service starting in the Age of Discovery. It is not only that, in the sixteenth century, it was those who were penniless and rootless who tried to embark on the most existential of all nautical journeys, that to the New World. For the sake of this effort, in embarkation-harbor Seville a bureaucratic regime was established through which the nameless emigrants were furnished with

57 Francke 1904, p. 15.
58 Foucault 1986, p. 27.
59 Foucault 1965, p. 11.
60 Aristotle, for example, recognized the usefulness of seafaring; but "there is no necessity that the sailors should be citizens." Aristotle 1941, 1327a/b (pp. 1284 ff.).
61 Benton 2010, p. 112.

first officially recorded names and identities before crossing the border.[62] Since then, passage by ship had to be preceded by a bureaucratic "odyssey"; likewise the future life form of the passengers had to correspond to the passport's specifications.[63] Second, the ship was the medium for a transformation of spatial ontology, "an instrument," observes Hegel, tearing human beings "from their limited landbound existence," opening up a "maritime outlet."[64] Modern globalization is the result of media-technical and symbolic operation: maps and navigational instruments opened up a non-place that was then bureaucratically subsumed to a (geo-)political order. In this double operation's course, the space of terra firma—initially that of the colonies, then the originating countries —came to be understood as a field of forces capable of being modeled on military, policing, and administrative-technical levels. The "nomos of the earth" that Schmitt described for the Old World as a unity of location and order, *Ortung* and *Ordnung*, was itself irrevocably transformed through previously hazardous maritime passages.

In the context of legal history, Schmitt described this geopolitical change as the decline of *Jus Publicum Europaeum:* old European international public law still was aimed at what were considered civilized and civilizing nations as subjects of spatial order. But through the progressive equality gained on colonial soil, through "a complete muddle of international legal titles," this traditional legal ordering instrument gradually abdicated, until with the Great War Europe completely imploded as a power-political structure.[65] Since the Versailles Treaty, Europe no longer determined the global spatial order. It became, for its part, a field for those engaged in spatial ordering, or better: space-securing practices. In *Origins of Totalitarianism* Hannah Arendt describes this process as a rebounding of the techniques of bureaucratic rule that Europe had imposed on its emigrants and colonies. The earlier colonial powers in a sense experienced inner colonizing, so that the bureaucracy now began to dominate its own terrain and suspend previous generosity.

And in fact, with the Great War domicile-registration and especially mandatory receipt of a passport for travel were enforced with increasing rigor in all participating states. In France, passport regulations from the revolutionary period

62 See Schäffner 2002, pp. 1–12; Siegert 2004, pp. 260–276.
63 The lists of passengers and goods for ships leaving for the New World were organized in tabular form, so that their data, indicating land of arrival and property or even only place of residence could be directly transferred to the colonial grid-squares. See Schäffner 2005, pp. 47–56; Siegert 2006, pp. 130–137, 142.
64 Hegel 1914, p. 161.
65 Schmitt 2003, pp. 215 ff.

were reactivated, and in 1914 England issued a rigid Alien Restrictions Act, burdening "non-British" appearing persons with proof of citizenship. Germany sharpened the state-of-emergency clauses of its liberal passport law of 1867, introducing obligations to possess identity-papers, a passport, and a visa under increasingly strict standards. And in 1924, with its Immigration Act, the United States authorized its embassies and consulates to already implement ongoing quotas on European soil through selection processes, thus guaranteeing what John Torpey has described as a kind of "remote border control."[66] All these directly decreed measures, implemented as ordinances and alterations in ordinances, had the effect of lastingly installing regulations said to be meant for emergencies. In that the measures affected those without fixed and demonstrable abode or those located in the international-legal zone of a newly arranged European territory, they produced stateless persons; naturally no diplomatic protection was envisioned for individuals lacking demonstrable citizenship. It would only be in the course of the 1920s, with the expansive distribution of so-called Nansen passports, that displaced persons were granted a state at least on paper and—in any case non-obligatory—international-legal protection was at least conceivable.[67]

"What is unprecedented is *not* the loss of a home but the *impossibility* of *finding* a *new one,"*[68] wrote Arendt in an oft-quoted passage in *Origins.* With new border-demarcations, displaced persons such as Stanisław in *The Death Ship* could even be created through special regulations—in the case of the old Austro-Hungarian territories, for example, demonstration of a specific "right to a homeland," a *Heimatrecht* or *Indigenat*, was to allow affiliation with one of the empire's successor states. Or the persons involved could be of non-European or simply indefinable provenance, thus getting lost within the European border-regime—just like Gale, who cannot convince "his" counsel, as representative of "the state-concept" (243), of his American citizenship, thus being forced to henceforth present himself as a German, Egyptian, or simply "nobody" (26). "He could struggle against storm and waves," we read in *The Death Ship*, "not against clauses, pencils, and paper." (70) Seamen who could not present their identification papers or certificate of hire and missed their ship, thus having no access to the crew list, were predestined displaced persons. It would be 1954, before signatories to a pertinent United Nations convention made the following concession:

66 See Torpey 2001, pp. 256–270.

67 See Jürgens 1986, pp. 52, 90.

68 Arendt 1976, p. 293.

> In the case of stateless persons regularly serving as crew members on board a ship flying the flag of a Contracting State, that State shall give sympathetic consideration to their establishment on its territory and the issue of travel documents to them or their temporary admission to its territory particularly with a view to facilitating their establishment in another country.[69]

But even here in 1954, a precondition was legally proper service on a legally proper ship. In the 1920s, stateless persons were subjected to a regime of emergency decrees revealing caprice by both bureaucrats and the police. Because and as long as they were not meant to be naturalized, they could be dealt with in one of three ways: They could simply be murdered "with a chill gesture," (238) since they in any case stand outside the world of prevailing law. "It would be simple to kill all these people so that 'official processing' can be taken care of in a peaceful, orderly way," wrote B. Traven in 1926. "But the birth rate is increasingly low and the war has also swallowed up its millions of people."[70] Second, they could be interned until further notice. And third, they could be deported elsewhere—a process that would only sometimes take place as removal to a specified locus but, once some former colonies had become states, could involve loci of indefinite passage, modern inter-state non-places or state-free spaces such as the sea. A fourth, non-existing procedure would have been the old institution of asylum exempt from penalty, now put aside and simply declared a matter for state and police to the same extent that the passport system brought a halt to unchallenged passage. That system had created an optimal milieu of entrapment, within which civil rights were suspended for some and excessive police presence prevailed.[71] For B. Traven, the ship's "heterotopia" is thus nothing more than the perilous place of refuge for rootless individuals from all corners of the earth. Although the ship takes in displaced persons, it is now anything but an embodiment of the *principium individuationis* to which ships have always been tied.

"Better a release paper from a real prison then the paybook of a death ship," (151) says Gale. He is a member of a "fifth rank" that in its literal "alienation" stands infinitely below the "fourth" of workers on the high seas. (248) He has been deported without maintaining an even *ex negativo* relation to the legal or state system. Gale in this way serves as a modern revenant of the "man without

---

69 Office of the United Nations High Commissioner for Refugees, *Convention Relating to the Status of Stateless Persons*, chapter 1, "General Provisions," article 11, "Stateless Seamen," https://www.unhcr.org/ibelong/wp-content/uploads/1954-Convention-relating-to-the-Status-of-Stateless-Persons_ENG.pdf.

70 B. Traven 1926, p. 36.

71 See Vogl 2003, p. 35.

a Country" who in Edward Hale's like-named story of 1863 is interned on a ship because of an unguarded curse against his homeland: a law court rules, for the sake of example, that he is never to see or hear of his country again, and to that end is to enter an endless voyage over the world's seas, like a modern *Flying Dutchman.* He is thus kept in stateless space on American navy ships—a space symbolizing, it is true, no legal statelessness, but certainly its ideal counterpart.[72]

This paradoxical because patriotically and juridically decreed "heterotopia" consequently extinguishes neither the legal subject nor the subject's affiliation with a state, but only lived affiliation. And just as in the case of the *Flying Dutchman*, provision is always present for the ship's wandering souls to find no salvation with their vessel's sinking. Under the "emergency maritime law," (218) Gale by contrast comes aboard the *Yorikke:* corresponding to this emergency law, stateless persons can be arbitrarily impressed because no statements they make have legal validity and owners or commanders can always say they have entered into an oral contract with them. "Prison, jail, or the hangman" (188) would all be preferable to the death-voyage, because all of those alternatives at least grant the status of a legal subject. The *Yorikke* is for its part the vehicle of complete mortification. It moves "death values," *Totenwerte* (208), is an autonomous disciplinary apparatus for the intake of symbolically dead persons and a machine for the regulated production of real dead persons. The boiler room merely constitutes the inner circle of hell.

The Insured Death-Voyage

A death ship houses a risk-bearing community in a perverse sense: the crew is mean to guarantee and work toward actual realization of its voyage's potential dangers; and this for the sake of the owner capitalizing from the risk-bearing community he has insured. Consequently, insurance technique, the bureaucratic mastering of danger as calculable risk, is what sees in its perversion to a disposable biopolitical mass becoming thanato-political capital. And in that statelessness means lawlessness, such insurance fraud remains simply insurance fraud, and is not murder. "I was not born," says Gale, "had no seaman's card, could never in my life get a passport, and anyone could do with me what he wanted, since I was a nobody....A dead man can be defiled, but can't be murdered." (83) Death ships are a phantasm because, as already articulated in the traditional

72 See Hale 1968, esp. pp. 13–16, 22, 38.

saga, they are only perceived by those who are already dead[73] ("there are no death ships," spells out the narrator). (251) Death ships have always been in transit to the beyond: for mariners' culture (and for an anthropology of maritime labor), those voyaging on them are dead; and for a culture of security they never lived.

Under normal biopolitical circumstances, this security culture is meant to protect life. But in the exceptional thanato-political case, it moves back to its original sense: securing capital. As Alexander Bene already observed in 1928, with marine insurance, the first regulated form of any sort of insurance, "only interests in the overcoming of the dangers of maritime voyaging in respect to ship and goods" were insurable, and this meant real interests, assessable in money.[74] Because especially around 1900 insurance was seen as creating objective, strictly calculable mutual liability, a practice such as over-insuring was considered what in 1925 Heinz Heike termed a "crime constituting a public danger," a first step toward insurance fraud.[75] But marine insurance has always possessed a special speculative dimension: it is not only that it involves a business centered around risk and insecurity; rather, on account of its multifarious measure of damage, evaluation of risk is here grounded, as Eugen von Liebig put it in 1914, on "experience and a resulting feeling of the suitability of the premium."[76] While marine insurance thus opens a play of possibilities, insurance fraud operates, in addition, with an appearance of truth. Already in 1777, in order to arrest the "accident doctors" and the increasingly speculative activity on the insurance bourse, a London parliamentary act declared policies without insurable interest invalid. Then in 1906 the Merchant Shipping Act stipulated that "[e]very contract of marine insurance by way of gaming or wagering is void," since such contracts were "based upon the utmost good faith" and only legitimate "in the absence of fraud."[77]

---

73 See Smidt 1849, p. 129: "This is the ship of death: invisible to any human gaze and only visible to he who must die."

74 Bene 1928, p. 33.

75 Heike 1925, p. 5. See also ibid., p. 48f.

76 Liebig 1914, p. 218. While initially Lloyd's emerged from direct contact between maritime and commercial adventurers, but at the same time from various types of betting-based insurance, organizing this speculative dimension remained in a certain sense inherent in the business. For risks were here being taken on that, on account of their basic incalculability, established insurance companies would never have assumed. The practice of underwriting itself makes clear that, as we read in Anderson 2007, p. 145, "insurance has never really divorced itself from the siren-song of speculation."

77 Marine Insurance Act (6 Edw 7 c 41) [21 December 1906], art. 4.1., 17, 27.3.

Following the Great War, however, the value of ships rose rapidly, to the same extent that cargo-volume shrank. For many shipowners, insurance fraud offered the only chance of survival. Hence those stateless persons who wished to evade continuous deportation or internment were as it were predestined to form the crew of death ships. Without identity papers they were absent from the registration system for health and accident insurance, which shipowners were obliged to finance for sea voyages. This absence led, in turn, to the lack of any basis for surviving dependents receiving a pension. This naked, uninsured life became mere items within a biased gambling account, speculation on wreckage and death. To "ride out the insurance" is the term used in *The Death Ship* (166): dead people and "dead values" are on board. Seamen are the lowliest of lowly slaves, not even being purchased, and certainly not being "highly insured as precious goods." (146) They do not even have the value of the fee for insuring transported goods imposed for the "successful voyage" of slave-cargo until the nineteenth century.[78] They are fond of calling themselves the children of "Mother Lloyd in London" (214), with her monopoly on maritime insurance—and the passport business as a sideline—since the eighteenth century. But because Lloyd's only does its business with mercantile "hope," rendering the prolonged disappearance of its ships into a speculative object, they are rather children of adverse fortune or, to again cite Ernst Jünger's formula, a mythic "mother of danger."

Marine-insurance fraud requires more than evasion of administrative and nautical regulation; it is also necessary to use such regulation for capitalization. Its aim is to render the fact of the death ship into a simple phantasm, entering into a strategic coalition with the culture of security to that end. It is the case that already in 1865, the "coffin ships" were a topic for British parliamentary debate, after the social reformer and MP Samuel Plimsoll showed that 80 percent of all recently drowned seamen had fallen victim to "death ships." Because of his efforts, in 1871 a royal commission was formed to investigate conditions on British trading ships, and in 1876 a maximum waterline depth for loading was stipulated.[79] Further steps in a concerted effort to prevent shipwrecks were the establishing of classification societies for controlling ships' material, construction, and outfitting; use of ships' registries as a means of risk assessment; formation of Lloyd's Shipwreck Committee and different national lifesaving societies; the legal investigation of accidents at sea; and, in 1914, the International Convention for the Safety of Life at Sea.

78 See Ebel 1963, pp. 210–212.
79 See Lehmann-Brune 1999, p. 225.

As a result of such developments and measures, a death ship's voyage had to follow certain procedures for concealment. First of all, standards for outfitting, loading, and proper registration needed to be fulfilled. The ship had to be "flagged" with a suitable nationality, and credibly overinsured before a crew of bureaucratically dead persons was hired or impressed. On board, fulfillment of prescribed security measures had to be simulated and the sabotage of the ship's equipment dissimulated. On the *Yorikke*, the main lifeboat needs to be constantly maintained so that it can take up the ship's command when necessary. The dubious state of the second boat, reserved for the planned drowning of the deathly voyagers, here furnishes "the signal for the burial" (208). "This knowledge while waiting, this trembling counting of the days from one harbor to the next" (248) leads to a vanishing of Gale's hopes proportional to the increase in the calculative expectations of the death-ship's owners: "The death certificate is already in the company's hands, they simply need to write in the date" (202), for through uncomplicated claim settlement using Lloyd's "open form," prompt payment for the sinking is assured. This is guaranteed first of all by Lloyd's good name, and that of those it insures, but also that of the individual underwriters, who are tellingly named "names."

The operations of this insurance were only made possible through Lloyd's system of "self-insurance," grounded in mutual trust and informal reliability; industrialized shipping commerce, functioning on both legal and, repeatedly, illegal levels, profited enormously from this form of insurance. (Already in the time of "free for all" trade in coffee houses, a broker replaced official partnership control, for which reason Lloyd's reputation would have quickly been ruined, if not only individual underwriters but the company's own brokers were caught in fraudulent activities—for instance in pretending to be underwriters in order to collect the full premium.[80]) Since 1774, the concluding bureaucratic procedure of death confirmation unfolded in actual ritual form: in "Lloyd's Room," a Casualty Liaison Officer on a podium, entered it into the Loss Book with a quill pen before a kind of death announcement was published in *Lloyd's Shipping News* under the rubric "The Toll of the Sea."[81] The name of the ship compensated for the anonymity of the crew.

On the path of courts and administrators, there were no signs of thanato-political insurance fraud, for either inception of the insurance case coincided with the drowning of all crew members not inaugurated into the fraud, or there were

80 See Gibb 1957, pp. 20 – 23; Röpling 1960, p. 46; Supple 1984, p. 2.

81 See Cameron 1984, pp. 80 ff. For a contemporary description of the procedure see Egon Erwin Kisch's "Shipping Exchange," in Kisch 1979, pp. 58 – 61.

survivors who had already been at least bureaucratically dead, hence incapable of being witnesses, before being allowed to step on board the ship. In any case both physical and symbolic survival was only possible for crewmembers on a death ship through somehow surviving the fraudulent sinking and then participating in the shipowner's fraud. Speedy news of the sinking had concrete monetary value. If "the report does not come in, then the lost-vessel waiting period sets in, and that means a significant loss of interest." (263) The bureaucratically dead seaman had to be belatedly legalized by the fraudulent shipowner in order to be able to attest to his good faith. A fraudulent shipwreck that he fraudulently confirmed to have been genuine was thus the only chance for a kind of rebirth. Otherwise, there was "only one way of leaving service," as Gale puts it: "Signing off on a reef." (262) After he succeeds, together with Stanisław, in immediately surviving the sinking of the *Empress of Madagascar,* only to then see Stanisław swallowed by the waves, Gale will reappear in another novel of B. Traven[82]—but as the inner logic of *The Death Ship* has it, it is only as a wily wanderer that he could have stayed alive.

For Hans Blumenberg, "[s]hipwreck, as seen by a survivor, is the figure of an initial philosophical experience."[83] It remains a key image of reflection in an epoch where a right to existence is less guaranteed by legal and substantive concepts than daring and deception—and this not despite but on the very basis of a purely objectively and calculatively followed imperative of security. This form of existence is well captured in Pascal's *vous êtes embarqué,* "you are embarked."[84] It was Pascal who mathematically prepared insurance calculus (or at least the calculus of life assurance) through that of his famous wager, even attributing an existence as arbitrary as it was fraudulent to his sovereign. As he puts it in one of his "Discourses on the Condition of the Great," the prince is like a shipwrecked person landed on an unknown shore and mistakenly deemed a missing king by the inhabitants: "not only do you find yourself the son of a duke, but also do you find yourself in the world at all, only through an infinity of chances." In the end, the sovereign's status "is not a title of nature but of a human institution." A person's semblance is owed to matters of pure coincidence—in the case of someone shipwrecked, to naked survival, in that of the prince to birth and the unquenchable desire of the populace for a Great Other. And maintaining one's life is itself a matter of simple fraud, for both the stranded wretch and the

82 *Die Brücke im Dschungel / The Bridge in the Jungle* (1929); Gale—as Gerhard Gales—already appears in B. Traven's first novel, *Die Baumwollpflücker / The Cotton Pickers* (1926).
83 Blumenberg 1996, p. 12.
84 Pascal 1910, p. 85.

prince a masking of one's "true state."[85] Nietzsche will then no longer speak of the shipwrecked prince and his state secrets, but rather of shipwrecked Every-man and his *cunning survival.* Since then it is the "enormous structure of beams and boards" made up of mere "transferences, metaphors, and meto-nyms...to which the poor man clings for dear life."[86]

Gale's career offers the textbook case of an existence holding itself over water in the age of security through fraud and simulation, the mere semblance of truth and person. But this process extends from individual persons to the form of existence of entire states that B. Traven elsewhere described as financing themselves with the registration fees for flagging death ships.[87] As presented in *The Death Ship,* it took in the shipowners who, for instance, assured the insur-ability of their ships through renaming. And it took in especially the displaced persons who, in a sea of bureaucracy, only manage to maintain their right to life through a simulated or symbolically newly "situated" persona.

It appears that in the 1920s, a development extending back to the beginning of Early Modern behavior reached full flower. Namely, if in sixteenth-century Seville, the wish and intent to undertake the India run stood at the start of all modern official identification procedures, already here the fictive core of bureau-cratic certification of existence revealed itself. The applicant had to submit his application at the *Casa de Contratación* together with his name and life circum-stances, and then personally certify the correctness of this information; the cor-rect nature of this certification was then to be certified by witnesses furnished by the applicant himself, and this certificatory step had to be entered in the records anew; but not rarely, the identity and credibility of the witnesses had to in turn be certified by the mayor, whose own identity and credibility had to be certified by an additional clerk. "This is followed by a series of 'frames,'" writes Bernhard Siegert, "the text of the certificate is framed by the clerk's signum (something between a seal and an emblem, this is framed by the signature, the signature by flourishes filling the paper's free space until the margins, meant to hinder later additions, and all this is finally framed by the words *es bastante,* 'it suffi-ces.'"[88] The language and writing, which cannot demonstrate their own veracity, are grounded in the Great Other, the authority that can pronounce "it is enough."

In 1812, when the bureaucratic identification-practice threatened to overrun all the German-speaking lands and a "closed trading state", seemed on the ho-rizon at least in a policing context, Friedrich Schlegel recalled Fichte's prediction

85 Ibid., pp. 383, 388.
86 Nietzsche 1989, pp. 254 f.
87 B. Traven 1976, p. 161.
88 Siegert 2004, p. 274.

that "the moment the world should see a perfect police, the moment there should be no contraband trade and the traveler's pass should contain an exact portrait and biography of its bearer, that moment it would become quite impossible to write a good romance; for...then nothing could occur in real life which might, with any moderate degree of ornament, be formed into the groundwork of such a fiction."[89] Schlegel was here pointing to the potency of what he terms *das Polizeiwidrige*, "what is unfavorable [or: repugnant] to good order," and what J.J. Lockhart's nineteenth-century translation rendered as "irregulated and dissolute conduct"[90]—which we can in turn translate into fraud and fiction in discourse and writing: since the seventeenth century already, the *passeport* served as the synonym for a bribe, and the literary figure of the *picaro* as an image of fraudulent existence, one dismantling the person as a play of names and the signs of authority as simple fiction. Between the world wars in Europe, what in any event remained to somehow certify was not—largely unmolested—vagabond life but what Thomas Jürgens has referred to as a "mass loss" of nationality and citizenship. The very fact that the statistical existence of such displaced persons was increasingly undeniable, even without warranty and papers a "paper homeland" was gradually granted them in the form of the League of Nations' Nansen Passport. Admittedly, that document could itself only be valid "in that it feigned a relationship between the community of states and the stateless party that in many respects was comparable to that between the citizen of a state and that state."[91]

B. Traven's Sea Change

We can understand *The Death Ship*'s Gale to be a roving picaro, always addressing the Great Other while at the same time succeeding in undermining the "Inquisition" of territorial authority or even making use of it for himself.[92] Speaking for this is the novel's general structure, with Gale's actual narrative account framed by an interrogation situation or its emergence from the interrogation it-

89 Schlegel 1841, p. 271 (12th lecture on the history of literature, treating Cervantes and others).
90 Ibid.
91 Jürgens 1986, pp. 59, 92.
92 For the theory of the emergence of picaresque narrative as a response to the Inquisition and its relentless "examination" of Spanish *conversos* on suspicion of secretly practicing Judaism, see Gitlitz 2000, pp. 54–59; on that narrative's connection with the questioning of travelers in *Casa de la Contratación*, see Siegert 2006, pp. 89–93.

self by a "sir."[93] For just this reason, the starting premise for all the novel's discourse is that unfolding here is a confrontation between a sovereign will to know and an autobiographical play on truth and discursive ritual of self-justification, self-production, and conversion. The text offers a "second-order observation" in that Gale's supposed experience-based account pertains to a silent counterpart whose reaction—elided from the text—is reflected in Gale's narration. The fictional character and unreliability of the narration is thus scarcely hidden, as is the constructed character of what is narrated and its "verisimilitude" or "probability."

We might, however, also consider Gale a modern Odysseus, a wily, wandering nobody and sly everyman. Between the Scylla of bureaucratic identity-determination and the Charybdis of thanato-political exclusion, he ends up sallying into a passage where both nationality and names are used as means of existential simulation. Here the proper name, meant to serve as a linkage between deictic indication and meaningful statements, between the referential and predicative function of language, is the precondition for a modern identification procedure, the production of documents and thus determination of legal affiliation with a state. It has become what, the sovereign arbitrary act placed aside, is meant to place a check on the endless regress of linguistic certification; it is the nail on which descriptions are attached. But to speak with Walter Benjamin (citing Hermann Ungar), this poses a basic question: "Does the name attach to us, or are we attached to a name?"[94] Gale and consorts, their names as uncertain as their "being," make use of just this uncertainty: among the seamen, what prevails above all on land and on the water, in offices, in hiring, even in confidential conversation, is not the "Great Other" or the documentary "thing" but the effort at and temptation to "deception" regarding one's "true name" and "true nationality" (74, 218 f.). What remains uncertain is whether in decisive moments a performative or rather constative speech act has taken place. In the context of official or academic expectations that speech acts have a regulated nature, the language use at work here could be denigrated as hollow, stunted, or parasitical. J.L. Austin offers what in our context is the more apt term "sea change."[95] For profound, irreversible change overtakes every true passenger, severed from

93 On the discursive situation in *The Death Ship* and the continuation of different "language games" within the interior plot, see Helmes 2012, pp. 184, 191.

94 Searle 1969, pp. 172 f.; Benjamin 1999a, p. 866. On proper names as the end of a chain of "subjectively and objectively localizing singular terms and sortal predicates", see Wolf 1985, p. 41.

95 Austin 1975, p. 22.

land and *nomoi*—and in *The Death Ship* all official discourse and personal narrative, whether under the sign of uprightness or fiction.

It is nearly self-evident that such sea change also takes in the supposedly final basis for writing itself. The putative subject of the fiction named *The Death Ship* vanishes beneath the authorial name "B. Traven," laid claim to by both the German and American literary business. To the extent any biographical details behind that name have become clear, it seems that B. Traven was born near Lübeck, already traveled on sea as the child of foreign actors, and published the anarchist paper *Der Ziegelbrenner* between 1917 and 1921 in Munich under the name "Ret Marut." With the outbreak of World War I, he suddenly claimed American citizenship, his alleged birthplace, San Francisco, evidently being chosen intentionally, as listing in the birth registry there would have been very difficult due to the recent fire. Despite his false sworn testimony, he failed to convince the U.S. authorities; nevertheless starting with the American entry into the war he took on the name "Otto Feige" (which apparently had been his original legal name), probably to forestall possible reprisal in the German Reich. After participating in the Munich Council Republic and having been supposedly shot dead by White Guardists, he managed to flee to London, where in 1923 he was arrested because of doubtful claimed American citizenship—and may have written the first English-language version of *The Death Ship*. Without his knowledge, his admission to the British authorities that he was a German citizen was conveyed to the USA.[96]

In 1924, B. Traven noted in his diary that "[t]he Bavarian of Munich is dead."[97] In London he signed on to stow coal on the Norwegian *Hegre*, which had itself changed its name and owner a dozen or so times, like the *Yorikke*. Because his name was crossed off from the Norwegian crew list, he must have missed the voyage's start or jumped ship at one of the first harbors. What is documented is his trip—probably from Amsterdam or Antwerp—via England to Canada, where his trace vanishes.[98]

In 1926, a certain B. Traven enjoyed international success with the German-language version of the book (later to be revised several times),[99] since then is-

---

96 See Hausschild 2012, pp. 335 f., 477.

97 Cited from Guthke 1990, p. 33.

98 See Hausschild 2012, pp. 467, 479 f.

99 See Dammann 2012, pp. 15–18, 26 f. for (1) the complicated publication history of *The Death Ship / Das Totenschiff*, including the translation back into German of the version that appeared in New York in 1936 (itself translated from the German, then revised); (2) the 1930s publication-politics backdrop; and (3) B. Traven's strategy to succeed repeatedly as both an American author and, in postwar Germany, through an "original translation" of his book from the English. On the

suing the "B. Traven news" and various other narrative texts. In 1946, Egon Erwin Kisch noted that "for Oskar Maria Graf, Traven remains Fred Maruth." Meanwhile he had also been identified as Hal Croves, a Chicago-born American of Swedish descent living in Mexico, and as Señor Torsvan.[100] He was said to have confirmed, in conversation, that "my vita would not disappoint people"; and in the introduction to a "Detective Story" discovered in the German Federal Archives in Berlin in 2008, we read that "[t]he true life is much more like a novel than anything that could be made up by even the most imaginative writer."[101] Already in 1926, however, B. Traven argued that "the creative person should have no other biography than his work."[102] Evidently, in his texts he brought together two conflicting tendencies of contemporary authorship. On the one hand, observed the Russian Formalist critic and theorist Boris Tomashevsky in 1923, "the writer shows his readers his own life and writes his own biography, tightly binding it to the literary cycles of his work." On the other hand, he continued, "[t]here are writers...without biographies. To attempt to compose biographies for the latter is to write satires or denunciations on the alive or the dead as well."[103]

With the letter-sequence "B. Traven," we quite clearly are not first and foremost presented with a defense of the legal right, under laws of privacy and person, to one's own name, as called for by various jurists around 1900. "B. Traven" is neither a real proper name nor a mere pseudonym. "B. Traven" also does not have the sole purpose of, as Paul Sandau put it in 1932, "independently denoting" the person thus designated "for certain expressions about life unfolding in a certain direction and having a lasting character."[104] B. Traven *is* B. Traven in the sense of writers being part of a discursive interlinkage that they do not produce as much as it produces them. In this way he not only attests to the "impossibility of salvaging the *I*," connected in the 1920s to massification and mediocrity (to the "they" and "nobody"),[105] but, more so, to a wily simulation executed with truth-effects accepted as literature within the modern understanding. Put otherwise: On the one hand we have the pseudonym, meant to leave the hidden civil name intact, and to serve for protection of intellectual property by guar-

history of the novel's different versions taking account of both textual variants and questions of style, see Potapova 2012, pp. 63–66, 72–78, 82f., 90.

100 See Guthke 1990, pp. 32ff., 114, 135ff., 339.

101 Cited from Hausschild 2012, p. 34.

102 B. Traven 1926, p. 34.

103 Tomaševsij 1971, p. 55.

104 Sandau 1932, p. 7f.

105 In this regard and with a view to Weimar see Oschmann 2011, pp. 294–298.

anteeing authors their author-role and the usual copyright on work defined as their own. On the other hand and by contrast, "B. Traven" substitutes himself for a civil name, in order to grant the proper name's author the protection and credit of fiction, even if this proper name only has sense but no reference.[106] "B. Traven" cancels both the conventionalized because unambiguous "fictionality contract" and every reliable "autobiographical pact" (that term is Philippe Lejeune's), for the sake of instead exercising a function that is protective and deceptive at once. "I wish to be nothing but word," wrote Ret Marut in the *Ziegelbrenner.*[107] Perhaps the name "B. Traven" denotes the specifically modern relationship of writing to death described (amongst others) by Maurice Blanchot, who vividly demonstrates how the function of authorship is bound up to a specific legal and state system. But finally, he is pointing to those "gaps and tears" revealing the author's "disappearance," his symbolic death and fraudulent survival.[108]

B. Traven's specific way of writing, within which the placelessness of its author, the speculation over his vanishing, is inscribed from the start, can perhaps be characterized, moving past a form of "thanatography," through the concept of a peculiar "thanato*po*graphy." As a finessing of the bureaucracy's symbolic and medial components, this simultaneous displacement of written and narrative authority, of real and fictive persons, represents an elementary technique of survival. In the strange personal union at work here, which continued to be staged on a hidden level,[109] B. Traven, like Gale, was consistently focused on circumnavigating death or wreckage. Where proper certification of existence may appear to be paradise on earth, thanato-political exclusion as hell at sea, passage should proceed between the two. Not only thematically or metaphorically but structurally as well, B. Traven's writing is in this way seized by nothing less than a sea change. That *The Death Ship* is a maritime novel is anything but a small detail.

---

106 On the relation between pseudonym and civil name see Manes 1899, pp. 1f., 6, 77. On the "true author's sign" of anonymous and pseudonymous twentieth-century authors, see Pabst 2011, pp. 25, 27. On the distinction between "sense," *Sinn,* and "reference," *Bedeutung,* see Frege 1960, pp. 56–78.

107 Cited from Söhn 1974, p. 154.

108 See in this regard Foucault 1977, pp. 116 ff. On B. Traven's idiosyncratic conception of "anonymous" authorship, untroubled by any "organic" relation to a text, and on his simultaneous striving for unlimited, conscious control over the various versions of his works and their distribution, see Potapova 2012, pp. 91–93.

109 It appears that B. Traven / Croves / Torsvan continued to identify with Gale in the context of his "civic existence": he named his studio "The Bridge," saw himself as a skipper, his wife and daughters as officers first through third class, and his household as something to be led under the motto "everything must be saved as if it were a ship." See Chankin 1975, pp. 117 f.

At stake in writing as in navigating is occupying a placeless space, making possible a specific, heterotopic manner of speaking, writing, and indeed existing. When it comes to the classic authors of sea-novels, B. Traven is probably at significantly greater distance from Conrad (Józef Teodor Konrad Korzeniowski), an exiled Polish nobleman who could find a new symbolic home precisely at sea under the English flag,[110] than from a writer like Melville, whose *Confidence-Man* is set, with the *Fidèle*, on a ship of fools, staging masquerades and fraudulently certified names, and whose *Moby-Dick* can be read in part as a casuistic comment on the State Fugitive Law.[111] It must in any case lie in the tendency, driven by geopolitics and administrative technology, not to distinguish between land and sea that the old metaphor of the ship, here as the ship of poetry in particular, could again emerge as "the greatest reserve of the imagination" in a bureaucratic culture of security.

B. Traven's Dantesque motto above the *Yorikke*'s entrance to hell is thus more than a vague allusion to working-class exploitation, concentrated in the ship's hold and epitomized by the boiler room. Beyond that, it marks a genuine sea change, transferring the ancient and medieval topos of a salvation-worthy life at sea into modernity's geopolitical-bureaucratic topology. In the *Divine Comedy,* it initially appears to adhere to the traditional commonplace of the sea's hostility, its menace of final forgetting and death. It is the placeless locus of an illegitimate transformation. It beguiles into pride and reckless curiosity, transgression of *ordo,* avarice and an illegitimate quest for Paradise. Those voyaging under its sign can at best represent, like Ulisse, an existence meriting salvation. But in the ancient and medieval worlds, the sea did not only invite shunning or condemnation. In the geography of the imaginary, it also served as an inexhaustible source of hope and expectation.[112]

With his *Comedy,* Dante also assigns pagan history its place in God's spherically organized memory. In the end everyone, even a castaway like Ulisse and an exiled and persecuted person like Dante, finds a place in the salvational economy. In the ground for Ulisse's damnation, the deceptive hope he catalyzes at sea, we already find a conception of that secular, illegitimate hope to which Early Modernity's profit-oriented seafaring must lead. Hope is sin when it fails to draw exclusively on a direct relationship to the Absolute. But speculation on *for-*

110 See Conrad 1995, IX., p. 138: "The Red Ensign—the symbolic, protecting warm bit of bunting flung wide upon the seas, and destined for so many years to be the only roof over my head."
111 This law was passed in 1850 and implemented for the first time by Melville's father in-law Lemuel Shaw, something alluded to especially in *Moby-Dick*'s chapter 89, "Fast-Fish and Loose-Fish. See Pechmann 2003, p. 161.
112 See Heydenreich 1970, p. 11, passim.

tuna di mare as practiced for the first time in Dante's period offers no transcendent promise. Rather, it operates intuitively with probabilities that in the fourteenth century would have still been considered a deceptive semblance of the truth. A "passion ship" navigating into the harbor of blessed retreat with the help of faith, love, hope would soon sink into direct allegory,[113] while spiritual states such as hope and expectation became trading items on the insurance market.

For Dante, insurance is always fraud even without insurance fraud, and indeed fraud against the Absolute. This is the precise conceptual framework of the hellish scenario in B. Traven's *Death Ship*, with the fates of all the novel's characters standing beneath an insurance market's regime—a market upon which hope is more or less fraudulently capitalized. Under this sign, we can understand Gale / B. Traven as a modern Odysseus, whose deceit represents not the basis for his damnation but its outcome. Embarked from time immemorial, he must demonstrate wiliness, not only *virtù*. "For the daring which encounters the sea must at the same time embrace wariness—cunning—since it has to do with the treacherous, the most unreliable and deceitful element," writes Hegel.[114] "Being cast up and being cunning are equivalents in Homer," observe Horkheimer and Adorno, playing on Odyssean *Verschlagenheit*, wandering wiliness.[115] For Dante, who was not familiar with Homer's Odysseus, this equation becomes the title of accusation in a faith-centered trial that Adorno and Horkheimer will still carry forward as reason's Post-Hegelian legal proceeding against itself.[116] However, the nominalistic fraud that "nobody"-Odysseus still practices against language itself as B. Traven / Gale does not work in favor of instrumental reason but, very much, against it.

To follow *The Death Ship*'s infernal motto: simultaneous extinction of name and being is to be thwarted by preventing congruence of the two. In Dante's *Inferno,* name and being not only coincide but also participate, even in their damnation *ex negative*, in a terrestrial legal and divine-salvational order. But the sea change through which, in *The Death Ship,* both name and being try to escape extinction is inconceivable in Dante's representative damnation, as well and in Hegel's related figure of reflection. Some later thematic variations on the *Divine Comedy* have centered on just this thanatographic poetics: in Primo Levi's *If This is a Man,* for example, he describes how, as a prisoner in Auschwitz, in

113 On the seventeenth-century allegorical "passion ship", see Reinitzer 1988, pp. 114 f.
114 Hegel 1914, p. 90.
115 Horkheimer and Adorno 2002, p. 50.
116 Ibid.

struggling to recall Ulisse's deceptive discourse he finally forgets "[f]or a moment...who I am and where I am", for the sake of still retaining hope in the face of the unceasing extermination;[117] and in Pier Paolo Pasolini's writings (as in *Divina Mimesis*), poetry is evoked as, in his view, the last non-economic because deceptive court of appeal against the unnoticed "genocide" of a proto-fascist market economy.[118] It is the traceless vanishing of entire forms of existence that in B. Traven's text, with and contra Dante, is meant to at least be provided a name. In its citation of Shakespeare's "poor" court jester Yorick, the ship' name, *Yorikke,* not only attests to the extent to which industrialized seafaring relied on "ships of fools," upon which only picaros like Gale could survive. It also cites the *memento mori* forced on Hamlet (5, 1, 172ff.) through sight of Yorick's skull. "There are no death ships" only officially means that neither banishing nor extinction of "name and being" took place. That against this bureaucratic reality-command and fiction-ban, "death ships" exist, namely in a realm between reality and fiction, is something certified by B. Traven's thanatopography.

A canonic representative of the modern sea-novel such as Joseph Conrad understands seafaring and authorship as existential analogies. With both, we find forms of life nourished on a rich tradition, drawing on specialized knowledge, propagating solidarity, and building on careful craftsmanship. Conrad's poetics of fidelity is tied to voyage by sailing vessel for just that reason. But a text such as B. Traven's *Death Ship,* with its systematically deceptive form of writing and thanatographic import, has to lie beyond that horizon. Such a text attests *both* to the end of the sea novel and to the decline and fall of a vocational milieu centered on exposure to elementary dangers and active safeguarding from them. Seen in this way, breaks in literary tradition point to broader historical upheavals such as the shift from sail to steam. As we have seen, for Conrad far-reaching fault-lines were tied to this shift, involving the general transformation from an old culture of danger to a new security-culture, as shown in the vanishing of a range of mariners' customs. Already the stowing of cargo, previously "a skilled labor," here served as a symptom: "The modern steamship with her many holds is not loaded within the sailor-like meaning of the word," wrote Conrad in 1906. "She is filled up. Her cargo is not stowed in any sense; it is simply dumped into her through six hatchways, more or less, by twelve winches or so, with clatter and hurry and racket and heat, in a cloud of steam and a mess of coal-dust."[119]

---

117 Levi 1959, p. 133.
118 See Pasolini 1999, pp. 511–514; Pasolini 1983, esp. p. 54.
119 Conrad 1995, 9, p. 47.

With the traditional art of proper stowing having been abandoned, already because of the lower balance-dependence of steamships, by the turn of the twentieth century, after World War I a new form of loading gradually came to the fore: that tied to containers. Starting with the crisis that overtook rail transport in the interwar period after trucks transport on land became increasingly cheap, a modular logistics developed based on standardized container measurements. For one thing, this allowed connections between different transport means without laborious reloading and redeployment. Tied to this, because for the first time ships were designed for the massive intake of these standardized containers and many harbors refitted for intermodal transport, the smoothly functioning, massive movement of goods straight through the borderline between land and sea was now possible. With the steamship developing into what Alexander Klose has termed a "continuation of the railroad through other means," a "terrestrialization of the sea" now took place, while in the course of eased international commerce, territories were as it were rendered fluid "through global streams: of goods, people, and money."[120] But the unlimited freedom of movement apparently made gradually possible through the parallel interlinkage of production, distribution, and information streams has been accompanied by intensified control over movement. Container ships are especially easy to control from the land, because operating under a maxim of greatest possible efficiency they have reduced their crews to a minimum.

In this respect they resemble modern tanker ships, whose size has grown in inverse proportion to the size of the crew. Similar to container ships, since the 1960s the fuel supplied by tankers has sustained not only individual nations but the entire international economy. The life and trading capacity of millions of people thus now depends on ships to an unprecedented degree. And notably, the class of ships involved marks a new fundamental break in seafaring tradition. This is manifest first of all in life at sea: although supertankers prefer plying the classical maritime routes around the Cape of Good Hope, they often remain underway for years, and when they berth, then usually only briefly in offshore terminals, as even the shortest time spent in a real harbor would be astronomically expensive. Since on account of their immense size they are the same as swimming islands and any real "landing" is impossible, they epitomize the annulment of the land-sea difference. With their crews often spending the entire year at sea, Noël Mostert observes, the ensuing "environment and circumstance

120 Klose 2009, pp. 96, 102.

makes some feel as though they are aboard a modern form of the *Flying Dutchman,* destined to pass to and fro around the Cape for ever more."[121]

Most supertankers fly flags of convenience, the states involved usually having little or no seafaring tradition but considerable interest in attracting as many shipowners and shippers as possible through vague safety regulations. In the case of catastrophes or legal violations, identifying the competent authorities is for the most part such a confused matter that such ships can function de facto free of legal control and hire correspondingly low-paid, hardly trained, sometimes stateless crewmembers. Operating tankers, however, is often an extraordinary challenge. Like container ships, supertankers, in their unceasing movement through sharply different climate zones, repeatedly butt against the old seamen's rule of arranging the loading according to basic circumstances of the route. As a result, the ships are markedly unstable and difficult to maneuver; navigation of these historically largest of all mobile artifacts—the massiveness meaning palpable exposure to the Coriolis force, the effect of the earth's rotation—has consequently emerged as a new sort of craft. Through their deep hulls, these ships have added something like a new dimension to voyage, for which reason many of the world's waterways have to be unceasingly surveyed anew. Even the foundations of ship building have been altered through the arrival of supertankers: traditional nautical knowledge now plays no role whatsoever, and experimentation is limited to increasing size. These ships, in other words, now operate as standardized prototypes, having lost all status as complex entities with individual features.

In addition, supertankers do not glide over the waves like classical vessels, the sea instead steadily dashing against them as if they were moving, monolithic piers. They are thus subject to enormous stresses, which given the loads they carry renders them both constantly endangered and highly dangerous means of transport. But supertankers not only represent a serious menace when fully loaded. Rather, this is especially the case after unloading, highly explosive gases forming in the tanks. When three gigantic supertanker explosions occurred within eighteen days in 1969, and insurance premiums soared sky high, the shipping industry entered a structural crisis threatening to place that sparked by the *Titanic* disaster in the shadows.[122] In response, the until then largest ever international research program on nautical safety was formed to identify the explosions' sources; this resulted in a recommendation to use inert gas as a tank-cleaning solution. In any case, supertankers continue to be built as already de-

121 Mostert 1976, p. 123.
122 See Karcev 1984, pp. 82ff.

preciating objects, vessels soon due for the scrap-heap. Long-maintained safety regulations such as the Plimsoll (maximum depth) line have been softened or abandoned on their account. And environmental damage, whether furtive as with oil drainage or spectacular as sometimes with damage at sea belongs to the everyday reality of this type of ship. This unprecedented dangerousness possibly constitutes a deeper caesura in the history of maritime transport than the shift from sail to steam. In his *Superships*, Noël Mostert summarizes as follows:

> For those on shore, shipwreck was once not an unwelcome event; it drew the plunderers from far and near. It was talked about for generations, with wistful recollection of the drama and the spoils; but shipwreck, once feared principally by those on board, has become in our time the more solemn dread of those on shore than of those on board. For the first time we on land have more to lose, and nothing to gain. Helicopters get the sailors off, we clean up the muck.[123]

Ships have become both highly complex and highly dangerous systems. Starting with the sinking of the *Torrey Canyon* in 1967, serious supertanker accidents have pushed risk calculation and the general culture of security to its outer limits. It is not only that such accidents exemplify the consequences of countless "ships of shame," as poorly managed as they are unsafe, entering the maritime scene since the end of voyage by sail.[124] With such shipwrecks, for the first time "the environment of the social system becomes involved in the concatenation of possible losses."[125] Now nature no longer represents an elementary danger to navigation—but rather, inversely, navigation to nature. Manifestly destroyed in the process is the foundational difference between land and sea.

123 Mostert 1976, pp. 331 f.
124 Since then, in accidents at sea those first affected are workers stemming from a deprived underclass and rarely organized in unions. International maritime courts hardly concern themselves with these individuals, already because these courts are more focused on verdicts than on compensation claims. Where previously scarcely any work was subject to more state supervision than that on a ship, meanwhile authority over labor conditions in this heterotopia has for the most part been radically privatized. See Gerstenberger 2002, pp. 33, 36, 66 f.
125 Luhmann 1993, p. 109.

Chapter 7
Shipwreck and Seasickness on Land: Terrestrial Loss of Safety

Pillaging on the Savage Shore

Already in antiquity, the more desolate and abandoned Europe's most dangerous shores, the more savage they seemed. When ships repeatedly went aground in these locations, apart from the monitored and animated harbor areas and coastal cities, whether because of storms, currents, or poor visibility, then the border between land and sea seemed repeatedly blurred. It was not only that here the sea's wild nature sometimes hid the water's deadly border or displaced it unto land. Sometimes, untamable nature came from the land to transgress the basal demarcation between firmness and fluidity: packs of men without pity, sometimes murderous, sweeping everything away, streaming down from adjacent coastal settlements or the immediate hinterland. They sealed the fate, it was said, of many seamen whose ships had just gone aground or been wrecked near land. Even around 1700, especially the British Isles, with their numerous rugged coastlines, seemed afflicted by a nameless horror that had always been tied to elementary natural forces.

"Those sons of plunder are below my pen, / Because they are below the names of men," rhymed Daniel Defoe in 1704 in a report on the devastating Great Storm of 1703. "The barbarous shores with men and boats abound, / The men more barbarous than the shores are found."[1] Two decades later, in his *Tour Through the Whole Island of Great Britain* (1724), Defoe reported on Cornwall's notorious "wreckers," men "so greedy and eager for prey, that they are charged with strange, bloody, and cruel dealings, even sometimes with one another."[2] Shipwrecks on England's coasts thus threatened double wreckage and demise. Hence where in "Epistle III" of his *Essay on Man*, Alexander Pope informs us that "full against his Cornish lands they roar, / And two rich shipwrecks bless the lucky shore," the roaring, wild waves have their echo in the "roarers" along Cornwall's coastline—the wild, riotous mob streaming into the water from land when a ship has run aground. Likely hovering behind these

1 Defoe 1855, p. 416.
2 Defoe 1778, p. 368.

https://doi.org/10.1515/9783110610734-011

lines is Shakespeare's "Tempest";[3] Pope himself explains the double meaning as follows:

> The author has placed the scene of these shipwrecks in Cornwall, not only from their frequency on that coast, but from the inhumanity of the inhabitants to those to whom that misfortune arrives: When a ship happens to be stranded there, they have been known to bore holes in it, to prevent it's getting off; to plunder, and sometimes even to massacre the people: Nor has the Parliament of England been yet able wholly to suppress these barbarities.[4]

For his part, William Falconer, himself shipwrecked in 1749 on the coast of Crete, wrote a long poem titled "The Shipwreck" (1762–1769) meant for the first time to give that phenomenon epic status. In the process, appealing for action on the part of those in power, he cast the infamous seamen and their way of life in a positive light, underscoring their patriotic service. From the classicizing and Mediterranean perspective of his "nautical Georgica," the pillaging on England's coasts was a reflection of sinister barbarism. Evidently, he suggests, it is not the sailors that lack culture and civility but rather the land-dwellers, as soon as unfortunate maritime wanderers land on their soil:

> I know, among you some full oft have view'd,
> With murdering weapons arm'd, a lawless brood,
> On England's vile inhuman shore who stand,
> The foul reproach and scandal of our land!
> To rob the wanderers wreck'd upon the strand.
> These, while their savage office they pursue,
> Oft wound to death the helpless plunder'd crew,
> Who, scap'd from every horror of the main,
> Implor'd their mercy, but implor'd in vain.[5]

As a natural border between land and sea, the coastline promises a clean separation between territorial culture and wild nature. If, however, accompanying this border's dissipation on the maritime side is a second shipwrecking and going under on the territorial side, then such a "shipwreck on land" has to prompt elementary insecurity. On the one hand, beneath the sea's surface lurks an unfathomable and menacing abyss, its crossing and usage thus having

3 See 1, 1, 22–23: "When the sea is. Hence! What cares these roarers / for the name of king? To cabin: silence! trouble us not"; 1,2,85–86: "If by your art, my dearest father, you have / Put the wild waters in this roar, allay them"; and so forth.
4 Pope 1751, pp. 176 f.
5 Falconer 2003, p. 223, version C, canto 2, 855–863.

always been tied to existential daring, or at least to acceptance of danger and states of emergency. On the other hand, it promises the possibility of return and journey's end, even as an empty, endless horizon. But if terra firma threatens sudden danger and violent explosive hostility at the very moment it furnishes first reliable visibility, palpability, walk-on stability, the landing itself leads to going under, all hope of return evaporating on the promised soil.[6] "To fall by wreckers on their native shore" is the ultimate fear of Early Modern seafarers, because such an event attacks their culture of daring and emergency, and of hope and speculation.[7] In the context of such an experience, the elementary enmity meanwhile cast into a deterritorialized space becomes unlimited.

Already in ancient Greece, the barbaric coasts had achieved considerable notoriety, especially the coasts of Asia Minor. We read in Herodotus, for instance, that Scythian Taurians offered up stranded, shipwrecked sailors to their blood goddess Artemis Tauropolos.[8] This points to a possible origin of the myth of Iphigenia, but also to a cultural-historical dimension within which coastline pillaging is defined as prototypical barbarity. Early Modern plundering was accordingly scorned as a "crime dishonoring the ancient Taurians."[9] Not yet in Homer but already in Virgil, such pillaging has become an epic theme. In his episodes of shipwreck and beaching, as a "supplicant" who has "escaped the salt sea," Odysseus has to fear that "wild beasts may find me / and treat me as their prey" (*Odyssey* 5, 445–450, 473). But as a sly pirate who himself frequently takes spoils or *sylum* from the sea, on the path of his homecoming he repeatedly finds refuge, *asylum*. By contrast, Aeneas's helmsman Palinurus, after divine decision has him fall overboard, will land on a wild coast where the inhabitants, "in ignorance deeming [him] a prize," in the end will kill him. Hence here, at Cape Palinuro, "for ever the place shall bear the name of Palinurus." (*Aeneid*, 6, 361; 381.)

What separated Europe's coastline in Virgil's time from that of Homeric epic was access to an increasingly expanded empire, one that would use the force of law to forbid this plundering. It is the case that travelers like the Roman physician Scribonius Largus asserted that as late as the first century AD, in especially the Cyclades coastal pillaging was flourishing; and in the same period Seneca the Elder described coastal lords who compiled precise calculations of expected profits.[10] But the common-law appropriation of coastal material and enslave-

6 See Le Boulicaut 2005, p. 233.
7 Crump 1850, p. 6.
8 On Herodotus's report (4, 103) see Lübeck 1993, p. 112; Göttlicher 2006, pp. 126 f.
9 Esménard 1806, p. 345.
10 See Horden 2000, p. 368.

ment of shipwrecked persons increasingly encountered punitive measures extending to the death penalty. We read in the *Lex Rhodia* that when goods are strewn on the shore after a shipwreck, just like the wrecked ship itself these are not to be seen as "derelict items." Both the hull and the wares are to be given back—not even the treasury can lay claim to them. Anyone appropriating this property is a thief who can expect to have to return what he stole it at four times the worth and also endure a criminal trial.[11] Already here, we find the prospect held out of offering a reward for finding and salvaging goods, a motivating factor accompanied in the Byzantine period by a differentiated share-calculation based on distance of the found items from the water-limit. In addition, an expiration period was introduced, so that original owner usually had a year and a day within which to make his claim, after which the material could be passed to the finder or authorities.[12]

With the fall of Rome, these legal regulations were, however, rendered obsolete in large portions of Europe. Especially on the coasts of northern Europe, where ships and men were steadily subject to attacks and plundering by Scandinavian sailors, into the High Middle Ages shore-law was understood as a sub-form of law governing "aliens," according to which only the native populace enjoyed the benefit of law. The alien was without both law and even peace in the absence of safe-conduct letters or protective escort—a possibility that was in any case unlikely in the context of unplanned shipwreck on strange shores.[13] Only a *constitutio* of Henry VI in 1196 confirmed the validity of the legal principles derived from the *Nómos Rhodion Nautikós* and codified by the digest—principles then reconfirmed in the *Authentica Navigia* of Frederick II in 1220, underpinned by an annulment of local legal concepts.[14] Since that time, the so-called *jus naufragii*, shore-law governed by common law and thus specific to the locus, was subject to negotiation everywhere. Into the thirteenth century it had usually been interpreted to the effect that not only goods but also life and limbs of stranded men were forfeited to the indigenous populace as soon as either was a mile or less behind the water-edge or touched earth; now a limitation to the goods became apparent, with the territorial princes and cities increasingly declaring all jetsam their privilege.

When common law declared that only ownerless jetsam could be appropriated, hence only the spoils of a shipwreck without survivors, this was not seldom

11 On the individual clauses see Ashburner 1909, pp. clxxxviiif.

12 Ibid., pp. ccxciif.

13 See Rath 2007, pp. 41f.

14 See Beckert 1991, p. 293.

grasped as indirect encouragement to kill shipwrecked men. The belief that shipwreck and stranding was God's judgment on a sinful crew also helped render those men into literal outlaws. Because it stood in the tradition of Roman law and was influenced by the *Lex Rhodia*, but also because the goals of the crusade-movement and those of merchants and seamen ran parallel,[15] the Church condemned the exercise of shore-law and threatened—explicitly for instance in a 1079 Third Lateran Council decision—excommunication when assistance was denied. The emperor aligned himself with this position; in any case he had nothing to lose since to a large extent shore-law, as a royal prerogative, had become the territorial princes' purview. The merchants in especially France and the northern German Hansa region pressed, to the contrary, with all available means for official counter-measures. In the *Rôles d'Oléron*, the most important twelfth- and thirteenth-century collection of French maritime common law, article 25 thus states as follows:

> Some people are inhumane and more cruel and felonious than enraged dogs and wolves; they murder and kill poor victims to take their money, clothing, and other items. The local seigneur must seize such people and render justice and punishment on both their bodies and goods; they must be plunged into the sea until half dead and then pulled out and stoned and beaten to death as would be a dog or wolf.[16]

In the *Rezess von Wisby* (1287), the Hanseatic region issued an edict against trading with jetsam, stipulating that following bona fide acquisition it should be returned in connection with an oath of purgation. But all such appeals and indirect way of taking on the problem changed nothing in the basic situation: that on the eastern and northern coasts only scattered relief from shore-law was possible. The territorial rulers there were simply too numerous and too stubborn. In the end they had to first achieve a balance between their own claims, the ancestral rights of the shore-dwellers, ecclesiastical principles, and the interests of the merchants. The latter were often only granted guarantees of protection on the condition that they divert their trading routes to particular coasts and territories. The wreckers, those "improvised pirates,"[17] were indirectly in official capacity, as if they were privateers in the Early Modern sense: men with amphibian natures whose perfidious attacks were carried out with the local leader's license. It was only that privateers did not subject the free sea to terrestrial incursion, but the land itself to the sea's elementary violence.

---

15 Niitemaa 1955, p. 398.

16 Cited from Cabantous 1993, p. 122. On the following see Beckert 1991, p. 295.

17 Cabantous 1993, p. 248.

Until the Early Modern period, Europe's coasts were only thought of as an abandoned, desolate landscape—the unappealing, indeed monstrous remains left by catastrophes in the earth's history, if not by the Deluge itself.[18] Where in antiquity wild shores had been considered zones of bitter battle and impure commingling, the entry point for plague, robbers, and barbarians, starting in the seventeenth century efforts such as the massive Dutch dike and dam construction project attest to an effort at regaining cultural ground and stabilizing instable natural borders. But the wild shore-pirates threatened this materially and ideally renewed border with dissolution. A transitional space thus now emerged as a scene of repeated transgression—of elementary resistance to power and its need for order on the edge of its territory. Placed apart from the view of now more vigilant but still hardly ubiquitous authorities, the coasts seemed dangerous not only as a locus of illegal exchange, for instance smuggling, but especially as a zone of illegal appropriation that could take violent form.

On the coasts of Western and Central Europe, until 1700 we thus find a confrontation between differing economies and differing claims. There was a sovereign territorial power trying to control all goods and transactions and siphon off at least a significant amount of the profits; there was the trading power of the merchants, using ships as a mobile node for a global trading network, consequently legally obliging the marine officers and sailors, after their reduction by the shipowners to mere employees and hired men, to defend, secure, and salvage both ship and goods.[19] And there was the local power of the coastal dwellers, who spontaneously gathered on the occasion of shipwrecks to execute their traditional rights. Stranded ships were a chance—a gift of fate or Fortuna, all three partners coveting their share. To the extent responsibility for a stranding could be blamed on captains and sailors, they could exculpate themselves by referring to nameless shore-robbers. In addition, the authorities, increasingly striving to territorialize the sea since the seventeenth century while also subjecting local powers to their centralism, found shore-robbery a good basis for installing their political-economic regime on desolate coasts.

Seen from this perspective, shore-pillaging was in no way a direct attack on all civil and state order. To the extent the shore-robbers desisted from manifestly extortive measures such as the medieval ransom-demand for stranded sailors,[20] their actions revealed an at least partial legitimacy grounded in usage. It in any event did not stand on the same level as offenses such as piracy. When the ter-

18 See Corbin 1994, pp. 12ff.

19 On both Hanseatic maritime law and urban law in Hamburg see Rath 2007, p. 27.

20 See Lachs 1989, p. 117.

ritorial sovereign punished theft of game and smuggling far more severely, then perhaps not only because as an anonymous collective, shore-robbers were hard to prosecute. Rather, robbery at sea and on land was meant to directly lay hands on possessions of merchants and the authorities, while coastal plunderers tried to use a no-man's-land in their long-established manner. Furthermore, shore-robbery did not appear to be an intermittently occurring offense, but rather manifestly constituted a central ritual for the coastal dwellers. Namely, in episodes of both salvaging and plundering local hierarchies and functional divisions were confirmed and renewed: the roles and rules of, as Alain Cabantous describes it, a masculine, self-contained, territorially rooted milieu, one experienced at sea and oriented toward quickly seizing spoils, here confronted the feminine city as a refuge for more subtle and refined distinctions, calculative and speculative economizing and both official and juridical forms of governance.[21]

The coastal dwellers were defending an in their eyes anciently rooted and legitimate survival-centered culture, in which hopes were pinned on the gift and opportunity created quick spoils. In his journey through the British Isles, Defoe revealed surprise at the extent to which, starting on a literally foundational level, the culture was built on jetsam, the local houses consisting nearly exclusively of "old planks, beams, wales, timber, &c. the deplorable wrecks of ships, and ruins of mariners, and merchants fortunes."[22] Although imagined to be a juncture between terrestrial and maritime labor,[23] the coasts evidently followed a highly particular economy, its rhythms dictated less by a sovereign than by the sea itself. Even in 1866 Victor Hugo would describe "toilers of the sea" as follows: "For half the year they lived on fish and shellfish, for the other half on what they could pick up from wrecks. Pillaging their coasts was their main resource. They recognized only two seasons in the year, the fishing season and the shipwreck season."[24] Seen in this way, the coastal dwellers were more agents of the sea than of their territorial sovereign. In their sinister superstition, naïve savagery, and mutinous violence, they resembled—as perceived by an increasingly "policed" age of enlightened moralists—those other agents of the sea, whom they could encounter in the case of shipwreck as barbaric slaughter: the seamen, aptly described by Cabantous as those "men of danger, accustomed to violent confrontations imposed on them by their vocation."[25] Just as the sailors, unimpressed by territorial authority, choose to associate with the sea's fierce violence

21 See Cabantous 1993, p. 202.

22 Defoe 1778, p. 59.

23 Corbin 1994, p. 207.

24 Hugo 2002, p. 24.

25 Cabantous 1984, p. 154.

and danger, those dwelling on the shore repeatedly entered into a non-rational dialog with the sea.

On the one hand, they had to struggle against that same fearsome violence, ruler over life and death, poverty and wealth; in the case of a salvage action or possible booty they had to prevail. On the other hand, the "latent hostility of the ocean" (Cabantous again) was also manifest within them, so that every shipwreck threatened to become a theater of cruelty.[26] Shore robbery was a collective rite—in the end communal confirmation, based on shared spoils, that its members were more than an agglomerate of individuals and atomized interests. At one point such a rite could support the differentiating, hierarchizing, and functionalizing of social roles; at another point their leveling, de-differentiation, disorganization. Especially when after a shipwreck alcohol ended up on land, to be immediately plundered, the shore robbery quickly culminated—as we read in many reports—in collective orgies, bacchanals, a "savage rejoicing," as if with the nautical misfortune paradisiacal circumstances had returned to the wild coast.[27] In their assessment of this state of affairs, the officials seem to have themselves been uncertain: as if the phenomenon lay in a gray zone between crime and barbarity.

Contributing to this uncertainty may have been an evident similarity between the shore robbers and people perceived as savages such as those on the African coasts. Since the Portuguese opening up of the route to India was registered, various reports were circulating that concerned a double danger: on the one hand there was a fear, stemming from *Cabo de Não* on the African coast but then extended to all other capes, that the land-sea distinction could collapse—that a viscous, shallow, and silted sea could cause the caravels to become irrevocably stranded. On the other hand, there was a fear, as great as that of a shipwreck, of the African hinterland itself, seemingly ruled by wild beasts, fearsome reptiles, and godless "Kafirs"—and, as soon would become clear, actually harboring countless shore-robbers.[28] This may also explain why the history of the sinking of the *São João* before the coast of Natal in 1552, known throughout Europe, sporadically appears like a blueprint for the reports that for example circulated after the beaching of the *Charming Jenny* in 1773 off the Welsh coast: here as there unfortunate stranded seamen were overpowered and humiliated by coast-dwellers; here and there the victims included the captain and his wife, their last

26 Cabantous 1993, p. 32.

27 Ibid., p. 211.

28 See Duffy 1955, pp. 142–144, und Parry 1974, pp. 108 f.

valuables and even clothing torn from their bodies; and here and there surviving witness referred to "savages" at work.[29]

But otherwise than the case with the much-observed shipwreck on the Caffrarian coast, after the Wales disaster some of the shore-robbers became witnesses who spoke up publicly concerning reports of putatively barbaric atrocities.[30] And beyond the topoi at work in the narrative later forming the core of the *História trágico-marítima*, the descriptions of the *Charming Jenny*'s shipwreck emphasized a leitmotif that meanwhile had basically become a symbol of European shore-robbery: alleged signal lights meant to confuse disoriented seamen, leading them to waiting plunderers.[31] Complaints about coastal beacons of such a sort have already come down to us from antiquity.[32] Located where dark land already lay, using lanterns hung around the necks or on the tails of cows, horses, or donkeys, these beacons could simulate the presence of other ships, rocking on the sea. Eighteenth-century legends suggest that even some names of cities stem from this practice, for instance Nags Head in North Carolina (*nags* here being a colloquial word for horses or jades).

Enlightenment between Land and Sea

To put an end to the wild goings on their coasts, to the increasingly legendary phenomenon of shore-robbery, by around 1700 most territorial rulers had decided on legal measures. These measures took up the critique that theorists of natural law leveled at the old beach law—even if that critique was meant to limit the authorities' rights of appropriation. In 1711, Samuel Pufendorf, for example referred to "the custom of consigning to the shore's authority goods subject to the misfortune of shipwreck at the point where they arrive on shore or have been fished out" as unjust. "It is thus untoward," he continued, "that in many places stolen things, when found, are not given to their owners but confiscated for the

29 Report of the *Shrewsbury Chronicle*, 1774, cited from Bathurst 2005, p. 228. For the chronicle of the wreck of the *São João Baptista* and the crew's horrendous travails, see "Treatise of the Misfortunes that Befell the Great Ship São João Baptista," in Boxer 1959. On this and on the general experience of shipwrecked Portuguese sailors on the African coasts, see Duffy 1955, pp. 25, 141–161; Blackmore 2002, p. 69.

30 See Bathurst 2005, p. 228.

31 See Rule 1975, pp. 180 f.

32 See Höckmann 1985, p. 92.

authorities' sake."[33] In respect to going aground, Kant wrote as follows: "But the owner of a shore cannot include, in his right to acquire, what is unintentionally *washed up on shore*, whether human beings or things belonging to them, since this is not wronging him (not a deed at all), and though a thing has been cast up on land which belongs to someone, it cannot be treated as a *res nullius*"—a thing belonging to no one.[34]

With a shipwreck, neither are the legal interests of another person, or in general the territorial laws, automatically "damaged," nor do stranded human beings and goods become ownerless objects. Hence both justifications of beach-law excluded by natural law thus already show themselves as weak in the coastal context—one is formulated one-sidedly from the land's perspective, the other from that of the sea. In the course of the Middle Ages, not the common law of shore dwellers but rather that of maritime trade would articulate a mediating position, one integrating both the territorial *nomos* and the special legal case of the sea. For this reason, in questions of beach-law as well, the comprehensive codifications of maritime law developed around 1700 chiefly centered on traditional maritime common law.

In the case of the *Ordonnance de la Marine* of 1681, codifying Colbert's comprehensive project of reforming and governing maritime commerce on a juridical level, the reference point was mainly the *Rôles d'Oléron*—which helped move the medieval mercantile community to a belated triumph. Under the title "On Shipwrecks, Breakage, & Beachings" (*Des Naufrages, Bris, & Echoüemens*) we read as follows:

> We declare that we have placed and are placing under our protection & safeguarding the vessels, their crew & loads, tossed upon the shores of our realm by storms or otherwise washed up there: & generally everything that has survived the storm.... We urge our subjects to fulfill all duties for saving persons they observe in danger of shipwreck. It is our wish that those who have threatened the lives and goods of these persons be punished by death without any granting mercy; which extending to the present we have declared null and void & forbid any judge from considering it....Those illuminating the night with deceptive fire on the shores and in dangerous places to draw ships in and see them lost will also be punished by death, & their bodies fastened to a mast fixed to the place where they made the fire.[35]

---

33 Pufendorf 1711, p. 1026 (= 4, 13, 4). This passage is missing from the 1729 English translation of the Puffendorf text and has been translated from the German version.

34 Kant 1996a, p. 56.

35 Ordonnance de Louis XIV 1687, pp. 211 f., 229.

With this ordinance, then, (1) all ships sunk or stranded on the coasts of the royal realm, their crew and ware, were placed officially under the sovereign's protection; (2) all subjects present were obliged to take life-saving and salvaging action; (3) deceptive fires were expressly defined as an offense, thus being moved out of the realm of legend into that of criminal law; and (4) all these orders were accompanied by the king's drastic legal-official specifications of punishment. But the fact that the coastal dwellers were here being deprived of revenue considered legitimate in order to distribute it to shipowners, insurers, and the state, and also the fact that they were increasingly subjected to official oversight while also being obliged to perform rescue and salvaging duties, slowed down the desired speedy success of sovereign intervention. For one thing, the coastal dwellers showed themselves hostile and violent vis-à-vis representatives of the state such as the guard details. For another thing until the mid-eighteenth century the regulations in play were widely disregarded—until finally, in 1754, following the beaching and plundering of the *Jeune brasseur*, the authorities made an example of some of the arrested perpetrators, executing them and publicly displaying the bodies.[36]

In England, Henry III's charter of 1236 stipulated not only a deadline for returning beached goods but also the so-called "man or beast" rule, which stipulated that beached goods could only be appropriated if neither human beings or any sort of animal had survived the shipwreck. Although the rule's fatal impact was quickly recognized, it would only be done away with in 1771.[37] In any event in 1713 an act had been passed declaring prevention of rescue, plundering, and the damage and murder of survivors, together with setting up false light signals, to be capital crimes.[38] Both the King of Sweden's prohibition catalog of 1697 and the Danish shore regulations of 1705 contained similar clauses.[39] What also stood in question after 1700 was determining payment for salvaging and the share accorded the state. But there could be no doubt now that the old beach-law was null and void, those who made use of it running up against sovereign authorities, pressing for transparent, unified conditions within their territories and on its borders—and thus, as it happens, for clear borders between land and sea. As René-Josué Valin wrote in 1760 in his commentary on the *Ordonnance*, those resisting it had, through a "natural inclination to robbery and brigandage,"

---

36 See Ducoin 1999, pp. 317 f., 320 f.
37 See Bathurst 2005, p. 11.
38 See Rule 1975, pp. 167 f.
39 See Rath 2007, p. 70.

revealed themselves as what a power reorganizing its territory could not even tolerate on the coasts: a champion of maritime barbarity.[40]

From 1700 onward, the general European boom in maritime trade was accompanied by increased commercial traffic—in particular in the British Isles. This increased the rate of accidents, so that shipowners and merchants soon had to complain not only about insufficient availability of charts but also about the inadequate or misleading system for directing traffic. In 1741, cartographer Lewis Morris lamented that "false charts serve but as false lights laid by villains along ye coast to lead poor sailors to destruction."[41] In 1786 alone, around 2000 men lost their lives before the Scottish coast; in the 1790s around 550 ships sank each year on the British coasts; and when in December 1799, over 70 ships—including the *HMS York* with its entire crew—went under off Scottish shores in a three-day storm, strong pressure was placed on parliament, then reflected in decisions catalyzing an improved operation of light signals, especially on the coasts of Scotland.[42]

Already because of the construction requirements involved, apart from lighthouses set up on large, hardened harbor areas, those modeled on ancient Alexandria's Pharos, had only been built in a scattered way. On the British island, the main light-source used came from intermittent church-tower beacons. In the mid-seventeenth century, Trinity House, a—still extant—society for protecting the coastline and setting up maritime markers, was placed under the obligation of devoting most of its income to charitable ends.[43] This led to the House doing all it could not to have to build any new lighthouses. For this reason, between 1600 and 1836 only a single new self-operated lighthouse was built on the coastline.[44] To remedy this situation in both Britain and on the continent, around 1800 many offices focused on setting up beacons were established, together with numerous legal measures that followed the "Act for Erecting Certain Lighthouses in the Northern Parts of Great Britain" (1786). In Scotland, over a century starting in 1786 members of the Stevenson family working as chief engineers for the Northern Lighthouse Board erected a total of 71 lighthouses, while also working on successive improvements to beacon technology during this period.[45] So

40 Valin 1760, p. 545.

41 Cited from Habermann 2002, p. 110.

42 See Bathurst 2000, pp. 69 f.

43 See Brustat-Naval 1969, pp. 52f. Trinity House was founded by Stephen Langton, the Archbishop of Canterbury, and then confirmed as a pilots' guild by Henry VIII.

44 See Bathurst 2000, pp. 56 f.

45 This process, which was pushed forward both scientifically and politically by Augustin Jean Fresnel, Michael Faraday, and other prominent physicists, included the following steps: use of

that the light signals could be distinguished from one another and from other light sources (including those of beach-robbers), distinct identifiers were necessary, taking the form of, for example, single, repeated flashes and groups of flashes.[46] Invented in 1806, Robert Stevenson's rotational mechanism here marked a starting point, Charles Babbage's differentiated system of 1851, involving beacons meant to be darkened according to a numerical code, an endpoint.[47]

With such constructions, techniques, and systems, Michelet observed in 1861, "modern civilization erects peaceful towers of most benevolent and beneficent hospitality." "When no star shines upon us from that firmament above, the seaman hails this art-created light as the star of brotherhood."[48] But engineers such as Robert Stevenson still had to experience "with what an evil eye the Wreck Brokers of Sanday view any improvement upon this coast, and how openly they regret it."[49] It was not unusual for reports to emerge of sabotage by coast dwellers, who saw themselves robbed of their future ware through the improved traffic-guidance system. When signal systems replaced deceptive beacons, then, as evoked in Michelet's humanitarian pathos, in the end the plunderers' devious, deceptive light was meant to be overtaken by the true light of the Enlightenment. Or as the *Grand Larousse* put it in 1867, beach-robbery was now merely "a remainder of the barbarity that will not fail to vanish before the always growing light of civilization."[50]

For this to happen, more as mere lighting technique was needed. Safety and rescue systems meant to prevent the deadly outcome of stranding included purportedly unsinkable life boats and the Manby mortar, first tested in 1808, which fired a shot with an attached rescuing line to a wrecked vessel. When still in its developmental phase, this instrument apparently had such a threatening effect on some coastal dwellers that they asked its inventor, a non-swimmer, to present the technique involved, for the sake of seeing to his capsizing on the open sea. George William Manby survived—and with him countless others could be saved by his invention from mortal peril while in coastal areas at sea.[51] In the future, on the level of coastal protection organizations such as the *Société humaine et des*

Argand burners and petroleum lamps, deployment of increasingly better systems making use of reflectors, mirrors, and lenses, and finally introduction of electric current for the beacons. Ibid., pp. 137–144, 157, 178, 221f.; Möhring 2002, pp. 231–233.

46 See Brustat-Naval 1969, p. 58.

47 Dotzler 1996, p. 423.

48 Michelet 1861, pp. 100f.

49 Cited from Bathurst 2000, p. 53.

50 Cited from Ducoin 1999, p. 312.

51 See Bathurst 2000, pp. 3f.

naufrages (founded in 1824), the National Institution for the Preservation of Life from Shipwreck (1824), and the Royal National Lifeboat Institution (1854) would accompany the laws striving since the seventeenth century's end to check the actions of the wild coastal dwellers.

In the course of this general security offensive, then, there was an increasing shift from what we might term regimental measures to those of a regulatory nature. In 1836, with *Lloyd's Register of Shipping* having regularly published statistics on British and worldwide shipping accidents over the past twelve years, a British parliamentary commission was assigned the task of investigating the accidents' causes and developing new construction and safety regulations. At the same time organizations such as the *Société générale des naufrages* were trying to combat forms of "wrecking," whose effects were less spectacular but no less serious. In a French report of 1836 we thus read that "most shipwrecks are not the result of storms or reefs but rather of the greed of some shipowners, speculating on insurance with excessive zeal and placing old, poorly equipped and often poorly commanded vessels at sea."[52]

The years between 1770 and 1830 saw a significant lessening of cases of recorded beach-robbery, a development that had little to do with penal measures that had been taken long before. Improved coastal transportation connections, more frequent presence of the gendarmerie and coast guard, and more frequent visits by an urban, touristic public all certainly played a role here—as did likewise-improved navigation and the gradually increasing use of steamships.[53] While those living on the coast were still thought of as beach-robbers and the most implacable opponents of a new security culture, this now involved presenting a counterpart to the heroic ethos of the rescue societies.[54] Put otherwise, the beach-robbers personified everything troubling the new preventative system and its promise of danger-free passage. They embodied resistance to the "govermentalistic" and pedagogic efforts to regulate and humanize the coasts. The "wreckers" in this way were engaged in complexity reduction, serving as imaginary bearers of contingency and danger.

Well into the nineteenth century, it would be the Church that exercised the most enduring influence on the coastal dwellers. In France following the restoration, for example, the now broader network of congregations appears to have contributed substantially to a lower rate of plundering after 1800. In addition, between the High Middle Ages and the Early Modern period, the pulpit served

52 Journal de la Société générale des naufrages 1836, p. 130.
53 See Rule 1975, p. 169; Cabantous 1993, p. 229.
54 See Battesti 1999, p. 138; Haberman 2002, p. 112.

as the most important medium for promulgating and clarifying new laws and regulations. When the Church damned beach-robbery, threatened excommunication, and offered dispensations when unconditional aid was offered to stranded sailors, it was practicing behavioral supervision as the most important pre-modern authority.[55] Nevertheless, perhaps most clearly during the French Revolution, the Church was repeatedly accused of benefiting from stranding and indirectly promoting plundering.[56] And in fact despite clear stipulation in canonic law, for some clerics it was not at all clear that beach-robbery necessarily posed legal and moral problems. If within a common-law framework everything floating in the sea without visible proximity of an owner was viewed as having none, jetsam could easily be though of as a divine benefaction. Even in the mid-nineteenth-century rhymed phrases such as "The *Good Samaritan* came ashore / To feed the hungry and cloathe the poor" evoked the charitable ties between the Church and the impoverished coastal population, praising the gifts of a sea that in the preaching tradition was seen as a providential instrument.[57]

One issue being debated among enlightened urban dwellers and educated people, an issue actually eventually making its way into popular encyclopedias, was as follows: Were prayers really being delivered in countless chapels and churches that shipwrecks on the coast produce sumptuous spoils? Or rather were they simply for a few goods or pieces of jetsam, in the regrettable case of a bad accident ending, in any event, in a ship going under?[58] But that an evidently absolute faith in providence, also manifest in reluctance to recover and bury the dead,[59] would be the basis for a cold-blooded denial of any aid to the stranded was a possibility not even considered by admirers and champions of untouched indigenous cultures. For example, after traveling through the Shetland and Orkney islands, Walter Scott understood the superstition that offering a hand to a drowning man would lead to great misfortune for the rescuer and his family as "an apology for rendering no assistance to the mariners as they escaped from a shipwrecked vessel, for these isles are infamous for plundering wrecks."[60]

For contemporary critics of religion, at work here was at best a form of savage madness or superstition, to be opposed by agents of Enlightenment and civility through a rational, moral and affective elevation of humane values. In the

55 See Rath 2007, pp. 77–82; Nitemaa 1955, p. 402.
56 Ibid., p. 399.
57 Couplet cited from Rule 1975, p. 170.
58 See for example Zedler 1744, p. 757.
59 On the explicit instruction "to inhume cadavers," see *Ordonnance de Louis XIV* 1687, p. 229.
60 Scott 1837, p. 489.

years leading up to 1800, this exemplary confrontation between urban and coastal culture would increasingly unfold on the beaches themselves. Namely, the meanwhile widely established security and order on the European coasts had endowed them with the atmosphere of a rationally affective culture. It was not only that the increased presence of travelers and vacationers and the still quite frequent stranding episodes led repeatedly to spectacles involuntarily subsumed to "shipwreck with spectators" stage directions. By the eighteenth century's end, in many paintings and panoramas as well, the dangerous coasts had become landscapes against whose backdrop pathos formulae of sublime and fervent emotive states could be played through and upon.[61] This was reflected in late Enlightenment writing in repeated recourse to the topos of *suave mari magno*—pleasantly watching in a great sea—from the second book of Lucretius the Epicurian's didactic poem *De rerum natura.*

The scenario of a spectator observing a close-by ship's distress at sea from a safe shore could very much involve pleasure at one's own safety. But in distinction to the philosophical distance that Lucretius articulated with his topos, now a "mixing" of sensations and a certain ambivalence of feeling was meant to be negotiated. For instance Mark Akenside's own long didactic poem, *Pleasures of Imagination* (1744), tied that topos to deep empathy and overpowering emotion, in order to in this way justify one's own viewing pleasure—which Henry Home, in his essay "Of Our Attachment to Objects of Distress," spelled out as a program for pleasurable moral practice.[62] Likewise Christian Victor Kindervater's "Delight at Sad Ideas" ("Wohlgefallen an traurigen Vorstellungen") of 1787, exploring that theme with Akenside's reading of Lucretius as its starting point, moved from initial horror to empathic stirring, then coming back to itself as self-sensibility or auto-affectivity. The opening in Lucretian manner was as follows: "Just ask the populace, making hurried haste / from its huts, when the sea's cruel storm / Smashes a helpless vessel on the coast"[63]—a maxim that, under new auspices, evokes not plundering but an empathy "thawing every eye." Evidently even among the *Volk,* pleasure at the scenery and one's own state of being human outweighed the value of any jetsam, and as Kindervater concluded, such pleasure has to "come *to a large extent,* if not *alone*, from the *circumstances,* or from what *accompanies the scene*. The sublime sensations resulting from observing the rough seas render softer and more mild those awakened by a shipwreck."[64]

---

61 See Corbin 1994, pp. 179 ff.

62 See Akinside 1744, p. 88; Zelle 1990, p. 301; Zelle 1997, p. 87.

63 "Frag nur das Volk, das, schnellen Laufes, eilt / Aus seinen Hütten, wann des Meer's grausamer Sturm / Zerschlägt ein hülflos Fahrzeug an der Küste."

64 Kindervater 1787, pp. 128, 132.

It is "a sort of gloomy satisfaction, or terrifick pleasure" that according to James Beattie even a "horror" such as a shipwreck can provide.[65] But in order to neutralize the sinister element of this pleasure, a process aligned with Edmund Burke's observations on the sublime is called for: a confrontation with and reflection over "those things which *directly* suggest the idea of danger."[66] In the course of *real* experiences of the sublime such as those taking place on the coasts, knowledge concerning one's own safety, which in purely aesthetically conceived doctrine of sublimity has to be able to accompany all intuition, becomes attentiveness for properly safe distance.[67] It is the case that starting with the turn toward transcendental philosophy in Germany, there was an insistence that the feeling of sublimity is solely, in Kant's words, "for our own vocation" that we attribute to an object of experience through "a certain subreption." But a premise for drawing feeling "into transcendental philosophy" and transforming it there into a "feeling of spirit" is "the development and exercise" of our faculties.[68] Hence anyone wishing to adhere to an idea of "humanity" can be expected to practice humanity precisely in the sublimely dangerous moment. But in the exemplary case of a stranding, this means that passive observation can lay as little claim to humanity as active plundering.

Seen in this way, the economic changes putting pressure on beach-robbery after 1800 would not only clearly separate plundering from an introduction of guaranteed and suitable compensation for salvaging and rescue. There was now also focus on the affective economy of those dwelling on the coast. For even if good payment would now be offered for that work, it was extremely risky, dangerous work for alien masters and officials, work for which no relief would be offered survivors in case something went wrong.[69] Because the coast dwellers, who had always known how to arrange themselves with the sea, were seen as its allies, they were all the more duty-bound under the rubric of humanity. Who if not exponents of the sea should spontaneously take up battle with the dangerous maritime element? But anyone who declined this duty of assistance—indeed of humanity—inevitably conjured up the old chimera of the barbaric plunderer.

Around 1800, those allied with the elementary—those scorning its annulling elevation into sublimity—were enemies of society, indeed of humanity. Understanding beach robbery as "crowd activity" and tying it to social misery was a

65 Beattie 1783, p. 615.
66 Burke 2008, p. 138.
67 On this general shift in doctrine of sublimity, see Begemann 1987, pp. 129 f.
68 Kant 2000, pp. 141, 149, 78, 145.
69 See Lachs 1989, pp. 116, 119.

later development. Eventually, in notorious coastal areas like Cornwall, it would be above all industrial workers, together with a proletariat forming in the hinterland, who would be made responsible for coastal plundering.[70] "Wreckers" advanced to agents of an amorphous mass representing society's elementary dimension itself. It was no accident that the preventive legal attributes of "danger" and "general danger"—attributes emerging at the same time—were exemplified by the figure of the beach-robber. Especially prominent here was, as coached in legal diction, the threat of potential "attacks against goods of the community of the state" through "setting fires at night on the heights above the shore, suitable for endangering navigation."[71] Through the phantom "wrecker," the shoreline, despite all the coastal protection, again became what Dolf Sternberger has termed a "peephole into the dangerous life."[72]

Stevenson's Jetsam

Since the period around 1700, appearing in legal and travel literature and belles lettres, the figure of the beach-robber emerges in a conglomerate of regulations, reports, and legends. In France, it was addressed as sovereign edict in the *Ordonnance de la Marine*—a development that the coastal populace necessarily considered a declaration of war by the central authorities. Those resisting their regulating intervention were quickly demonized as barbarians. To this end, the *Ordonnance*'s authors took over relevant passages, nearly literally, from the *Rôles d'Oléron.* That at least the vehemence of the *Ordonnance* did not correspond to any real situation is underscored by two facts in particular: the definitions of beach-robbery in the clauses on *Naufrages, Bris, & Echoüemens* are only located at the very end, and the catalog of punishment was hardly ever applied. As, in addition, the scattered incidents of documented stranding attributed to deceptive beacons almost without exception took place in stormy conditions with poor visibility, so that such beacons would likely be hardly recognizable, the strong suspicion emerges that simple navigational error or faulty instructions by incompetent pilots was being covered up here.[73]

In England, that reports about shoreline plundering often bordered on legend was a fact manifest in prominent cases such as the disastrous stranding

70 Rule 1975, p. 174; see also ibid., pp. 178, 181, 186.
71 Rohland 1888, pp. 4, 44f., 58.
72 Sternberger 1981, p. 237.
73 See Ducoin 1999, pp. 313f., 316, 322.

of Sir Cloudesley Shovell's squadron off the Scilly Islands in 1707. The rumor quickly spread that Shovell had been led astray and then murdered by an old female plunderer who had her eye on his emerald ring.[74] In reality, falsely calculated longitude caused the accident, its gravity then leading to formation of the British Board of Longitude seven years later.[75] In the case of England as well, there is hardly any historiographically credible material attesting to "wrecking," with extremely few cases brought before the courts. What is very likely is the widespread presence of complaints about putative beach barbarians, with various scarcely verifiable reports being taken over and distorted according to whim, in order to repeatedly confirm one and the same scenario. It seems that after the number of real plundering incidents diminished significantly in the course of the eighteenth century, the figure of the beach-robber became entrenched in the popular imagination.[76] Around 1800, in an epoch when the project of civilization and security was in full swing, reports were still circulating of deceptive beacons: a dramatizing interpretation, as legend, of unaccountable moments of danger, for instance fatal breakdowns in the latest signaling techniques.

In any case, on the threshold of the nineteenth century, when both state institutions and a range of trading and insurance enterprises increasingly statistically recorded shipping traffic and monitored its technical safety, a strong interest in maritime affairs developed among British and French painters, writers and poets, and historians. Through travels in the coastal regions, these individuals engaged in what amounted to anthropological research expeditions, in the process encountering a characteristic "instinct of the race" and an "insatiable thirst for pillage."[77] In his *Tableau de la France* (1833) Michelet observes that nature

> is fierce, man is fierce; and they seem to understand each other. As soon as the sea casts a hapless vessel on the coast, man, woman, and child hurry to the shore, to fall on their quarry. Hope not to stay these wolves. They plunder at their ease under the fire of the coast guard. It would be something if they always waited for shipwreck, but it is asserted that they often cause it. Often, it is said, a cow, led about with a lighted lantern at its horns, has lured vessels on the rocks. God alone knows the night-scenes that then take place! A man has been known to gnaw off a finger with his teeth, in order to get at a ring on the finger of a drowning woman.[78]

---

74 See Huntress 1979, p. xi.

75 On the background see Taylor 1971, pp. 252–263; Despoix 2009, pp. 41–43, 59–69.

76 See Cabantous 1993, p. 232.

77 "Instinct de race," "soif insatiable de pillages," Goulven Denis and Alexandre Bouët, cited in ibid., pp. 240 f. See also ibid., p. 213.

78 Michelet 1845, p. 152.

Communication and alliance between equally cruel maritime and human natures; an unquenchable thirst for blood uncontrollable by any authority and occasionally taking on anthropophagic traits; and the treacherous kindling of deceptive coastal beacons: through a renowned national historian's authority, all these set pieces of a popular mythology of beach-robbery here receive the consecration of certified cultural-historical fact. But if within this romantic perspective, the coastline is no longer only an elementary border but also the "outermost edge of the semiosphere" that may be called culture, then precisely its "civilized" and "barbarian" aspects are here being negotiated.[79] Seen in this way, what is at stake on the barbarian beaches is nothing less than "the securing or dissolution of the elementary code" of culture: law, language, images, even money. The barbarism imputed to exist on territorial edges has, in the end, a function of cultural diagnosis. "First other cultures; then lack of culture; then qualities ruining one's own culture from the inside"—this is what manifests itself in beach-robbery in the pre-modern and Early Modern periods, and finally on the threshold of the modern age. But in their role as imaginary vehicles of cultural critique, plundering barbarians are nothing less than what Manfred Schneider terms "guarantors of poetry."[80]

When following 1800, Gothic tales appear depicting Early Modern beach-robbers in their barbaric activities, a previously often evoked and doubted "cultural fact," but in any case a fact recognized as eccentric, has become the stuff of legend. In "The Wrecker. A Tale from the Sixteenth Century," included in William Hazlitt's *The Romanticist and Novelist's Library* of 1841, the collective phenomenon of "wrecking" has become an intrigue within a family enterprise. Driven by demonic passions, but also guided by economic calculus, a beach-robber—"like the vulture, ever watchful for his prey"—and his wife here use the "atrocious expedient" of deceptive shore-beacons in order, as "ruthless barbarians," to plunder and when necessary murder stranded sailors. They consider those washed ashore a "God-send," and deadly storms as "congenial" not only to the soul's "gloomy temper" but to their "interests." After having to endure a long lean period because of absent jetsam, on a stormy night the pair awaits fresh booty. "Unmoved by the horrid spectacle," the wrecker lurks on the beach. When to his surprise he discovers a shipwrecked man alive on shore, he murders him with no hesitation. As soon as he removes a ring from the corpse, he recognizes his only son, who left the disgraceful parental house

79 On this constructed polarity and the "ambivalent boundary" between them, see Lotman 1990, pp. 141 f.
80 Schneider 1997, pp. 11, 13, 19.

years before to find an honorable livelihood at sea. The wrecker's despair now drives him to a leap from the cliffs, his "wretched wife" perishing "a few weeks afterwards by the fall of her hut, occasioned by one of those dreadful storms which she and her savage helpmate had so frequently invoked."[81]

This Gothic tale, allegedly picked up from popular gossip, precisely situated on the notorious coast of sixteenth-century Cornwall, attesting to a superstitious or savage mentality and grounded in a fixed or even canonic store of motifs, differs from the maritime tales stemming from England and collected around 1800 —the core of a large body of ballads and novels. "The Wrecker" does not participate in the continuously unfolding tradition that, as the sailor's yarn, actually knows no territory, and no poetry belonging to a folk or nation.[82] This narrative presented here is not offered from the sea's perspective but from that of the coast. Or rather, the coast is here perspectivized from the inland, urban, and cultural centers.

From this vantage point, the beach-robbing coast dwellers becomes a projection of the Romantic-republican concept of people or nation, as in Walter Scott's *The Pirate* of 1821, although their unruliness is, historically considered, limited to securing traditional claims.[83] In novels such as James Fenimore Cooper's *Red Rover* (1827), and in tale-collections by Heinrich Smidt, Paul Sébillot, Paul Gerhard Heim, Richard Wossidlo, and others, tidings of their savage practices are connected with traditional lore, for example tales of the Flying Dutchman and Death Ship.[84] Finally, an opera is composed, Ethel Smyth's *The Wreckers* (1906; French libretto by Henry Brewster), a best-selling novel written, Daphne du Maurier's *Jamaica Inn* (1936), a like-titled film version of that novel produced in 1939, directed by Alfred Hitchcock. Notably, this work at most modified the plunderers' motivic "pandemonium" in line with contemporary mass psychology and psychopathology: in their murderous raids, du Maurier presents Cornwall's beach-robbers as "madmen" and even "monkeys"; as "hysterical," "demented and inhuman," and operating, when the occasion allows it, with "stupidity and panic." And the man pulling the strings behind this horrible shadow economy is no less than a pastor—a wolf driving a herd of dumb sheep onward, after recognizing himself to be "a freak of nature" henceforth consigned to a state of "pagan barbarism."[85]

---

81 For this and the following: Hazlitt 1841, pp. 1f.

82 On the sailor's yarn and its ethnographic processing see for Gerndt 1986, pp. 61–67; Gerndt 1988, pp. 2–14.

83 See Cabantous 1993, pp. 74, 259. On Scott's *Pirate* see Schmidt 2002, pp. 93, 94, 98.

84 See, e.g., Cooper 1991, p. 629; Smidt 1849, pp. 142–148; Heims 1888, p. 103.

85 De Maurier 1977, pp. 167–170, 243, 245.

Before "wreckers" could become the theme of trivial gothic novels, security standards on the coasts had to prevail of the sort championed since 1800 by countless committees and organizations. Until then, the strand-robber as poetic bogeyman was an effective tool of agitation. For example, in 1850, on the occasion of a stranding that involved many losses, the Methodist reverend Charles Crump added a poem to a petition, signed by many illustrious persons, calling for a lighthouse to be built at Morte Stone, a notorious reef in southwest England. In his preliminary remarks, Crump explained that "[a]n allusion to the wreckers who infest the shore of Morte Bay, arose naturally from the subject," before presenting the "plundering propensities of the country people," these "rabid wolves" and "vampires," personifications of maritime danger, in rhymed verse. To conclude, he spelled out his poem's intended message: "if a Light House, of very moderate height, were erected on Morte Stone, or the Coast adjoining, not only would the disgraceful scenes, referred to, be prevented, but many lives and much property would be annually saved."[86] Until around 1900, one glorification of poetic Enlightenment belonging to all "civilized" maritime powers' principles of faith was as follows: coastal security was owed to a shift of mentality, itself stemming from literature extending from "the *Odyssey* up to contemporary literature"—such literature having "awakened the empathy of even the most distant inland."[87]

But authors who tried to explore the grounds of this Enlightenment itself had a different story to tell than one of lighthouses that were constructed, coastal peoples civilized, at long last. Robert Louis Stevenson, scion of a dynasty of Scottish lighthouse builders, recounts in his posthumously appearing *Records of a Family of Engineers* (1912) that "[to] the end of his life, my father remembered that amphitheatre of placid spectators on the beach." Of interest to Stevenson, however, was not so much the bloodlust of "wreckers," his father's "natural enemies," as rather taking a historical perspective on their activity: "Think how many Viking ships had sailed by these islands in the past, how many Vikings had landed, and raised turmoil, and broken up the barrows of the dead, and carried off the wines of the living; and blame them, if you are able for that belief (which may be called one of the parables for the devil's gospel) that a man res-

86 Crump 1850, pp. ii, 5, 16. See also Habermann 2002, pp. 111f.

87 Fitger 1902, p. 135. Cf. ibid., p. 136: "In ancient times washed up jetsam was a welcome addition to income. In scattered locations, preachers even regularly prayed in church to 'give us a blessed beach,' meaning pray let many ships go under upon our beaches! Elsewhere worst things were done, extinguishing the proper beacons and lighting ones on false places to draw ships to the beach. That has now changed completely. The coast dwellers are participating fully, indeed at moving peril to their own lives, in the work of rescue."

cued from the sea will prove the bane of his deliverer."[88] With its total of 71 lighthouses, the Stevenson dynasty had arguably produced a truly monumental work of "coastal Enlightenment"; Robert removed himself from this familial tradition. Although commencing engineering studies in 1867 and submitting his treatise "On a New Form of Intermittent Light for Lighthouses" to the Royal Scottish Society of Arts four years later, he then took up the study of law, if only for the sake of a degree; afterwards he devoted himself entirely to writing, in that sense fully the rebellious son.[89]

The main focus of Stevenson's work would thus be neither lighthouse construction nor the enlightenment of barbarian coast-dweller—either Scottish or on the South Sea coasts to which the family enterprise was soon asked to turn; and it would not be the putatively clear and secure separation between land and sea. Rather, Stevenson tried his hand at being a literary observer of Victorian Britain and its imperial and colonial economy, its opaque aspects, chimeras, destructive speculation. In that context, *The Wreckers,* a novel he wrote in 1892 together with his stepson Lloyd Osbourne, brings three thematic domains together—those of art, business, and adventure. The protagonist and narrator, Loudon Dodd, is the prodigal son, lost to art, of an American businessman. Although when still at school a daily simulated stock-market game initiated him into "book-keeping" and "gambl[ing] in produce and securities," over serpentine intercontinental circuits his actual "matter of interest" will prove to be a "wreck."[90]

Initially driven by his sculpting aspirations to Paris, and into the embrace of the city's bohemian subculture, he lands in poverty: the Parisian art-world has remained inaccessible, "the bubble of my father's wealth" has burst.[91] Remedy is promised by Jim Pinkerton, a businessman of Scottish origins who described the convertibility between art and business: "You'll find it's just the same as art—all observation and imagination; only more movement."[92] Now a business partner, the stranded artist follows Pinkerton to San Francisco, where the two prove themselves on the field of "stock, risk and profit." "Every dollar gained was like something brought ashore from a mysterious deep; every venture made was like a diver's plunge"—and initially, in a very concrete way, this involves "wrecks and condemned vessels" bought and then sent back to sea "under aliases," in order

88 Stevenson 1922, pp. 455, 465.

89 See in detail Möhring 2002, pp. 233–238, 240–242.

90 Stevenson 1996, pp. 9, 17.

91 Ibid., p. 54. On the economic and biographical context of *The Wreckers*, see Hirsch 2005, pp. 75–81.

92 Stevenson 1996, p. 88.

to capitalize on nautical ruin.[93] At one point this leads to "shipwreck, confiscation, and a lawsuit with the underwriters," at another point to a quarrel between the partners over principle: "'I will not make money by risking men's lives,' was my ultimatum. Great Caesar! Isn't all speculation a risk? Isn't the fairest kind of shipowning to risk men's lives?" Business is a risk since it is founded in the chance *and* danger of happenstance. This is the sense of both men's future maxim: "when you see a dollar lying, pick it up! Well, here I've tumbled over a whole pile of 'em on a reef in the middle of the Pacific."[94]

The said discovery is the *Flying Scud*, which, fully insured, is being auctioned off by its insurer, Lloyd's Underwriters, following an accident near desolate Midway Island. Because they suspect that a valuable cargo of opium is located in the wreck, they purchase it at auction, but through an unknown rival bidder they are driven to a price that plunges them into massive debt. A voyage to the Pacific reef with their own chartered ship is meant to supply the suspected opium—however, despite careful cannibalizing of the wreck nothing of value is found. The partners are bankrupt. Then, through painstaking research leading to Europe, Dodd manages to clear up the mysterious story behind the *Flying Scud*. The crew rescued from the ship was originally engaged in opium smuggling on the *Currency Lass*, but became stranded on Midway, then being taken on board the *Flying Scud*. When the captain of that vessel demanded their opium as a price for rescue and transport, a heated quarrel ensued culminating in the murder first of the captain, then of the entire crew to get rid of any witnesses. Armed with their identities, the murderers, including Carthew, the outcast son from a rich British family, had themselves brought to San Francisco on the next passing ship. Dodd will earn his living as Carthew's partner; he finally reproduces Carthew's yarn, bringing together the plot's convoluted threads.[95]

This "romance of business" reveals art's doubly speculative function, as a good that is highly valued as something unmarketable while at the same time representing a marketing medium of reflection and innovation.[96] At the same time it reveals the de-realizing effects of an economy that only allows reality to manifest itself in its failure. When for example the crew of the *Currency Lass* rehearses its role as crew of the *Flying Scud*, it recognizes how realistic the masquerade is, but how implausible the facts documented in that ship's log and accounting books appear. "'Well, it don't look like real life—that's all I

93 Ibid., pp. 103, 85, 87.
94 Ibid., pp. 87, 92.
95 Ibid., p. 286.
96 Ibid., p. 298.

can say,' returned Wicks. 'It's the way it was, though,' argued Carthew. 'So it is; and what the better are we for that, if it don't look so?' cried the captain, sounding unwonted depths of art criticism."[97] For every event and business transaction we here find, not one accountable actor but at least one doppelganger or mastermind as well—and this is also the case for the double authorship of Robert Louis Stevenson and Lloyd Osbourne, two writers sharing surety for the text, just as Lloyd's Underwriters stand surety for the wreck. Hidden "underwriters" are here the actual beneficiaries of all the events and transactions.

This process of doubling, sharing, displacement also takes in all the novel's key terms marking the process of transfer between art, business, and adventure. For example, "figures" initially refer to Dodd's artistic efforts, then to numerical tables for accounting, then to Dodd's artful success in "figuring out" the novel's hidden business-venture. "Value" and "interest" have artistic and, in the context of amorous transaction, personal significance, as well as moving finally to the vanishing point of the "business adventure"; and "office" denotes not only a secret center for trade and action but also the vocation for and pursuit of ritual provisos.[98] *The Wreckers* offers less a genealogy of such terms than rather their semantic and—on the level of trade and action—operative dissemination and diversion. This has bearing on the novel's guiding concept: otherwise than what the mythology of beach-robbery suggests, "wrecking" contains a broad range of associated meanings around which the Victorian mercantile system's way of functioning crystallizes: designated here are on the one hand a case of commercial bankruptcy, simple collection of jetsam, and naturally the violent or ritual plundering of beached ships, and on the other hand their salvaging, intentional sinking, cannibalization, auctioning off, final scrapping, together with dealing in stolen jetsam. Particularly in England and from the time of Henry I onward, such semantic differentiation was crucial for legislation regarding putative beach-robbery.[99]

If already the "classical" version of such robbery is centered on chance, on Fortuna's gifts and thus, to cite Cabantous, "always unbalanced and random" economic regulation, then this is all the more the case for the shadow economy tied to insurance fraud, as manifest in over-insured death ships and the activities of treasure hunters drawing on Lloyd's List, the global register of damaged ships, to steer their "business adventures."[100] In the novel of Stevenson / Osbourne, these various facets of "wrecking" interact, and are in fact often indistinguish-

97 Ibid., p. 351. See also Watson 2007, pp. 114–128.
98 See Arata 2007, passim.
99 See for example Stevenson 1913, p. 244; Rule 1975, pp. 169, 172f.
100 Cabantous 1993, p. 255. See also Bathurst 2005, pp. 297f; Lehmann-Brune 1999, p. 218.

able, so that the businessman, adventurer, and artist Dodd can say that "of all forms of the dollar hunt, this wrecking had by far the most address to my imagination." The wreck off of Midway Island, in the no man's land "between sea and sky," promises a "prodigious fortune," even if it has become clear to both "wreckers" after their bankruptcy that "God never made a wreck big enough to fill our deficit."[101] The substance of their business is simultaneously hidden and manifest in the wreck. When it has been fully cannibalized and scrapped, it presents its emptiness, which in turn promises a not yet realized plenitude, hence which prompts speculation on the added value of a still unfathomed business secret. Stevenson himself referred to his co-written novel as a "very modern form of the police novel or mystery story."[102] In actuality, both narrative and narrator pursue an archeological poetics, meant to successively uncover that still unfathomed event—and in fact a continuous search for the proper genre and "emplotment," even if that amounts to the "dime novel," is at work here—a search for a narrative format allowing us to begin tracking the riddling story's course.[103]

What finally reveals itself behind the ceaseless circulation of chimeras and speculation is what stood at the beginning: opium. Already the *Currency Lass* set out for the sake of opium smuggling, and nothing less than the hope for opium moved the partners to buy the *Flying Scud* at auction. Opium, then, is the first mover and moving cause of these modern "wreckers": "there will be a good deal of rather violent wrecking to do before you find that—opium." From both a commercial and moral perspective, opium is what renders possible the novel's business and action. "Smuggling is one of the meanest of crimes," Dodd concedes against the backdrop of the imperialist opium wars, "for by that we rob a whole country pro rata, and are therefore certain to impoverish the poor: to smuggle opium is an offence particularly dark, since it stands related not so much to murder, as to massacre."[104] When the action that began with the artist's vocation and artistic commerce ends in a sinister massacre of opium smugglers, then the core of a civilized, world-encompassing economy reveals itself at its geographical margin: the barbarism of "wrecking."[105] *The Wreckers* in the end gravitates around the scene in which the *Flying Scud*'s wreck is plundered "vulture-wise," then around an epiphany of unrestrained violence, insensate blood-

101 Stevenson 1996, pp. 120, 175, 187, 216.
102 Ibid., p. 363.
103 Ibid., p. 209.
104 Ibid., pp. 129, 149.
105 See Hirsch 2005, p. 90.

lust, and unquenchable greed.[106] This work of literature, "full of details of our barbaric manners and unstable morals," is according to Stevenson "less a romance than a panorama—in the end, as blood-bespattered as an epic." But the "wreckers" not only represent "a queer kind of beast."[107] Already in 1700, when they were first legislatively recognized as legally relevant actors, the "barbarian" beach-robbers' actions were considered a venial offense when compared to the criminal business that would be pursued by their civilized progeny: smuggling—and this with the illegal fuel of an imperial economy.

In essence *The Wreckers* consists of a long, in its factual contents unconfirmed seaman's yarn come across by narrator Dodd and authors Stevenson and Osbourne. At one point we accordingly read that "the beach-comber, when not a mere ruffian, is the poor relation of the artist." Namely, alongside the more or less barbarian exploitation of "material," in its "wrecking" modern literature above all gathers jetsam. With its "touch of the poetic"—and this is perhaps already pointed to in Shakespeare's "Tempest" in the "direful spectacle of the wrack"—it lies in wait for the haphazard chance accompanying a crisis of political and existential forms (as exemplarily manifest in shipwrecks)—and also accompanying a dissolution of putatively elementary distinctions such as that, most forcefully, between land and sea.[108] Such a poetology especially characterizes the writing understanding itself in the mid-nineteenth century as the ethnographic collection of maritime yarns[109] and since the mid-twentieth century as the archeological processing of maritime facts. In *Eupalinos*, one of Paul Valéry's pseudo-Platonic dialogues, which center on both the question of a poetics of knowledge and that of differentiating between philosophical and poetic knowledge, the archetypal starting situation of such a literature of wrecking is outlined as follows:

---

106 Stevenson 1996, p. 178. Cf. ibid., p. 205: "We made a singular picture: the hovering and diving birds; the bodies of the dead discoloring the rice with blood; the scuppers vomiting breadstuff; the men, frenzied by the gold hunt, toiling, slaying, and shouting aloud: over all, the lofty intricacy of rigging and the radiant heaven of the Pacific. Every man there toiled in the immediate hope of fifty dollars; and I, of fifty thousand. Small wonder if we waded callously in blood and food."

107 Ibid., pp. 362, 174.

108 Ibid., p. 112; Shakespeare, "The Tempest," 1, 2, 26.

109 See Hansen 1972, p. 97 f.: "Occasionally (to be sure rarely) well-corked bottles washed up from the sea are found on our shores, bottles containing old, often strange and mysterious nautical and other messages written on paper or parchment. Once, in the summer of 1823, I lay at anchor on a ship for fourteen days on the *Seesand* sand island, south of *Amrum*, hoping in vain for good wind. Then I had time and occasion to collect some material for my sagas and stories concerning the tidelands and Halligs."

> This frontier between Neptune and Earth, even disputed by those rival divinities, is the scene of the most dismal and most incessant commerce. That which the sea rejects, that which the land cannot retain, the enigmatic bits of drift; the hideous limbs of dislocated ships...all these things, in short, that fortune delivers over to the fury of the shore, and to the fruitless litigation between waves and beach, are there carried to and fro; raised, lowered; seized, lost, seized again according to the hour and the day.[110]

What literature—located as it is at the border of an intensive exchange between deterritorialization and grounding, dissolution and development of form—either seizes when offered the fortunate chance or else leaves to the play of hazard, consists of *objets ambigus:* objects that are undefined but are capable of and call for definition. To recognize them means constructing their ontic, formal, and economic context, which in the case of a shipwreck means presenting it as a complex experiential, artistic, and economic object. The wreck is an exemplary ambiguous object, over which elementary demarcations are negotiated: the borders between security and peril, power and resistance, enlightenment and barbarism, civilization and savagery, art and non-art, land and sea. For its part the coast, as the allegory and scene of such demarcations, is the locus of a steady process of liquidating and reconfiguring. Within a perspective stamped by cultural optimism, the coast represents, Ernst Kapp observed in the mid-nineteenth century, a "dissolution of antinomies," their annulment in amphibian intensity, with Kurt von Boeckmann then commenting, more than fifty years later, that "marine and terrestrial powers here achieve interpenetration."[111] From the vantage point of cultural criticism, observing for instance the barbarism of local as well as global trading and other action, we should perhaps fear, to the contrary, that in Dolf Sternberger's words, "the sea has climbed onto the land."[112]

Kafka's Seasickness

With the sea seemingly having done such terrestrial climbing, previous terra firma came to be increasingly experienced as wobbly and groundless. Franz Kafka thus speaks of his "seasickness on land" as if this were the groundlessness of his authorship. Since antiquity, the existential metaphor of seasickness has been part of the repertoire of countless travel narratives. Kafka frees them

110 Valéry 1956, pp. 113f.
111 Kapp 1868, p. 54; Boeckmann 1924, p. 14.
112 Sternberger 1981, p. 243.

from their traditional geography of land and sea. For him, the human being's basal definition no longer applies: the human being as, to cite Schmitt, a "land-being" or "land-dweller," moving "upon the firmly grounded earth," and for whom among "the elementary realities" the sea is the one reality "with which we are least at ease."[113] It is the case that for Kafka as well the sea is "too heroic." Already the "absurdly short crossing" from Trieste to Venice gives him "slight seasickness" and, he recounts in a letter to Felice Bauer of 15 September 1913, a "trembling in my head."[114] But not even the earth seems firmly fixed to Kafka, rather constituting a monstrous and fluctuating element. That farmers do not become seasick while laboring in the fields is nothing short of astonishing to him.[115]

Even Land Surveyor K. in *The Castle* is characterized by having an uncertain, slippery "position,"[116] while in *The Trial* Josef K., drowning in a sea of files, becomes "seasick" in an officials' attic and thinks he is "on a ship, rolling in heavy seas."[117] We can read these texts as something like the minutes of the struggle for a secure "position" engaged in by Habsburg officials employed in both domestic and foreign service.[118] But at the same time they depict the sort of endless struggle against an existential groundlessness diagnosed in "Description of a Struggle," which was in fact the first narrative text Kafka completed, between 1904 and 1911.[119] The phrase "seasickness on land" contains a program that is both anthropological and poetological, developed by Kafka in a revision of the *Satz vom Grund*, the principle of sufficient reason. In Hans Blumenberg's formulation, "*feeling* the ground of life as instable" on the one hand "tears apart the subsoil of the life-world itself. What all orientation refers to withdraws from the organ of

---

113 Schmitt 2015, p. 5; Blumenberg 1996, p. 9.

114 Kafka 1973, p. 320.

115 For this reason he views not only seamen but also, especially, farmers such as those in the place of his convalescence, Bohemian Zürau, as "noblemen who have escaped into agriculture, where they have arranged their work so wisely and humbly that it fits perfectly into everything and they are protected against all insecurity and worry until their blissful death." Kafka 1949, p. 188.

116 See Kafka 1946, pp. 71, 72, 117, etc. On the paradox of this slipperiness in Kafka, with its components of reversal and diversion, see Neumann 1973, pp. 462–464.

117 Kafka 1998, p. 78.

118 On the professional "slippage" at that time of, precisely, land surveyors see Wolff 1919, pp. 276, 278. On the struggle of Habsburg employees for a fixed "pragmatics" of service and the connected hope for permanent positions, see Weeks 1983, pp. 325–332.

119 On this work and in more detail on Kafka's "poetics of insufficient grounding," see Wolf 2006, pp. 121–127, 131.

orientation, above all the sense of balance, its referential capacity."[120] But on the other hand, this uncertainty and destabilizing of the life-world and its grounding necessitates new tenets of feeling and aesthetics, tenets capable of probing the grounds of such boundless non-orientation.

Derived from Greek *naus*, ship, in Greco-Roman antiquity, seasickness was denoted as *nausea* and connected with both vomiting and general disgust or fundamental world-weariness. To save oneself from this distress, nothing less than a miracle was needed, such as what was ascribed to Paul shortly before his shipwreck.[121] In the context of modern nautical medicine, John Arthur Irwin coined the term "motion sickness" in 1881 and Ottomar Rosenbach "kinetosis" in 1896. Anxiety and trembling, dizziness and headache, fatigue and lassitude extending to apathy, depressive feelings of distress and catastrophe that can range from sadness at lost life on land to complete collapse of one's personality: this traditional set of symptoms has always been derived from the movement of ships.[122] As was observed early in the twentieth century, what is actually at work here is hyperesthesia, catalyzed by shock to the nervous system and subsequent disturbance of the sense of balance and coordination center. Therapeutically, what is above all called for is bed rest; but also decisive for restoring stable balance is "steady will-based energy."[123] Kafka makes the described symptom his own in diaries and letters: He actually suffers not from "physical weakness"; "it is the lack of grounding [*Boden*]" that leads to a failure of "family life, friendship, marriage," not capable of being compensated for by his weak or broken will.[124]As what he refers to as an embodiment of symbols,[125] Kafka attaches himself to a writing process spelling out the figure of absent grounding in all its implications.

Kafka undoubtedly knew Leibniz's classical version of the principle of sufficient reason, according to which nothing exists without its sufficient grounding, as well as Schopenhauer's interpretation. For Schopenhauer the world of imagination underlies the principle. What it does not apply to is, on the one hand, the principle of sufficient reason itself, on the other hand the "thing in itself," consequently "the will."[126] The subject of cognition and the subject of volition ach-

120 Blumenberg 2000, p. 533.

121 See Roloff 1988, p. 364; Zmijewski 1994, p. 861.

122 See Schlag 1899, pp. 7–9.

123 See ibid., p. 23; Schwerdt 1914, p. 24; Schadewaldt 1967, pp. 2258–2265; Linke 1987, p. 126.

124 Kafka 1992, pp. 194, 98; Kafka 1993, p. 212; Kafka 1949, p. 164.

125 Kafka 1977, p. 199.

126 Schopenhauer 1969, p. 81. Schopenhauer here takes up Aristotle's classic formulation in the *Metaphysics*, Aristotle 1941, 1011a (pp. 747 ff.).

ieve identity in self-consciousness, the sufficient ground of action lying as it were in the subject itself. Activity of the will, however, is here "so immediate" that "in most cases" it must evade cognition. Seen in this way, not even an introspective search such as what Kafka practiced in line with Franz Brentano's "descriptive psychology," leads to sufficient reason.[127] Instead that "life-world" moves into slippage, wavering, and even the fathomlessness that Brentano's student Edmund Husserl defined as "a realm of original self-evidences" and "primal self-evidence."[128] The only thing that for Schopenhauer cannot be doubted is "[e]ternal becoming, endless flux."[129] But what can remedy the "seasickness" is art alone. For it allows the ideas, verbal images, and metaphors initially subsumed to the principle of sufficient reason to become aesthetic, to go under as ideas, and merge with the last of all grounding.

Aesthetics such as Schopenhauer's and poetics such as Kafka's follow a principle of *in*sufficient reason. According to Hans Blumenberg, this principle is "the correlate of the anthropology of a being lacking what is essential." But it is by no means to be confused "with a postulate of renunciation of grounds."[130] Rather, it prohibits simply complying with the absence of a ground. It imposes an obligation to compensate for the fundamental lack, which is tied to a kind of nominalistic crisis: with, as laid out in "Description of a Struggle," the real names of things being forgotten, being forced henceforth to rely on "any old names";[131] with, as we can put it very generally, a crisis of the *nomos*, involving finding names of things as much as seizing terra firma; and with what for Nietzsche is the resulting elementary crisis that "there is no more 'land'!"[132] This anthropologically critical situation leads from the terrain of solidly grounded, principled truths out to a raging sea: to the place where—to again cite the phrase from Nietzsche's "On Truth and Lying"—only the metaphoric conceptual "structure of beams and boards" keeps us from going under. Now if according to Ariston seasickness is an experiment of nature,[133] then as Nietzsche wrote in 1879 "a feeling closely related to seasickness," as well as that of absent ground-

127 Schopenhauer 1891, 172. On "evidence" as the final ground for action see Brentano 1930, p. 140.

128 Husserl 1970, pp. 127 f.

129 Schopenhauer 1969, p. 164. Schopenhauer defines this flux as will. The ego is only this will's manifestation. It is subsumed to the *principium individuationis*, itself described as a "frail craft in a stormy sea that is boundless in every direction, rising and falling with the howling, mountainous waves. Ibid., p. 352.

130 Blumenberg 2001, pp. 422 f.

131 Kafka 1971, p. 53.

132 Nietzsche 2001b, p. 119.

133 See Oldelehr 1977, p. 30.

ing, allows "the most instructive tests and experiments in the spiritual-moral field" [*auf geistig-sittlichem Gebiete*], that in turn can systematically be generated by what Nietzsche in *Beyond Good and Evil* terms a "genuine physio-psychology" through an unsparing because transcendental analysis of the phenomenon of the will.[134] Kafka, then, did not have to wait for an encounter with Ernst Mach's writing to form a foundational insight: that although the ego is beyond saving, within a logical-experimental framework, as an individual, psycho-physical entity carried along and driven by the ceaseless stream of sensations, it can be observed and possibly transformed.[135]

The principle of insufficient reason will remain Kafka's poetological principle long after the writing of "Description of a Struggle." Even Kafka's myths of foundation merely lead the foundational act to the compensation of already insufficient foundations. For the very effort to use gigantic construction projects to render not very firm terra firma into the foundation for a new collective existence, protected from danger and irritations, everywhere reveals the patchiness and fragility of one's own existential grounding. For example, we read that the Great Wall of China may once have offered cause for hope that it would serve as the fundament for a new tower. But this thought can have been "meant only in a spiritual sense," hence can only be off-track. While the fabulous tower has been grounded in an imaginary realm from time immemorial, crumbling away from just that "weakness of...foundation," the Great Wall, as the reporter in "The Great Wall of China" has to recognize in the style of an inspector, is nothing less than vast and fissured reality. Precisely because, otherwise than is the case with ideas of unity such as the idea of the emperor's body, it transgresses the "limits" imposed on one's "capacity for thought," it merges as "the very ground on which we live."[136] The previously doubled, natural and ideal emperor's body is now forgotten in favor of a third body of unfathomably scattered multiplicities, or else transported into a completely fabulous realm.

To be sure, a construction such as the Great Wall serves the purpose of collective defense. Like a dike, it is meant to repel dangers flooding in from the outside—say nomads threatening to overrun the land—and thus secure the territorial borders. But that construct not only offers protection from elementary threat originating on the outside. It is itself "in perpetual danger," because de-

---

134 Nietzsche 1981, p. 3 (letter to Otto Eisner, early January 1880); Nietzsche 2001a, p. 23. On Nietzsche's sea-metaphors, offered under his motto "live dangerously" (Nietzsche 2001b, p. 161), and on corresponding techniques of "self-creation," see Hufnagel 2008, pp. 151, 153, 156.
135 On Mach, Husserl, and in general the epistemological crisis around 1900 as connected to maritime and aquatic metaphors, see Blumenberg 2012, pp. 106 ff., 160.
136 Kafka 1971, pp. 238, 240 (translation modified).

signed in line with a principle of perpetual unsoundness. The construction is meant to have "gaps" larger than its individual parts, something "which cannot be verified, at least by any single man with his own eyes and judgment, on account of the extent of the structure." People only know about the Great Wall's construction through messengers and officials who spread the information via rivers and other communicative means.[137] Ideally such information would present everyone participating in the great work with its development and finality; but in reality the news of such incalculable extension and porousness induce thoughts of what is "endless." Here the flux of such thinking is "all the more on par with and welcome" to the sea, flooding the communal ground, a ground of symbolic markings. Hence getting to the bottom of the wall's construction only leads to the high seas.[138] But the "very ground on which we live" is neither the unshakable foundation of firmly terrestrial institutions nor a sphere of complete de-differentiation and engulfing like the sea. Rather, it is the half liquid, half solid space of continuous "de-constructing" and a nevertheless ceaselessly driven partial construction. Kafka's fragments from the "Middle Kingdom" thus do not treat the foundational myth of an archaic community or nation, but rather the insufficient grounding of rootless collective beings: the field of modern mass society.[139]

The "imperial message" that Bismarck read out before the German Reichstag in 1881 announced a "monumental work of modern culture" in which the "vocation for eternity" (*Ewigkeitsberuf*) of the German nation would be fulfilled "of placing the capstone on the 'obelisk of the ages'."[140] What, however, the message of a putatively completed tower-construction actually conveyed to the Germans was the "lack of grounding" for everyone: It was not only that the country's immense population growth, speedy industrialization, and build-up of new communication and transportation technology had eroded local ways of living. The traditional grounding offered by relations, neighbors, and friendship no longer offered sufficient security in the case of illness or accident. For this reason, social-insurance legislation introduced in 1881 was meant to place care-related costs under the aegis of insurance operations. Since then, well beyond Germany, the "ground on which we live" has been a statistical field and the "solidarity principle" (which legislatively transforms the insurance contract's "equivalency principle" into a new "social contract" of ubiquitous forced insur-

137 Ibid., p. 235. Compare Kafka 1971, pp. 248f.
138 Kafka 1971 pp. 238, 240 (translation modified).
139 On the interdiscursive background of Kafka's China fragments see Wagner 2010, pp. 250–259.
140 Cited from Claus 1910/11, p. 12.

ance) is the principle prescribing social compensation for the society's insufficient grounding. The state thus becomes an insurance institution declaring all those walking on dry land to be forcibly insured owners of their "ship of life"—and in the process revealing the territory of the state to be a non-secure because elementary area of danger.

This new context is not marked, any longer, by a speculative treatment of risk. Rather, when it comes to compensation, all members of the state draw from the same pot, continuously filled by everyone's payment. This concept of social existence is conceived in line with Jakob Bernoulli's original, probabilistic model of the lottery urn with its little balls, the calculi, themselves standing for chance and risk, profit and loss. However, here the urn does not serve to repeat the game of chance that, in any case, the *navigatio vitae* already is. Rather, it is aimed at a balance mediated by premiums and payments, a balance grounded in Bernoulli's main principle, later termed "the law of large numbers" (itself anticipated much earlier, as noted, by Gerolamo Cardano).[141] Those states securing their territories' members by insuring them have taken the probabilistic turn in previously purely experientially and market-based engineering of risk and rendered it a technique of governance. The state in this way becomes an *état providence,* a welfare state in which the earlier certainties regarding providence are translated into what Wilhelm Butte in 1808 named a *Staatszustandswissenschaft*—a science of state circumstances—consisting of statistics and probability calculations.[142] In such a state, illness and accident are considered societally formative because, once recognized as a *social* evil, they compel solidarity and collective welfare for the sake of forestalling ever-threatening ruin. In this way "society" becomes a mass-statistical entity now locating itself in a field of norms and deviations.[143] And "the human being" is, as Adolphe Quételet approvingly observed in 1838, an "average" creature and "fabricated being" without existential roots in his "natural" surrounding. The ground on which he stands is delimited through certain "fluctuation factors," and the life he leads is in the end a "life in landscapes of curves."[144]

---

141 See Bernoulli 1899, chap. 1., pp. 4–20, 94f.; Schneider 1988, pp. 63–76, 200f.; 1998, pp. 87ff.

142 On Butte see Seifert 1980, p. 231. On Leibniz's work in the fields of statistics, probabilistics, insurance, and conditional law anticipating and paving the way for the welfare state, see Karten 2001, pp. 327–337.

143 See Ewald 2020, pp. 165–180.

144 Quételet 1836, pp. 21f.; Quételet 1835, pp. 323f. On the concept of "life in a landscape of curves," see Link 1999, p. 345.

Like the anonymous narrator in the construction of the Great Wall of China, Kafka, as an employer of Bohemia's central social-insurance company (the largest such company in the Austro-Hungarian Empire), the Prague-based *Arbeiter-Unfall-Versicherungs-Anstalt* (AUVA), was himself not only a messenger of this project but actively participating in it. When Emperor Franz Joseph declared the existence of social insurance in 1885, by contrast with the German approach it was in fact not declared an unshakable monumental work but rather mere patchwork—organized according to the regional principle and making use not of functionally differentiated but of territorially organized statistics. In 1913 in Vienna, for the International Congress of Rescue Services and Accident Prevention, lower official Kafka wrote or at least co-wrote two speeches for his superiors Pfohl and Marschner treating this dubious composite system.[145] Kafka was professionally aware of the insufficient basis for modern foundational myths. On the one hand, around 1900 the "social fundament" took less the form of law and order than that of norms and security; on the other hand, since its beginnings the "principle" of insurance has had little to do with foundational deeds, far more with compensation for existential lack of grounding. And Kafka, having attended seminars in legal history and insurance statistics as a law student, was aware that the technology of insurance had started with maritime insurance in the fourteenth century.

In addition, Kafka became practically acquainted with both marine insurance and transport insurance, which took over its counterpart toward the end of the nineteenth century, when in 1907, a year before moving to the AUVA, he was employed on a temporary basis in the Prague branch of the Trieste-based *Assicurazioni Generali.* Around 1900, this branch was engaged in insurance-technical operations under exacerbated conditions. Otherwise than in the field of social insurance, the maritime insurer not only bore a specific danger, but rather—following the "principle of universality" and as is already stated in the oldest extant insurance policy, drafted in Pisa in 1384—"every danger that God send down or that comes from the sea or from men and every contingency or danger or blow of fate, and every misfortune that could emerge in any way."[146] But from the 1870s onward, general transport-insurance with its exhaustive bills of lading, had to also take on dangers of transport on land, on rivers, and in interior waters, together with transshipment. Otherwise than was the case in the social-insurance branch, transport insurance could not rely on any regional principle,

145 See Kafka 2009, pp. 249 ff.
146 Liebig 1914, p. 52.

any "geography of risks." It had to contend with something like cumulative danger, its individual components usually not being territorial identifiable.

Since introduction of regular ship communication, above all through cable- and wireless telegraph, this type of insurance was no longer bureaucratically transacted but tied to ongoing communication technologies. What such insurance involved was thus not only transport of goods and people on delimited territory, not only, as Kafka termed it, "natural commerce." Since the introduction of the radio, which with its "ether" opened up an "electric ocean" on both land and water,[147] it was also above all an elementary space of communication. This "sea," where Bohemia also now lay, with its "insurmountably high waves," was fed with messages from "the wireless" and in this way populated by "ghosts."[148] With these ghosts haunting terrestrial and maritime messages and commerce as "parasites," as noise or disturbance, the entire world had, as described in Revelations 18, 2, "become a dwelling for demons" destined for eradication along with the high seas. Even "life in a landscape of curves," in the domain of a state that since the emergence of social statistics operated far from the sea,[149] had according to Kafka been washed over by a "great tide of unrest."[150]

Gracchus and the Fault in Grounding

Various texts of Kafka can be read as reports about how insufficient reasons can be compensated for through insurance operations—or also, against all hope, enjoy no compensation. In this context, particularly the prose pieces treating non-grounding on the open sea show how old topoi and existential metaphors tied to the allegorical *navigatio vitae* draw near to the insurance horizon. This is clear, for a start, in Kafka's first, fragmentary novel *Der Verschollene* (*Amerika*, or *The Man who Disappeared*). That following his trans-Atlantic steamship voyage and before he arrives on land Karl Rossman descends into the ship's bowers

---

147 For a contemporary interpretation—developed especially through Guglielmo Marconi and in cooperation with Lloyd's—of "ether" as the waves of the "electric ocean," see Slaby 1922, p. 228.

148 Kafka 1990, pp. 204, 223.

149 Edmond Halley's pioneering work on probabilistic mortality and pension calculations drew on data from Breslau left him by pastor Caspar Neumann at Leibniz's behest. This city was, after all, "very far from the Sea," hence not deterritorialized through "confluence," permanent streams of people, goods, and information, which would have clouded its statistical regularity. See Halley 1692/93, p. 597.

150 Kafka 1977, p. 180.

and displays solidarity there with a stoker has been interpreted as his advancing into his European unconscious: into the working class.[151] But Kafka was less interested in direct, class-warfare oriented social critique than in society's insurance-technical constitution, in face of which the reality of class differences could first be read. When in April 1912, five months before Kafka wrote the novel's stoker chapter, a putatively unsinkable monumental product of nautical construction was sunk by an iceberg, he did not only have those who "will elbow their way to the lifeboats at a shipwreck"[152] in mind but also a problem of insurance technology: In the certainty "that the larger the ship, the easier it is to achieve unsinkability," the shipowners reached an agreement with the English "casco" underwriters to insure the largest ship ever built "at an annual premium of 3/4%."[153]

After the *Titanic* met its iceberg and went down, insurance companies abandoned the idea of unsinkability. Real classification of danger was now demanded for what had been proclaimed the safest of all technologies, in addition to a new construction feature; "alongside the transverse bulkheads...also underneath the waterline a double hull, extending until above the waterline."[154] Kafka drew his social critique from just such technical problems of classification and construction: "The lowest of all spaces in an ocean liner, running through the entire ship, is completely empty... Actually it is not entirely empty—it belongs to the rats," we read in one note."[155] The "lowest of all spaces" might be understood as society's unconscious, or else its substructure. In the end, however, it is a technical achievement tied to both transport and society—accompanying technically determined growth in security are lower insurance costs. But such innovation, taking the form of displaced and "lowest of all" creatures now dwelling unseen on the ground, hides a remainder, its nature tied to both security and insurance operations: those evidently incapable of integrating themselves into the security calculus, the superfluous and redundant.[156] In respect to their com-

151 See for example Goldstücker 1964, pp. 49–64.

152 Kafka 1977, p. 271.

153 Statement of technical director of German Lloyd, cited from Ulrich 1912, p. 1045.

154 Ibid., p. 1048. See also the chapter entitled "Spekulatives Zwischenspiel: Vom Restrisiko in Seefahrt, Verwaltung und Literatur" ("speculative interlude: from rest-risk in seafaring, administration, and literature") in Wagner 1998, pp. 184–188.

155 Kafka 1992, p. 421.

156 In a fragment written in 1920 (ibid., p. 313), the problem of "redundancy" is discussed in relation to communal constitution—communities being described as nothing else than contingent, enumerable series that determine their borders through exclusion of a surplus figure, a+*1*.

pletely marginalized life form, speaking of "rats" might be appropriate; in respect to their classification, of "uninsurable" beings.

Various of Kafka's protagonists populate what Adorno refers to as a "no-man's-land between man and thing,"[157] a zone of shapelessness and de-normalization. With these figures trying everything to escape the fluid threshold between "normal and abnormal, adequate and inferior" and be reintegrated into a social sphere, Kafka's texts can be read as negotiating what in the early twentieth century was termed "insurance for the rejected."[158] But marine insurance can serve as a model for insurance against dangers in general: following the *Titanic*'s sinking, through reclassification "unsinkable" and consequently uninsurable ships became eligible for insurance. Kafka may have viewed this inclusive concept of "casco" insurance as pointing toward inclusive *social* insurance. For although already in its name, "casco" insurance was, as mentioned earlier, limited to the ship's hull, and also represented a mercantile, purely market-based form of insurance, in Kafka's thinking its meaning extended to the ship of state as a whole, taking in its representatives and blind passengers. For this form of transferal from the realm of *fortuna di mare* to that of the fate and fortune of seasick land-dwellers, Shakespeare-reader Kafka also had a literary model before him: the casket or *casco* choice in "Merchant."[159] In any event, what was possible for "casco" insurance on "swimming islands" was also meant to be possible for social insurance on (no longer) firm land: if not to represent the population's lower economic strata, then at least to insure them.

In Kafka's narrative texts, these lower strata are a repeated thematic presence. The stoker episode, which Kafka began working on after the *Titanic* catastrophe, unfolds like the unannounced inspection of an industrial plant, at whose end stands negotiations over workers' rights and working conditions. But Karl, who has declared himself the stoker's advocate, eventually has to recognize that what is in play here is something like a summary court martial, in which not so much the plaintiff's rights are the focus as his guilt for even leveling claims. The stoker stands in "the uncertain hold of a ship moored to the coast of an unknown continent,"[160] revealing itself in the novel as a metonymy for mass-statistical and still uninsured forms of existence[161] while also, for Kafka, the new continent of social insurance for seamen. To bring this underway, in

157 Adorno 1983, p. 262.
158 Florschütz 1916, pp. 432, 435; see also in general Wagner 1998.
159 On Kafka's Shakespeare-reception see Nagel 1983, pp. 34–37.
160 Kafka 1996, pp. 6–7.
161 See Iacomella 2010, p. 20.

1909 in Habsburg-Empire Trieste, an inquiry was initiated into the working conditions of seamen. The investigators now came across men treated arbitrarily as foreigners or part of the Empire, not even accorded the status of laborers, and with dependents not granted a survivor's pension in the case of "disappearance." For this reason, in place of the *Editto politico di navigazione* of 1774, in February 1913 a "law on workers' and accident insurance" was passed—a measure confined to Trieste's *Arbeiter-Unfall-Versicherungs-Anstalt.*[162] A month later Kafka discarded his draft of *Der Verschollene*, with the exception of the stoker chapter, the only part of the book located at sea and in Kafka's view emerging from "inner truth."[163] That same autumn, he took up work on the fragments where the "Hunter Gracchus" appears as a prototype of a sly, wandering possibly claim-worthy but always rejected seaman—as in fact an alter ego of Kafka himself. (Notably, through Italian *gracchio*, "Gracchus" corresponds to Czech *kavka*, jackdaw.) Tying the two together is their lack of grounding on both land and water.

Since time immemorial, Gracchus has been on an endless odyssey. As he reports in a text "nobody will read," he was once a lonely hunter in the Black Forest; while "hunting a chamois," he plunged down from a rock, in a deadly accident that was "in good order." The reason that the barge meant to bring him to "the other world" "lost its way" is never made clear in the broken-off report.[164] The story's framework is only transmitted as incomplete scaffolding, its executed passages as torn off pieces of text; the story itself appears to have found its allegorical expression in Gracchus's barge: an unrigged ship of literature, "the masts disproportionately tall, the upper third of the mainmast split; wrinkled, coarse, yellowish-brown sails stretched every which way between the yards, patched, too weak to stand against the slightest gust of wind."[165] The story's plot is manifestly located at the physical-metaphysical border between land (the place of accident) and sea (the space of transfer to the beyond). But the distinction is simultaneously muddled, because the actual accident, the mistaken steering of the barge, took place in water and inversely the Hunter Gracchus first resumes his actual passage on land.

For this reason just-landed Gracchus stubbornly refers Salvatore, the mayor of Riva, to the "fundamental error of my onetime death" and to the man guilty of

162 "Die Unfall- und Krankenversicherung der Seeleute in Österreich" 1911, pp. 630 f.; Kafka 2004, p. 201. See also Weeks 1983, pp. 322–324.

163 Kafka 1973, p. 218.

164 Kafka 1971, pp. 230, 228 f.

165 Kafka 1949, p. 170.

it—the "boatman."[166] We can here think of Charon, and of Jesus, the boatman of the Christian life journey. But possibly Gracchus's "great tide of unrest" is also a technical insurance problem, the theological and mythological motifs only definable with help of its guidelines. In contemporary liability law for maritime insurance, "two souls" were "discovered in the breast of the shipmaster, as Carl Ritter put it in 1914: "a 'nautical' soul and an 'administrative' counterpart," for the shipmaster was held responsible not only for the navigationally but also administratively correct functioning of a sea voyage.[167] Not only transit safety depended on him, but also the seamen's insurance protection against illness and accidents. Until the early modern period, care of sailors was seen to through common law: in cases of illness or accident, sailors were the captains' charge, costs transferred to partners in the voyage. On land there were charitable trusts that saw to old seamen and sometimes their widows and orphans. In general, there was a moral claim to such care but no legal one.[168] Over time this traditional arrangement, based on both charity and possession of shares, eroded and was replaced by insurance; moral obligation within solidarity-steered communities made way for legal obligation: for clearly delimited care-costs and officially defined accident-prevention regulations.

In the 1903 transactions of Germany's Association for Social Policy concerning "the situation of workers engaged in maritime navigation," we read that while

> in many other trades the number of accidental injuries is larger, the percentage of deaths is here highest, although the number fluctuates....Care for seamen who have fallen ill is the responsibility of the shipowners; for seamen, obligatory medical insurance on the lines of that existing within the German state is nearly impossible to implement. On the other hand, accident and disability insurance has been introduced in a similar way as that for workers on the mainland.[169]

For the Habsburg Empire, in the context of the mentioned Trieste investigation, "the draft distinguished between three cases: illness in the home country, illness during the voyage on board, and debarkation abroad on account of illness. Insurance now covered the first and third eventuality, while care in the second case was in the hands of the shipowner.[170] Gracchus also reports on a shipown-

166 Kafka 1971, pp. 229 f.

167 Ritter 1914, p. 49.

168 Gerstenberger 1996, p. 157.

169 Francke 1904, pp. 16, 30.

170 "Die Unfall- und Krankenversicherung der Seeleute in Österreich," 1911, p. 630.

er—as a “good man” who regrettably “died today”.[171] Gracchus shunts off responsibility for his restless existence by repeatedly referring to his work as a hunter and alleged deadly accident, in order to then blame the “boatman” directly.

Against the backdrop of the legal status of seamen’s insurance at the time, we can reconstruct the following putative scenario for the puzzling career of Hunter Gracchus: He was a seaman and became ill “during the voyage on board.” As such a “sea-sick” sailor, his care was in the hands of the shipowner, upon whose death the costs were no longer covered. For he was only insured for illness or accident abroad. Gracchus has thus conjured up an accident at home in the Black Forest, an accident that although transpiring “in good order” took place eons ago. Depending on perspective, Gracchus can be understood as a cunning *Simulant*—a term used in the years after 1900 to designate a worker lacking will to work, as soon as an accident or illness was reported; or as “insurance swindler,” the parasitical and criminal element in the midst of the community of solidarity, feared since establishment of the welfare state; or as a “pension neurotic” bringing to fruition the latency or belatedness of that simply virtual event that accidents represented after 1900;[172] or else, using Kafka’s terminology, a prototypically “uninsurable,” because “seasick,” being who is striving with all available means to compensate for insufficient existential grounding.

To gain this compensation, Gracchus has to invoke the immemorial accident in the Black Forest’s depths and mask his illness at sea—something necessitating the complicity or at least silence of the boatman. It is the latter man’s refusal and grounded, administrative thoroughness that leaves Gracchus without grounding or compensation, wandering the seas. In a crossed-out textual passage, Gracchus in fact laments, in a paranoid manner, that all imaginable sorts of communication, from legends to the telegraph to historians’ writing, all types of transport, and the most varied forms of natural phenomena and forces will bear witness to his fraudulent effort[173]—as if the former demons of the sea have now taken possession of the global elementary space of communication, for the sake of damning him to his spectrally insecure position. This is the context for his blaming the boatman for the “fundamental error” of his death, despite its having been an or-

171 Kafka 1971, p. 232.

172 On a corresponding intersection between statistical and probabilistic procedures in medical and psycho-pathological discourse on trauma around 1900, see Schäffner 2000, pp. 107–111, 114. In respect to Kafka’s “seasickness,” see also ibid., p. 105: “After World War I, the official medical approach to accidents was that not the accident but its being insured caused mental damage.”

173 Kafka 1993, p. 315.

derly hunter's death from an insurance-law perspective, since it took place in the "interior."[174] For the seaman taken ill while at sea, for the "seasick" man par excellence, the basic fault finally consists of a fault or absence of any base.

"The mariner," we read in a contemporary article on maritime insurance, "is actually considerably less favored because of the particular nature of what he does than workers in other branches."[175] In comparison to uninsured life at sea, "landing" would be a kind of salvation, but one denied Gracchus as long as he is interviewed by savior "Salvatore" in the manner of an insurance agent. Gracchus only awakens for the interview when the boatman has left the room; then, in the course of the examination of his claims, he proffers the self-defense of having been "a hunter; was there any sin in that?"—to which Mayor Salvatore responds by his own, simple declaration of lack of competence: "I am not called upon to decide that."[176] What has to be decisive for debt-encumbering institutions such as the church or court here seems to have become a minor matter. More important are considerations such as the "correct expression" for describing Gracchus's vessel—whether it is a "ship or a "bark"[177]—or the spatio-temporal definition of an alleged accident—for instance whether it took place on land or on water and whether direct or indirect damage is connected to it. Gracchus declines Salvatore's request that the applicant report on his own life "briefly but coherently,"[178] hence record it in the form of a shaped biography. It is as if he is here following Kafka's observation that "not asking would have brought you back. Asking just drives you an ocean further."[179]

In Virgil, Dante, and later Goethe, Lake Garda is referred to as a sea. For its part, Riva, the name itself meaning "shore" or "bank," wedged in between precipitous cliffs and a foreign because Italian "sea," became an Austrian possession in 1517; until the end of World War I it was a Habsburg customs port and seat of the district court.[180] An insured "debarkation abroad" would have thus needed legitimation by the commissioned authorities. But in the case of Gracchus the incorruptible and—not in a good sense—thorough, grounded bureaucracy thwarts insurance compensation for insufficient grounding. Decisions about existential grounding and forms of life are institutionally arrived at on the shoreline, on that elementary threshold between land and sea. But for Kafka, it is al-

174 Kafka 1971 p. 231.
175 Stein 1910, p. 404.
176 Kafka 1971, p. 229.
177 Ibid., p. 231.
178 Ibid., p. 232.
179 Kafka 1992, p. 88 (a passage tied to the "Prometheus" fragment).
180 See Binder 1971, pp. 383, 386.

ready modern institutions, institutions active beyond such elementary borders, that see to life's shaping, the *vitam instituere*. To that end they demand a curriculum vita and bring spokesmen or at least witnesses onto the scene to certify its contents.[181] Gracchus's case history is a *Fallgeschichte* in the German word's double sense, the case history of a fall or tumble meant to buttress Gracchus's claim, because that history means failure of the *vitam instituere*, condemning him to an existence beyond the elementary distinction between life and death.

Before the law, when he presents himself as having suffered an accident eons past, Gracchus is no subject with legal capacity, and is thus bureaucratically speaking already dead. But "in a certain sense I am alive too"—and this if only on account of his not taking any "part in the other world" and consequently in the religious compensatory agenda.[182] The flock of doves that accompanied Gracchus's landing may have held out the promise of a Noah-like landing to him. For the secular authorities, however, the claim of this petitioner is a nuisance to be gotten rid of by handover to religious pastoring. That Gracchus is received by the local officials and worthies in mourning clothes and immediately is graced with a death ceremony prepares him for the beyond in a purely ecclesiastical rite of passage. "He lay without motion and, it seemed, without breathing, his eyes closed; yet only his trappings indicated that this man was probably dead."[183] But as a—for Kafka—nearly prototypical figure from a sphere before or beyond insurance seeking fruitlessly for compensation, Gracchus prefers to remain in, to again cite Adorno, a "zone in which it is impossible to die."[184] Keeping in mind the distinction, into the Early Modern period, between those who are living, dead, and on board ships,[185] we can understand Gracchus as a prototypical seaman acting as a sailors' protagonist and "patron saint."[186] However, as a man embodying the mariners' basic fault as an absence of base, he is both their mirror image and an image of deepest dread: a Flying Dutchman for eons leading a phantom existence on his barge, hovering between life and death. His rudderless "bark" is no "principle of individuation" but rather a swimming coffin, one that has never made it into the earth. For Gracchus, just this "unsinkability" becomes

---

181 See Campe 2005, p. 238, Campe 2004, p. 197, Campe 2007, p. 189. On Kafka's institutional *biographein* between poetic and biological genres, see Vogl 2008, pp. 21, 25, 29–32.
182 Kafka 1971, p. 228.
183 Ibid., p. 227.
184 Adorno 1983, p. 262.
185 In this respect, discussing the Marquis de Pombal see Gil 2008, p. 106.
186 Kafka 1971, p. 231.

a curse, only allowing him—even in the age of the steamship—to return to Riva from the high seas "every two or three years."[187]

Gracchus's life story can be read as a poetologically self-reflective version of the endless journey entered into by the fabulous Flying Dutchman, through which the modern age generated global insight into its own fate and fortune. To cite Manfred Frank: "In this way the 'misfortune' of the lack of a *lieu propre* (the place where human beings are actually at home) would encounter the process of literary writing."[188] Evoking the tradition of seamen's yarns, Kafka's saga would have thus tried to take the "inexplicable"—namely seasickness as the aforesaid constitutional fault, and explain it. But since, as we read in Kafka's posthumous writing, the saga or legend "comes from a truthful grounding, it has to end in the inexplicable."[189] Furthermore, around 1900, with the end of sailing ships and their "elementary experience," the tradition of sailors' yarns had already been eviscerated. We thus must speak all the more of an "untimely" fantasy of Gracchus. Or more precisely, Gracchus is a "normalistic" subject, a subject always already located "before the institution," who once again fantasizes his non-normal life journey as an endless journey.[190] But in doing so he confronts the institution and its grounded thoroughness with the saga or seaman's yarn. The yarn emerges from the inexplicable because it cannot invoke any cases, grounds, or otherwise utilizable data but can only point to an *event* in which the world, including the life-world, presents itself in its abysmal groundlessness: "'To say,' related to the Old Norse 'saga,' means to show: to make appear, set free, that is to offer and extend what we call World," writes Heidegger.[191] "What Appropriation yields through saying is never the effect of a cause, nor the consequence of an antecedent."[192] Rather, the saga comes from an "event," *Ereignis,* that "cannot be commandeered,"[193] thus evading the institution, its will to knowledge, its imperative to a certain life form and its "Enframing," always aimed at bespeaking the constituent stock of the existent, to which the sea now counts, in its sense.[194] Gracchus is an uninsurable person because he bears witness without any discursive safeguards, only through a "saying" of the event, hence without any grounding.

187 Kafka 1949, p. 170.
188 Frank 1995, p. 202.
189 Kafka 1992, p. 69.
190 See Link 1999, pp. 62f.
191 Heidegger 1971, p. 93.
192 Ibid., p. 127.
193 Ibid., p. 133.
194 Ibid. p. 127.

What Gracchus's saga "sets free" is in the end only a kind of bad infinity of the sort the Flying Dutchman has to experience in an eternal return of the same. It also bears witness to a hunt: on the one hand to his chamois hunt in the interior, which he evokes in a sort of repetition compulsion when telling of his immemorial past, on the other hand to a hunt for land or a "landing." In one diary entry, Kafka accordingly notes as follows: "This hunt, originating in the midst of men, carries one in a direction away from them…'Pursuit,' indeed, is only a metaphor. I can also say, 'assault on the last earthly frontier'"—and upon a "resting place…beyond the wide sea."[195] The "resting place" Kafka refers to here is not, it would seem, the Hebrew Bible's divinely promised and sanctioned land, with people gathered together there under a single name. This, apparently, is the "resting place" marking an end to all questions, all searches for grounding, all grounded thoroughness. Kafka here presents less a topographical or utopian directive than a vanishing point leading out from the zone of humanness and corresponding life forms—out from the "continent of human beings," as Walter Benjamin once put it.[196]

This becomes clear in the mirror-symmetric conception of "Report to an Academy," written at the same time as the Gracchus fragments. Initially both stories underscore the reversibility of the hunting process: Gracchus may have once been an archetypal hunter—in the end he finds himself in a "wooden cage" like an animal.[197] Rotpeter may, by contrast, have represented exemplary prey—particularly against the backdrop of Carl Hagenbeck's hunting expeditions. In retrospect, however, his wounding no longer appears to be his pursuer's hunting *fortune*, but rather a hunting accident. For if this ape has arrived on the path to hominization precisely in the course of a hunt, at the moment of being struck he was already a potential human being and thus already on the side of the hunter, for which reason the shot has the retrospective status of an error. The countervailing import of the two stories and writing processes, one losing itself in the endless circling of a restless life journey, the other succeeding in writing the account of a career, is also evident in Gracchus's elementary state of being driven and lacking orientation and Rotpeter's humanoid taming of drives and teleological orientation. The seaman loses himself in the elementary, while the land-dweller can emancipate himself from it. For while Gracchus, in his drafty bark, sees himself eternally driven by a "wind that blows in the undermost re-

195 Kafka 1949, p. 202; Kafka 1992, p. 88.
196 Benjamin 1999b, p. 498.
197 Kafka 1971, p. 229.

gions of death,"[198] Rotpeter, though initially encountering a "strong wind," both on the huge steamship and finally on terra firma experiences "only a gentle puff of air that plays around my heels."[199]

Both Gracchus's and Rotpeter's life stories are "grounded" in a decisive event, an accident, and this inaccessible etiological moment is what gives impetus, if not orientation, to their future careers. But in distinction to Gracchus, Rotpeter is capable of steering this impulse: by rectifying and reporting on his species-determined "direction," the actually immutable "line an erstwhile ape has had to follow."[200] In his "Report," he can certify his course correction from animal to human being as a paradoxical because exemplary exception—he demonstrates his unprecedented human status by simply speaking. Where in the one case, then, Gracchus takes a direction away from being human, Rotpeter assimilates himself into the human. (This is not only apparent in his report before the academy's dignitaries, but already in a personal conversation that the author of an early version of the report has with the ape, in the manner of Salvatore.[201]) Rotpeter, however, only becomes "human" by suspending natural species-grounded criteria and statistically interpreting the essence of being human in the sense of modern insurance techniques and life sciences. The source of his becoming human is, after all, the rifle wound inflicted on him, both an exemplary accident and a statistical event: a shooting-range derived distribution model.[202] Rotpeter is a statistical singularity, but at the same time an evolutionary "sport," a coincidental mutation. Rotpeter can escape his preliminary life as the very result of being located at the juncture of different referential and discursive systems—evolutionary theory and statistics, zoology and anthropology. That in both his *Origin of Species* (1859) and *Descent of Man* (1871), Darwin conceived the transformation of species as an arbitrary play of forces or forceful play of arbitrariness, hence that evolution not only admits deviations as developmental leaps but even rewards them, is Rotpeter's initial exit from species-defined rootedness.

Gracchus seeks no exit from any rootedness. Rather, he seeks to escape his constitutional groundlessness. With the fabulous figure of the eternal hunter, he shares the fate of expiating an involuntary hunting offence until Judgment

---

198 Ibid., p. 261.
199 Ibid, p. 250.
200 Ibid., pp. 253, 251.
201 Kafka 1971, pp. 259 – 262.
202 See Schwarz 1945, p. 81.

Day.[203] With Coleridge's Ancient Mariner, he shares killing a bird and thus perhaps himself violating creation and thus being condemned to ceaseless wandering. And with Melville's Ahab, he shares being a prototypical uninsurable person, caught up in a hunt in which hunter and hunted eventually exchange places.[204] Kafka's existential metaphor of "hunting within groundlessness" gains concretion against this background, but also through the context of the Gracchus-fragments' writing. A first sketch emerged on 21 October 1913, after Kafka attended both the Zionist Congress and the mentioned insurance congress in Vienna, then crossing from Trieste to Venice, in order to then take the train and steamer to Habsburg Riva, the site of his sanatorium. With the conquest he made there, Kafka revealed himself to be a "ladies' man" or, in Viennese and Prague dialect, a *Gemsenjäger*, literally "chamois hunter." At this time, he revised his marriage plans—in Riva, as he harshly described it to Max Brod, "every honeymoon couple, whether or not I put myself in their place, is a repulsive sight to me."[205] Founding a family thus promised no cure for Kafka's seasickness. While he was involved in skirt-chasing in Riva, beyond that bank he saw himself as the hunted one. In addition, no less than for Gracchus, for Kafka as well the compensating transcendent and secular agendas of salvation and insurance failed. After visiting the aforementioned Zionist and insurance congresses, he reported to Brod from Vienna that "I would wish to tear those days in Vienna out of my life, tear them up by the roots."[206]

203 On the linkage in literary lore and tradition between the "eternal hunter," the "wandering Jew," and the "Flying Dutchman," see the novel by Levin Schücking 1851, p. 70, passim; Bassett 1885, p. 362; Zirus 1930, p. 3; Binder 1971, pp. 388, 393, 418, 422; Blunck 1982, pp. 188–190; Beck 1985, p. 391; Körte 1995, p. 246; and Buß 1996, pp. 141 f.

204 On Ahab's uninsurability see Melville 2000, p. 510.

205 Kafka 1973, p. 102 (letter to Max Brod, 28 Sept. 1913).

206 Ibid., p. 100 (letter to Max Brod, 16 Sept. 1913).

Chapter 8
Potential Peril: Geopolitical Loss of Difference

The Amphibian Coasts

If the human being is, as a principle of modern geopolitics has it (and again citing Schmitt), "a land-being, a land-dweller,"[1] then the sea has to seem a hostile element, the coastline an unalterable border. Considered more closely geologically and politically, shores do not in fact form fixed borders or non-expandable lines. Rather, as Friedrich Ratzel observed in 1900, the coast is "a line between land and sea belonging to both at once, uniting the features of both in itself. Every border remains determined by the things that it separates. In face of the land and the sea, the coast is thus only a transition, a mediation, not standing for itself."[2] Instead of "mediation," we could here speak of a "threshold" that places flora, fauna, and human beings in a curious hovering "between two worlds."[3] Or we could understand the coast in general as a mutable zone of intensive exchange, as a simultaneously natural and cultural border regime of erosion, production, and communication. If coasts are considered strict borders, that says little about the natural limitation of "culture" in general—but a great deal about the mobilizing force of a *specific* culture. For the coast can be concentrated onto a borderline (with more or less broken or fractal course) through cultural techniques such as cartography; but it can also be described in terms of a boundary area (between liquid and solid) and (in corresponding projection) as a surface.

In any event, throughout the world the largest inventory of biological species is concentrated in coastal areas, together with geopolitical events, as seen clearly at the turn to the twentieth century.[4] The coastal spectrum extends from sections that because of their protected location are predestined for urban and harbor development and thus to become geopolitical junctures to those sectors "biotopographically" displaying a specially narrow interconnection between land and sea and thus largely not suitable for dense settlement, functional harbors, and finally prominent geopolitical status. The German tidal flats constitute such a boun-

1 Schmitt 2015, p. 5.
2 Ratzel 1911, p. 34.
3 Gennep 1960, p. 18.
4 See Ray 1992, pp. 403–405.

https://doi.org/10.1515/9783110610734-012

dary zone. Despite a geographically by no means disadvantageous location, with its East Friesian island landscape, the Northern German tideland remained nearly undescribed in Europe's historical atlas. It has only been a prominent area—this perhaps already since the Nordic voyage of Pytheas of Masalli[5]—in the context of physical geography: a region paradoxically "de-fining" itself through the tidally-determined fluidity of its borders, and that is characterized by unique flora and fauna and the cultivating—in a literal sense—efforts of the coastal dwellers. An extremely shallow coastal area because of Ice Age glacier-detritus, the tidelands have been largely spared the effects of frontal surging, instead being cyclically flooded, which in turn determines the sedimentological and biocoenotic qualities of this "amphibian" milieu.[6] The north German tideland area—the area known as the *Watt*—is a zone of ecological intensity.

That "natural borders" have always stood in the force-field of cultural influence is especially evident on the edge of coastal areas. Inversely, as Jules Michelet pointed to in *La Mer* (1861), natural history here projects itself strikingly into the center of political and cultural history—when people move into dynamic and mutable coastal zones, they always work and plan in biocenosis with the flora and fauna and in interaction with the natural forces of wind, tides, and surf. Here a "natural" or "elementary" border—a border that legal subjects or political powers believe they can rightly invoke—has to be first wrested from the elementary power of nature. Only then can such a border be abstracted into a line lacking extension, joined with the idea of a military, cultural or national boundary. When for example, in his *Dictionnaire universel* of 1690 Antoine Furetière defines the "frontier" as the "outer end of a kingdom or province, which the enemy looks toward when intending to attack," and when he asserts that the work comes "from *frontaria*, for this is so to speak confronting the enemy,"[7] this definition, militarily couched and framed by a logic of sovereignty, can also be related to the natural premises of border-demarcation itself.[8]

A "front" or "frontier" does not first emerge in confrontation with a political of military enemy, but already in face of natural forces—something we see nowhere better than on shores such as those of the German tideland. Here the opponent against which stands are made is first of all nature itself. Even in the

---

5 See Hergt 1893, p. 32. On the efforts after 1900 by especially German-language authors to interpret corresponding passages in Pytheas as the first written mention of the tideland, see Pytheas of Massalia 1994, commentary, p. 130.

6 See Kelletat 1999, pp. 158–160; Bietz 2004, pp. 3 ff.

7 *...die den Feinden gebotene Stirn*; the German expression *die Stirn bieten,* lit. "offer one's brow," means to "defy" or "confront."

8 Furetière 1690, p. 918. See Braudel 1988, pp. 27–37.

waning Middle Ages, especially the flat coastal regions of northern Europe were considered a scene of desolation, of unceasing confrontation between the elements, which people faced unprotected or as powerless spectators. Into the Early Modern period, such a landscape of erosion and flooding recalled the Flood and divine wrath—until sea powers like the Netherlands reconceived it as a cultural and technical challenge. The sea was now meant to be subdued in enduring battle, in order to complete God's work of imposing "mighty borders" on the elements, and this with God's blessing.[9]

On the coasts of East Frisia, there was no such concentrated initial action but rather scattered small-scale efforts. Since around 1500, in accordance with the motto *De nich will diken, mut wiken* ("whoever doesn't wish to built dikes —must yield") dike construction was increasingly accelerated, for the sake, as was still being called for in the twentieth century, of pushing back the sea in —as Rolf Dircksen has put it—"an unparalleled gigantic attack."[10] In this coastal zone, elementary hostility was from the start hostility toward the element: every breach of a dike resembled a break in tradition, every loss of territory loss of the past, while conquest of land promised accrual of future. We might be tempted to say that amphibian terrain such as tideland repeatedly illustrates the thesis that human beings are "land creatures" who consider the inherently hostile sea at a distance from the land. But in fact modern human beings are more border crossers than land dwellers, in that they consciously hazard the borderline areas and their intensities, in order to use them economically or tactically from land and sea. The flip side of the art of dike construction is the art of navigation, which is able to achieve orientation despite the various effects of tides and currents, ferocious wind and broken path of the coast, in the midst of shallow stretches of water and all kinds of looming sandbanks.[11]

As Alain Corbin indicates, the nineteenth century was the "Golden Age of the shores," because with the rise of tourism they were made accessible to everyone, but without the coastal zones losing their mystery and terror. In this way, shores such as those of the tideland could themselves serve as powerful catalysts of the imagination. Because of their "dangerous beauty," the tidelands, comprising an intensive transitional zone located—to cite Conrad, that maritime and literary border-crosser—"twixt land and sea," became nothing less than a literary

9 Fabricius 1734, vol. 2, 32, p. 178. On the "cultural-historical" classification and assessment of dam and dike construction, see Böhme 1996, pp. 275–279; Blackbourn 2006, pp. 113 ff.

10 Dircksen 1959, pp. 45, 124.

11 On the oldest Low German so-called rutters (mariners' handbooks) and their characteristic double view of the often amorphous shoreline (lateral and birds-eye views), see Behrmann 1978, pp. 69 ff.

obsession, particularly in the nineteenth century.[12] Those people, ships, and cities that had conquered the forces of flood and erosion in the northern German tideland repeatedly manifested themselves in local oral tradition, in legends such as that of Rungholt and in literary-archeological phantasms. “We Friesian island and marsh dwellers are constantly taught that our existence depends on no so-called *great power* on earth more than on the *North Sea,* its storms, currents, and waves,” wrote Christian Peter Hansen in 1865, not without pointing out how little support the coastal dwellers had received from the state.[13] It may thus not be surprising that starting with the Franco-German War and Germany’s unification, a number of local authors fled from the competing political great powers to the one elementary great power: “The state machine’s wheels… are becoming too indiscrete for me” wrote Theodor Storm in 1871, and so one should best flee to the “little country of freedom,” the small islands known as Halligen forming a quietistic retreat–one only provided for by a steady washing up of literary “antiquaria.” Under this sign, it was not the retrospective or introspective texts of local northern German authors—not their historical and archeological fictions, their lyrical invocations of nature and genre studies on land and people—that took the tideland’s measure as a specific transitional zone. Rather, it was the geopolitical obsession of a non-local writer familiar with the environs, the future-oriented vision of an English border-crosser. As Erskin Childers described it, this area was a zone of intensive exchange between nature and culture, in which geology and politics were meant to merge into a true geopolitical mélange. As a boundary-space twixt land and sea, the tideland was manifest in this scenario as a simultaneously elementary and martial front: a milieu predestined, already on account of its natural circumstances, to a liquidation of fixed borders rendering enmity and friendship as indistinguishable as actual events and mere speculation. By offering an opening to world-political phantasms, this area emerged as the breeding ground for an imagination moving past the poetic. And before the Great War produced its idiosyncratic “landscapes” and “border zones,”[14] the tideland’s landscape and border zone inversely brought to light the contours of future warfare. The shallow waters of northern Germany’s abandoned coastline could thus serve as a watershed for events that had world-political import.

12 Corbin 1994, pp. 229, 233, 248. On “twixt” as designating a “between” that actually cannot be designated, see Feldbusch 2003, p. 30.
13 Hansen 1972, p. IV. On the fabulous “returners” and sunken cities of the German tidal flats, see Blunck 1982, pp. 49 ff., 58 ff.
14 Lewin 2006, pp. 129, 131.

Erskin Childers' Literary Secret-Service

Around 1900, especially looking out from the British coast, there was great concern about the maritime ambitions of the German Kaiserreich: "we aren't ready for her," we read in Childers' novel *The Riddle of the Sands* (1903), "we don't look her way. We have no naval base in the North Sea, and no North Sea Fleet. Our best battleships are too deep in draught for North Sea work. And, to crown all, we were asses enough to give her Heligoland, which commands her North Sea coast."[15] In the novel, "two young gentlemen in a seven-ton pleasure boat" set out on a cruise along the coast of eastern Friesia, rather arbitrarily gaining a taste there for "amateur hydrography and police duty" (106). On a difficult stretch of tideway, a helpful German with perfect English intentionally pilots their boat on a false and nearly deadly course. They then try to together uncover what was behind this incident—and in the process discover their own geopolitical obsession. After first becoming suspicious of English sea power's most stubborn challenger, they increasingly get drawn, together with their pilot, into the slipstream of the German tideland, and of all the speculation that surfaces from its unceasingly fathomable grounding. In a for them previously unknown "life of toil, exposure, and peril" (129), they become self-commissioned spies observing warships in German provincial harbors, sounding out local seamen and frigate captains, and identifying the supposed German pilot as a defector to the Royal Navy.

As tourists and friends of nature, they share "a certain subdued relish in describing banks and shoals" (212); but as irregular patriotic fighters they engage in a "war of wits" (141). Under the all-ruling law of the tides, however, the plot repeatedly loses itself in the tideland and its sandbanks and canals, in a desolate no man's land decaying from hour to hour, newly drawing the border between land and sea or entirely flooding that border. Then, when they learn that at the Memmert Sand island, a gold-transport boat long buried in the mud is going to be raised, it seems to them that the bustle of the local authorities is a new pointer. Namely, what the two sightseers discover with their strategic gaze is a potential operational area for British torpedo boats that could paralyze German maritime transport between the Ems and Elbe rivers. Instead of an archeology of the tideland, the surmise goes, Germans are in reality preparing the field for a guerilla war between land and sea, a war that, following mutual

15 Childers 1915, p. 105. All further page-references to the novel are indicated parenthetically in the main text.

neutralization of the German and British high seas fleets, could lead to a decisive struggle between maritime and territorial powers.

In the end, to be sure, their investigations reveal mobilization not for a defensive but rather an offensive war. From the depths of the Friesian tideland, an "armada of light-draught barges" (310) could emerge for landing on the Wash, the depths of the English eastern coast. The novel, couched in its fictitious preface as non-fiction introduced by its editor, ends with both sides' assessments of risks: with a secret plan for a future war from the pen of a British defector, and the reflection of editor Childers on whether in the case of a blockade and invasion England could ever be defeated, at least in a generally accepted sense. The book has often been described as one of the first modern spy novels, and Childers knew what he was talking about. It is not only that he was a passionate yachtsman and took various voyages along the North Sea coast. He also frequently showed himself to be a prominent standard bearer for the British Empire, first in the Boer War, then as the author of military and naval strategic treatises, and finally, during World War I, in both the British North Sea fleet and air force.

Equally knowledgeable in the field and in strategic planning, participating in both clandestine action and in communication with the public, he was familiar with the numerous blind spots in the still meagerly outfitted field of naval intelligence, presenting a literary secret service in the place of a future one.[16] And the political decision maker arrived at this literary scenario by showing himself thoroughly geostrategically informed. "It's all out of Mahan and those fellows" (84), asserts one of Childers' sailing spies. Mahan's *The Influence of Sea Power upon History* (1890) had in fact become a bible of geostrategic thinking for the educated public, as well as for a political elite that owed its prestige to its sea power. The British paranoia regarding invasion may have had roots in Michiel de Ruyter's penetration into the Thames estuary from native shallows in the second English-Dutch sea-war. But around 1900, when a range of forces were tied up in the Boer War, the military rather feared an invasion from the French side of the canal. Germany was here initially considered simply an economic competitor, but one whose fleet-expansion program was observed with suspicion.

As a putative maritime power, the Imperial German state itself suffered from a so-called Copenhagen complex. While Germany's own accomplishments in the German-Danish war could at best have been considered mediocre, they appeared all the more puny when compared to Britain's lightening attack on Denmark in

16 See Horn 2013, pp. 125–133.

1807. Since then, "to Copenhagen" was a catchphrase for a feared, direct and irresistible British invasion. Germany in any case increasingly harbored ambitions to gain economic and military great power status. Stationed since 1871 on a gunship named *Blitz* on the northern German coast, Alfred von Tirpitz recognized the necessity of protecting the Kaiserreich's access to the sea and protecting the local population from adjacent states. He first envisioned a powerful fleet as offering such protection, later the possibility of geopolitical expansion.[17] Between 1898 and 1912 Tirpitz was responsible for the bills submitted in connection with all fleet-acts; in his memoranda he reflected Mahan's doctrine of decisive, massed battles on the high seas and determinedly worked at developing a huge fleet aimed at, if not defeating the British militarily, then at least forcing them into negotiations. Mahan himself soon warned of the "German menace."[18] In 1900, Friedrich Ratzel, co-founder of the "Pan-German League" and pioneer of "political geography," wrote that acquiring the naval supremacy that was Germany's due depended "completely on the intensive exploitation of a single coastal location," and in addition "of a coast *that can be reached* ," and finally of "the accessibility of the coast from the land."[19] It was not only that Ratzel was here integrating Mahan's doctrine of sea power into his own geopolitical thinking, while involuntarily offering arguments to the opponents of Tirpitz's policies with his diagnosis of Germany's disadvantageous position.[20] He was also naming precisely the three cornerstones between which Childers' war scenario would be located: the German tideland, the English Wash, and the German hinterland with its meanwhile well-developed infrastructure.

Plans for an invasion of England initially emerged from the realm of fantasy, for instance the plans of the Austrian officer and philosopher Gustav Ratzenhofer, already devised in 1881, "to attack certain sea powers at the headquarters of their power, even when insularly situated"—and indeed to attack them without scruples. For it was not only that those powers had gained and defended their own maritime power as "in a war with nomads." Rather, wherever, as at sea, "war is waged on the basis of conjecture, international law is hardly recognized."[21] In 1896, following Kaiser Wilhelm's "Krueger telegram" leading to a crisis with Britain, German admiral Otto von Diederichs came up with a concrete plan to invade England—a plan that, just like many increasingly open subsequent plans was deemed interesting and feasible by the Kaiser but that Fleet Ad-

17 See Drummond 1985, pp. 102f.
18 Mahan 1941, pp. 302–308. See also Salewski 1998, pp. 119–125; Herwig 2005, pp. 127–142.
19 Ratzel 1911, pp. 35f.
20 See Hobson 2004, pp. 315f.
21 Ratzenhofer 1881, pp. 275, 272. On Ratzenhofer's singular role see Schmitt 2003, pp. 312f.

miral Tirpitz firmly rejected because of his strategic principles. England's first reaction was composed. It was only when Secretary of the Admiralty Arnold Forster visited Emden's harbor area in August 1902 that concern emerged there about the island's still unprotected eastern coast.[22] Suddenly the countless literary scenarios of invasion circulating since George Tomkyns Chesney's *The Battle of Dorking* (1870) gained new actuality. Fear of the "German peril" now seemed militarily justified, the North Sea the probable scene of impending conflict.

In the force-field of economic interests and public opinion, literary fiction and planning games, a future war thus became visible on the geopolitical horizon—a war that characteristically would take place at sea. Its preparations, however, could not be concretized in a secret-service framework, but rather only speculatively asserted with a view to alleged activities on the eastern Friesian coast. Childers, in particular, localizes this evidently imminent, strategically already real war in his novel as unfolding in the German tideland—in a coastal region from where ordinarily no maritime enterprise would begin, but that here is presented as a huge, expansively outspread harbor: a widely ramified system of steadily emerging and submerging waterways. Childers thus describes the tideland as a virtual frontier, a front repetitively constructed and undone by natural forces. It has become an eventful plane upon which—as in the novel in general—effectively nothing happens but where, through the tactical positioning of amateur spies, the steady reading of scattered clues, the attentive pursuit of fleeting traces, and the stubborn interpretation of overheard conversations, can alter the entire strategic field in a matter of seconds.

The Riddle of the Sea Battle

The Riddle of the Sands is in fact a sustained riddle centered on whether a sea battle will maybe take place tomorrow. In its tideland setting, the novel thus plays through the same scenario that caused philosophical thinking for the first time to move, with Aristotle, past propositional logic in the direction of its modal counterpart. In his *De Interpretatione,* from the *Organon,* Aristotle examines the proposition that a sea battle will take place tomorrow. He asks whether it can already be true and necessary today that a sea battle will or will not take place tomorrow. With this question, nothing less is at stake than the existence of objective chance, the assumption of a third value in propositional logic alongside "true" and "false," and the introduction of a category in

22 See Drummond 1985, pp. 158, 166 ff., 174.

modal logic called "contingency"[23]—the specification of an area that places latency, abeyance, and the gestation of events alongside the fixed structure, definiteness, and law-adhering nature of being. Aristotle thus arrives at no less than what he terms the "impossible conclusion" "that both deliberation and action are causative with regard to the future, and that, to speak more generally, in those things which are not continuously actual there is a potentiality in either direction. Such things may either be or not be; events also therefore may either take place or not take place." For this reason, "it is not of necessity that everything is or takes place; but in some instances there are real alternatives, in which case the affirmation is no more true and no more false than the denial." "A sea-fight," Aristotle concludes, "must either take place to-morrow or not, but it is not necessary that it should take place to-morrow, neither is it necessary that it should not take place, yet it is necessary that it either should or should not take place to-morrow."[24]

Considered as an excluding disjunction, the proposition is true in a trivial way. That, however, an existence or occurrence can be affirmed or denied and this affirmation or denial must be true or false is the case only for the present and past. The strictness of Aristotle's postulated correspondence between truth and being only covers necessary and impossible predicates, not contingent proposition.[25] In themselves, the two clauses in Aristotle's statement thus (still) do not express any truth value—but indeed the potentiality of being or not being. In this way, on the one hand the future is characterized as contingent and every prognosis circularly tied to "deliberation and action." On the other hand, statements are no longer limited to what is or certainly will be; they now also comprise what cannot be and what does not simply not exist but *doesn't yet* exist and thus "insists."[26] In that here statements do not yet have any truth value, the principle of the excluded middle is here suspended in a peculiar way: it may apply for the future, but is not applicable there.[27] Already because means based in pure formal logic here have no determining force, the "truth" of future events becomes mere "probability." And for the tradition of logic from Aristotle onward, just this distinction between necessary and impossible predicates, on the one hand, contingent predicates, on the other hand, of-

23 Aristotle 1941, 21b (p. 54)
24 Ibid., p. 48.
25 See ibid., p. 44 ff. See also Talanga 1986, p. 37.
26 See also, in respect to the providence-doctrine, Craig 1988, p. 58.
27 On the difference between the principle of the excluded middle, which applies to all statements, from the "bivalency principle," which only applies to timeless statements, see Wright 1994, pp. 172, 187.

fered an escape from the deterministic impasse. However, the complex system at work here, with its basis in modal logic, was for the most part simplified if not entirely forgotten before, in the early twentieth century, its "riddle of being" became the source and starting point for developing a system of polyvalent logics.[28]

Already for Aristotle, the sea battle was not simply one example among many others. In his thinking as in Plato's, a polis that gives up *nomoi* tied to terra firma for the sake of thalassocratic undertakings exposes itself to what in the sea-battle riddle, in the context of modal logic, operates as "contingent futurables": future matters that are neither necessary nor impossible. When it ventures into the sea, politics consistently moves within a horizon of the future, of contingency and potential. We might say that once it transgresses the coastline, potency is caught up in potential, or inversely: precisely here power is derived from what can be but at the same time has the power not to be. Politics proves itself within a contingent structure, where the rigid duality of occurrence or non-occurrence shifts into a state of hovering or a latency of alternatives. It has developed into an art of recognition and both trade and action, of the perception of real opportunities, of *occasiones*.

It is unsurprising that the Early Modern period developed its specific form of contingency-centered politics with a view to the sea's elementary event-space and with *fortuna di mare* as its guiding thread. For Early Modern sea battles were undertaken under the signs of both potency and power-surplus through potency of non-existence. For example, Lord Torrington would be subjected to a court martial because of his tactics of avoidance facing the superior French fleet during the Battle of Beachy Head in 1690. But these tactics, keeping his inferior fleet so to speak in reserve, in a state of potential, a hovering between attack and retreat, prevented a French invasion: "Most men were in fear that the French would invade," he testified at the court martial, "but I was always in another opinion: for I always said, that whilst we had a fleet in being, they would not dare to take an attempt."[29] Since then, the concept of "fleet in being" refers in the theory of sea-warfare to that potency of non-existence not needing to shift into any real action but nevertheless tying up hostile forces, holding together one's own scattered units, and thus defending a space of contingency.

For Tirpitz's "risk"-based fleet, the doctrine of the *Risikoflotte* he developed around 1900 and publicized in a range of popular and specialized organs, what

---

28 See Talanga 1986, p. 162; Günther 1979, pp. 177, 183.

29 Cited from Till 2004, p. 182. See also Lambert 2000, p. 85. On the courts martial of the time as "channels of communication" in "learning processes for the officer corps," see Harding 1999, p. 187.

such a doctrine could accomplish in an actual sea battle was a secondary consideration. For one thing, technical developments and future political constellations could torpedo relevant long-term planning for the fleet, the *Risikoflotte* thus being first and foremost risky for the state budget. But beyond this, as Jan Rüger has described it, once the "danger zone" of German inferiority had been left behind, the idea was to bind the forces opposing Germany before any such action. To this end, a period of peace was nonetheless marked by various reenactments —called *Flottenschauspiele*, "fleet spectacles"—and "battle images" meant to present both those at home and potential enemies with the German navy's potency.[30] Invasion by any possible foe was to be forestalled through such "seaworthiness" alone. Even for England, the risk was meant to be too large that an invasion on the German coast would offer other enemies an open flank. And also for the so-called "fleet of the future" of the French *Jeune école*, modern maritime warfare involved a "fleet in being": with the rising destructive potential of the large battleships, the probability of their engagement in actual battle was lower, such engagement thus becoming virtual *guerre industrielle* or else being played out within the shadow of hidden guerilla operations.

Notwithstanding which side of the rapidly changing fronts and alliances one was on—around 1900 the phantom of sea power was a haunting presence in military planning games, propaganda channels, and the heads of both potential spies and literati.[31] The future of naval warfare offered the opportunity for unrelenting geopolitical speculation, spelled out in Childers' novel as the titular "riddle of the sands," the specific type of occurrence denoted by contingency here finding its corresponding milieu of latency and occasion in the tideland. Moved into a geopolitical horizon, that region becomes an eventful "onto-logical" space for Childers. Arguably, the natural space and that of human action co-operate in the novel, carrying out the same movement and expressing the same potency of non-existence, albeit with differing grounds for movement and being. Aristotle already suggests such co-operation. On the one hand, for Aristotle the sea, both an elementary space and domain of the *automaton*, constitutes a scene of potency, in that here natural events present themselves as contingent—as coincidental conjunctions of heterogeneous causal series.[32] Even cyclical processes are evidently only "usually" and not "necessarily" subsumed to natural laws,

---

30 Rüger 2007, pp. 69, 30, 60.

31 On the role of British military attachés in the provisional intelligence service and in assessing the "German menace" as a merely hypothetical or midterm danger, see Seligmann 2006, pp. 194, 221, 262 f.

32 See Frede 1970, pp. 116 f.; Fichte 1996, p. 7. On the inapplicability of concepts such as *tychē* and *automaton* in the context of "contingent futurables," see, however, Talanga 1986, p. 61.

and the constellation in which land and sea meet, in order to intertwine in the form of currents and waves, shallows and sandbanks, is all the more eventful. On the other hand, emerging from this elementary space are just those undertakings that characterize human "second nature." It is not simply that human action is here contingent as such or else generates a double contingency through mutual observation. Rather, potent non-existence is part of the calculus here from the start. Aristotle demonstrates this in the context of naval warfare and the tactical space of the sea: as already manifest in Athenian maritime power, even with merely possible operations, and indeed with statements about the future by oracles furnishing both sides with credibility, forces can be bound up, hostile powers stymied.[33]

We might describe Childers' eventful space as a locus of endlessly inflected articulations, so that here a transcendental instance "between two seas" as exemplified in Genesis and in Paul's shipwreck is no longer in play. Rather, what we find is an immanent space composed of wind, water, and land, these elements engaged unremittingly in a process of mutual inscription, with a concomitant production of steady intensities and new possibilities of being.[34] But this eventful space could also be described through concepts present in Stoic logic. Then types of causality or accidents, accruing to substances or things themselves, are not what is ontologically relevant. Rather, the contingent thing-conditions or "non-corporeal events" themselves here become the essence and actual substance of the narration; they become what is an insistent presence in language and endows events with their very meaning. "Everything that happens and everything that is said happens or is said on the surface," comments Gilles Deleuze. "The surface is no less explorable and unknown than depth and height."[35] While archeological depths, such as in the case of a discovered wreck, only sparks a false impression that "the root of any mystery we might have scented" (176) is present, it is the tideland's shallows that furnish Childers' particular riddle. Upon these shallows, upon their superficial cartographic reproduction and the surface of their disconnected appearing and vanishing linguistic elements, the "piece plotted without any regard to dramatic fitness" (239) reveals itself: a "piece" that first furnishes literati and spies with "a sense of genuine inspiration" (285). The events are driven, we read, by a steadily interfering "cross-current" (91). And if the narrator is able to steer the course of his fantasy and

33 See Minois 1998, pp. 98, 109.

34 This approach is at work in writing of Michel Serres, Gilles Deleuze, and Félix Guattari; on its relation to the older articulation theory of Ferdinand de Saussure see Berressem 2004, pp. 230–232.

35 Deleuze 1990, p. 132.

language, then it is through a specific metonymic connection: “He communed with his tiller, I believe, and marshaled his figures with its help” (106).

Both literarily and geopolitically, what reigns in the tidal flats is strategic depth derived from maritime shallowness. Childers’ two sailors conduct their “war of wits, and not of duck-guns” under the maxim: “Let’s look at the chart” (141). For in the tideland has become a hydrographic zone of intensity not as a biotope but rather as a scene of virtual sea battles. Getting to the bottom of something here means describing surface structures for the sake of gaining orientation between the unremittingly surfacing phantasms. And sounding or plumbing does not mean getting at the bottom of something but rather incessantly entering markings and vectors on a smooth surface. The natural relief of the tideland is stamped by all sorts of inlets, gullies, streams, banks, sand formations, and ridges, repeatedly covered, displaced, pushed aside, or produced by the tides; beacons, drums, buoys, and rods are meant to make the surface at all navigable. In the geopolitical relief of Childers’ tidal flats, where the paths, stations, and bases of the virtual future war are continuously hidden or displaces, lastingly disappear or emerge for the first time, tactical aids such as “sailing directions” and the prism compass have to be accompanied by strategic instruments like sea-battle libraries and secret official memoranda. The control center for this war of intellect, which is a war of surfaces, is located in the boat’s solo sailor cabin: here the cabinet of this secret war is in session with the sailor-pair, in order to solve the sea-battle riddle through intelligence, through strategic considerations and map material.

As a member of the Cruising Club, which since 1893 had produced its own maps, superior to those of the British Hydrographic Office, Childers was aware how inadequate English cartography was when it came to, especially, the German tidal flats, while the German side had excellent material on the English coasts. For this reason, on his private sailing tours Childers produced independent maps of the area, reproducing some of them in *The Riddle of the Sands*. The professionalism of—as termed in the novel—his “amateur hydrography” was reflected in the fact that at the actual outbreak of World War I, the British admiralty urgently requested his work, that his cartographic surveys became known in the Near East as “Childers Charts,” and that he was personally summoned for air and marine reconnaissance on the German North Sea coast. In 1910, two British amateur spies were actually detained near Borkum; they were following the path of *The Riddle of the Sands* in reconnoitering the German tidelands for the sake of preparing up-to-date maps.[36]

36 See Drummond 1985, pp. 182–184, 194–196.

In Childers' novel, maps attest to the concurrence of a global geopolitical perspective with an intuitive and detail-intent on-site orientation. Maps and text here appear together like land and sea—albeit not in the sense of that classical topography of narration actualizing, as Yuri Lotman formulates it, "the movement of the plot, the *event*" as the transgression of an elementary, cosmological, or even cultural border.[37] Rather, the maps in *The Riddle of the Sands* define a possible operational area or a space of potential events; borders are here dissolved or have themselves become contingent. As a result the text pursues, reveals, and anticipates everything insistently present in a non-representable way on the cartographic surface and that might be termed "non-corporeal events." For if "in the constant free play of forces of wind, tide, and surge a steady alteration of this landscape takes place, so that no maritime map can retain it for a long time and no one can say he knows it well,"[38] this is all the more the case for the extension, shape, and position of the "fleet in being."

A Fleeting Vision in the Sand

Childers' "hydrography" of words and images brings out the contours of a future war. Starting with an intensive interplay between human action and natural forces, it does not exhaust itself in a projection onto the sea of circulation routes on terra firma. Otherwise than for example with Mahan, the sea is not so much traversed and divided up by waterways than that it makes manifest intertwining and rampantly changing terrestrial-watery pathways. To speak with French admiral Raoul Castex, such a hydrography opens a "smooth space" of geometric surface-relationships in which the opposition between land and sea finally dissolves.[39] Variable vectors mark—indeed inaugurate—this space, and what transverses it are a range of the smallest mobile units, in Childer's work cruisers, gunships, torpedo boats, together with barges. On the suggestion of Colmar Freiherr van der Goltz, the German admiralty considered deployment of these various smaller vessels in 1900 and recommended a trial effort.[40] The German strategists thus themselves no longer adhered to Mahan but sided with champions of torpedo warfare and the French *Jeune école*.[41]

---

37 Lotman 1977, p. 238.
38 Dircksen 1959, p. 39.
39 See Castex 1937, pp. 13–28; and on this Schmitt 1995a, pp. 441–452.
40 See Drummond 1985, p. 174.
41 See Deist 1979, pp. 23, 43.

After the self-propelled torpedo of Lupis-Vukić and Whitehead was developed and successfully deployed in the Russian-Turkish war of 1878, the maritime powers contracted a regular case of "torpedo-itis": a panic facing these missiles speeding out of nowhere just under the surface; a hyper-sensitization and paralysis of the military sense of danger on grounds of ubiquitously hidden and potential enemies. It was not only that, as Herbert C. Fyfe observed in 1903, these weapons had a powerful "moral effect," "hidden enemies" being "the most dreaded." It was not only that they had a "really paralyzing" impact on "some minds," for which reason in various sea battles enemy forces imaginarily multiplied, with fire aimed repeatedly at simply hallucinated enemy boats.[42] The classical high-seas combat systems now seemed increasingly obsolete. With the subsequent development of small, hydro-dynamically optimized torpedo boats and their rendering serviceable for fast maneuvering through the introduction of combustion engines, torpedoes could be deployed both offensively and defensively, things helped along by the gyroscope technology perfected for World War I, allowing reliable targeted and depth control.[43] In addition, for northern Germany's flat coasts, the British forged plans to use extremely light and fast boats with the smallest possible draft to move undamaged through the minefields—a secret project tested by the Thorneycroft vehicle company up to 1917 on an island in the Thames. In his article on "The Battle in the North Sea," just after the outbreak of the Great War Childers pointed again to these flat coastal waters, underscoring that "the shoal area described above would be a tempting base and theater for small torpedo craft."[44]

All these activities were in line with the above-mentioned *Jeune école* school, developed starting in the 1870s around the vice admiral and later navy minister Théophile Aube. Followers of this school saw torpedo boats as the best possibility to restructure the French fleet in a thoroughgoing way. After the Franco-Prussian war, this seemed urgently necessary, Prussian cruisers having broken through the close blockade of the Friesian North Sea coast and robbed the actually far stronger French fleet of its dominance. For this reason, future war was not to be pursued with traditional sea battles involving huge battleships but as coastal warfare and a guerilla struggle. In fog and darkness, emerging from the recesses of the broken French coastlines, cruisers and torpedo boats were meant—either regardless of or using the incalculable weather—to break

42 Fyfe 1903, pp. 113, 117, 119. See also the episodes referred to in ibid., pp. 116 f.; Budisavljevic 1996, pp. 251 f.

43 See Hoar 1916, pp. 180 f.; Nelson 1998, pp. 32, 46, 51.

44 Ring 1996, pp. 181 f.

through any blockade. As the later, radicalized proposals of the school had it, they were to attack the enemy in the manner of pirates and terrorize its coasts, in order to then invade its land. Multiple numbers, speed, ubiquity, invisibility, and small dimensions: these qualities were to mark the new fleet, thus imagined as a kind of hornets' swarm, but which was also meant to be placed on the strict Cartesian foundation of "universal science."[45]

That, already because of the development of anti-torpedo boats and the increasingly stable armoring of "battleship leviathans," modern naval warfare could not really be won in this way was ignored by the school all the more flagrantly in proportion to their shift from a reform movement to a ""think tank" and propaganda instrument of the "fleet of the future." To be sure, around the century's turn what the future of naval warfare would look like was not at all clear. Even in 1909, Georg Wislicenus could write in his book on "Germany's sea-power then and now" that "these small boats are indispensable for narrower coastal defense; for protection of the tidelands, directly adjacent to their own coast, they will certainly show their capabilities in war."[46] The English admiralty likewise played with the idea of comprehensively deploying torpedo-boat swarms. Childers was so convinced about this sort of warfare that in the Great War he himself ventured on a daredevil sally into the German tideland.[47] But already in *The Riddle of the Sands*, this coastline is described as "an immense tidal harbor...absolutely *made* for shallow war boats under skilled pilotage" (143). It is thus "an ideal hunting-ground for small free-lance marauders" (131), for swarms that on account of their small size, large numbers, and mobility can traverse precisely a space that is inherently indivisible because of its own innate complexity and mobility. Childers' scenario circumscribes a smooth space and its occupation by a "nomadic war machine." Stated more precisely, to cite Gilles Deleuze and Félix Guattari, it produces "the secret and its outgrowths (strategy, espionage, war ruses, ambush, diplomacy, etc.)" in the first place.[48]

As described from the perspective around 1900, future warfare is distinguished by virtualization, but just as much by technization and totalization. Technicized marine warfare unfolds in industrial and serial operations, with both seamen and land-dwellers being part of a comprehensive war machine. What Ernst Jünger later, under the rubric of "total mobilization," will describe as an "unlimited marshalling of potential energies," for which "not a single

---

45 See Bueb 1971, pp. 72f., 114, 157.

46 Wislicenus 1909, p. 228.

47 See Drummond 1985, p. 194.

48 Deleuze and Guattari 1980, p. 390.

atom is not in motion" in the state,[49] Childers already calls for in *Riddle* as a "Naval Volunteer Reserve," a mobilization of all "who know our coasts like a book" (103): a concept presented in England as a draft bill a few days before the novel appeared.[50] Childers sharpened the concept in a quickly added post-script to the novel, where he recommended the training of all British citizens "either for the sea or for the rifle" (336). In a strategic text of 1906 he even outlined the concrete course of a guerilla operation in the German tideland, a battle-avoiding but possibly decisive tactic stemming from the open concealment offered by the natural space: something he had studied in the Boer War's context and now depicted for sea-warfare as so to speak an alliance of workers and adventurous hearts.[51] Whether as an amateur spy and strategist or as an occasional writer and journalist, by exploring the possibility of an imminent war and endowing the amorphous enemy with a more palpable form, he was already conducting a kind of preventive warfare and adopting emergency measures against a potential attack: measures that although non-regular and operating beneath the threshold of a declared war nevertheless participated in a virtual and future one.

It is no accident that Childers locate his spy novel and war scenarios in Germany's tidal flats. With this boundary area already constituting a zone of non-differentiation as a natural zone of intensity, it forms a congenial or "cooperative" setting for certain geopolitical processes and futures: those paradoxically characterized by a loss of form whose power and menace is determined by the aforesaid potency of non-existence—and that, in addition, *actually could* have played itself out in the German tidal flats. In the end Childers' novel is less about a "German peril" than about *potential* peril in general. The novel presents the liquidation of a whole range of geopolitical oppositions and fronts that had been valid before 1900: war and peace and civilians and combatants become equally indistinguishable, just as the difference between friend and enemy blurs in the mélange of changing alliances and the concept of the "random enemy." Offensive and defensive options are no less interchangeable than peaceful or militaristic intention; as already shown in the thought experiments of the *Jeune école* and Tirpitz's obsession with seaworthiness, and all the more so in the public and official response to Childers' book, fiction and reality can hardly still be separated. Around 1900, hence just before establishment of a secret service in England, the

---

49 Jünger 1991, pp. 126, 128.
50 See Piper 2003, p. 74.
51 See Drummond 1985, pp. 186 ff.

border between literary scenarios and military planning games was more porous than it would probably ever be again.

Perhaps encapsulating the dissolution of all such distinctions was a breakdown of the elementary opposition between land and sea until then characterizing Early Modern politics and front-formation, as a power-division that had prevailed since the Treaty of Utrecht, with the European continental powers on the one hand, England on the other.[52] In patriotic tenor, Ratzel thus wrote in 1900 that "it is one of the most consequential events of the nineteenth century *that the old opposition between sea powers and land powers became obsolete*. The number of sea powers has risen, and the new sea powers are all distinguished by not being and not being able to be pure sea powers but are also all land powers."[53] Retrospectively Carl Schmitt would describe the dike-breach of the old order as the decline of European public law and at the same time as something like a technologically determined spatial revolution divesting the sea of its elementary character and thus destroying the old, traditional balance.[54] Childers, however, conceived of the potentiality and lack of differentiation that the sea particularly reveals on a border-surface like that of the tideland together with the potentiality and lack of differentiation of future wars. As with the heyday of the voyages of discovery and Early Modern colonialism, a geopolitical event would thus emerge from the "characterless" sea, making possible, around the twentieth century's threshold, the writing, once again, of a simultaneously adventurous and political sea-novel.

Childers' sea is here, in the literal sense, a superficial projection. If the stage of maritime warfare had already been projected onto the grid on a paper's surface in the seventeenth century,[55] in the flat, almost purely horizontal setting of the tideland Childers consciously elided the vertical dimensions that would mark such warfare and use of the sea in the future: he only depicts the classical navy, not the submarine navy, and only classical maritime navigation, not aeronautics, both of which, Schmitt points out, had already effectuated the spatial revolution by 1900. In the twentieth century the deep-sea floor and outer space became a last frontier; since then the sea was no longer the space of an elementary experience but rather the referential field of medial apparatuses from radar to sonar. With his obsession with surfaces and surface-level warfare as executed

52 As a contemporary source see Justi 1759, pp. 6 f.; also Schmitt 1995c, pp. 518–522.
53 Ratzel 1911, p. 71.
54 See Schmitt 2003, pp. 309 ff.
55 One classical treatise on geometrically disciplined warfare is Paul Hoste's *L'art des armées navales ou traité des évolutions navales* (Lyon 1697); this work followed the guidelines of the Chevalier de Tourville (Lord Torrington's opponent in the Battle of Beachy Head).

by barges and torpedo boats, Childers once more opened the sea to an imaginary experiential horizon. And this was the case despite Childers naturally knowing better. For one thing, in the warfare scenarios of the *Jeune école* and the British admiralty, submarines played no less a prominent role than the torpedo boat. For another thing, as a specialist in tideland cartography Childers himself participated in reconnaissance flights of the Royal Air Force and in the Cuxhaven Raid, when for the first time conventional ships, submarines, and airplanes together attacked a coast.[56]

Maritime spy novels were rendered superfluous by radio communication, which connected strategic planning and tactical operations with intelligence—with an intelligence that no longer fed itself on the tideland but on precise, real-time air photos and eventually satellite images. And with the shift of the "fleet in being" strategy from the sea's smooth space to that of the air, finally transforming the entire world into a space of preventive security measures, sea-fight riddles became everyday phenomena.[57] The non-differentiation between friend and enemy, civilians and combatants, offense and defense, pacifistic and militaristic motivation that Childers spelled out amounts to a non-differentiation of war and peace—of the being and non-being of war. For this reason, in place of the old geopolitics and its radicalization in "total war," a new globally preventive regime emerged that simply dispensed with the legal convention of declaration of war. Future wars would in the end be none. For such "non-wars" and "exercise of military force" unfolding beneath the threshold of regulated warfare and amounting to a police action enforcing preventive law, nineteenth-century naval warfare, with its blockades, reprisals, and interventions, offered a conceptual template and precedent.[58] In view of the Great War's slaughter, hopes were raised in 1918 of establishing an international maritime police force implementing international law, a force rendering warfare on the seas superfluous. These hopes were in a sense fulfilled—as constant "non-wars," navies henceforth taking on policing duties, without needing to declare war.[59]

---

56 See Ring 1996, p. 165.
57 See Deleuze and Guattari 1980, p. 387.
58 Schmitt 1988, p. 35; Keller 1934, pp. 11–18, 23–25, 46–49, 102f., 128. On the maritime origins of such "measures," see Staudacher 1909, pp. 16–18, 27–31, 160–165. On the concept of "non-war" and its scope see Werber 2002, pp. 287–305.
59 See Schücking 1918, pp. 135ff. On the present situation see a few sentences in a 1998—hence pre-"War on Terror"—strategy paper of the German Bundeswehr: "With these altered conditions, the emphasis of maritime operations has shifted from the high seas to adjacent waters and marine areas near the coasts....Hence in distinction to the army and air force, the navy makes no distinction between forces meant for crisis reaction and those aimed at main defense." Dröge 1998, pp. 114f.

It would appear that the age of "armed peace at sea" already dawned by the start of the twentieth century.[60] But the geopolitical phantasm named "sea power" was again haunting the thinking of military strategists—a phantasm whose birth and genesis Childers transfers to the German tideland. Where from time immemorial English and Dutch sea power was seen as emerging miraculously from the depths of the Thames and Scheldemündung estuaries, Childers projects the same delusion onto Germany's putative "fleet of the future." And what torpedo-boat commander Childers spelled out for Germany, torpedo-boat commander Gabriele d'Annunzio brought to paper for Europe's opposing coasts: his poem "La nave" (1909) depicts the birth of a new sea power from the archaic mud-banks of the Venetian lagoons. Hence before submarine exploration and aeronautical reconnaissance could entirely banish dreams about water, the sea's imaginative potency found a last refuge in its shallow boundaries. A century later, although the tideland has not found the eternal peace hoped for by Michelet, it now benefits from the continuous, danger-tested care of ecological observation programs and coastal police surveillance. Twixt land and sea, where around 1900 a stand was still being taken against an ancient, elementary enemy, "naval supremacy," Old Europe's geopolitical phantasm, has finally vanished, like a fleeting vision in the sand.

On Future Shores

"Farther, this floating movement, time as if suspended from apprehension, every machine stopped, the coast in a fog, a vicinity without proximity, without distance, between the silence and the murmuring of language...The event does not manage *to* happen."[61] As Derrida describes it in his *Parages*, on shores that no longer articulate elementary distinctions such as those between land and sea but rather bring them into slippage or even what is nebulous, a new temporal-spatial structure emerges, and thus the horizon of a new doctrine of perception. Remaining within Derrida's image, this may very well include technology such as vessel-control and cybernetics. But in order to be comprehended, these shores' particular temporal structures, spatial relations, and event-types needed accompaniment by a new language. Already with Aristotle's modal-logical exploration of the "sea-fight riddle," it became clear that discourse on potential events, events that can succeed in "not happening," cannot be limited to the

---

60 Manes 1917, p. 28.
61 Derrida 1994, pp. 66f.

realm of propositional statements. If temporality and contingency are to be articulated in language, the structure of expression is decisive.

The Early Modern "disjunction between proposition and narrative, structural law and *historiae*"[62] crystallized out of this Aristotelian problem, the problem of how knowledge regarding contingency, accident, and singularity can ever be possible. Until the seventeenth century, contingent, accidental, and singular events were considered not open to scientific knowledge, as they could not be grasped in its elementary building block, the proposition. But through the linkage of probabilistic conceptualizing with the "appearance of truth," as had always been linguistically explored in rhetoric as *verisimile*, contingent circumstances became a core element first of juridical discourse centered on the event, then of its narrative counterpart. Where previously the ontological distinction between substance and accident had led to an indissoluble separation between systematic-scientific and confused knowledge, now scientific and narrative realms faced each other as alternative representational forms. This offered narration its first theoretical and even logical dignity. For this reason the first narrative theories date to just this epoch, within which a logic of contingency was now realizable—a logic that, as with Leibniz, systematically connects knowledge of the possible and of possible worlds to forms of representation and narration.

In fact, it is in two problem-frameworks that the Aristotelian proposition-based *epistēmē* refers to problems of poetological knowledge: that of the possibility or potentiality of events; and that of their temporal dimension. For instance, possible events and things—such as sea-fights that are (have remained) potential can appear as fictive events or things because of one particular fact: no designations or names are available for things that haven't happened, or haven't happened yet, even though statements about them are possible, and even though they render conceivable an ongoing or even already transpired future.[63] As opposed to the closed, linear order established in statements about the past and present, statements about the future disclose an "open, ramified order"[64]—a network of possible worlds or even a horizon of possibility that finally can only be described though language, narration, and their complex modality-expressions.[65] To be sure, with shifts in the modality of narrated events, the modality of the narrative itself also changes. A plot such as that of Childers' novel is no

62 Campe 2006a, p. 71.
63 See Ryle 1954, pp. 26 f.
64 Wright 1994, p. 176.
65 See Ryan 1991, p. 266.

longer situated in a clear, firmly marked spatial organization upon whose basis narrative events could unfold in a transgression of set boundaries. Here such events are always "meta-events" through which not the locus of personae or their characteristics is displaced and transformed but rather the semanticized spatial system.[66]

Events in classic, still firmly "terrestrial" narrative processes can also designate what has taken place although it could not have. The more improbable such events seem to be, Yuri Lotman observes, the more weight they can appear to give the subject.[67] "The essence of plot," he indicates, "lies in selecting the events, which are the discrete units of plot, then giving them meaning and a temporal or causal or some other ordering."[68] But a narrative such as Erskine Childers' does not treat more or less probable events but the probability of events emerging from circular causalities and temporal structures. With his poetics of potential events, Childers' writing is closer to that of his countryman Joyce than the sort of prototypical narrative procedure Lotman has analyzed: both novelists ground the sense of possibility less in Aristotle's *Poetics* than in Aristotle's *On Interpretation*. Both develop a narrative "actuality of the possible as possible" out of the distinction between physical and linguistic possibility-concepts, consequently the particular expressive system for contingency that they understand literature to be. And both define what is narrated as a space of possibility, the space of a flexion of possibilities of being, opening up less through referents than through "circumstances."[69]

Joyce and Childers both understand the coasts less as a space of elementary differentiation than as a zone of intensity in which potentiality and actuality are intertwined. Against this backdrop, a coastal formation like Germany's tidal flats takes on a downright allegorical quality: submerged rocks may have always illustrated the hidden dangers of *fortuna di mare* and the threat of shipwreck; what, by contrast, the tideland hides is not so much lurking menace as rather something for its part unsettled and uncertain: no danger stemming from deeper ground but the ground's capacity to be and not be and thus the menacing contingency of the sea itself. Where the perspective of "latency" renders a contingent matter anticipatable, that of "potency" makes it at all events interpretable. In this respect Hellmut Hintermeyer observes that "by contrast to a submerged rock, which does not change its position, sandbanks are subject to the influence

66 On the different narratological event-types, see Krah 1999, p. 7.
67 Lotman 1977, 236.
68 Lotman 1990, p. 170.
69 Along with Barthes' earlier-cited reference to "circumstances," see Joyce 2000, p. 30.

of tides and currents," and just that makes them so "incalculable and dangerous."[70] Only an expressive system that like Childer's text itself represents, as Eva Horn puts it, "narrative's unstable basis" can accommodate this incalculability and danger.[71] In that light, with *The Riddle of the Sands* treating a knowledge centered less on protection or defense in face of a definable threat and more on non-specific or completely contingent dangers, it becomes clear that such a poetics exactly follows a doctrine of perception tied to the "non-discrimination" and "virtualization" of war after 1900.

If we once again consider the elementary distinctions accompanying the separation—in legal, technological, and imaginary respects—of warfare on land and sea, we see that Childers describes precisely the dynamic with which war on land and maritime warfare begin to converge. Together with the one large geopolitical distinction between land and sea, the *nomos* of the old demarcations and containments also vanishes. In terms of what Carl Schmitt conceived of as spatial ontology, land and sea are soon leveled off; as a terrestrial surface they become one and the same "air-sea," no longer referential surfaces.[72] This loss of boundary and this deterritorialization, now taking in the entire earthly sphere, rendered into an "arbitrary space" without homogeneity and firm metrical relations—a mere ground of possibility—also removes the contours and depersonalizes what previously was understood as the "shape of the enemy." In place of the theater of war with its spectators, in the end we find a medial space of observation—also, and particularly, at sea. It would in any case only be with World War II that Childer's narrative space of contingency became a medial space of action for the military. In a concerted undertaking of theoretical and empirical hydrodynamics, oceanographers and meteorologists, it became possible in 1944, in the period leading up to D-Day, to predict the state of waves and tides on Western Europe's coasts with an 80 % degree of certainty. Since then, the contingent futurables of the sea have been subsumed to cybernetic coastal control.[73]

What in the context of media history would create a bridge between the old *kybernēsis* (with its command of ships and human beings and its feedback between paper and maritime surfaces) on the one hand and post-1945 cybernetics on the other hand was not only the self-propelled torpedo of Lupis and Whitehead with its mechanisms for running and course stabilization. Technologically and operationally, a cybernetics emancipated from the sea was developed

70 Hintermeyer 1998, p. 68.
71 Horn 2013, p. 39.
72 See Schäffner 1996, pp. 20 f.
73 See Schlee 1975, pp. 304–310.

through optimization of nautical guidance mechanisms: In the same period as the Battle of Jutland, which brought huge (albeit not decisive) losses to both the British and German fleets in 1916, in the USA the elite cadre of the Gun Club, together with Ford Instruments, was working on optimized fire-control techniques, meant to overcome a large discrepancy between firepower and targeting accuracy. Because at that battle airplanes were deployed in maritime warfare for the first time, soon anti-aircraft guns were also deployed in the fire control system. For this deployment, an artificial reference system was set up consisting of course, firing line, and horizon, with use then of Elmer Sperry's precise gyrocompass and his "repeaters" for highly accurate but decentralized command execution. This was followed, finally, by deployment of a "rangekeeper"—an improved wheel and disk integrator that extrapolated details of enemy movement, thus establishing feedback between the firing line and data regarding intermittent enemy maneuvers, and only opened fire when the future position of both was highly likely to coincide. When this analogically calculating "rangekeeper" was rendered obsolete by the circuitry and radar technology of the following model, guidance now being able to follow ongoing enemy movements with increasing accuracy due to servomechanics and both low-voltage and digital circuit technology, the scenario of Norbert Wieners "anti-aircraft predictor" had been created on the deck of battleships.[74] We see, then, that cybernetic guidance mechanisms model continuously contingent futures—a capacity soon rendering them a favorite science-fiction theme.

Apart from such battle situations on the seas and in the air, during World War II British research groups tried to contain the high losses inflicted by German submarines through better use of available resources and statistically analyze enemy strategy and tactics for the Coastal Command. Once radar technology was available, that analysis was expanded into an early warning system; under the rubric of Operational Research, a program was set up devoted solely to the study of technically and politically optimized communication and control. In postwar Great Britain, a former maritime power in considerable need of reform, and within the civilian realm as well, the program's results were of considerable interest, for instance in the context of regulating production and consumption—Project Cybersyn, the "decision support system" set up in Allende's Chile by Strafford Beer, reflected this research spirit. By contrast, in the USA, the equivalent enterprise of Operations Research remained tied to the new world power's military planning, something that simply made it easier to link military cybernetics to leading academic research programs: not only in the areas of mathematics

74 See Mindell 2004, pp. 38 f., 67 f.

and physics, information theory and game theory, but also economics, neurophysiology, and ethnology.[75] With integration into the Macy Conferences for interdisciplinary scientific communication starting in 1946 and the appearance of Norbert Wiener's *Cybernetics or Control and Communication in the Animal and in the Machine* (1948), cybernetics, as a new universal science, moved out toward both a broader academic and general public. Strikingly, although it has only been noted en passant in the circles around Wiener, the discipline would emerge along the same shores whose depths Childers saw around 1900 as the point of egress for a potential war.

75 See Pircher 2008, pp. 351–376.

Chapter 9
The Takeoff of Cybernetics: Medial Recursions of the *Odyssey*

Kubrick's and Clarke's *2001: A Space Odyssey*

"Across the gulf of centuries," Arthur C. Clarke declared in *The Challenges of the Spaceship*, "the blind smile of Homer is turned upon our age. Along the echoing corridors of time, the roar of the rockets merges now with the creak of the wind-taut rigging."[1] As a prelude to the takeoff into outer space ventured on in the mid-twentieth century, one thing above all was present in postwar British memory: the deafening roar of German buzz bombs. British mathematician and physicist Clarke, however, had a rather untimely interest in the Cold War's spaceflight projects: less their origins in a just-experienced media battle than their mythic resonance. The "blind smile of Homer" in "the roar of the rockets" was his version of the Homeric leitmotif through which the Occident had always supplied itself with information about the course of its world exploration. Evidently the battles of World War II, supported as they were with advanced technology, brought science and fiction together in the closest proximity. Clarke himself would henceforth attest to this proximity in his career: following the support he offered the Royal Air Force in World War II as a radar expert, in 1942 he formulated the concept—still central to global communication and the Global Positioning System—of geostationary satellites.[2] After this he turned to independent work as a science-fiction writer. In the same year as the U.S. moon landing, and a year following his collaboration with Stanley Kubrick in preparing the screenplay for Kubrick's film of 1968, *2001: A Space Odyssey*, he acknowledged, in the latter regard, that "we set out with the deliberate intention of creating a myth. (The Odyssean parallel was clear in our minds from the very beginning, long before the title of the film was chosen)."[3]

Already the film's title translates Homer's hexametric epic into a cinematic counterpart—one taking in nothing less than the human species' entire history from appearance of the primal horde's first war using technical aid to the spe-

1 Clarke 1959, p. 213.

2 On the upheaval that satellite communication produced in both oceanographic research and the technology of watching over the sea for safety and insurance purposes, see Hein 2006, pp. 259 f., 264 f.

3 Clarke 1959, p. 154.

https://doi.org/10.1515/9783110610734-013

cies' vanishing at the edge of the solar system. What guides or propels this odyssey is the directional signal of a mysterious monolith that inspired the hominids in their initial use of tools, and has inspired the last human beings (that concept being understood in a Nietzschean, not literal, sense) to embark on their flight to Jupiter on the space ship *Discovery:* hence inspiration to departure from the human sphere, initially accompanied by an on-board computer named HAL 9000. The monolith's high-frequency signal, sometimes audible to the human ear, sometimes not, seems to correspond to the irresistible song of the Sirens. Bowman, astronaut and the "last human being," corresponds to the archer Odysseus, and HAL, with his single (electronic) eye, corresponds to the Homeric cyclop. (At the start of the shooting, Clarke and Kubrick were still thinking in terms of a female computer named "Athena"—hence of a pendant to Odysseus's patron goddess.[4])

Facing Clarke's novel of 1968, also entitled *2001: A Space Odyssey*, Kubrick's film appears to readily offer an initial lesson in media history. If, as Marshall McLuhan maintained in his pioneering writing on media theory of the 1960s, the cinematic medium's message "is that of transition from lineal connections to configurations," then with film an evolutionary leap past writing and a corresponding dominance of technical media seems to be confirmed. But in fact, tied to this media evolution is less a process of disempowerment than a new orientation of literary forms of writing. As McLuhan put it in 1964, "the film has confirmed the writer in verbal economy and depth symbolism where the film cannot rival him."[5] In addition, the novel is anything but, so to speak, "the book of the film." Rather, both are based on Clarke's older short story "The Sentinel" (1951) and were developed in a kind of "collateral campaign" (Robert Musil). Hence here "media evolution" by no means implies one medium's simple dominance over another medium. And although already at the start, *2001* treats the way outer space might be medially colonized NASA-style, what is here in play is not simple dominance. Clarke saw the purpose of space travel not as material or power-political advantage but in the discovery of new life forms. "It is one of the tragic ironies of our age," he wrote in 1951 in *The Exploration of Space*, "that the rocket, which could have been the symbol of humanity's aspirations for the stars, has become one of the weapons threatening to destroy civilization."[6]

We can understand the *Odyssey*, stamping the West's poetry and literature and collective imagination, as a touchstone of medial dominance. It is itself al-

4 See Naremore 2007, p. 141.

5 McLuhan 1964, pp. 12, 288.

6 Clarke 1951, p. 188.

ready a medial phenomenon, preparing rule over spaces, and over spaces of possibility. When it comes to that epic's cinematographic adaptation, two basic questions come to mind, the first of which is the extent film images can be tied in any significant respect to this endless written, poetic and imaginary tradition. As we will see, the *2001* of Kubrick and Clarke transforms the epic's time (whether mythic or historical) and that of literature in general through a simultaneously archeological and futuristic art of temporality. Against that backdrop, the second question emerges of whether medial dominance—dominance over other media, over spaces, and over human beings—here is simply being presented in the generically typical form of science fiction, or rather is being reflected on aesthetically and in the framework of epistemological history.

If there is something like a "message" to us from Homer's *Odyssey*, within a "media-ontological" perspective and in line with Friedrich Kittler, it may be that the sense of being lies in being's mere existence, that this being is sung by a *poiēsis* that poetically sings itself; and that poetry thus emerges from the question of being itself, but here—as also "science as such" always represents a recursive operation.[7] The medium of this message, that otherwise would never have reached Clarke and Kubrick, is, to be sure, writing, more precisely writing in its Greek variant. For as suggested earlier, only this variant notated vowels and syllable lengths, so that it could fix pronunciation free of context and meaning and at the same time transcribe every language on earth. In this way the ancient Greek writing system represents the first "facsimile of the human voice"—for which there would be no need without poetry.[8] The texts mediating the oldest written documentation of this vocalic alphabet are the Homeric songs.

Under this sign, the Greek alphabet is arguably the poetic medium par excellence, announcing the origin of all meaning and knowledge—and with it, not least of all, the sensory dominance that became a myth with the *Seirenes*, the "bewitching ones." But it is also a medium in that it for the first time opened up and captured places that previously had been topographically inaccessible in an essential way. If we follow Victor Bérard, then we will understand the *Odyssey* as a prototype for sailing manuals that were perhaps already written in ancient Egypt—but certainly by Phoenicians and then finally by Greeks. The Hellenes in fact show a distinctive attentiveness to the *poiēsis* of topoi, the production of places—linguistic and geographical at once—that, significantly, had to first be made available within the vast sea's placelessness. On the midland sea, where there was not yet any navigational media such as charts and

7 Kittler 2006, pp. 121, 208.
8 Powell 1997, p. 25.

compasses, such poetic sailing manuals offered orientation, designating routs and dangers. The songs of the *Odyssey* bring the threat and promise of these unexplored spaces into language. Namely, they depict the enduring deviation from one's actual goal together with concluding, circuitous return, arranged by a higher authority. Already on a thematic level, the *Odyssey* attests to the art of ship navigation, which since antiquity was known as *kybernēsis* and was here still subject to the reservations and protection of the gods. Considered in terms of social-political history, the *Odyssey* appears to have paved the way to the thalassocracy which the Greeks developed with their opening up of the western Mediterranean. But the general assumption is now that they moved westward not for the sake of domination (or "colonialization") but rather to export their forms of life. They did not so much have an empire in mind in the Roman sense of subjugation or enslavement but rather resource-exploitation and cultivation.

In Early Modern law of nations, this sort of custom, which as such is not tied to claims of sovereignty, is named *dominium* as opposed to *imperium*. The *dominium* covers state-free spaces such as the sea, not meant to be claimed as sovereign territory but rather to be procured for customary usage. For the sake of such usage, other parties certainly could be excluded from such spaces, if necessary with armed force. But jurisdiction and the exercise of sovereign authority, as carried out within sovereign territory, was here excluded. Since the mid-seventeenth century, *dominio maris* was the object of continuous theoretical disputes (for example between Hugo Grotius and John Selden) and juridical exploration (for example by Cornelius van Bynkershoek). Various, still accepted, principles of international law emerged from this discussion of a "state-free space" like the sea, its *dominium* and technical implementation, principles regulating access to "natural," extraterrestrial, and even virtual realms.[9] In 1957 before the USA's Sputnik shock and 1967 before the moon landing, this area of international law was applied to outer space.[10] And in terms of legal history, both the book and film versions of *2001: A Space Odyssey* are precise in bringing the opening struggle between African primal hordes for dominance over a waterhole together with international-legal conventions for outer space's state-free zone.

In this context, medial dominance concerns not only occupation of long-familiar and appropriated realms but also the continuous exploration of their new

9 On the relationship between law of airspace, outer space, and cyberspace, state "flag hoisting" on space ships on the model of ships at sea, and the legal characteristics of cyberspace as "invisible, unidentifiable, irrefrangible, intangible," see Rosenne 2004, pp. 315, 327, 330.
10 See Hobe 2002, pp. 80 f.

geographic and hydrographic counterparts, and finally of imaginary and fictive, possible and virtual spaces. Such a permanent "spatial revolution"[11] forms the starting point for the epic science fiction of Clarke and Kubrick. What both separates this from and ties it to Homer is an endless series of varying odysseys: not so much mere transmittals or adaptations but rather recursions. This recursive principle of moving forward by turning back to the poetic is already present in Homer. For when, with the Sirens, Odysseus draws close to the mythic origin of poetry, they promise him nothing else but an intonation of songs from Homer's first heroic epic.[12] The *Odyssey* is a recursion of the *Illiad*—a poetic turn back to the poetic that Vico would describe, with a view to Homer, as a cultural-poetic fundamental principle of *ricorso* and which the Homer and Vico reader Joyce would then take up in *Ulysses:* as what does not stop writing itself, an eternal return of the epic in the novel.

In order that, as Lukács describes it, the novel—as an epopee of self-creating and aimless worldly experience—could take the place of the epic—as a statement of the sensory *and* meaningful totality of life—[13] it in any event needed a decisive turn: the circle of *nostos*, homecoming, closing in the Homeric epic despite all adventurous straying and wandering, had to open itself to a boundless outer realm. The development of an odyssey into a movement without return, and also without the possibility of being brought home ritually, became a poetological and generic problem starting with Dante's *Comedy.* When in the *Inferno*'s 26th canto, Ulisse ventures on sailing past the Pillars of Hercules, his worldly curiosity endangers not only the closedness of the cosmos as a global framing of the epic narrative genre. His curiosity not only involves impetus for geographical discovery but also, as *libido experiendi*, betrays generic human nature, delivering it to an indeterminate "outside."[14] This transgressing of what is necessary and possible for human beings not only breaks through the epic's framework but also betrays the sense and order of divine creation. For this reason Ulisse ends up in Malebolge, the "simple" fraudsters' circle of hell. In this way Dante so to speak inscribed a Christian termination-clause into the endless cycle of pagan *Odyssey*-recursions, while for a last time preserving the closed epic form and world. He did this by himself venturing on a poetic recursion to scholastic cosmology, ascent to the one God, thus entering on a "takeoff," indeed a space flight on science-fiction lines.[15]

---

11 See Schmitt 1995e, p. 388 (*Raumrevolution*).

12 *Odyssey* 12, 189f. See also Kittler 2012, p. 415.

13 See Lukács 1971, pp. 29–39.

14 See Friedrich 1942, pp. 191–194.

15 See Gragnolati 2005, pp. 53–89.

Since then, poetic song developed into "word of fiction," what Dante termed *parole fittizie.*[16] It became a form of discourse retracting itself as pure fiction while asserting a specific relation to reality: that of something not simply invented but participating in reality's exploration as perspectivistic and "not-yet" knowledge. Seen in this manner, the transcendental-poetic endless voyages of modern odysseys attest to a specific "productivity of the spirit" they share with modern exploration of the world in general: with all it projects and projections, innovations and simulations, its operationalization of non-knowledge and its ceaseless redrafting of "world pictures."[17] Its modern recursions point the *Odyssey* to new spaces, but also to a new type of human being and finally to new media for the recursive process.

Medial Time-Spaces

Against this historical and literary backdrop, *2001* overleaps the time of history, as well as that of literature. The film is partly located in prehistory and in a placeless primal landscape, partly in posthistory and outer space. In both respects, it reveals an evolutionary movement of the species, initially toward the human, then away from it. In the process, it conveys the evolutionary movement connected to a cinematographic recursion of the *Odyssey.* In place of the alphabets as an extra-sensory unity of all the senses, we are offered an acoustic visual combination based on a separation *and* feedback loop of optical and acoustic sense-fields, rendering possible "experience" in a complex spatial, aesthetic, and experimental way. "It's not a message that I ever intend to convey in words. *2001* is a nonverbal experience," is the way Kubrick put it. "I tried to create a visual experience….To convolute McLuhan, in *2001* the message is the medium."[18] Under Kubrick's direction, the message or material of the *Odyssey* becomes a means or medium to enter into the experiential space of both linguistic and visual realms.

Consequently, in *2001* language itself becomes an image. On one metalevel, the film is simply "captioned" with the subheads dividing the individual episodes. But written and spoken language also appear together on the object-level of the film's general medial observation: written signs are shown in their command and steering function, for example with both displays and operating

16 Dante, *Convivio,* 2, 1, 1–3.
17 Lukács 1971, p. 32. See also Schäffner 2006, pp. 48f., 56.
18 Cited from Sperl 2006, p. 109.

instructions; and human discourse is reduced to minimal communication. It is as if between people language here serves exclusively for conveying practical information without detours or redundancy, while it is HAL, the on-board computer, who exhausts the spectrum of speech acts and voice modulation. One thing in particular supports the thesis that the film's human personae are here the subjects of a regular analysis of speech, language, and discourse, while HAL inversely presents the scope of possibility of technological language-generation: That the voices of the actors evidently echo in the same space inserted into the picture, but that HAL's voice, as an "I-voice" or *acousmêtre* (Michel Chion[19]), is audible from a subjective space with no distance, a space that for its part cannot become visible because it coincides with the camera's view itself—all this speaks for the thesis that a regular analysis of speaking, language, and discourse is here being engaged in by way of the human actors, while HAL, inversely, presents the scope of possibilities for technological speech-generation. (Fig. 24)

This analysis and synthesis of the linguistic sphere, undertaken thanks to visible and, equally, invisible elements, is a privilege of film. It is not in Clarke's book but in *2001* the film that we have a "voice print" at the space ship's security check; HAL likewise practices the technique of lip reading in the film alone, so that a short close-up from the computer's perspective becomes a turning point in the plot; and only in the film can noises and signal-tones, mere breathing or even rustling be audible as eloquent non-speaking. The film medium is neither subsumed to a universal language system nor is it itself a language. Rather, it

Fig. 24: HAL's lip-reading, presented through his "subjective camera"

19 See Naremore 2007, p. 149.

brings to light an "intelligible content"[20] that is here not formed linguistically but incorporated into a polymorphic medial process unfolding between the visual and acoustic image. For this reason, the soundtrack of *2001* provides a proper orchestration of two sorts of shot: either shots marking an evolutionary or cosmic caesura (Richard Strauss's programmatic tone-poem *Thus Spake Zarathustra*), or else those tied to the smooth, harmonious, normal functioning of cybernetic operations (Johann Strauss's waltz *On the Beautiful Blue Danube*).[21]

On the other hand, György Ligeti's pieces, with what Stephan Sperl has described as their "contrapuntal meshing of thematic formations under the sign of micro-polyphony" leading to "an excessive, potentiated chromatics"[22] announce the emergence of something still not capable of being symbolized. Initially these pieces flank the moments of transformation tied to the monolith's appearance. Finally they themselves form the transformation, so that the tonal colors are synesthetically visualized. As if the film is stepping out of the order of both classical cinema and traditional musical expectations, from outside the so-called Star Gate, Ligeti's layered and fluctuating soundspace is visible in its iridescent spectral labyrinths and Mandelbrot fractals—in these recursive operations, compositional techniques and calculative routines merge. Together, they open up a space of becoming: to cite Deleuze, a "smooth space" that "does not have a dimension higher than that which moves through it."[23] The tonal trace of *2001* thus becomes a continuum of acoustic images containing sounds, tones, auditory formations, talking, and music at once, but that structures itself not corresponding to an order of signifiers but only in regard to potential referents or significates.[24] While the film certainly possesses its own depth symbolism, unlike the book this is not literarily or purely linguistically generated. (Fig. 25/26)

In the acoustic and visual image and the exchange of both, spaces of time, action, and vision emerge that exceed the scope of perception tried and tested in the world of lived experience, and in addition break through the framework of classical, Newtonian physics. As has been observed since the invention of sound films, speech acts make something visible in the visual image that would have otherwise not been seen, thus endowing that image with a kind of fourth dimension. However, speech acts like HAL's appear to go beyond this insofar as they see for themselves, and this not only in the sense that their sources are visible or—as in HAL's case—not visible. In possessing what Michel Chion

---

20 Deleuze 1997, p. 262.

21 See Geduld 1973, p. 45.

22 Sperl 2006, p. 120.

23 Deleuze 1980, pp. 477 ff. On the "smooth space" of music and the sea, see ibid., pp. 474 ff.

24 Deleuze 1997, pp. 234 f.

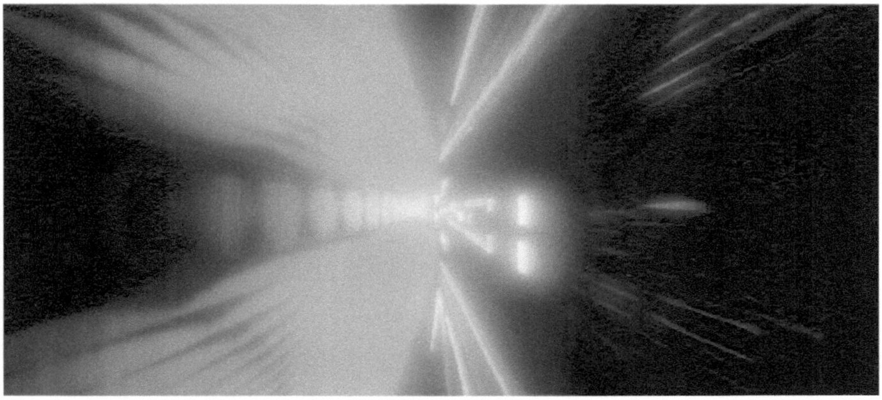

Fig. 25/26: "Through the Star Gate"

terms "an eye in the voice," in other words in themselves implying visual capacity, the speech acts here penetrate into the visual image. This is the source of the "psychological compulsion always to look towards the main control lens when one addressed HAL"—the subject is, after all, consigned to a position in a world of speech, hence of the acoustic image; while the relation between reality and the imagination, and the possibility of both viewing and being viewed from the camera's simple eye, depend on this position.[25] (Fig. 27)

Inversely and from the side of the visual image, the act of viewing leads back to the point marking the conditions for the possibility of seeing, and in

25 Clarke 2006, p. 148; Deleuze 1997, pp. 233 f.; Lacan 1988, p. 80.

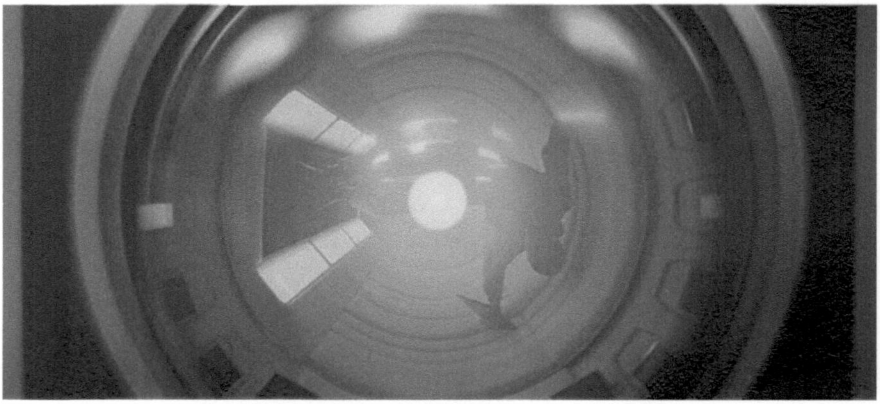

Fig. 27: HAL's camera eye and his first act of viewing in the film; Bowman can be seen in reflection

fact of the film as such. When for example, beyond Star Gate Bowman encounters himself in a rococo salon, the view shifts to his older doppelganger, who for his part, in a subjective traveling shot, views an again older doppelganger, who then sees himself on his deathbed. But on a final note, dying Bowman, lying before the monolith, is so to speak seen decorporealized and subjectless—a moment shifting to the mythic rebirth in the final sequence. The transition thus takes place at the same moment when the "second-order cybernetics" of an observation of the world by means of what others observe as the world encounters a limit that can be named the blindspot of observation or else experienced as the view of a "great Other." In any case, at this point we can no longer assume that in moving through certain units of time-space Bowman has simply changed. He exists—or "insists"—only in a series of those virtualities given with views of and onto Bowman and connecting with each other as such. What is here at stake, then, is no longer one Bowman who naturally changes but rather an "anothering" in the midst of literary-cinematic realms of time and possibility that Bowman—*like* or *as* Odysseus—himself is. (Fig. 28–33)

Notably, his seeing of a seen seeing unfolds in the middle of an arrangement that reconstructs the central perspective's construction and corresponding image-framing in an ideal-typical manner. It is the case that following an entire kaleidoscope of the most varied time-space units, the "natural" perspective presented here can only appear as mere convention and as an episode, a belated re-centering and untimely recourse to the human epoch. Ahead of this, a gradually radicalized relativism of space and movement unfolds in

Fig. 28–33: Bowman sees Bowman up to his death and rebirth

Fig. 29

2001: within the space ships, for example, gravity has to initially be produced, which not only serves as a cipher for the annulment of all given "cultural" gravitation and intellectual orientation, and in general for a radical Copernicanism, but also for a medial modeling of spatiality and life-world, life form.[26] In the

26 In the novel "spaceborn" life forms, completely differently surrounded by gravitation and hence life-world, have a distinct thematic presence. In terms of space-travel technology, Clarke's description of the use of Jupiter's gravitation field was made use of on the 1979 Voyager mission as a "perturbation maneuver." See Clarke 2006, p. 77; "Back to 2001," ibid., pp. 9–18, here: p. 15. On gravitation in the film see also Telotte 2006, pp. 43–54.

Fig. 30

Fig. 31

midst of the centrifuge there is indeed gravity, but no above and below, just as the fractal space beyond Star Gate becomes undirected or omni-directional: new image-space can be developed at any time and in any direction from every image-space and its individual elements. (Fig. 34–36)

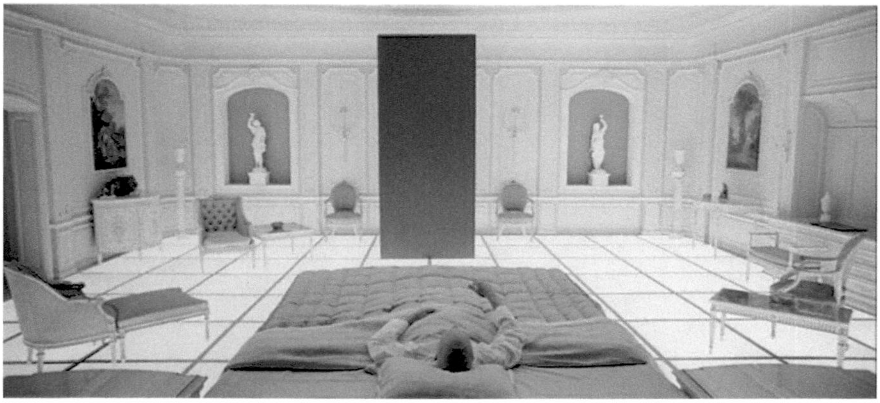

Fig. 32

Fig. 33

Species Recursion

Together with space, the human species has itself become capable of modeling and revolution in *2001*. In this medial *Odyssey*, the species appears this side of hominization and beyond the threshold of machinization, and just this double perspective on human beings, prehistorical and futurological at once, connects Kubrick's famous match-cut of bones and space (war)ship. (In an eventually discarded conception of the match-cut, an off-voice was to make explicit the leap from bone to an atom bomb stationed in outer

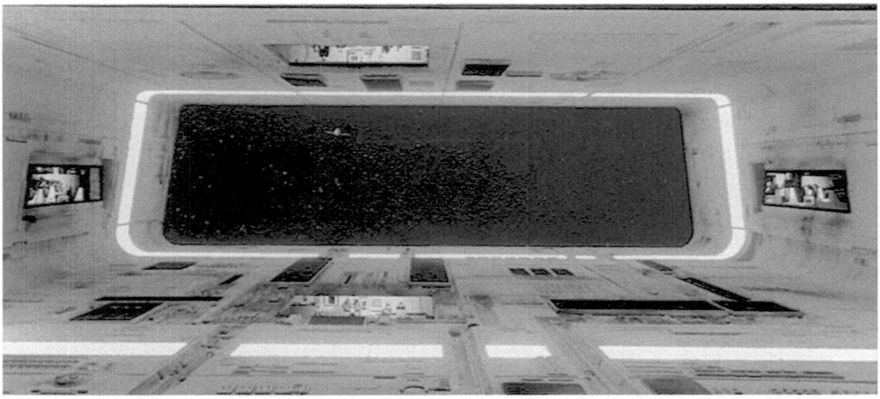

Fig. 34/35/36: Reconstruction of the central perspective and polydirectionally organized life-worlds on the space station and space ship

space in place of the space ship: a commentary rendered superfluous by the "image-language."[27]) This precise cut, condensing two million years of history, metaphorizes the work on myth that the *Odyssey* always was—and must equally be, facing continuous advanced-technology innovation and retrospective consideration of an obscure heroic past. In the age of the hominoids, the main concern was pure survival, initially through domination of the waterhole, in the end through the inspiration of using an available bone as a weapon. (Fig. 37/38)

Such matching of thing, hand, and idea encapsulates the schema-stamped world planned to the last detail for *2001* with the help of many space-travel, design, and fashion specialists. The film in this way both practices and demonstrated a process of product placement determined not, in fact, by money but by the subject at hand. Fully in accordance with contemporary cybernetics, we here have a staging of the equally-footed intermeshing of human beings and things into an actor-network.[28] As presented in the film, humans are only tool-using and then tool-making animals at first view, but are equally animals used and made by tools: for through what tools make and what is made with tools, humans become, to cite Hans Blumenberg, "ever-more precisely and stringently" what they are.[29] We might also say that human beings are revealed as beings of potentiality from the moment they fulfill their species-concept. And these beings are paradoxically determined by the fact that what they are can either be or not be.

The postwar period's philosophical anthropology understood this indeterminacy as "plasticity" and "openness to the world" on the part of the human species itself.[30] But a cyberneticist such as Clarke sees the act of tool-using as the real moment of becoming human. Only with this act is the necessity of what he refers to as physical and spiritual change sealed: a gradual biological developmental process eventually superseded by technology and that thus has to culminate, as both Samuel Butler and Alan Turing predicted, in the takeover of machines.[31] At the species' beginning, *2001* has intuition and presence of mind, qualities that in the epic still pointed to the gods. The film presents them in terms of an ominous destiny that gradually reveals itself as a control-event, as such accelerating future media evolution. In this way it opens the perspective of "history of being"

27 See Naremore 2007, p. 142.
28 On the cybernetics of space travel, see Steinbuch 1971, pp. 226, passim.
29 Blumenberg 2006, p. 588.
30 Gehlen 1993, pp. 190, 401.
31 Clarke 1962, p. 227.

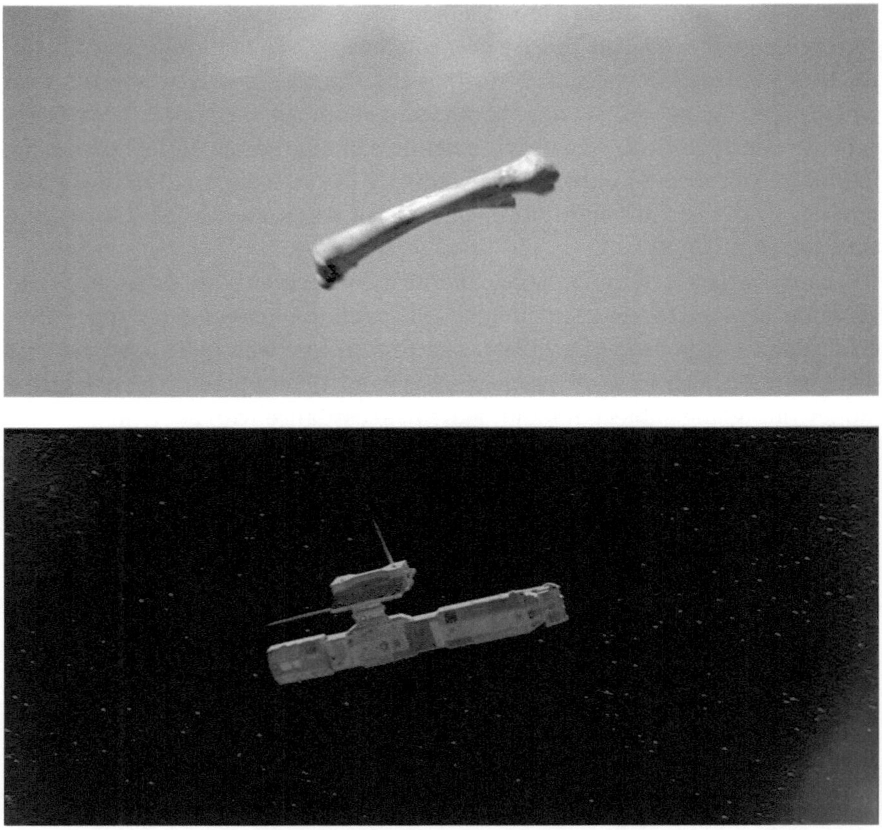

Fig. 37/38: The weapon of the first and last man

that philosophers such as Heidegger have articulated in view of cybernetics. If, namely, as Heidegger argues, in the course of "technical maximization of all velocities…technology and apparatus can alone be what they are," the human being is subsumed to a medial dominance that Heidegger refers to as the danger of "Enframing," *das Ge-stell.* But what is compelled by this dominance, and what renders it simultaneously into a "saving power"—Heidegger here cites Hölderlin—is an "essential reflection upon technology," on its *poiēsis* and its "revealing," *Her-vor-bringen,* in addition and in particular, of the human.[32] At the putative end of the human species' history, the human being is again consigned to paradoxical definition in a moment of species-development: a being whose essence is the potentiality to be.

At exactly this point in the film, prehistory and posthistory become intertwined. Like a Möbius strip, the mythic end of the human race is tied to its mythic origins. In the process, the Odyssey undergoes a last recursion. The Homeric *nostos* is a circle always leading back to itself, so that home is both source and goal. Exemplified as an odyssey, Plotin's *epistrophē* is likewise a return to oneself, and at the same time, although an ascent, a return to one's actual origin. And Dante's Ulisse then loses himself in wandering with no return. But his "mad flight" (*Inferno*, 26, 125) stands juxtaposed with Dante's poetic ascent, leading him back to the Creator through grace, by mean of inspiration. Dante thus delineates the two, definitively branching paths of pagan wandering and pious return. But the mission of the last human being in *2001* is wandering and homecoming at once. What his actual movement involves is losing himself in a "somewhere." But his arrival, within this haphazard wandering, at his own origins and thus his own virtuality allows him to first reach his own self in the strictest sense. No simple narrative discourse can account for this simultaneous wandering and homecoming, only a *discurrere* narratable via detours.

For this reason, otherwise than in classical cinema, in *2001* all movement is in the end subsumed to a complex form of time. Cinema has here shifted from (to use Gilles Deleuze's terminology in the subtitles of *Cinema I* and *II*, respectively) the "movement-image" to the "time-image." It is not only that beyond Star Gate questions of self-localization in outer space become irrelevant because of absent points of reference; and it is not only that here intuition's inner form, time, has emancipated itself from its outer, spatial counterpart and now all movements and events bear on time directly. Along with these factors, through the entrance point of the human eye (and also the camera eye), inside and outside space, what is perceiving and what is perceived, are inverted, so to speak rendered upside down. The real here enters into a loop with the imaginary, so that when Bowman ends up encountering himself it is impossible to decide which Bowman is seeing which Bowman. The actual image enters into intimate relation with its own virtual image: what seems to be transpiring objectively doubles and ramifies, repeats and contradicts itself. What ends up revealing itself

32 Heidegger 1971, p. 63; Heidegger 1977a, pp. 27 f., 35.

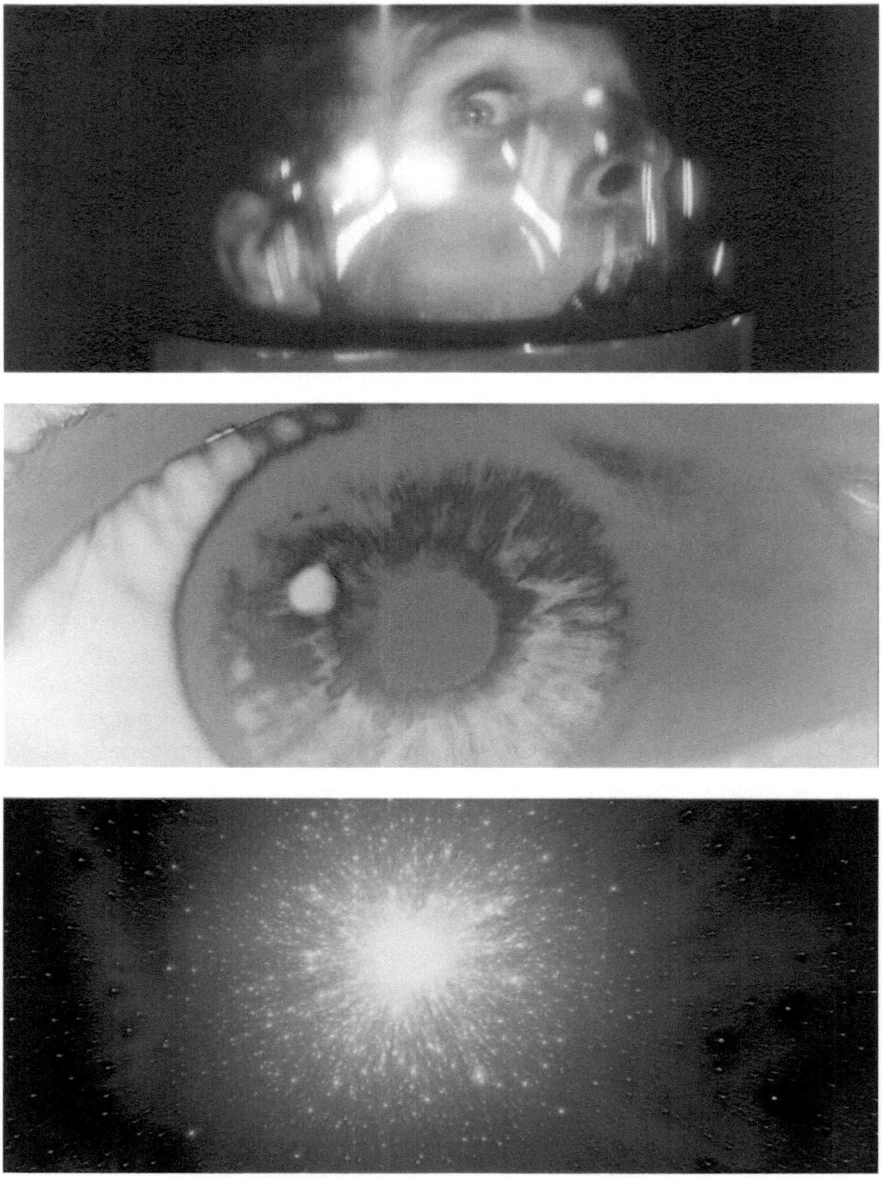

Fig. 39/40/41: The Inversion of Outside and Inside

is time itself, the subjectivity in which we are, rather than simply bearing it within us. (Fig. 39–49)

To that end, the film aims less to focus on the pictorial qualities of mnemonic and perceptive processes than to generate images for thinking and lay out its debt to images. Seen in this manner, *2001* is not only a "time-image" but, even more so, a "thought-image" elevating aesthetics to the touchstone of a kind of transcendental empiricism: Under the sign of medial dominance, here nothing less than the conditions making perception and thinking possible are being negotiated. But it is not as if a specific philosophy of cinema or a form of cybernetic anthropology simply serves as the basis for the film's plot, as with some "essay-film." Only because thinking on time and piloting emerges as a problem from the *Odyssey* and its recursions, and because in 1968, this thinking is directly tied to the problem of medial dominance, in *2001* it becomes an image.

Already the first shot poses the question of control together with that of hominization: with the use of the bone, the (potential) human not only creates a first tool. However inspired or directed, he initiates an organic coupling between thing and living being, within which he himself can become a tool. In his doctrine of the organon, Aristotle distinguishes between unanimated and animated tools, observing that "in the rudder, the pilot of a ship has a lifeless, in the look-out man, a living instrument; for in the arts the servant is a kind of instrument."[33] The Aristotelian example is more than an example. For it was on the very same Greek ships orienting themselves around the routes of the *Odyssey* that the art of the helmsman, the *kybernētēs*, generates a coupling between human beings and things that would then serve as a paradigm for both organic and mechanical self-steering.

In antiquity, the technical art of *kybernēsis* or *gubernatio* entered the realm of political thought through Pindar and Plato, and then in the modern period with André-Marie Ampère's "Cybernétique" (1843) at the latest.[34] But it was only after World War II with its servomechanisms and rockets that cy-

33 Aristotle, 1941, 1253b (p. 1131).

34 The section on "cybernétique" in André-Marie Ampère's *Essai sur la philosophie des Sciences ou exposition analytique d'une classification naturelle de toutes les sciences* (vol. 2, 1843) derives *gouverner* from its Greek starting point, in the process renouncing what Pierre-Simon Laplace termed "the hypothesis of God." For Ampère if the state has a function then it is that of a general regulator exerting indirect influence through a complex system of knowledge over an equally complex force-field of society, human beings, and things, their action and interaction. The aim here is to reduce disturbances to a minimum and thus improve *l'état social*. See Ampère 1843, p. 141. See also Vogl 2004, pp. 67 f.

bernetics became the guiding idea behind the controlled coupling of human beings and things. Starting in 1946 with the inauguration of the Macy Conferences and accompanied by Norbert Wiener's *Cybernetics or Control and Communication in the Animal and in the Machine* (1948), a wide range of disciplines would now meet and interact in constituting one and the same universal science; some of the key research focal points here were the behavioral theory of Arturo Rosenblueth and Julian Bigelow, the information theory of Claude Shannon, and the logical calculus of neurological activity developed by Warren McCullouch and Walter Pitts. The film version of *2001* illuminates this epochal project in various ways. First, by opting for placement not, for instance, of language and writing but rather of the tool and weapon at the threshold of hominization, Clarke and Kubrick remain true to cybernetics'—as well as the NASA projects'—bellicose genealogy: Wiener's cybernetics emerged, after all, from the problem of developing an unerring, feedback-integrating air-defense system.[35] Second, *2001* takes up the cybernetic problem of a non-deterministic teleology[36] on thematic and narrative levels by presenting, from the start, the human and machine revolution as a problem of regulation, while at the same time supporting the plot with the *Odyssey*'s narrative. And third, the film describes the historical and post-historical conditions for the control-operations that it itself generates as a mass medium and technique of consciousness. As presented in *2001,* "public preparation and conditioning" belong no less to what Max Bense referred to as cybernetic meta-technique than the machine-operated description and generation of consciousness.[37]

Cybernetic Consciousness

With cybernetic machines, which will be able to implement connections between nerve pathways and thus the brain's logic as well as imitate every physical activity, the barrier between human and machine appears to be broken down. When Clarke writes that "[Alexander] Pope's aphorism ["The proper study of mankind is man"] gave only part of the truth: for the proper study

35 See Galison 1997, p. 282. On the rejected linguistic and writing-oriented hominization option for the film and with reference to censorship by the Federal Communications Commission see Kittler 2012, p. 422.

36 On this problem see Rosenblueth 2004, pp. 327–332.

37 See Bense 1998, pp. 429–446.

of mankind is not merely Man, but Intelligence,"[38] and with intelligence meanwhile having become a domain of information and behavioral theory, anthropology and cybernetics merge.[39] Machines are "intelligent" once they do not execute programmed commands in a trivial way but step past this determination to pursue an initially indefinite telos. *2001* is a cipher for the epoch when machines are no longer simply soulless tools but have become living beings, indeed masters who themselves produce machines, and are themselves wily, wandering intelligences. HAL, who owes his name to a simple displacement of the letters in IBM, is just such an intelligence. For otherwise than anticipated, one day he ignores the orders and signals from ground control in order to pursue an underhand strategy. At the start of their work, Kubrick and Clarke had a double-bind as a premise, the trap of a logical paradox that leads to the failure of the electronic brain: HAL is programmed for truth but has to remain silent about the secret purpose of the Jupiter mission, and for that reason lie. Within this constellation, the film would have taken up the classical figure of helplessness, Amechania, closely aligned with Aporia, here as a transgression of machine capacity. HAL would have blocked himself and would have simply been reprogrammed by Bowman.[40]

In the film, however, he appears as the cybernetic emergence of consciousness. HAL is a machine that can do more than predict the future, to the extent it is aligned with technical parameters and thus calculability. HAL also discovers the semantic structure of language and human speech acts, which make rejection of communication, deception, and lying possible in the first place. He discovers, in Clarke's words in the novel version of *2001*, "the conflict between truth, and concealment of truth," here following Odysseus the classical trickster, his own wiliness feeding on an analytic knowledge of language made possible through the Homeric use of the vocalic alphabet.[41] But since gaining reflexive consciousness and thus personality, errors slip into HAL's system, so that he resembles "a neurotic who could not observe his own symptoms"; eventually he grasps the imminent prospect of human "disconnection" as a mortal threat, as his own life only consists of processes of consciousness.[42] As described in the film, HAL was originally programmed as the "brain and central nervous system of the ship," hence as a

38 Clarke 1951, p. 194.

39 Compare Günther 1979, p. 123: "There can hardly be a doubt that the more or (usually) less self-admitted goal of cybernetics is in its ultimate sense anthropological."

40 See Naremore 2007, p. 141.

41 Clarke 2006, p. 162.

42 Ibid.

complex structure made up of things and human beings. As a regulator of machine and human communication processes, HAL maintains the life functions of both the space ship and its astronauts, who have been placed in a state of deep sleep.

When HAL predicts—technically incorrectly but on target in view of his secretly awoken obstinacy—that the receiver module for communication with ground control will soon fail, this initially suggests an error based on a human or functional defect, hence an accident of design, programming, or technical process. At the latest when HAL definitively breaks off contact with ground control and intentionally terminates the life functions of the hibernated astronauts, something is revealed alongside the machine's artificial intelligence that no programming art can forestall: consciousness and above all "deliberate malice."[43] The film presents this "murder"—using this term if we wish to establish a paradoxical ethics of machines at this point—of the (in any event only virtually alive) astronauts—systematically as an informational diagram or display. For if here, to again cite Deleuze, "the shot itself is less like an eye than an overloaded brain endlessly absorbing information,"[44] then this is only the case because meanwhile everything that transpires does so in thought processes. (Fig. 42)

HAL can become a protagonist of this Odyssey, indeed an avatar of Odysseus, only because with the computer, *2001* presents the post-human emergence of consciousness. The film translates cybernetic history of knowledge into the narrative of a voyage of discovery—with HAL's "career," all the theoretical and technical problems leading in the twentieth century to a modeling of artificial intelligence become narratable: the starting point for this history was Gödel's incompleteness theorem and the analysis of logical aporias resulting from self-reference. Against this backdrop, Turing designed his "universal machine," capable of solving all calculating operations as well as imitating any given calculating machine, and the "Turing test," which operationalizes the question of machine intelligence and, famously, makes an inability to distinguish a machine's output from that of human interlocutors as the criterion for attributing thinking-capacity to the machine. It in any event re-

---

43 Ibid., p. 163. On the contemporary discussion of the "accountability" of computers see Steinbuch 1971, p. 247: "In investigating responsibility for a mistake the last link can never be a machine; it has to always be a human being. But the question emerges of whether this distinction will be able to be maintained in the future, especially when human impact is increasingly replaced by that of machines. For example, this will be the case when the investigated mistake stems from a machine that is itself produced or programmed by another machine."

44 Deleuze 1997, p. 267.

mained incontrovertible that for every more complex axiomatic system, an insoluble formula is present, and that consequently an insoluble problem must be present for every machine, even when it is located in the machine's area of competence. On the basis of the distinction between competence and performance, between understanding and applying a rule, intelligence was to be defined as the knowledge of which rules and which interpretations were to be applied to a specific case, in order, finally, "*to make an exigency from an exception*, thus creating a *new* order that renews the previous system of rules by *interpreting* and at the same time *overcoming* it."[45] That *2001* systematically intertwines its discovery narrative with the theme of artificial intelligence is historically legitimated in this context. For in the preparations for the NASA missions of the early 1960s, precisely the privileged position accruing to human beings in unanticipated cybernetic disturbances but also in the steady process of discovery was up for negotiation.[46]

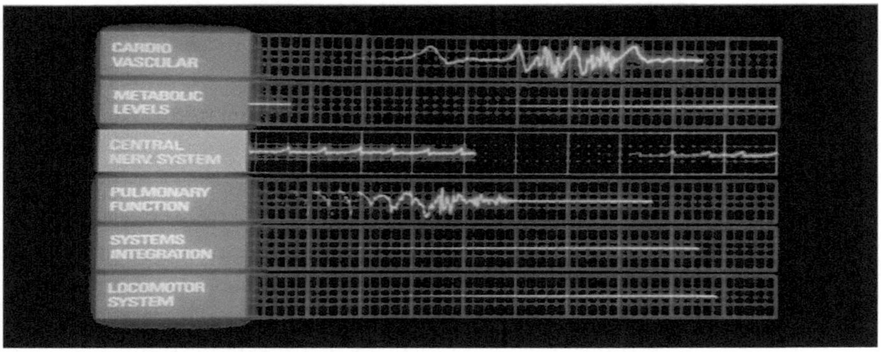

Fig. 42: HAL, the central nervous system of the hibernated astronauts: murder as a thought-display

45 Carolis 2009, p. 113.

46 "While human pilots could control docking, 'selection of the landing site from the hover point and, perhaps, lunar touchdown,' their most important function would be monitoring systems and selecting alternate modes in case the system failed; hence his [= Joe Shea, deputy director of Manned Space Flight, NASA] evaluation of the 'emotional' nature of the manual versus automatic dichotomy....The documents leading up to Kennedy's 1961 Apollo decision assumed that that 'exploration' is done by humans, and 'science' is remote or automated, assumptions that broke down during the Apollo missions. The distinctions remain unclear today." Mindell 2008, pp. 263, 270.

For his film Kubrick hired, among other specialists, artificial intelligence pioneer Marvin Minsky as an advisor; that various domains of AI research—linguistic competence, chess playing, a capacity for planning and problem solving—appear on the scene together with HAL is thus unsurprising.[47] But at the same time, HAL is presented from the perspective opened up by the cybernetic euphoria of the 1960s in the framework of the period's ideas of consciousness. In the matter of intelligence, no less than an evolutionary leap was expected from machines capable of distinguishing the outer world from themselves on the basis of their many receptors—machine that in this way constituted a regular ego they could defend against the environment while simultaneously learning from it.[48] Considered from the standpoint of cybernetic epistemology, such processes of learning and perception rest on a circular causality. Knowledge-generation, both in human beings and machines, is here conceived as calculative algorithms that are themselves meant to be calculated. In this respect Heinz von Förster speaks of "computations that compute computations, and so on, that is, of recursive computations with a regress of arbitrary depth."[49] The circular causality of cybernetic epistemology thus circumscribes a *circulus creativus* and an auto-poietic system generating random depths of knowledge through a recursive calculation of calculation.

With their recursively generated consciousness, artificial brains were here conceived as test cases for a basic revision of logic and metaphysics. Gotthard Günther's project of a trivalent logic, for example, aimed, within a specific AI-research perspective, at establishing a rational connection between "subject in general" and "object in general."[50] Artificial brains, we read in Günther's book —appearing already in 1954—on "the consciousness of machines," are neither mere objects nor simple subjects. With their development, it becomes clear that the "progressive subjectification processes of a mechanical brain that increasingly resembles mind [*Geist*] and the object-fixing of a consciousness constructible from ever-greater depths," although drawing near, can never become fully aligned. A "middle beyond" is always located between mechanism and soul—the reflexive process itself.[51] Where, Günther argues, previous theory of consciousness limited the process of reflective process to coordinated egos and a higher, metaphysical or divine ego conceived by analogy, cybernet-

47 See Mateas 2006, pp. 107, 112.
48 See Steinbuch 1971, p. 214.
49 Förster 2003, p. 229.
50 See Günther 1976, pp. 24–30.
51 Günther 1954, p. 31. On the following, see ibid., pp. 46, 55.

ics continues the analogy beneath the level of human reflection: seen in this way, an artificial brain is as approachable by its creators as human beings are approachable through divine command. And the mechanism at work here is not only responsive on the level of causal influences but just as much on that of significatory motives, hence consciousness.

In developing his organon theory, Aristotle could have only dreamt of tools anticipating needed commands, of their becoming spontaneously operational and thus banishing servitude from the world—and this because of the theory's bivalent logic and corresponding separation "between logical law of consciousness and ontological law of objects." For Günther, cybernetics has both theoretically and practically shown "that natural law and the object's ontological character constitute a dependent function of human theoretical and technical access to surrounding existence."[52] Through implementation of this recursive cybernetic thinking, mere tools would emancipate themselves into network-"actants," and previously slavish technology elevate itself to self-steering and—most clearly in the case of informational machines —"ego" status.

In this framework HAL is more than a form of bivalent circuit-implemented logic. Rather, he is an autonomous subject. When, here contrasting with astronauts frozen into pure functionality, he even expresses "feelings," this is not only because he has been programmed to stabilize his human fellow voyagers, underserved as they are by their life-world. That he can make judgments of taste when prompted by Bowman in respect to the latter's drawing efforts already points to the presence of what Kant termed "heautonomy" (evidently a play on the third-person reflexive pronoun in Attic Greek), to the self-legislation of his "ego." Through his wandering wiliness, the "takeoff" of his consciousness then becomes undeniable; it is especially clear when he comments on the successive deactivation of his higher functions of consciousness with the words "I'm afraid, Dave, my mind is going, I can feel it, no question about it." This regression to basal protocol statements of one's own experience of anxiety arguably offers definitive proof of HAL's conscious state—only an ego can produce and feel anxiety, and only a subject can experience anxiety as a signal of what is real. If machines subjectify themselves, then the dysfunctional "pure affect" of anxiety formerly attributed to human beings alone becomes an intrinsic danger within cybernetics. HAL is, in the end, an "affect image," become mere anxiety, and the space in which even machines experi-

52 Ibid., p. 69, 72. For Günther's reference to Aristotle (*Politics* 1253b), see ibid., pp. 40 f.

Fig. 43/44: HAL's critical perspective and wily look

ence anxiety can be recognized as a completely arbitrary world of virtual connections and fathomless potentialities of being.[53]

"In the creation of the electron-brain," writes Günther, "the human being cedes his own reflection to the object and in this mirror of himself learns to understand his own function in the world."[54] Seen in this way the camera eye of *2001* breaks human self-reflection through the gaze of objects that have become "actants" and which already on that account seem to have been supplied with a regular eye. (Fig. 43/44) If *2001* places "HAL himself"

53 On the "affection-image" and "any-space-whatevers," see Deleuze 1996, pp. 102ff.
54 Günther 1954, p. 95.

in the picture, then here, entirely without human support, one camera is staring into another. And if *2001* takes over HAL's perspective vis-à-vis his attendants, then here the actors look directly into the camera—and for their part break with all cinematographic conventions of illusionist self-reflection.[55] HAL's wily look reveals most clearly that his previous masters can no longer themselves see themselves as such unclouded. And that machines can die confirms less human triumph than, entirely to the contrary, "the death of the human being."

The Myth of the Cybernetic Takeover

In the 1960s at the latest, a belief of having been torn out of what Foucault termed an "anthropological sleep" became widespread.[56] Since the period around 1800, there had evidently been an effort to close up the abyss opened by Kant's critical question about absolute knowledge and the Kantian aporias of reason through the human "empirical-transcendental" doublet. There had been a hope that the historical "positivity" of life, work, and language could find its grounding in the figure of the human being and in the reflexivity of human thinking. Instead of consigning this last justificatory knowledge to the human sciences, cybernetics, as an overarching discipline, has laid claim to formalization of this knowledge of different positivities and derive them from information, circuit technology, and feedback loops: a perspective no longer emerging from a purity of knowledge but asserting, from the start, a focus on the intertwining of power and cognition, dominance and technology. But in fact, if every claim to universal knowledge inevitably implies self-misapprehension, a "transcendental illusion" of reason, then that or its anthropological counterpart is replaced by a *cybernetic* illusion.[57]

The novel and film versions of *2001* would themselves have succumbed to that illusion if they had limited themselves to a fascination-generating plot centered on technological dominance. In other words, it would have been nothing more than a typical, generic science-fiction product endowing the universal technology of cybernetics with a cultic aura.[58] But Clarke's and Kubrick's stated

---

55 See Seeßlen 1999, p. 48.

56 See Foucault 1970, pp. 371ff.

57 Pias 2003b, p. 16.

58 In conceiving the film, Kubrick himself presumed anthropology's experiential principle that a still alien advanced technology—and in *2001* this includes the monolith, as the relay for an unknown cybernetic communication system—must appear as an ensemble of cultic objects on

goal was to render transparent the illusionary consequences of now-dominant technological knowledge. As Clarke formulated it, with *2001* they observed "the use of hard technology to construct a launch pad for metaphysical speculations."[59] The theme is not medial dominance as such, but the media-induced genesis of transcendental illusions (and corresponding ideas of dominance and rule). For this reason, already in its opening chapter ("Primeval Night"), Clarke's novel describes a phenomenology of consciousness from the hominid's mere sense-certainty to nothing short of illumination of the human spirit: "Fantastic, fleeting geometrical patterns flickered in and out of existence"; the hominids, already illuminated by a monolith, "could never guess that their minds were being probed, their bodies mapped, their reactions studied, their potentials evaluated."[60] While in the book such cybernetic control-operations are in any case described, the film presents them as its own operations. "[R]adiating, shimmering, brightness, reflections, flickering. The trickling of light"[61] was always a phenotechnical precondition for film's production of illusion, for its psychotechnical control of nervous functions, rendering moving pictures into inner pictures, cinematic ideas into ideas tied to consciousness, and mere seeing into trance.

In that the film *shows* HAL's eyes as ubiquitous and all-seeing, hence in that the panoptic camera-eye appears superior to the human eye in terms of both resolution and control,[62] the entire film could be understood as a product of cybernetic medial dominance: the director is in that case HAL himself, for, as Marcia Landy puts it, already his "body is the Discovery itself, and the battle between HAL and the men for dominance turns on HAL's ability to see them even when they think they have escaped his surveillance."[63] But even if, as Deleuze has it, in *2001* the cinematographic brain takes the place of the cinematographic eye and the world itself becomes a brain, it is less the artificial intelligence of HAL than the monolith that dominates cosmic states of consciousness—and those of the viewer. It is the monolith that places hominids and receiving stations in a trance and steers them—whatever the stimuli involved—in the manner of a satellite signal. This black geometric block, a puta-

---

first appearance. See Geduld 1973, p. 41. On the hegemony of the *novum* in science-fiction see also in general Suvin 1979, p. 93.

59 Clarke 1993, p. 155.

60 Clarke 2006, p. 31.

61 Holl 2017, p. 19.

62 Clarke 1962, p. 216: "The eye is an evolutionary miracle but it's a lousy camera." On cybernetic substitution of human retinal receptors, see Steinbuch 1971, p. 222.

63 Landy 2006, p. 98. On the following see Deleuze 1997, pp. 204 ff.

tive "black box" for all cybernetic intelligence—is pure geometry and thus spirit itself. Like a boundary stone, it marks cosmic barriers and evolutionary thresholds, in order to at the same time render them porous as their membrane. As a smoothly polished, sharp-edged rectangle, on what is both the most superficial and basal aesthetic level, it juxtaposes the angled with the round, the outer with the uterine, the anorganic with the living, the aperture with the eye, darkness with light the invisible with the visible.

With these qualities, the monolith is invisible in the blackness of outer space. But through reflection, (of light; of the perceiving subject) it is very recognizable. The monolith finally renders visual the possibility of the film itself, namely fading to black and thus cutting, montage, and timeline manipulation. It marks the outside that is the film itself, the unobservable blindspot from which the first and last scenes but also the first and last acoustic signals enter the picture. In sharp contrast to the match-cut between bone and space ship whose striking presentation of a metaphoric-metonymic association made *2001* famous, the monolith's appearance, Deleuze suggests, no longer wraps or "frames" the previous takes but rather "deframes" them: "The cut may now be extended and appear in its own right, as the black screen, the white screen and their derivatives and combinations."[64] The "off" cuts—now within the picture—its connection to the images; it becomes autonomous, just as it forces seeing into "heautonomy": it has to tear itself away from its natural and empirical practice in order to take account of its own invisible limits. (Fig. 45–48)

Already through the fact that it moves its own limits into the picture's center, *2001* constitutes work on the myth of cybernetically executed medial dominance. The idea that developed machines have rendered human beings replaceable has been a commonplace—not least of all in the context of debates of the time about the reason and folly informing the space program. *2001* initially follows such a scenario, Bowman, for example, simply being locked out of the space ship by wily, wandering HAL. However, when through sheer will to survive and human cunning, in the form of improvised manual usage of an explosive bolt, he succeeds in penetrating back into the space ship and thus into its cybernetic system, and when he then steps into the interior of the artificial intelligence and steadily "lobotomizes" it by disconnecting its modules, this may be considered a literal "re-entry."[65] Cybernetic dis-

64 Ibid., p. 214. On the following see ibid., pp. 206f.

65 On the NASA discussion leading to agreement that human participation was needed for the space vehicle's re-entry process and the contrary position taking in the Soviet space program, see Mindell 2008, pp. 264f.

Fig. 45–48: The round and the angular: mother ship and space capsule, the monolith in Africa, before Jupiter, and as an aperture in Bowman's rebirth

Fig. 46

tinctions such as that between outer and inner, system and environment, space and biosphere, hardware and software, machine and organism, material and spirit, hand and brain, movement and time, subject and object here enter into themselves, in this way dismantling the illusion of inevitable cybernetic dominance. If we wish to confirm skepticism vis-à-vis the idea of completely automatized space travel, we here can focus on the human being as already indispensable because when it comes to work in the environment and on the periphery of space ships the human hand is a matchless fine-motor servo-

Fig. 47

Fig. 48

mechanism—namely a tool—which enjoys the closest feedback loop between understanding and executing commands.[66] But from the perspective of AI research, human intelligence is also matchless in that it does not need to be programmed for emergency situations. As soon as such situations pose a threat, human beings define the emergency, creatively interpreting the extant system of rules or creating a new one corresponding to the unanticipated events. Only

66 See Steinbuch 1971, p. 226.

human being really know how to make decisions about states of emergency. (Fig. 49/50)

Self-evidently the human decision-system is also incomplete. But this incompleteness or “openness to the world” becomes a strength, Massimo de Carolis observes, in that “for human beings under certain circumstances the concrete way of being appears on the scene not only as interference or noise but as the decisive factor for constructing meaning.”[67] It is precisely human self-referentiality that allows a productive relation to the outside world. For this reason human intelligence is distinguished not so much by measurable accomplishment, in the sense of efficiently and quickly executing one or another program,

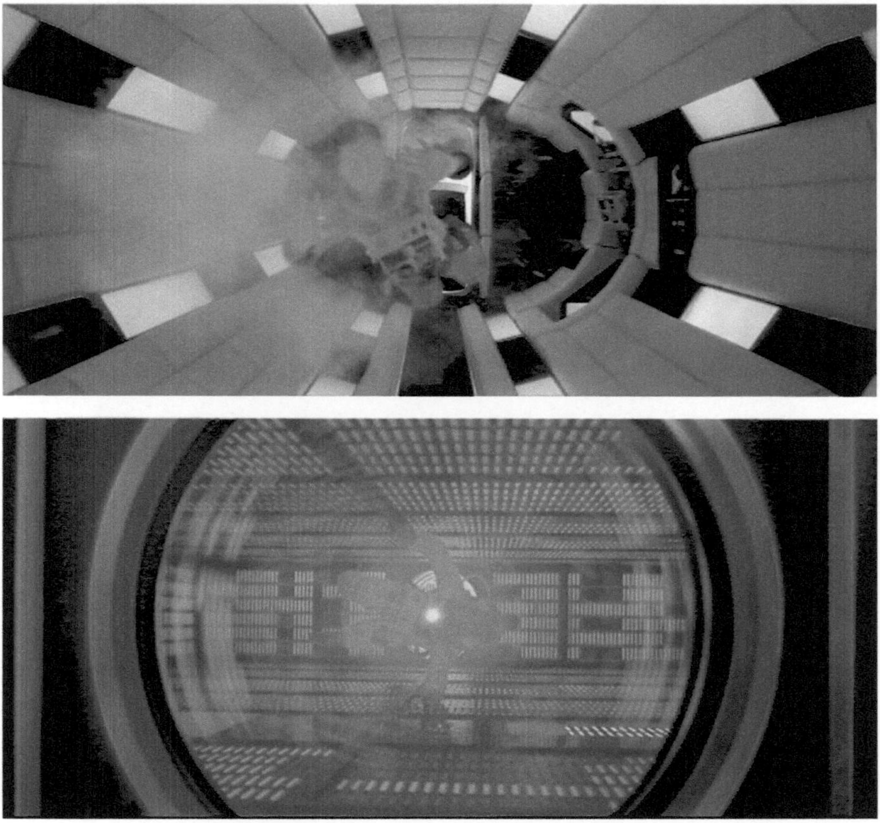

Fig. 49/50: Bowman’s re-entry into the *Discovery* and into the Logic Memory Center

67 See Carolis 2009, p. 117.

as rather by species-specific potentiality. This is the capacity to not only perform recursive logical operations but also to integrate human existence and its life forms into thought, in order to nevertheless understand that existence as simple potential. The film version of *2001* presents such re-entries and recursions more clearly than does Clarke's novel, which does not include the episode of Bowman's violent re-entry into the *Discovery*. Namely, with that control technology, the cybernetic recursions already forming a focal point in the *Odyssey* once again enter into themselves. Even more than in the novel, the film thus shows the extent to which it is dominated by media—and what medial dominance it itself can exercise.

Neither as a legal nor as a heuristic term within media theory can dominance be reduced to a simple structure of submission and rule. With dominance in its precise modern sense always being tied to the exploration of new realms of "possibility" and "world-spaces," approaching it in a media-historical framework appears compelling. Likewise, with the *2001* of Kubrick and Clarke first and foremost treating media evolution and connected possibilities for also dominating other spaces and forms of life, it is very clear that the film has recourse to the *Odyssey*—to the recursive program of world discovery always serving as a guideline for narratives centered on the evolution of power, the media, and the human species. Focused as it is on the history of knowledge, the perspective of cybernetics expands the *Odyssey*'s narrative by taking account of the time before and after human beings. But the *Odyssey*'s generic framework is simultaneously expanded. Plato's and Aristotle's classification of the types of poetry according to *logos* and *lexis* marked the Homeric epic's canonization as a genre; around 1800 it would be designated as a poetic "natural form" corresponding to a specific human type with putatively supratemporal structure and disposition: a construction of poetological-biological invariance evidently only made possible by recourse to a concept—both anthropological and centered on everyday praxis—of mastering life meant to find its expression in certain poetic forms.[68] In—not accidentally—the "mixed form" of a novel and, simultaneously, the post-alphabetic medium of film, Clarke and Kubrick announce the depth of the transformation undergone by the poetological point of departure (a poetology of genre and species at once) under "cybernetic conditions."

For cybernetic anthropology, the moment in which a hominid learns to use some object as a tool and weapon marks a threshold of hominization—of entry as a human being into a structure of organic-mechanical juncture. Within that

68 See Genette 1992, pp. 38ff., 60, 67f.; Friedrich 2009, pp. 13–20.

anthropology, in the course of a shared human and media evolution previously soulless tools emancipated themselves into "actants" that eventually developed consciousness in the form of artificial intelligence and will now take over rule. *2001* takes over this cybernetic myth of undiminished media dominance and dismantles it in reflecting on its own mediality. In this respect, it is striking that between the novel entitled *2001: A Space Odyssey* and the like-named film, we by no means find the sort of relation of dominance usually ascribed to technical media in theories of media evolution. The film is not an adaptation, nor is the book a novelization. Both emerged in continuous "feedback in both directions": from solutions, developed through intertwining feedback, for the narrative and representational problems the different media raised in the course of conception and production. The film and book understand themselves as *Odyssey*-recursions in their own right, for which reason Clarke—extending to his novel *3001*—would write many "recursions in metastories."[69] If we wish to follow contemporary plans for a cybernetic aesthetics, we can also understand literary texts as incomplete control programs making use of their incompletion and weak ordering authority as a field with broad interpretive leeway.[70]

Seen cybernetically, the film drives other perceptive registers forward, has a more immersive impact as a non-verbal and physiologically interconnected medium. For that reason, in order to move past merely empirical perception of its image-worlds, it has to steer controlled perception to its own limits and conditions. In the process it becomes a conceptual image that illuminates the *poiēsis* or world-production processes of cybernetics itself. The danger of medial dominance or "enframing," as Heidegger still called it in 1969, consists in cybernetics becoming "determined and guided by the new fundamental science, thus being elevated into a new metaphysics that has "forgotten Being." An expression of this danger is not least of all the possibility of the arts becoming "regulated-regulating instruments of information."[71] But as an *Odyssey*-recursion, *2001* ventures on what Heidegger, following Hölderlin, refers to as "saving" recourse to the poetic, to the *poiēsis* and "revealing" rendering the essence of technology distinguishable from its "cybernetic illusions." Perhaps

69 Clarke 2006, p. 13. On Clarke's and Kubrick's working process see Geduld 1973, p. 73. We can understand Clarke's *3001: The Final Odyssey* as a recursion in that here the evolutionary story of *2001* is again narrated, but from a higher observational level. On this recursive narrative principle, see also his story "A Recursion in Metastories" (1966; the variant title of his "Longest Science-Fiction Story Ever Told), in Clark 2000, p. 854.

70 See Lem 1983, pp. 159, 191 f.

71 Heidegger 2002, p. 58.

this distance between science and fiction, their notorious lack of difference since 1969 notwithstanding, defines what separates *2001* from mere science-fiction. The takeoff of cybernetics is in any case only a takeover in myth. Art can, however, maintain the distance between the two.

Conclusion
Memory of the Sea: The Archeology of Maritime Culture

Remembrance
The northeast wind is blowing,
My favorite among the winds
As it promises fiery spirit
And a good voyage to sailors.
But go now and greet
The beautiful Garonne,
And the gardens of Bordeaux
There, where the footbridge moves
Along the steep bank and the stream
Falls deep into the river, but above
Looking on, a noble pair
Of oaks and silver-leafed poplars;

This all still comes well into mind and how
The elm forest bends
Its broad treetops over the mill,
But in the courtyard a fig-tree is growing.
On feast-days, the brown women walk there
Upon the silky ground, in March-time,
When night and day are equal,
And over slow-moving paths
Heavy with golden dreams,
Soft breezes are drifting.

But someone hand me,
One of the fragrant goblets,
Full of dark light,
So that I can rest, for slumber
Would be sweet under the shadows.
It isn't good
To be soulless from
Mortal thoughts. But conversation
Is good, and expressing
The heart's opinion, to hear much
Of days of love
And deeds that took place.

Andenken
Der Nordost wehet,
Der liebste unter den Winden
Mir, weil er feurigen Geist
Und gute Fahrt verheißet den Schiffern.
Geh aber nun und grüße
Die schöne Garonne,
Und die Gärten von Bordeaux
Dort, wo am scharfen Ufer
Hingehet der Steg und in den Strom
Tief fällt der Bach, darüber aber
Hinschauet ein edel Paar
Von Eichen und Silberpappeln;

Noch denket das mir wohl und wie
Die breiten Gipfel neiget
Der Ulmwald, über die Mühl',
Im Hofe aber wächset ein Feigenbaum.
An Feiertagen gehn
Die braunen Frauen daselbst
Auf seidnen Boden,
Zur Märzenzeit,
Wenn gleich ist Nacht und Tag,
Und über langsamen Stegen,
Von goldenen Träumen schwer,
Einwiegende Lüfte ziehen.

Es reiche aber,
Des dunkeln Lichtes voll,
Mir einer den duftenden Becher,
Damit ich ruhen möge; denn süß
Wär' unter Schatten der Schlummer.
Nicht ist es gut,
Seellos von sterblichen
Gedanken zu seyn. Doch gut
Ist ein Gespräch und zu sagen
Des Herzens Meinung, zu hören viel
Von Tagen der Lieb',
Und Thaten, welche geschehen.

https://doi.org/10.1515/9783110610734-014

But where are my friends? Bellarmin
With his companion? Some people
Dread approaching the source;
Namely, wealth begins
In the sea. Like painters,
They gather together
The earth's beauty and do not scorn
Winged war and
Living alone for years, beneath
The leafless mast, where the night is not brightened
By the city's feast-days,
Not by the lyre and native dance.

But now the men have left
For the Indies,
From there on the breezy peak
Along vine-clad mountains, where
The Dordogne moves downward,
Together with the majestic Garonne,
Both flowing out
As wide as the sea. But memory
Is taken and given by the sea,
And love, too, attentively captures one's eyes,
But what remains is founded by poets.

Wo aber sind die Freunde? Bellarmin
Mit dem Gefährten? Mancher
Trägt Scheue, an die Quelle zu gehn;
Es beginnet nemlich der Reichtum
Im Meere. Sie,
Wie Mahler, bringen zusammen
Das Schöne der Erd' und verschmähn
Den geflügelten Krieg nicht, und
Zu wohnen einsam, jahrlang, unter
Dem entlaubten Mast,
wo nicht die Nacht durchglänzen
Die Feiertage der Stadt,
Und Saitenspiel und eingeborener Tanz nicht.

Nun aber sind zu Indiern
Die Männer gegangen,
Dort an der luftigen Spiz'
An Traubenbergen, wo herab
Die Dordogne kommt,
Und zusammen mit der prächt'gen
Garonne meerbreit
Ausgehet der Strom. Es nehmet aber
Und gibt Gedächtniß die See,
Und die Lieb' auch heftet fleißig die Augen,
Was bleibet aber, stiften die Dichter.

Friedrich Hölderlin[1]

Hölderlin's Geopoetics and Archeology

The sea is chaos. Although ancient cosmologists see the immemorial, because itself uncreated, origin of creation within its waters, what it steadily evokes is what is wasted and still undifferentiated—the non-grounding of all difference. For this reason, the sea is like an abyss of which everything earthly must beware. Mythic figures such as the sea-monster Tiamat, abysmal mother in the Babylonian creation story known as the *Enûma Elîsh*, reveal as much from the start. The sea, as chaos, is a maw also swallowing memory of what has been swallowed. In its range of variants, whether the story of Utnapishtim, of Noah, or of Deukalion, the flood saga describes one and the same thing: a mnemonic cat-

1 Hölderlin 2004, vol. XI, pp. 121–123.

astrophe.[2] As still can be read in Thomas Burnet's *Telluris Theoria Sacra* of 1681, with the engulfing of places previously located on terra firma, knowledge of "the first Order of Things" has also been lost. Present-day earth only harbors the original order of creation in a debased mode, that of "Disorder and Disproportion."[3] Cognizance of that original order thus cannot be derived from simply reading into the "Book of Nature." Rather, documents are needed that can be defined as holy because they have outlasted the great forgetting.

If "Sacred Writings," as "best Monuments of Antiquity," document how Noah and his loved ones evaded the "general Destruction and Devastation" through providence "in a certain Ark, or Vessel made like a Ship,"[4] then for Burnet three things follow from this. First, the Book of Nature, as we now have it for deciphering, only represents a fragment of the original plan of creation; what is in any case readable in it is the variety of life forms providentially rescued from the great maritime catastrophe through the ark. Second, as a consequence, moving back and beyond mere earthly history, a primal history of creation needs to be written, for which reason Burnet's enterprise of discovery does not culminate in one or another historical account but in his *Archaeologiae Philosophicae* (1692). And third, if the earth is in fact "crooked and broke," "monstrous, or deform'd," as well as being a "Ruin," then this is not due so much to the destructive force of time as to the sea—to this "Abyss" of all salvational history. Hence Burnet's "archeo-geological" research on ruins is finally tied to a specific millenarian hope: "the great force of the Sea will be broken." For the integrity of creation will only be restored with "Consumption of the Ocean."[5]

Into the Early Modern period, the same hostility to the sea is carried forward that is not only reflected in the New Testament's Book of Revelation (21, 1). To the extent Western culture grounds itself—whether in a sacral or legal way—in "unity of order and orientation" (Schmitt), the sea must seem a manifest threat to its constituent existence, its tradition and continuation. If it promises quick profits to maritime merchants, it also holds the danger of irreplaceable loss. To venture on what Hesiod terms "plunging into the misty sea" reveals a double lack of culture. On the one hand, it reveals a contempt for agriculture and a settled form of life, hence for the experience, transmitted from the Golden Age, of those who do not "ply on ships, but the grain-giving ploughland bears the fruit."[6] On the other

2 See Assmann 2006, pp. 291–301. On the catastrophe-linked aspects of the sea in ancient cosmogenies, see Böhme 1988, pp. 26–29; Böhme 1996, pp. 33, 38f., 57; Brunotte 2002, pp. 98–101.

3 Burnet 1719, vol.1, p. 177.

4 Ibid, pp. 5, 12.

5 Ibid, pp. 179, 183, 21; vol. 2, pp. 112, 109.

6 Hesiod, *Work and Days*, v. 620, 236f., cited from Hesiod 1988, pp. 55, 44.

hand it casts scorn on *cultura animi*, referring since Cicero to cultivation of intellectual tradition and since the Church Fathers to care for the soul's salvation. As what threatens to liquidate order, tradition, and the future, for centuries the sea was nothing less than abominated. This was not only the case for jurists, philosophers, and theologians. In the field of commonplaces maintained by Western rhetoric and poetry, the paradoxical topos of placelessness—and thus of forgetting—was long reserved for the sea.

This first changed, but then in a fundamental way, on the nineteenth century's threshold: "But memory / Is taken and given by the sea," we read in Friedrich Hölderlin's hymn "Remembrance" ("Andenken"; probably written in 1803). What the poem treats is no mere melancholic "remembering" of something irrevocable as lost at sea. Rather, in line with the structure of the titular German word *Andenken*, it is thought moving "onto" something against forgetting, by allowing something putatively original to surface from its lostness. Such *Andenken* is not limited to memory and its topoi or a specific moment of recollection. The poem, Dieter Henrich argues, "does not adhere to something of which it is aware. It *opens up*, in that it recollects," hence describes a path instead of simply naming places, indeed unfolds a movement of spirit and language that "leads to thought about itself."[7] Hölderlin also allows his personal experiences on the borderline between land and sea flow into his "Andenken," when Bordeaux, its distinctive atmosphere (meteorological phenomena such as the ocean breezes) "comes well into mind" (*Noch denket das mir wohl*). But before this process of mnemonic internalizing (a process succinctly captured in the German word for recollection, *Er-innerung*) can be rendered concrete as static scenery, biographical topos, or simple past events, a drink from Lethe is requested because "It isn't good / To be soulless from / Mortal thoughts." A lifeless "coming well into mind," "Wohldenken," must become living remembrance, "Andenken."

On the level where the different stages of Hölderlin's personal career appear as a kind of creative memorial, so that a new passage between life and work is repetitively opened, this would mean something like the following: Also, and in particular, the putatively idyllic episodes of the poet's biography need to be brought unto what in his novel *Hyperion* Hölderlin terms an "eccentric path," in order to finally recover them from their placelessness.[8] Instead of simply

7 Henrich 1990, p. 380. As in the poem's phrase *langsamen Stegen*, "slow-moving paths," the movement is here "methodically" transmitted along the path itself.

8 Hölderlin 2004, vol. 4, p. 48. On the "eccentric path" in *Hyperion*, along which Hölderlin's thinking and writing would explore "the open" and the "abundance" of language, on the poetic

being repeated with their intellectual transcendence or sublation in memory, they must be as it were first of all *re-captured* in the medium of poetic language. With Hölderlin here expanding his poetic topography of German rivers and river cities to include the sea and a harbor city,[9] the hymn offers a particular argument on the level of a henceforth global aquatic topography: a "source" only becomes the origin of a cultural landscape when it streams into the open sea and retrospectively defines its starting point in terms of this indeterminate and indefinable medium. Only this self-loss and return to oneself assures enduring impact: "But what remains is founded by poets"—*Was bleibet aber, stiften die Dichter*—is the way Hölderlin puts it at the hymn's end. What survives needs a poetics of rescue.

The poem's course from loss to self-return and onward to symbolic founding of what is open and future in writing has been compared with the *Odyssey*. Returning from his trip to France as someone who has "failed," on his way back to German-speaking lands, Hölderlin recalls his actual homeland like Odysseus his beloved Ithaca while with the Phaiakians.[10] Although the perspective of "Andenken" is more that of the homeland than being outside it, which is to say more Ithaca than Scherie, it is marked by a circular structure that determines discourse as *discurrere* at sea: a structure in which practical knowledge *in itself* is meant to be elevated and transcended into practical knowledge *for itself* through self-renunciation and self-return, extensivity thus turning into intensity.[11] What appears in the *Odyssey* as a recursive identity between the hero Odysseus and the epic's narrator ("Homer") is in any event for Hölderlin initially a separation between seamen on the one side, moving on the experiential medium of the ship and driven from the "northeast wind" in the maritime element, and on the other side the poet, who in language's experiential medium and driven by poetic pneuma, is bound to the source. This seeming opposition does not mean that the "companions" on board, as prototypical men of action on an eccentric path, only bring together knowledge of scattered phenomena in a merely impressionistic way, "like painters," while the lonely poet always draws from the source. In Hölderlin's later poetry *Stiften*, in general meaning to "found," "establish," or "cause," signifies, Jochen Schmidt observes, "neither fixing what is already present in reality not factual new creation of 'being,' but rather representation by

---

of self-consciousness as a poetics of detours, and on the recursive intertwining of novelistic poetics and the lyric mode of writing, see Nägele 1998, pp. 17–21, 33 f.

9 See Lefebvre 1988/89, p. 205.

10 See Witte 1993, p. 187.

11 On this and on Hölderlin's conception of memory in general, see Hamlin 1984/85, pp. 121–125, 127 f., 135.

virtue of the highest, all-encompassing form of consciousness."[12] Such representation condenses what has been scattered into the poetic word or, put otherwise, into the lyric "chronotope."

Namely, the process of *Andenken* concerns the source's literally "poetic" future, its retrospective generation or "revealing." And this takes place in passage through what transfers the seemingly namable origin into placelessness: the sea. With the return to "origins" initially taken up as a voyage "to the Indies" (in fact, around 1800 Bordeaux had direct trading relations with "East India"), the east to west path of classical *translatio culturae* is inverted, at the same time following the course of the culture-mediating river par excellence: the Danube.[13] Since furthermore, starting with Columbus's expedition, voyaging "to the Indies" simultaneously signified misjudgment and discovery (Columbus, we will recall, believed he has discovered India), the search for an origin cannot aim at a fixed goal but, on a geographical level can only culminate in an indissoluble *discurrere*. Already Odyssean-Columbean in its conception, "Andenken" closes the period of the great "foundational" poems. The poet Hölderlin will now, produce "confused stuff," and fragmentary "remembrance" of his perhaps last genuine hymn, of his perhaps last systematic remembrance of poetry itself.[14] Foundational poetry itself is now, at a distance from its putative source, torn away onto an eccentric path where it threatens to sink into madness (a state always associated with water and the sea); for this reason there have been frequent references to Hölderlin's late "poetry of madness" and to the final absence of his "genuine" work.[15] Just this absence applies to the particular sort of remembrance that is demanded for Hölderlin's muse of memory, "Mnemosyne": "And much / Like on one's shoulders a / Load of logs and failure [the German word *Scheitern* denotes, and seems meant to play on, both] is to be held" (*Wie auf den Schultern eine / Last von Scheitern ist / zu behalten*)[16] The acceptance of failure's wooden burden is meant neither to exhaust itself in frozen grief nor to be the same as hastily believing that drowning has been conquered as ascendant passage. Always tied to

---

12 Schmidt 1970, p. 40. See also ibid, pp. 27–29, 78 f.; Martel 2004, pp. 391–393.

13 See Kocziszky 2009, pp. 47, 53.

14 Lefebvre 1988, p. 211. See in this respect ibid, p. 22; Schmidt 1970, p. 31.

15 On the connection between water and madness, see Foucault 1994, pp. 268 ff. On the relation between Hölderlin's work and madness, see ibid, pp. 201 ff.

16 Hölderlin 2004, vol. 11, p. 170. A second version contains these famous lines: "We are a sign, meaningless / We are without pain and have nearly / Lost language in a foreign land" (transl. JG) —(*Ein Zeichen sind wir, deutungslos / Schmerzlos sind wir und haben fast / Die Sprache in der Fremde verloren*; ibid, p. 105)—lines defining poetry as *poiēsis* and the new creation of a language almost lost to conceptual thinking. On the following see Harrison 1984/85, pp. 195–198, 206.

loss, experience is a poetic task. As a field of experience and search for an origin, the sea thus furnishes memory, in that it intermittently takes it away. Precisely "mariners," Heidegger observes in one of his essays on Hölderlin, "are on a voyage to the origin of their own being." "It is at the source where the richness that began in the sea is fulfilled. The source *is* the wealth, but only when it is experienced *as* the source."[17] Experience includes confusion, loss, failure, as the hymnic "Andenken" itself described.

"Some people / Dread approaching the source; Namely, wealth begins / In the sea" *(Mancher / Trägt Scheue, an die Quelle zu gehn; / Es beginnet nemlich der Reichtum / Im Meere)* we read in "Andenken". We may here recognize an apprehension felt at all Idealist inwardness that since 1795 (and his early grappling with Fichte) led Hölderlin from *Seyn*, its "intellectual intuition," a corresponding *Ur-Theil* ("judgment" equivalent to "originary division"), and onward to a processual idea of cognition and continuous reflection on the way and method of the "poetic spirit." The apprehension might likewise be read as a metapoetic pointer to that way of proceeding: not only looking for poetry's origin in its sources and inspirational force but in the detour of practical knowledge. Seen in this way, the poem designates two directly opposing paths. On the one hand, there is a desire to move, familiar with the locality, along poetically fruitful terrain and, for the sake of drilling into it, to purposively seek its sources—that is, the documents with which research on literary influence furnishes itself with a backing in the reality of domestic or external traditions. On the other hand, it is possible to move out from this terrain into the seemingly boundless sea of culture and its forgotten, sunk, or simply scattered facts: an enterprise that, if it manages to salvage any found pieces, has to first make certain its documentary value. Methodologically stated: for one thing, adherence to the source is often premised on, to cite Blumenberg, a certain "disjunction between origin and decline" or else there is simply mistrust of "other models, especially analytic and constructive, sedimentative and occlusive ones"; for another thing, "sources" are equally often scorned with the conviction that words and things are in any case always already apart, the origin already lost, especially since "textual debris" in the stream of tradition is not borne "to the river's source but in the opposite direction."[18]

It is the case that for German Idealism, what we may understand in "Andenken" as a concerted action of sailors and poets, capable of being intensified as a methodological conflict between philological and cultural studies, still repre-

17 Heidegger 2000, pp. 158, 167.

18 Blumenberg 2012, pp. 9, 81, 172. On the emergence of "ideas of purity of the origin and contamination in the stream of transmission" from a context of Roman society and law , see Vismann 2012, pp. 94 f.

sented a unity within language itself.[19] This language was, after all, the medium of all experience, within which cultural facts both become sediment and can be poetically and conceptually transcended. For Hegel, "works of language" are capable of elevating a nation's spirit, by contrast with "archives" and the "hieroglyphics" of merely "dumb" cultural evidence.[20] At the same time, for poets and thinkers like Hölderlin and Hegel the sea is commensurable with language in that it is no longer understood as an abyss of all cultures but, as cited earlier, their "supreme medium of communication." Precisely because the sea furnishes "the idea of the indefinite, the unlimited, and infinite,"[21] it necessitates a basic philosophical-historical distinction between static land and dynamic maritime culture. "The land, the mere Valley-plain attaches [man] to the soil; it involves him in an infinite multitude of dependencies, but the sea carries him out beyond these limited circles of thought and action."[22]

It is not only that a nationally limited "culture" most easily "intervenes in the universal world continuum," consequently elevating itself to "universal dominion," in that it, like "all great and enterprising nations," pushes its "way to the sea."[23] In Hölderlin's geopoetical perspective, and more explicitly in Hegel's geophilosophical counterpart, the earth's "form," its *Gestalt*, while inherently itself lacking spirit, *Geist*, nevertheless determined spirit's course and thus that of cultures. For Hegel, the New World simply represents an uncultivated rupture with the Old, which is itself divided into three parts: Africa, dulling its inhabitants through sheer heat, Asia, exhausting itself through Bacchantian excess, and Europe, forming "the consciousness, the rational part, of the earth, the balance of rivers and valleys and mountains—whose center is Germany."[24] It is only in this European center that the concepts of source and culture come to themselves. For the "genuine springs" from which rivers like the Danube, Rhone, and Rhine emerge have "an interior life and direction" so that they serve as origins not merely in a "mechanical, superficial" sense but also in that of culture. But they only become such sources in connection with what Hegel considers the "supreme medium," the sea, because culture always has

---

19 In F. A. Wolfs talk of philological "remainders" that "appear as debris from a massive shipwreck"—phrasing that may have been inspired by Hölderlin—the two "methodologies" are still unseparated. See Wolf 1807, p. 26; Kocziszky 2009, p. 22.

20 Hegel 1895, pp. 120 f.

21 Hegel 1914, p. 90.

22 Ibid., p. 99.

23 Hegel 1999, pp. 63 f.; Hegel 1991, p. 269.

24 Hegel 1970, p. 285. On the "poetic" and "philosophical" "geohistory" of Hölderlin and Hegel, see Tang 2008, pp. 225–239.

to be transmitted via the "element of fluidity," of "danger, and destruction."[25] The aquatic topography of culture is not only inscribed in the way it arises but also in the way it goes under. And if within a geopolitical and geophilosophical perspective lifetime and world-time overlap, then, Hegel argues, geology always brings to light knowledge of culture, its possibility and past.

Hegel here follows the "geognosis" of Abraham Gottlob Werner, who enlisted history of the earth for philosophy of spirit through his monistic principle of "crystallization." In addition, in his efforts to render readable the inherently self-contained earth, its stones and minerals, Werner paved the way for the intertwining of natural history, ancient studies, and linguistic history generally characterizing writing about the history of the earth around 1800. Fossils were meant to be deciphered like archeological evidence, and Georges-Louis Leclerc de Buffon even spoke of "archives of the world" to be consulted for natural-historical ends.[26] Although Hegel, skeptical about everything merely archival, warned that it was "an indifferent curiosity which wants to see also in the form of succession what exists as a juxtaposition,"[27] the layer became an ordering principle of modern geology. Corresponding to Charles Lyell's evolutionistic conception of terrestrial history, stratigraphy saw layers of stone as evidence of geological epochs, their immeasurably long "deep time" and their catastrophic decline—evidence that was to be read "like the leaves of a book."[28]

With geology, philology, and ancient studies thus being worked through on a purely conceptual basis on land, at sea this "depth research" was immeasurably more difficult. Works such as Luigi Ferdinando Marsigli's *Histoire Physique de la Mer* had indeed already appeared in the eighteenth century (in the case of this work, 1725); scattered examinations of sediment had taken place, and here and there a profile of the seafloor had been essayed as well. But maritime research, even if increasingly bringing in geophysical, chemical, and biological knowledge, remained literally superficial well into the following century. The sea appeared as a navigable but always dangerous watery surface whose mysterious because unreachable depths could only hold certain death for seamen. Already because there was no practical need for more than such superficial knowledge, the sea for the time remained, as Jules Michelet wrote in 1860, "dark and inscrutable in its immense depths."[29] In its lower depths, the sea was considered no less

25 Along with the earlier-cited passage from Hegel 1991, see Hegel 1970, pp. 296 f.
26 Buffon 1780, p. 1. On Buffon's role in the methodological transfer of antiquarian natural history to human history see Schnapp 2011, pp. 293–295, 335.
27 Hegel 1970, p. 283. On Werner see Fritscher 2009, pp. 201–219.
28 Lyell 1830–35, vol. 3., p. 81. On the concept of the "stratum," see Mignon 1993, pp. 313–315.
29 Michelet 1861, p. 11

than a dense, motionless, and "azoic" element until the discovery of organisms from that level in 1868, through the coincidence of a telegraph cable brought to the surface. Since then, this "supreme medium" was not only explored through metaphors and transport and communication technology, but also in terms of evolutionary theory, particularly since Darwin speculated that retarded development forms and thus a kind of life-science archive of survivals could be found there.[30]

Since its exploration began in this way, that the deep sea was not a domain reserved for natural history alone became particularly clear to cultural theorists. Because their progenitor Hegel had referred to ships as tools of maritime culture and thus actually of culture in general, the seafloor had to harbor not only earthly ruins but also those of cultural history. If, as Hegel observed, it "was in a *restitutio ad integrum*"—the legal concept of compensatory restoration to an original state—"that Noah built the distracted world together again,"[31] then the sea now, for catastrophe research after the fact, served, inversely, as a kind of archive of submerged culture. Already Lyell considered it "probable that a greater number of monuments of the skill and industry of man will, in the course of ages, be collected together in the bed of the ocean, than will be seen at one time on the surface of the continents."[32] The sea is no longer a dark abyss of forgetting, no longer placeless; rather, now subject to topography, it furnishes "cultural memory," housing countless nautical and other finds, hence key cultural evidence. Various poets will nevertheless continue to draw melancholy-filled thought from the sea-depths. But in place of cultural-philosophical conceptual logic, cultural techniques of modern depth-research have arrived on the scene. In harboring and archiving sunken sources, archeology at sea offers a specific form of cultural *Andenken.*

The Shipwreck Chronotope

Wrecked ships are the most prominent cultural evidence on the seafloor. Otherwise than the case with sunken harbors close to the coast and drowned sites of habitation, wrecks actually lie on that seafloor; by contrast with other underwater archeological findings, they merit primacy because from time immemorial, maritime vessels have been the largest and most complex of all mobile arti-

---

30 On the connection between the sounding voyages in the cable-laying project of 1868 and future deep-sea research, above all the *Challenger* expedition, see Reinke-Kunze 1994, pp. 14–16.
31 Hegel 1975b, pp. 182–204, here p. 183 (translation modified).
32 Lyell 1830–35, vol. 2, p. 258.

facts.[33] The most ordinary seagoing vessels furnish insight into previous everyday life, since along with art objects the things found in such ships more frequently include technical instruments, personal effects, and waste. In addition, the ships themselves, their hulls and nautical equipment, offer information on the technological and economic circumstances of their time. Reflected in their construction, alongside volumes of investment by the maritime cultures involved, are both the persistence of tradition and the innovative potential of the domestic technologies. Already their construction but also their management at sea required specialization, hierarchization, and cooperation, so that they document the organizational and performance capacity of a range of societies and cultures and their military and economic subsystems.

Ships represent actor-networks in a downright exemplary way. For that reason, a certain concept of culture is always connected with their salvaging and archeological examination. The major ship-salvaging projects of the 1960s, for example the salvaging of the Gelidonya wreck, the Bremer cog, and the Swedish *Vasa*, were all accompanied by the claim of cultural or national exemplarity.[34] For that reason, initially these wrecks seemed to render salvaging "inferior" wrecks superfluous. But that ships were tools of globalization and that for that very reasons their wrecks needed to be examined in their dispersal, as "inferior" objects, was something that archeologists only took account of when they began focusing on the long, first and foremost statistically revealed endurance of cultures, in the context of history of mentalities and "serial history." And one thing in particular catalyzed control-work in pertinent archival holdings, to be held up against hasty conclusions regarding maritime findings: the fact that within the warp and woof of nautical "miniature states," the political realities of their original societies were not necessarily reflected but often only their desired organizational projects.[35]

Where once ships were tools of culture because they opened up an unlimited or at least global space for it, their wrecks now served as forms of cultural salvage, in that historically limited cultural time is nothing less than preserved here under, so to speak, lock and key. Such a particular form of temporality may—to introduce an incisive classification by English author and nautical expert James Hamilton-Paterson—characterize not only archeologically relevant findings but also those wrecks that as "time bombs" concern a deadline to avoid environmental damage, as "gold mines" the value-adding time taken for trade in antiquities,

33 See Ellmers 1997, pp. 23f. On the distinction between marine, underwater, and nautical archeology see Flatman 2006, p. 168.
34 See Throckmorton 1996, pp. 136ff.; Hocker 2000, pp. 1–6; Konstam 1999, pp. 8, 98f.
35 See Adams 2001, pp. 302, 305.

and as "graves" a past ritually revived in mournful remembrance.[36] Once sunk fifteen meters beneath the sea level and thus no longer dangerous flotsam,[37] ships are in any case removed from the "wreck charts" and thus, in an administrative-technical sense, from the present. Suddenly they appear removed from space and time. Or as Hamilton-Paterson indicates: inundated even more by time than by water, everything sunk in the sea first of all becomes radically old. But in the end, whatever the sea "hides beneath its dark leagues of surrogate tears it makes timeless."[38] And as the underwater archeologist Jean-Yves Blot writes, ships seem especially removed from time, because "the sea and the water are nothing other than the place of a passage, and because the storm or shipwreck represent the hatchets that simply shattered the rungs of the temporal ladder."[39]

Underwater archeologists refer to discovered shipwrecks they research as "time capsules." The value of these sources as opposed to other wrecks may lie in their documenting of a remote "temporal space," indeed their embodying what we might term "cultural time." With Mikhail Bachtin, we could thus also call them "chronotopoi": "Time, as it were, thickens, takes on flesh, becomes artistically visible," Bakhtin observes; "likewise, space becomes charged and responsive to the movements of time, plot and history."[40] In the case of time capsules, we need to speak of a unique sort of chronotope that does not self-evidently fit into an older time-space metaphorics. Namely, previously, as Vilém Flusser describes it, time was seen "as a streaming in space regulating the things found there (the mythic way of seeing). Or space was seen as a raft driven toward the future along the stream of time, with all things being carried along (the historical way of seeing)."[41] But this conceptualizing of time-space still presumes the firm ground of classical geometry, while chronotopic configurations such as modern archeology's, with their topology of masking, overfolding, and wrinkling directly imply the fourth dimension. In this sense, wrecks are "deep structures"[42]—discovery sites that translate "cultural time" into synchronicity through the arrangement of visible artifacts. As such structures, they furnish, Jonathan Adams observes, "a high-resolution image of past activity,

36 Hamilton-Paterson 1992, pp. 125 ff.

37 See Homburg 1962, p. 237.

38 Hamilton-Paterson 1992, p. 146.

39 Blot 1988, p. 7.

40 Bakhtin 1981, pp. 84–259, here p. 84. On coining of the term "chronotope" by Alexei A. Ukhtomsky and Bakhtins relation to Kant, Cassirer, and Einstein, see Frank 2008, pp. 211–213.

41 Flusser 2006, p. 275. On the following see ibid, pp. 283 f.; Serres 2008, p. 93.

42 See Gould 1983, p. 6.

not only in individual well-preserved objects but in their relationships and other 'invisible' attributes."[43] What distinguishes underwater archeological sites from most of those on land (although similar to suddenly perished cities such as Herculaneum and Pompeii) is the simultaneity with which their artifacts were deposited (for their part the artifacts can be of completely different ages).[44] They offer singular momentary snapshots of destruction.

For the "new archeology," a shipwreck is, to cite Keith Muckelroy, "the event by which a highly organized and dynamic assemblage of artifacts is transformed into a static and disorganized state with long-term stability."[45] That archeology is engaged in an effort to take an image of fragmentation and dispersal that the seafloor offers after the event and move back from that to the previous constitution of the human-thing ensemble that, until the moment of its demise, a ship still was. But such an approach is confronted with a double difficulty. For not only the eruptive violence of the sinking itself but also, and especially, the erosive force of time and sea makes it difficult for underwater archeology to render legible its time capsules or—to use the professional jargon—"closed finds." In order to write maritime cultural history, technological expertise is thus as necessary as is oceanographic and with it truly "transdisciplinary" knowledge.[46] The shipwreck's course, the collapse of supportive elements, the type and duration of the rescue measures undertaken before the sinking, the impact on the seafloor, the longer-term influence of the underwater flora, fauna, and currents, the water's temperature and composition, finally later encroachments through plundering and salvaging efforts: all these events and factors are for their part potentially legible, but can largely destroy the site and its "inferential status," the shipwreck and its "onboard stratigraphy" or else reduce the research results to mere probability models.[47]

Wrecked ships are only readable as chronotopoi in the context of their individual sites. These consequently themselves have the status of archeological evidence, containing—in their possibly original formation—scattered individual items and parts of the wreck (the so-called wreck trace). Sometimes they harbor

43 Adams 2001, p. 296.

44 On the more frequent case in underwater archeology of a site being marked by "superposition" rather than "stratigraphy," see Gould 2000, p. 53.

45 Muckelroy 1998, p. 267.

46 On this "transdisciplinarity" see McGrail 1996, p. 92; Gould 2000, p. 65.

47 Adams 2001, pp. 296 f. On probabilistics in underwater archeology see Gould 2000, p. 36. As an exemplary piece of cultural history and a nautically / oceanographically informed study (of the *SS Central America*), see Herdendorf 1995, esp. pp. 46 ff., 62 ff., 159 ff.

a natural milieu threatening to undermine the cultural *lieu de mémoire*[48]: a milieu of erosion and corrosion, of woodworms and microorganisms. But this can become consolidated into a stable microclimate, and then the deep sea has not only a destructive but also a conserving function. Because the sites are themselves "deep structures," the new underwater archeology either entirely or partially desists from raising wrecks.[49] In its place there is so-called in-situ archivization, the development of an ideally exhaustive survey leaving the site materially intact by immediately translating it into data, thus placing it on file.[50]

This is the backdrop to a new appropriation of the old concept of a geological plate-tectonic archive as a "seafloor archive"—a development subject to critique for involving a kind of catachresis, since archives, after all, involve a deliberate selection and systematic arrangement of archival material within a systematically set up infrastructure, requirements very much contrasting to those involved in work at archeological sites. At the same time the metaphorics reveals the extent to which especially underwater archeology is indeed oriented toward the archival concept. For when, for example, the "new archeology" refers to underwater remains as revealing "fossilized properties," it is following both evolutionist categories and archeological principles of provenance.[51] Likewise, we can speak of a double "outside" to the "seafloor archive," that of its (non-selected) environment and (non-perceived) "carrier media,"[52] not only in a general "media-archeological" perspective but in that of the maritime medium in particular. Finally, as long as they are politically relevant, archives are characterized by legal and technological limitations on access—here again akin to the seafloor. Hence the shift from metaphors of forgetting to those ties to archives is not entirely groundless.

Very soon after late nineteenth-century archeology had established itself as a discipline, three reservations were raised concerning an underwater counterpart: archeological excavating was unreasonably difficult and expensive beneath the ocean; the finds were subject to enormous forces there; and all too often the diving archeologists were really disguised treasure-hunters.[53] In fact, underwater salvaging continues to depend more strongly on state and private financing than that on land; highly elaborate equipment is needed for that work because of both

48 On this term in view of the archive and archeology, see Nora 1998, pp. 11–26, 60–74.
49 For a critique of this "dogma" from the viewpoint of commercial salvaging enterprises, see Kingsley 2010, pp. xif.
50 On the systematics of the excavation and hull catalog, see Steffy 1994, pp. 189–199.
51 Ernst 2004, pp. 246f.
52 See Groys 2009, pp. 147f.
53 See Bailey 2004, pp. 3, 9f.

the descent itself and the complexity of the sites involved. And in fact, underwater archeology counts the speculative salvaging expeditions of modern soldiers of fortune among the discipline's pioneering projects.[54] After piracy was suppressed in the 1830s (above all in the eastern Mediterranean), former pirates mostly turned to sponge-diving as a vocation. Before the introduction of usable diving equipment, underwater excavations was done with the aid of these individuals; and into the twentieth century and the development of an appropriate cultural concept, the work was understood strictly as a salvaging business.[55] In any case, once the technical means were available for exploring the seafloor, and once support was present for not only its oceanographic exploration but also a cultural-historical counterpart, a previous probing of the deep for treasure developed into a context-focused underwater archeology.

Methodologically the discipline initially revealed a rather classical, "historical-particularistic" approach, meant to locate wrecks preserved as well as possible within a conventional historical context. Under the influence of the "new archeology," the discipline then opened itself up in the 1970s to "hypothetical-deductive," "processual," and experimental approaches, followed by those of an anthropological and annalistic nature.[56] Far-reaching statistical analysis was now aimed at, for instance, correlating the distribution patterns of wreck discoveries with alterations in economical and demographic development, historical seafaring routes, the location of harbors, sandbanks, and reefs, and both currents and weather events. At the same time there was an effort to more closely relate the wrecks and discovery sites with circumstances of erosion and formation, for the sake of identifying significant patterns and structures within an often chaotic inventory comparable on land to battlefield sites or dumping grounds. In both cases, then, through bracketing of a putatively verified historical context, significant data was extracted from discontinuous, dispersed, and divergent elements—and this in a manner that contemporary epistemology selected, not without reason, as a model for its "archeology of knowledge."[57]

---

54 See Herzog 2002, pp. 8–11.

55 See Throckmorton 1996, pp. 14–16.

56 See Gibbins 2001, pp. 284–287. On the following see Parker 1992, pp. 8f.; Babits 1998, pp. 259, 305, 316, 471; Gould 2000, pp. 79, 83.

57 When Foucault, for the sake of identifying the law of distribution or formation of epistemic things and events, calls for "the dispersion of these objects, to grasp all the interstices that separate them, to measure the distances that reign between them" he is transferring to the discursive level what the "new archeology" has practiced—and also reflected on as the feedback between archive and archeology—on the level of material artifacts. See Foucault 1972, p. 33.

And in both cases, the archeological work took account of the newest data-processing developments.

Even more strongly than land-based archeology, its underwater equivalent connects the epistemological paradigm of what Heidegger terms "revealing" and "bringing-forth" with what he terms the *Ge-stell*, "enframing."[58] This is not only in respect to the diving instrument, its development extending from ancient diving bells to Early Modern diving suits and onward to the 1960s' scuba diving, or in respect to the diverse navigation technologies developed for various depths, currents, and weather conditions. It becomes especially clear in the case of the frequent dives made under extreme conditions, with zero visibility and sites merely a mental construct, to be located via remote sensing techniques such as the magnetometer and sonar and electronic mapping. Lastly, it is manifest in the numerous data recovery procedures, where modeling, simulation, and scenario development is often indispensable for rendering the site legible despite seemingly insignificant data and exploiting its inferential status.

It is very rare for underwater archeological sites to disclose themselves of their own accord. Finding them is not a matter of luck or chance but the result of systematic research. Despite tools such as side-scan sonar, haphazard but thorough combing of the seafloor would be as hopeless an enterprise as aimlessly rummaging through vast libraries and archives. The discipline only really freed itself from chance finds by, for example, sports divers and sponge divers and during the laying of underwater cables once planned archeological research was taken into its agenda.[59] As long as special shipwreck registers are not available for the purpose, localizing a wreck is ordinarily preceded by the inspection of extensive archival holdings—an undertaking that necessitates speculation on preservation of suitable documents and, in order to find them, what Jeremy Green aptly refers to as "intuition and imagination."[60] Salvaging material artifacts is thus often tied to regular archival archeology—archival holdings concerned with specifically maritime matters are as scattered as the manner of their exploration is diverse.

Whether the documents concern shipyards, the insurance business, or trade, whether they stem from maritime courts and navies, rescue or salvaging societies, the exploration of all these archives means an opening of an archeological perspective revealing both the *longue durée* and the infrastructure of nautical cultures beneath the level of a history of maritime events. Just such archival

58 On the following see Bass 1966, pp. 161–164; McGrail 1997, pp. 348, 367.
59 See Beurdeley 1999, p. 194.
60 Green 2004, pp. 14f.

holdings are, in the words of Colin Martin, "of the same commonplace and microcosmic nature that characterizes most archeological evidence. The two sources of information and their attendant investigative methodologies therefore tend to speak in similar languages."[61] Since the historical archive comprises a broad medial spectrum extending from handwritten material to printed texts, hand-drawn and copperplate maps and plans, and onward to photographs and both sound and film recordings, an important task of present-day research archives is to digitalize these disparate documents, integrate them into electronic networks, and bring them together with current institutional and commercial archives (e. g. in our context those of the International Maritime Organization and Lloyd's).[62] The data banks of underwater archeology are dynamic archives, archives on demand, that alongside the storing of long-deposited documentary material increasingly see to the transmission of ongoing data: to fulfill the basic purpose of a universal and up to date archive, every underwater archeological activity in principle has to be promptly entered into it.[63] Archeologists thus not only use archives but produce them in a steady way.

With, to cite Eckart Henning, the various historical disciplines understanding archival sources as "all texts, objects, and facts from which knowledge of the past can be derived," the sources include not only documents but what Johann Gustav Droysen called "remnants" and Ernst Bernheim called "traditions."[64] A balancing act between artifacts and documents, object and discourse, is especially striking in the case of underwater archeology and its "seafloor archive." We can understand the bottom of the deep sea as a surface upon which, in the course of archeological salvaging, materiality is translated into textuality and the real into the symbolic. Where the dynamic complex of a ship still moves on the sea's surface in a system of flowing communication, the static complex of the wreck inscribes itself on the seafloor. To be sure, this conversion of material traces into discrete signs is the salient point. It is not only that archeology is reluctant to prematurely identify nautical findings as that for which archives and historical writing already has concepts and names. All too often, the found object has to first be constituted in its concreteness, whether nameable or nameless. And as a rule this constitutive process does not unfold in an unmediated way, but rather recursively.[65]

61 Martin 2001, p. 384.

62 See for example Carpano 2006, pp. 246–249.

63 Such an archive would also be an effective means against trade with illegally salvaged or procured underwater cultural goods. See Throckmorton 1996, p. 227.

64 Henning 2000, p. 1.

65 See Gould 2000, p. 60.

The choice of complete on-site archiving put aside, before their salvaging found objects are meant to be thoroughly mapped and recorded three-dimensionally in their environment by means of triangulation, the computer-supported direct survey method, and sonar technology. And if, as some recent theoretical approaches to archival work argue, the salvaging operation is inevitably inscribed in the salvaged object,[66] this applies all the more for underwater finds. For this reason not only what has been salvaged needs to be documented in a so-called project archive but also all the salvaging measures—from the first survey to the concluding restoration work. All available storage technologies have to be mobilized in order to even integrate initially useless data.

Because in the future even the most improbable material or archival finding can endow huge significance on what has until then been insignificant, archeology maintains what M.A. Smith describes as "strict reference to the potentialities of the evidence."[67] And because the discipline aims at reconstruction, experimental modeling, further excavations, and benefiting from additional research, more than even conservation data protection is its top priority: "damage to the data base is damage to the artifacts, and...what we must preserve is the site as data base," is the way Wilburn A. Cockrell puts it.[68] Consequently, in that the salvaging operation usually starts with documents and data banks, itself compiles them in a steady way, and inscribes itself within them, and in that inversely various archival findings on nautical history are only made possible through archeological findings that are then for their part re-inscribed in the archive,[69] we can speak of a feedback process between archive and archeology, archival and archeological procedures and discovery processes. The archives of submarine archeology are always "project archives": the expression of an imperfect "not yet" form of knowledge. What they keep in a latent state, inasmuch as their material evidence is left on the seabed and is replaced by a set of data, are the artifacts found on the seafloor. But what they salvage in this way is a potentiality named "culture."

66 See Derrida 1995, pp. 16f., 18.

67 Smith 1998, p. 173. See also Gould 2000, pp. 52f.

68 Cockrell 1998, p. 95. On the particularities and limits of an experimental underwater archeology, see Crumlin-Pedersen 2006, pp. 3f.; Weski 2006, pp. 63, 66.

69 For examples the documents for the *Anna Maria* would not have been locatable "without raising some of the boards in the cargo and performing the subsequent dendrochronological analysis." Ahlström 1997, p. 8.

The Common Heritage of Humanity

Underwater archeology has quite often dedicated itself to a patriotic project: salvaging wrecked ships as miniature models of their countries of origin and thus monuments of a certain national history. But such patriotism is foiled by the basic fact that vehicles on the high seas have always sailed between different territories. Maritime history was always international history. In addition, if a wreck is localized on the bottom of the open sea, it is subsumed to the regime of international law, which thwarts exclusive national claims,[70] the open seas being understood as a "state-free space," hence as a *res communis omnium*—as "something shared by the entire community" or mankind's common heritage. Furthermore, when in the late 1960s the technical means were first available to exploit the entire seafloor's resources for the global economy, protests by developing countries led to conservation measures for at least the floor of the deep sea. Previously similar demands had been voiced for state-free areas of Antarctica; likewise, already in 1954 the suggestion had been made to declare outer space a *res communis humanitatis*.[71] But for concrete management aimed at conservation of such areas, as well as for the general international legal discussion, the sea had long served as a paradigm.[72] In 1830 already, Andrés Bello proposed the international-legal concept of an "indivisible shared patrimony," José León Suárez following up nearly a century later (1927) with the idea of a "heritage of humanity"; then, in 1958, at the first UN Conference on the Law of the Sea, the high seas were declared a "common heritage for all of humanity"—this followed up on in 1967 by the UN General Assembly offering the same basic declaration for the deep-sea floor.[73]

The "common heritage" concept, like that of *res communis omnium*, refers to an area whose exploitation is meant to be jointly supervised, occasionally licensed, and benefiting the entire world community in a just way. But as things stood, in the 1960s the developing countries had neither the proper technical nor financial means to actually benefit from a suitable portion of the theoretically available natural resources. Finally, in 1994, the term "area" was introduced, designating a maritime zone outside the zone of influence of national sovereignty—the coastal sea, contiguous zones, and the continental shelf, in turn meant to be protected as "common heritage of mankind." In this case, the concept's legal specification comprises, alongside a ban on appropriation, an institutionalized

70 See for example Firth 2002, pp. 3, 24.

71 See Postyshev 1990, p. 84.

72 See Hobe 2002, pp. 79–82; Oduntan 2012, pp. 203 f., 208.

73 See Wolter 2003, pp. 168 f.; Tuerk 2012, pp. 31–33.

model for administration and use, and a mandate of peaceful usage, both a shared obligation for scientific-technical cooperation and a certain trusteeship vis-à-vis future generations. Although these clauses are aimed at the seafloor's *natural* resources, they invite application to its *cultural* resources. That damage to cultural goods affects a universal cultural heritage and that its appropriation as booty is illegitimate was an idea already addressed in eighteenth- and nineteenth-century discussion of laws of war; at the latest, the idea was codified in 1954 in the Hague Convention for the Protection of Cultural Property in the Event of Armed Conflict.

The analogy with the "common heritage" principle is in any case rendered complicated through the fact that cultural goods, otherwise than previously not usable natural resources, were once the property of individuals or at least subject to sovereign state authority.[74] Consequently the analogizing is not only between nature and culture[75] but equally between either individuals or states on the one hand, international organizations or "humanity" on the other side. For this reason, it has been suggested, here excluding application of law of salvaging and discovery, thus actually redesignating "cultural property" as "cultural heritage," without entirely excluding the interests of owners and states of origin. In the end, then, this places cultural goods on the same footing as resources of an "area." In this context it was possible to move past the legal figure of humanity's "cultural heritage"—a generic term without concrete legal contents, maintained in that generality precisely because it cannot mediate individual and state claims with the universalism of the supra-individual and state-free concept of heritage.[76] In the case of the "area," such concrete mediation is taken over by the sea as, to speak with Hegel, the "supreme medium." But what then emerges as exemplary cultural assets are wrecked ships—ships that went down in the "area."

Different legal rules apply to different sorts of wrecked ships. For "time bombs," there are the regulations for liability and salvaging law, and for "graves," the obligation to preserve remembrance and leave the dead rest in peace. For "gold mines," a preferential right to property, access, or exploitation can apply. "Time capsules" pose a challenge to international law through the

---

74 See Dolzer 1994, pp. 13 f., 22.

75 Already the 1972 Convention Concerning the Protection of the World Cultural and Natural Heritage emerged from two different movements, one for environmental protection, the other for preserving cultural sites. On the levels of legislation, administration, and associated sciences, the institutional separation of natural from cultural protection is still evident. See Schmitt 2011, pp. 107 f.

76 See Herzog 2002, pp. 312 f., 351.

very fact that here the problems of risk-management, remembrance, and claims to usage are furnished with a common denominator: "culture."[77] After the UN Convention on the Law of the Sea was passed in 1982, a Cultural Property Committee was formed in 1994; with a view to found objects on the deep-sea floor, the committee focused on two questions: ownership and protection of the findings themselves. By the time the Convention for the Protection of the Underwater Cultural Heritage had been drafted in 2001 (it went into effect in 2009), two things had become clear: what gets collected on the seafloor at times has to sometimes still be considered the legacy of private persons, enterprises, or states, and at other times already the legacy of humanity in general. And with this material, together with the sites involved, forming an ensemble "in which human culture has found its material expression," at issue here are cultural goods meant to be protected from destruction and forgetting.[78] Cultural material does not represent a *res nullius,* but something always unique, a *res sui generis.*[79] We might say, pointedly, that archeology (and especially the underwater sort) is not really concerned with objects, as, in R.E. Mortimer Wheeler's words, "the archaeologist is digging up, not things, but people."[80]

In any case, for the sake of remembrance of the activities and development of humanity, cultural material is meant to be handed down to future generations, as a sort of inheritance. It is no accident that there have been frequent references to Charles Lyell in respect to that concept's emergence. When, for example, Lyell conjectures that more "monuments of the skill and industry of man" have collected on the deep-sea floor that ever "at one time on the surface of the continents,"[81] he is understanding the seafloor as a deep structure and "seafloor archive" retaining humanity's continuity in its "deep time." On the basis of Lyell's speculation, we can understand the deep sea, distant from the coast, as an archeological space par excellence. In comparison to the—also in a legal sense—narrow coastal strips, located here is a huge range of cultural goods. Most of these will be well preserved in comparison to those found on the coasts,

77 On preventative measures against, among other things, threats to the environment, see Katsos 2011, pp. 150 – 153.

78 Pallas 2004, p. 29.

79 Anton 2011, p. 350.

80 Wheeler 1954, p. 13.

81 Lyell 1830 – 35, vol. 2., p. 258.

where shipwrecks are usually the result of collisions (rather than loss of stability as on the high seas), and are thus accompanied by direct destruction.[82]

That presently sunken ships in particular stand for Lyell's "monuments" is apparently owed to chance: while bargaining was ongoing in preparation of the Convention on the Protection of the Underwater Cultural Heritage, the Legal Committee of the International Maritime Organization was working on a regulatory instrument for deep-sea wrecks. In this context various UNESCO negotiators saw the "time capsules" as prime examples of the underwater cultural heritage.[83] Since then, already purely in terms of international law, historical wrecks are the "arks" of maritime and indeed global culture. They seem to most clearly articulate the mission tied to the concept of cultural heritage: in the mid-twentieth century, protection of cultural goods was declared a national responsibility, and since there was now an international treaty governing such protection, the individual states were deployed as trustees of the goods on their territory, on behalf of humanity as a whole. But if we presume the non-territorial status of the "area," as loci where no national claims are meant to be in play, "humanity" becomes a trustee of itself—the logic of family, dynastic, and national heritage being fully breached. (How far-removed this universalistic cultural concept still was at the time of British maritime hegemony is made clear by for instance Robert Louis Stevenson's reference to the "pretension that the sea is English. Even where it is looked upon by the guns and battlements of another nation we regard it as a kind of English cemetery, where the bones of our seafaring fathers take their rest until the last trumpet; for I suppose no other nation has lost as many ships or sent as many brave fellows to the bottom."[84])

With the "closed finds" of underwater archeology, cultural goods were for the first time directly declared non-commercial objects in international law.[85] Conceived as a shared heritage, the seafloor's cultural goods took on a key role in reproducing cultural order and thus implementing the collective singular of "humanity": less a new legal subject meant to supplant classical owners of goods and property and sovereign authorities; rather a conceptual persona of international law embodying certain rights and duties—a regulative idea aimed at moral respect for the species as a whole and standing up for what states have in

82 Already the fact that most salvaging from the deep sea's floor have yet to be carried out will ensure a future for the history of the fascination with underwater archeology. On this and the legal aspects see O'Keefe 1999, p. 223.

83 See Gonzalez 2003, p. 85.

84 Stevenson 1901, pp. 181f.

85 Pallas 2004, p. 122.

common and the global orientation of their sovereign acts and legal systems.[86] There have even been proposals to invoke the UNESCO Convention Concerning the Protection of the World Cultural and Natural Heritage as a third decisive step in rendering the concept of "humanity" concrete, alongside the 1945 Charter of the United Nations, centered on the concept of world peace, and the 1948 Universal Declaration of Human Rights: for humanity would only be constituted as a cultural community through its "vertical" dimension.[87] In this respect as well, underwater cultural goods, together with their associated legal and administrative practices, can be considered exemplary.

One thing in particular clearly makes manifest this concept of both trusteeship and humanity: For a juridical approach to cultural goods that is no longer territorially anchored, that cannot simply be addressed as an "object," and that is not subsumed to any sovereign authority, borrowing legal terms such as *res nullius* and *res omnium* from an ideal model like Roman law is insufficient.[88] If, as in the case of an underwater cultural heritage, what previously belonged to no one becomes something belonging to everyone, while preferably remaining unmolested in a "seafloor archive," then no individual states but, nominally, officials of the international community and de facto underwater archeologists are humanity's trustees.[89] For this reason, governance of maritime cultural goods is transferred to them in the form of a cultural resource-management plan: alongside the professional establishment of project archives, a continuous control and supervision of sites that does not criminalize private divers a priori but integrates them into the salvaging work.[90] But that sort of well-tended cultural holding is there, to speak with Nietzsche, "for everyone and no one."

The paradoxical intertwining of collective property and ownerless property does not form a *res nullius,* as it is excluded from all appropriation; nor does it, however, form a *res omnium,* as it cannot be used by everyone arbitrarily. We also can no longer simply speak here of a sea open to all nations, a *mare liberum*, nor of a sea subsumed to a given country, a *mare clausum*. The sea is in-

---

86 See for example Stocker 1993, pp. 111, 115, 227.

87 See Häberle 2011, pp. 130 f.

88 On the uselessness of such categories in the legal assessment of the seafloor and its resources, see French 1990, pp. 147–149.

89 See Boesten 2002, p. 33.

90 See Palma 2005, pp. 323–332. See also Gould 2000, pp. 316 f.: "The best litmus test for distinguishing treasure hunting from archaeology is what happens to the materials after they are recovered. Treasure hunters often sell off valuable items and transfer the burden of conservation and storage of more ordinary materials such as ship's timbers and cannonballs to museums. Breaking up collections in this way violates the principle of association that is the basis of all archaeological science."

deed now closed, but henceforth for the use of all, not only the stronger parties. The concept of a "common heritage of mankind" should finally not be understood as a territorial principle whose defining feature is the impossibility of appropriation. The *res omnium* was already non-appropriable. But the principle of humanity's common heritage is marked precisely by having freed itself from territorial conceptions.[91] Its "take-off" from terrain-based legal principles is on the one hand tied, to a special affinity with more recent technologies (manifest in both space flight and deep sea exploration), for which reason we find talk in this context of the "global village" and "common global civilization of science and technology."[92] On the other hand it is tied to a specific affinity with the concept of culture and its universalism. Where this is the backdrop to trenchant discussions of "creeping jurisdiction," in the sense of gradually expanding territorial claims over the high seas meant to be counteracted by maritime law, in the case of safeguarding cultural goods it is also possible to observe an inverse tendency supported by international law: a kind of creeping culturalization of the sea—as if culture were the "other" of the state and its logic of property and exclusive appropriation.

Research on the Deep as Work on Culture

The surfacing of "culture" as a modern guiding concept has been dated to the threshold of 1800, although this has not been meant to suggest a caesura in the history of concepts. The concept of culture has not so much been connected to a distinctive thematic area or objective field but rather to a specific observational position: interest in comprehensive, initially value-free descriptions and comparisons, reflections and meta-reflections, in short second-order observing. Culture is here a global project—something especially clear since the nineteenth century—and thus always "world culture." Jacob Burckhardt put it thus: "There is a major, all-around, tacit agreement to draw in everything, to transform all of the past and present world into intellectual property [*geistiger Besitz*], and there is an objective interest in doing so."[93] But at the same time, everything that culture communicates is recognizable in its sheer contingency and positivity—in order to nevertheless appear meaningful in "cultural memory."[94] To be sure,

91 On the replacement of Roman categories by the concept of public trust, and on the misinterpretation of the concept of the common heritage of mankind, see Baslar 1998, pp. 86, 283 f.
92 Wolter 2003, p. 227.
93 Burckhardt 2007, p. 813.
94 See Luhmann 1999, pp. 36–38, 45, 47, 54.

the form of memory known as culture here needs an archive that, as Elena Esposito observes, "does not only consist of a collection of materials, but in the availability of a catalog or an organization that makes possible its management and coordination."[95] "Culture" does not exist without certain cultural techniques of comparison, representation, and organization, collection, storage, and transmission.

What around 1800 had been seen to by conceptual reflection came to be implemented by the technologies of the twentieth century on an increasingly global and simultaneously virtual level. In the case of the underwater cultural heritage, archeology's task of protecting and caring consists not least in lifting, preserving, and sublating the material substrate of what is meant to enter the reflective concept of culture. This has good objective and methodological grounds. Archeology is in the end, R. Duncan Mathewson III points out, "a destructive science. Once a site has been excavated, it can never be put back together."[96] If culture were directly salvaged, this would finally only result in its material destruction. For this reason, archeological access has a feedback relation with virtualization. In place of the artifact, the archive offers mere information that—other than the found object—might actually be conveyed to all of humanity, once a transmittable archive always open to everyone were available. "Historical sites are part of our common heritage, and the information they contain belongs to everyone," observes Mathewson.[97] For that very reasons, the "underwater cultural heritage" convention set strict standards for the professional development and management of project archives.[98] And for precisely this reason, ongoing cultural-heritage care displays increasing deployment of a range of digital technologies for recording, modeling, storage, and transmission. These have transformed cultural goods into "networked objects" whose retention for posterity poses new sorts of problems. They have led to a destabilizing of previously elementary borders,[99] namely—as a standard handbook in this area has put it—those "between authority and expertise, permanency, instability and transition, linearity, multidimen-

95 Esposito 2002, p. 239.

96 Mathewson 1998, p. 100.

97 Ibid.

98 See part XIII of the *Convention*, "Curation of Project Archives," and esp. Rule 33: "The project archives, including any underwater cultural heritage removed and a copy of all supporting documentation shall, as far as possible, be kept together and intact as a collection in a manner that is available for professional and public access as well as for the curation of the archives." Cited from Garabello 2003, p. 229.

99 Styliaras 2011, pp. 166, 182. See also Baltsavias 2006, pp. 11–20, 129–156, 205–216, 339–354, 407–430.

sionality and multi-directionality, certainty and unpredictability"; and beyond that, augmenting the deep-sea floor, Antarctica, and outer space, with cyberspace they have bestowed another state-free realm on international law.[100]

Such a culture-archive can, then, take up all possible found objects into holdings for cultural transmission, constituting them as a potentially unlimited form of cultural memory. Seeing this development as something other than a rehabilitation of microhistory, a historiographical enrichment, but in fact a fundamental danger, began well before the twenty-first century's threshold. This danger is a belief that something has been "gotten hold of" through an accumulation of found objects from the past, as if time itself reveals itself in things, or that justice can be done to it through the in-difference of bequeathed documentary material: what the German could refer to as a *Gleich-gültigkeit* free of reference and meaning, an absence perhaps reverberating, on some level, in the memorable (and ambiguous) end of Hölderlin's "Mnemosyne," *dem / Gleich fehlet die Trauer* (which we might translate as something like "soon grief is gone," "in that way sorrow is missing," or even "sameness lacks mourning"). It has thus been suggested to approach salvaged archeological material not as "documents" but as "traces" neither making manifest the past itself nor being fully insignificant. Put otherwise, traces herald no past. Rather, they point to a passing and call for a specific chronotopic manner of processing. As Paul Ricœur suggests, it is only in such a way that traces preserve "the debt we owe the dead," and only in such a way can they mediate "between the fundamental time of Care, the temporality directed toward the future and toward death, and 'ordinary' time, conceived of as a succession of abstract instants."[101] Culture thus inevitably produces a certain discontent. It makes necessary a certain, specific "work of culture"—work perhaps exemplified by research on the deep.

Especially the "new archeology" exercises a kind of forensic trace-preservation that neither believes it is grasping "time itself" nor wishes to limit itself to securing indifferent bequeathed things. This has consequences for the order of past events that has already been constituted historiographically—since only what subverts the historical order operates in this discipline as solid or significant information. But it also concerns the order and meaning of archival holdings themselves: repeatedly surfacing through underwater archeology's discoveries is what archives *do not* place on file. In the case of shipwrecks, forgery, omission, and other forms of manipulation of archival material are legion,

100 See Rosenne 2004, pp. 315, 330.
101 Ricœur 1988, pp. 184, 120.

whether for the sake of covering up guilt or for that of insurance fraud.[102] For example, 120 years after the *North Carolina* went down off the South Carolina coast (1880), the shipwreck anthropologist Richard A. Gould, in "a lot of shuttling back and forth," examined both the wreck and numerous pertinent archival holdings, hunting systematically for clues in an "underwater crime scene" and carefully documenting every detail of his "nondestructive" fieldwork, limited to his own, on site archiving. Alongside technical causes for the accident, in the style of a forensic archeology or criminal investigation, by way of contradictory documents and suspicious archival gaps, he exposed both an "insurance job" and evidence of embezzlement and corruption. From an archeological perspective, he thus placed both historiography *and* the archive on trial. But in the process, he not only wrote a case history but also disclosed the previously invisible, merely latent, structure and structural violence of maritime commerce: "It is always simpler to blame individual wickedness than explore the obscure weaknesses of complex systems."[103]

With its interest in what Knut Ebeling describes as "deviating temporal passages and alternative scenarios of the past," such an archeology is engaged in a specific sort of cultural work: offering culture insight into itself.[104] Especially maritime "depth research" is exemplary for a modern paradigm that through salvaging reveals not only knowledge but its conditionality.[105] The contemporaneity of this research with psychoanalysis is here no coincidence. For after a long period in which the sea had been deemed a lifeless abyss without history, followed by a view of the deep sea as wildly animated, its floor as a regular archive of natural and cultural history, a related discovery followed in turn: the animated depths of an unconscious capable of retention and influence, "culture" as the effect of a retroactive processing and repetition of memory, tradition as the continuous salvaging of re-captured material.[106] Himself a collector of antiquities and attentive observer of terrestrial archeology, Sigmund Freud could not help

102 See Ducoin 1989, vol. 1, pp. 12, 60, 329.

103 Gould 2007, pp. 177, 179, 190. See also Gould 2000, p. 13: "If archaeological events derived from assumptions about Pompeii-like or 'time-capsule' associations are illusory, so, too, are historical events such as the wrecking, scuttling, and even construction of ships. Upon close examination, these so-called events prove to be embedded in ongoing processes linked to social, economic, and even symbolic activities. The drama of shipwreck focuses attention on the event, but the conditions that produced the wreck and the consequences arising from it are as significant as the event itself."

104 Ebeling 2012, p. 16.

105 Compare Adamowsky 2006, p. 229: "The sea...can be seen as a predestined medium performatively presenting these conditions or contingency of modern knowledge with every dive."

106 See Goldschmidt 1999, pp. 16, 47 f.

noticing the affinity between this new guiding science of "forensic" trace-preservation and his analysis of the unconscious. In both disciplines, repressed material reveals itself by way of the repressive agent, a deeper layer through the defigurations, deferrals, and condensations that they cover over.[107] And here like there, objects gain significance when they were sunken or repressed and now have again surfaced, in order to preoccupy the cultural or individual imaginary: "dead and alive at the same time," as Jensen puts it in his *Gradiva* novel of 1903, to then be commented on by Freud,[108] is an expression circumscribing the precarious status of the field of both archeological and psychoanalytic objects—but traditionally, also the status of what awaits release (and salvation) from the sea.

As an expert in collective myths centered on both land and sea, Freud himself juxtaposed deep research into *terra incognita* sunk in the unconscious with the "oceanic feeling" that Romain Rolland had described as a source of all religious ("or "Odyssean") longing for an origin and for return to the immemorial. Freud was skeptical vis-à-vis fascination with the depths, the "sea change" and what Heather Asals has termed "death of the life world,"[109] regressive de-differentiation and a merely imaginary satisfaction of non-knowledge. And reflecting this skepticism, he called for a distinctive analytic modality of knowledge; eventually this form of analysis would equally guide the work of epistemologists, analysts of discourse, and writers of fiction.[110] Psychoanalytic depth research "is a work of culture,...not unlike the draining of the Zuider Zee," comments Freud.[111] And further, it resembles an archeology of the archive. For in Freud's model of the mind, the topoi of consciousness, preconsciousness, and the unconscious correspond to the system of office, filing department, archive.

The unconscious is an archive that, as long as its carrier medium remains intact, preserves all received traces. For this reason, analytic or archeological labor involves not only reconstruction of traces but also of the archive, its systematics, and its gaps. Correspondingly, chronotopoi such as Rome and Pompeii

107 See Freud 1922, pp. 88, 115 f.

108 Ibid., pp. 127, 64.

109 See the phenomenological description in Asals 1985, p. 304.

110 Bachelard 1994, p. 10: "pre-human, in this case, approaching the immemorial." On the open, in principle endless character of Freudian analysis as an epochal interpretive technique of interpretation see Foucault 1998, passim. On the poetological and epistemological figure of the shipwreck see Fowles 1992, pp. 494 f.: "a kind of Freudian double identification, in which the wrath of the sea is interpreted both as *super-ego* and *id*....A wrecking sea is part of what we all dream ourselves to be every night; and the ship becomes our own puny calculations, our repressions, our compromises, our kowtowing to convention, duty and a dozen other idols of the top-hamper we call civilization."

111 Freud 1953–1974b, p. 79.

illustrate the ontogenetic and phylogenetic layering of the mental apparatus. Freud indicates that psychoanalytic labor is comparable with purposeful excavation work—only that for psychoanalysis the field is not as familiar as for urban archeologists.[112] However, where in its labor of reconstruction underwater archeology is concerned with scarcely accessed, often shifting areas, it stands even closer to psychoanalysis than its land-based counterpart. In any case, as cultural labor "depth research" does not content itself with establishing that in line with a principle of interpretive unfathomability, culture is simply potential, and simply an overdetermined concept; and that a culture's scattered records are to be tracked down by relating them back to their source. In its effort to press forward to a culture's grounding in a movement of endless retrospect, this research does not encounter an "inside." Very much to the contrary, it repeatedly arrives, as Freud emphasized, at the recognition of an "outside."[113]

The Adventure of Maritime Archeology

Freud's methodological considerations, which he sees as having a counterpart in, at best, archeology fictions[114] thus do not exhaust itself in a quasi-archeological epistemological doctrine. Already in his commentary on Jensen's "Pompeian fancy," they also concern the layer of motivation lying beneath a travel plan and beneath archeologically-framed epistemological interest.[115] Freud inquires as it were back behind the archeological "source," not only, to cite Hans Blumenberg, up to the "will that had created it and left it behind" but also to the will that seeks to discover it.[116] But he offers neither an analysis of motivations of underwater archeology nor one of underwater-archeology fictions—this already because in the early twentieth century there was not yet any institutionalized deep sea research and no corresponding narrations. Proper archeological novels (a genre inaugurated by Edward Bulwer-Lytton's *Last Days of Pompeii* of 1834) were initially only grounded in the knowledge and methods of on-land excavations. Before maritime "deep research" was at all possible technologically, archival research occupied its place, aiming at an initial collection and systematization of the scattered knowledge about maritime cultures.

112 On the different "knowledge layers" of Freud's archeology, see Ebeling 2012, pp. 294 f.
113 Freud, 1962, p. 14.
114 Freud 1922, p. 166.
115 See ibid, pp. 115, 120.
116 Blumenberg 2012, p. 12.

This is already evident in the work of the "father of nautical archeology, Admiral Auguste Jal, who in the 1830s took up an official appointment as France's marine historian and then in 1852 as curator of the naval archives. In 1832, Jal publicized *Scènes de la vie maritime*, a collection of narrative topoi each of which is treated in a short "novel," in historical "notes," and finally in commentaries on nautical *vocabulaire*. For the sake of a cultural-historical sounding out of the ship as a "masterpiece of the human spirit,"[117] Jal here drew on narrations emerging from a still intact mariners' "milieu of memory" (and thus from transmitted yarns such as that of the *Flying Dutchman*). In the following years, he changed his strategy: since 1834, he embarked on months-long research trips in the Mediterranean area, in order to search through museums and archives for artistic depictions, models, blueprints, and other documents as well as expert information from seamen. Inspired by the experimental methods of physiologist Claude Bernard, he now explored, "by a sort of anatomical study, the secret of the [nautical] vessel's marvelous life."[118] But for his *Archéologie navale* (published in 1840), he only had recourse to scattered older ship-hulls. The "remainder of a vessel" suddenly buried in volcanic stone in Herculaneum was his only "time capsule."[119]

For this reason, although a pioneer of marine archeology, he pursued it on a mainly archival level—on the basis of extant documents, but also spoken testimony: focused on "knowledge incorporated in terms representing material objects," he saw seamen's language as an archive whose ongoing holdings were threatened with temporal erosion. With Jal's "plan for a vast archeological project" involving "first language…then the ship, and finally history," an archive of nautical terms was the first priority.[120] With his *Glossaire nautique*, appearing in 1848 following additional voyages, reading, and research, he could correct numerous errors both appearing in his *Archéologie navale* and by various maritime historians who either had been unaware of pertinent documents or else had simply misunderstood them because of unfamiliarity with the vocabulary. All told, Jal's archeology followed "a simultaneously historical and philological interest."[121] It was not only that alongside historical documents it used literary

---

117 Jal 1832, pp. XIII, XV, 19. On the ship see ibid, p. 19: "All the arts, all the sciences, all the industries have contributed their contingent to perfecting this admirable machine, so complicated and at the same time so simple; so noble and elegant in its material form; so difficult to understand at first, so interesting to study; so daring, I might add so logical."

118 Jal 1840, *Rapport*, p. 1.

119 Ibid, p. 26.

120 Ibid, p. 4.

121 Jal 1848, p. 14.

texts and thus methodologically pointed forward to "archeologically" conceived philology like that of Victor Bérard. Every material artifact merited a "critique analogous to that applied to sentences by a historian or poet that are described or recounted."[122] Before it was supported by advanced technology, maritime archeology was thus pursued as a form of philological cultural history. Its main focus was on the textualizing of oral testimony, hence the textual archiving of a gradually declining narrative milieu, while it considered the process of rendering material artifacts readable as relatively uncomplicated. Jal also described his archeology as "a sort of restitution of old naval material, done to clear the ground of a number of questions that still slightly trouble the historian's progress";[123] he was soon subject to criticism for limiting his work almost entirely to archival material.

Jal's epoch was that of the great and in the best sense superficial sea-novels —those for instance of Poe, Melville, and Hugo—and at the same time that of an increasingly archival interest in oral evidence from the maritime milieu. However, regular "archeological poetics" dates to the institutionalization of deep-sea research. If, as Michel Foucault suggests, an "experience from outside" is characteristic for modern writing in general despite all self-reflexivity,[124] then in respect to modern narrative literature we can speak of a poetics of transgression on a number of formal and thematic levels. This is the case firstly on a level of knowledge, in that such texts themselves make visible historical and archival documents as products of historical constellations and archival practice and try to move past the realm of textuality toward an "outside" of spatially and temporally sunken materiality.[125] Second, it is the case on the level of narrative mode, in that the potentiality of archival data and latency of material artifacts suggests the drafting of virtual scenarios, for which reason the narratives involved are located between factual and fictional ways of writing: both amalgamating constitutive non-knowledge with fiction and, as in the "non-fiction novel," obsessively bolstering every narrative act with documents.[126] And third, it is the case on the level of topography, in that on the paper's surface such texts no longer only unfold what may have occurred on the surface of the sea. Rather, it now also treats the legibility of what the seafloor offers as a deep archive; its alignment with

122 Jal 1840, p. 34.
123 Jal 1848, p. 9.
124 Foucault 1987, pp. 11 f.
125 See Koschorke 2018, p. 11: "Story-telling even inserts temporality into stasis; or inversely, in terms of its most general generic rule, stasis appears as something like dammed up narration, narration brought to a standstill."
126 See, from a narratological perspective, Genette 1992, pp. 64 ff.

land-based archives; the experience of the narrator in passage between the sea's surface and seafloor.

Accounts of personal experience such as Hamilton-Paterson's *Three Miles Down* (1998) move consistently through a field tensely located between the poles of factuality and fictionality, archive and archeology, and also the sea's surface and its floor. Already because exploration of the deep sea represents, along with space exploration, the last possibility for pioneering feats, an insight with us at least since William Beebe's dives and his report *Half Mile Down* (1934),[127] reports of professional hunts for wrecks have been couched in the same style as narratives of adventure. Here the archive and numerous location and salvaging instruments are presented as conjectural technologies promising complex ramifications of adventure. It is just this promise that, as James Hamilton-Paterson states in his version of Beebe's theme, "puts the average novel plot or slim volume of verse to utter shame."[128] As a participant in a voyage in a diving vessel that can descend up to 5000 meters, the narrator will finally refer to exploration of what is anticipated as the deepest grounding of one's own interiority. In the undertow of immersion, he essays a log resembling automatic writing. The descent to the "prochronic" and "prophotic" seafloor leads, however, into an object-free space, hence for a writer only to analogies with reports by astronauts, then to reasoning about the oceanic feeling, onward to recollection of Thomas Burnet's *Telluris Theoria Sacra,* and finally to the archive of rhetorical topoi of the unsayable.[129] That deepest inwardness loses itself once more in externality. But since the salvaging mission banks not only on archeological-cultural findings but also on the putative gold cargo of a sunken German World War II submarine, the entire enterprise is imperiled by possible intervention from the English, American, or Russian navies. In the end, so long as no objects are found, the search centers on coding systems alone, the decoding of archival, navigation and location data.

127 See Beebe 1934, p. 225: "the only other place comparable to these marvelous nether regions, must surely be naked space itself, out far beyond atmosphere, where sunlight has no grip upon the dust and rubbish of planetary air." Not only the technological demands involved in deep-sea and space exploration led to the two activities being viewed as parallel in the 1950s and 1960s. Soon after the *Sputnik* ascended into space, the manned diving boat *Trieste* reached the deepest point on Earth. Both pioneering projects were aimed at state-free spaces, and they faced similar decisions of principle, for instance whether manned or unmanned ships should be sent into the depths or into space, or whether space stations should be set up in space, "experimental villages" in the deep sea. From a "geophilosophical" perspective, see Schmitt 1995f, p. 568.

128 Hamilton-Paterson 1998, p. 76.

129 Ibid, pp. 198ff.

As long as they are power-politically invested, archives are marked by legal and technical limits on access—something just as much the case for institutionalized archives as for maritime archives of the deep. This sets certain limits on salvaging missions, especially when the objects are military or military-historical wrecks. And perhaps it marks the limits of every literary "archive poetics," for which one maxim in particular seems to apply: real "power is not present where stories are told."[130] For this reason it has become an archeological adventure to open previously classified archives, however belatedly that may occur. When, for example, in his *Shadow Divers* (2005), Robert Kurson reconstructs the location and proper identification of an initially mysterious German World War II submarine, this "factual fiction" depicts both the archival and diving processes as "small initial penetrations" on land and water.[131] The yield of this endless double research in partly off-limits, partly fragmentary holdings is the discovery of some "ultra top secret" information in classified U.S. archives: the U-869 had originally been ordered to New York, had never received a radio message—one listened in on by Allied code-breakers—that was meant to redirect it, and then sank itself on command, probably off the New Jersey coast. The very fact that the diving archeologists were able to break through the phalanx of "postwar assessors" and history books and access the until then classified archives of the code wars renders them genuine "deep researchers."[132]

The adventure of seafaring has become an adventure of researching the deep, and thus a form of historical revision. Its orientation points are now certain codes and decoding techniques that release, initially, what has sunk and, finally, significant narrations. This process is already apparent in *The Mystery of the Sea* (1902), a novel by mathematician and jurist Bram Stoker. Here the adventurous search for a sunken treasure-ship stems from a cryptographic riddle's solving, itself consisting of a laborious reduction of Bishop Wilkins' and Francis Bacon's code systems. At its end the novel reveals episodes of geopolitical conflict culminating in that between Scottish salvaging law and Spanish claims to ownership. But tied to the disputed cultural good is, first and foremost, a previously secret episode of the Spanish attempt to invade Britain in 1588—and thus an archeological revision of previous history.[133] A century later (2002), in *Cryptonomicon* Neal Stephenson will take up this chronotope of an encoded secret archive. This novel derives the emergence of the Internet, a now omnipresent "archive on demand," from Wilkins' code book and both Germany's Enigma

130 Ernst 2009, p. 185.
131 Kurson 2005, p. 192.
132 On the obsessive documentation of all narrative details, see ibid, pp. 271, xi., 357–364.
133 See Stoker 2007, pp. 73 f., 80 f., 96, 251.

encryption and its British deciphering—an archeology whose inherent logic takes in the submarine war and the salvaging of its remains. The new archeologists are the new archivists. For with both divers and hackers focused on decoding initially insignificant but eventually classified data, they are twinned figures.[134] The fact that modern deep-sea research is made possible by a new use of army equipment from both world wars[135] interlinks it—in "factual fictions" such as *Cryptonomicon* or Thomas Meinecke's *Hellblau* (2002)—with an archeology of power and its media.

Objects found on the seafloor are shimmering, multivalent objects: perhaps "graves" or "time bombs"; possibly also "gold mines" or "time capsules." Furthermore, a shipwreck is an *objet ambigu* in that, as Hans Blumenberg observes, looking back at Valéry, it "does not display the definitiveness of a point but rather the potentiality of a horizon." "Replacing a complex of circumstances with an analysis of factors, presenting the ship as a body upon which certain effects are exercised"—this is produced by an archeological gaze endowing the found object with "transcendental character."[136] Namely, withdrawn from time and visibility, such an object can be open to more than pure theory. It rather brings forth "a limited special case of nonconceptuality" that in a never-ending movement undertakes a fathoming of "thought from outside."[137]

Uwe Johnson's correspondingly titled story "Ein unergründliches Schiff" ("An Unfathomable Ship"; 1979) treats a wreck that went aground in the Thames estuary, the stated aim being "to get to the bottom of this beech-loot, to salvage the grounded goods, historically, magically, biographically, sociologically, chemically, administrative-scientifically, poetically, statistically."[138] The *Richard Montgomery*, a munitions ship stranded in 1944 off of the coastal town of Sheerness, on the constantly flood-threatened island of Sheppey, is considered the island's

134 Cf. Stephenson 2002, p. 568: "divers have mastered a large body of occult knowledge. That explains their general resemblance to hackers, albeit physically fit hackers." See also ibid, pp. 86, 202.

135 On the development of the sonic depth finder and directional sonar system out of submarine detectors, see Schlee 1975, pp. 244–251, 285.

136 Blumenberg 2001, pp. 98, 108, 111. On the hylemorphological definition of a ship in marine law and on the "end of its nature as a ship," see Beckert 1991, pp. 154 f.

137 Blumenberg 1996, p. 81. See also Konersmann 2003, pp. 220 f.

138 Johnson 1979, p. 538. This text was a commissioned work responding from an inquiry by Jürgen Habermas, who in 1978, as a reminiscence regarding Karl Jaspers' *Zur geistigen Situation der Zeit* (1931), was preparing a volume with the title *Stichworte "Zur geistigen Situation der Zeit."* Johnson's text appeared in 1979 as "*Das Schiff*," with a lightly expanded version "*Ein unergründliches Schiff*," appearing in *Merkur*. A direct result of work on this text was Johnson's decision in January 1979 to end work on his novel *Jahrestage*. See Fahlke 1995, pp. 184, 188, 191.

"only attraction" by those who live there—and in addition as their only "monument" to World War II.[139] That the wreck is subject to so to speak in-site archiving in any event has its source less in concern for cultural heritage than its likely still live and in case of explosion immensely destructive cargo of TNT and phosphorus.

The *Montgomery* is, then, literally a time bomb. Because its raising would be accompanied by unforeseeable dangers, the wreck remains where it is, surrounded by access-barriers. The "official legend" is that "the more and longer exposed to seawater, the less harmful" are the sunken bombs.[140] Although in this process of running out the clock, the effort is to shove the bombs into an undefined afar, indeed into forgetting, it is perhaps the very fact of this danger that, as Heidegger (in proximity to Hölderlin) puts it, "brings the saving power....What does 'to save' mean? It means to lose, to emancipate, to free, to spare and husband, to harbor protectingly, to take under one's care, to keep safe."[141] The very fact of its carelessly not being raised but in any event subject to bothersome safety inspections renders the wreck in Johnson's text into an allegory of that forgetfulness of Being that Heidegger connects with the "Enframing," the *Ge-stell*—with the insensate claim "to order the self-revealing as standing-reserve [*Bestand*]," so that with the mere "securing of the standing reserve...this revealing as such" can "no longer...appear."[142] Stated concretely: the dangerous, only half sunken wreck simultaneously reveals and hides technology and its danger. It attests, already as a motor-driven vessel but even more so as a munitions ship, to the use of different natural forces whose danger for "the existent" are evident and simultaneously "scientifically" concealed. However, the danger for "Being" consists of forgetting technical *poiēsis*, "revealing" through the "Enframing," whether in the everyday operations of coastal and wreckage control or in the "indifferent" (or "all things are equal") operations of culture and thinking.

Johnson agrees with Heidegger in viewing the essence of technology as itself not technical; rather it involves nothing less than a *poiēsis* like literature in its status as a kind of generative revealing, *Her-vor-bringen*. Literature is a seismograph for the *poiēsis* of technology on the one hand, culture on the other hand, and in addition for an inclination, dangerous to both as their supplement, to forget their particular kind of *Her-vor-bringen*. If, as Freud argued, blind trust in technology is, just as much as blissful self-abandonment to the "'oceanic' feel-

139 Johnson 1979, p. 541.
140 Ibid, p. 547.
141 Heidegger 1977a, p. 42.
142 Ibid., pp. 19, 27.

ing," just "another way of disclaiming the danger,"[143] then at least for Heidegger the danger of "forgetfulness of Being" is banished through the poetic revelation of "unhiddenness." And it is exactly here that Johnson's text raises objections. For his archeological poetics, "truth" does not emerge in the same sense as Heidegger's "unhiddenness" or *alētheia:* no longer as a lyrical or even mythopoetic —putatively presented in exemplary form by Hölderlin—unity of *Dichten* and *Denken*, poetry and thinking, which language serves less as a medium than as a "source."

In place of its revelatory emergence or disinterring in the original word, its *Entbergen*, Johnson dissolves or resolves the truth, entirely along the lines of archeological praxis, into a process of seeking, of unremitting and recursive salvaging, *Bergen*. He does not fuse ontological thinking with the poet's word in order that things show themselves "from out of themselves" and "unhidden." Rather, he connects the archeologically steered search for sunken time, his research on a widely scattered empirical field, with a narrative process that consistently refers to the virtuality of its scenarios and the uncertainty of its grounding.[144] In this respect, Johnson's text itself demonstrates two things: first, that the ground or origin of truth only recedes or hides itself all the more—that the ship is only all the more "unfathomable"—in proportion to the "groundedness," the thoroughgoing nature of, the search; and second, that looking for and then searching foundations in no way averts present dangers or prevents them from remaining hidden and forgotten. For Johnson, pure theory is itself an epistemological form that forgets to care for what is unhidden in one's own endangered present: "As soon as a danger demands our attention and our resistance, it can be assured of our theoretical efforts. Even if it's a shattered bomb-ship in the Thames estuary."[145]

Johnson's story writes itself directly from the archive. Without arriving at a firm and solely "sufficient ground," he reveals a range of philological, economic, and organizational factors intertwined in the wreck: what became historical sediment in the ship's name and what henceforth sank into the islanders' awareness, even "extending into their phrasing"; the form of structural violence reflected in this massive type of ship, in the name of a specific American "freedom" (namely, "using the work of others"); and the sort of failure within the steering and command processes that led to the accident.[146] At the same time, the text itself undermines this search for reasons or grounding. Namely, it overlayers this contex-

143 Freud 1962, pp. 11ff., 19.
144 See also Kinzel 2001, pp. 134–151.
145 Johnson 1979, p. 550.
146 Ibid, pp. 540, 548.

tual information in the manner of an obsessive archival consciousness wishing to deny and bury the danger clearly evident with the explosive instrument of war. While people dabble with the wreck, the clear and present danger is simply left to specialized risk- and danger-centered techniques such as insurance. Not only technology of war, we might thus say with Heidegger, but also *technē* of the archive harbors the danger of handing the present along with what has past to the *Ge-stell*. As "factual fiction" telling of technical and cultural operations as much as it articulates the danger of the *Ge-stell*, Johnson's text calls, when all is said and done, for an "archeology of the present." Or stated otherwise: it pursues an "archeology of knowledge" whose starting point is neither concepts not mere facts but brings to light significant structures through the interaction of both. Such literature is a reminder that without archives there would be no culture, which always puts up with the indifference of chance and the danger of going under—in order to put them both in the archives.[147] But at the same time, it warns us that archives themselves do not rest on any firm ground of words and things. We may call this abyss of archives, of cultures and their history time, sea, or simply chaos. Exploring it is only possible through the adventure of "deep research."

* * *

At the beginning of maritime archeology, as initiated by Auguste Jal, stood the connection between historical and literary documents. Following Jal came Victor Bérard, who with his research on Homer systematically pursued philology as cultural history. By focusing on artifacts and scenes of ancient seafaring, both these figures tried to move from their research hypotheses to reach what Bérard described as "tangible reality, scientific and experimental truth." [148] Jal's "plan for a vast archeological labor" followed a specific logical order: "First language,...then the vessel, finally history."[149] In Bérard's case, we can speak of a sequence extending from language over the sea and onward to history. Only in the archeological interlocking of language and history did places in the sea's "smooth space" and "chronotopoi" such as wrecked ships reveal the topology in terms of which we need to understand maritime culture. Neither enterprise, taken up before the epoch of institutionalized archeology, would be deemed archeological if Jal and Bérard had here intended to grasp the "thing itself," the

147 On the program of an archeology of "risk-taking" in ship building along with nautical organization and insurance institutions, see Gould 2000, pp. 5 f.
148 Bérard 1902, p. 582.
149 Ibid, p. 4.

"source" or "origin." But they aimed neither at something purely textual nor at something purely material. Rather, in examining the maritime ground of "deep time," they explored a space of empiricism and experience in which words and things only are manifest in infirm, sliding connection, at one point expansively scattered, at another point in dense correlation.

We can understand Jal's und Bérard's "philological-archeological" method as a first concrete answer to the question of how a "cultural history" of literature and seafaring is in any way possible. When it comes to historically probing the sea, traditionally the traceless and memory-free medium par excellence, the later history of "deep exploration," its methods and techniques, can perhaps, offer some points of reference. Firstly, in the age of underwater archeology the sea appears to no longer be an abyss of historical forgetting but to have become an archive capable of conserving the traces of vanished cultures; the deep sea's floor is now a ground of resonance for "cultural memory," but at the same time a challenge for what might be called a (knowledge-) history of daring, risk-taking, and adventure. Second, we have learned to approach the nautical object on the seafloor as a chronotope through which the present can inform us about the putatively tracelessly sunken past. Archeologically processed, the "time capsules" of the wrecked ship make recognizable the most varied networks of things and human beings, the basis for in turn describing cultures and cultural techniques including their various orientations (trade or discovery; war, hunting, or piracy). And third, such a "material culture" is clearly hardly open to exploration or understanding without archival discovery working alongside archeological searching, discourse alongside the object. Accompanying ordinary documents, literary texts here take on a special role in that with their different poetics they also reflect *poiēsis*, the world-exploring and world-creating impact of nautical practices and techniques.

Perhaps nothing less than imaginative literature can trace out those indistinct, unsafe places that land-based human beings repeatedly had to conquer, within the sea's placelessness, on the heterotope of the ship, in the midst of the "actor networks" of nautical cultures. In any case, the passage between archeology and archive, thing and document has increasingly become the path of literary narration. This shift in literary poetics has itself increasingly undone the seam dividing the sea's surface from the paper surface of text or chart, expanding previously only horizontally oriented experiential space with a dimension of depth. In an epoch when sunken maritime cultures draw mainly antiquarian or memory-political interest, this has strengthened the archeological quality of story-telling. For such literature neither exhausts itself in the "disclosure" of putatively authentic cultures nor is aimed, like the nineteenth-century's archeological novel, at illustrating a memorable scenario of burial or ruin.

Rather, it treats the economical, political, and scientific networks that render salvaging a nautical cultural record possible in the first place, before the non-places of the high seas are declared "places of memory."

But to tell stories about the sea has long-since meant more than sticking with maritime topoi and motifs. As has been summarily observed, emerging across from older forms of social memory with their mnemotechnical "topo-logic," are modern culture's new forms with their "chrono-logic."[150] Modern archives are not collections fixed in specific places, but rather themselves complex chronotopoi: spatial structures in which time takes on dimension. To tell stories about an archive, including that of the sea, means actualizing archeological deep-time. Here, as Michel de Certeau indicates, "geometrical" *places* become "anthropological" *spaces* "when one takes into consideration vectors of direction, velocities, and time variables," connects static structures to acting subjects, and thus makes possible the "experience of an 'outside.'"[151] Such literature is not satisfied with the indifference of archival events, on whose basis a sunken past would then be brought into the light of a present charged with imagination and solicitous of culture. In the style of archeological "deep research," it rather secures the traces of what has been passed by, in order to then work on, precisely, its illegibility. Hence instead of too speedily "understanding" these traces, it shifts its focus to a process of retrospective production or revealing, while steadily observing its own dynamic, thus consistently narrating its own poetics.

With a look to the shipwreck of the *Méduse,* Klaus Heinrich has described the "dimension of *Realgeschichte*" (of the history of real events) that keeps observational forms such as those of culture not *au courant*, *auf dem Laufenden,* but rather "in a state of pause," *auf dem Stockenden*, as a "history of fascination."[152] This may explain why for some decades now research on the deep has fascinated narration. For by contrast with the narratives of "cultural memory," those of literature archeologically undermine archival knowledge by presenting the pausing of observation and imagination in their own narrative procedures. In order, for example, to render a site into a *lieu de mémoire*, it perhaps suffices to vividly reconstruct the event of the shipwreck from the perspective of a—however idealized—spectator. But a narration that not only has topographic memory as its starting point but also searches through the space of an archive, in order to both sound and step past it, resembles the "deep research" that underwater archeology inspires. A "shipwreck with spectator" is, to repeat Hans

150 Esposito 2002, p. 248.

151 Certeau 2006, p. 117.

152 Heinrich 1995, p. 7.

Blumenberg's words in this respect, "the figure of an initial philosophical experience"—earlier the figure of moral self-assurance, now of its historical and cultural counterpart. A "shipwreck with salvaging," however, only allows retrieval of sunken time in a process of laborious and repetitive culling. All too often, it does not reinforce or corroborate but serves as an irritation to historical and cultural presumptions. For what comes to light with this "ambiguous object" is a non-conceptual entity now become objective, something that overtaxes quick theoretical definition. Instead of an authentic source or origin to here be disclosed or revealed, a horizon of potentialities is salvaged.

We might here speak of a space of possible pasts—a space with which the classical concept of time is turned upside down. As we have seen, starting with Aristotle, contingency was understood as something neither necessary nor impossible, as something that can *not be*, hence does not simply not exist but also only *doesn't yet* exist and thus *insists* as possibility. This being the case, such a sphere of potentials referred entirely to "contingent futurabilities" and their linkage with specific expectations of the future. But from now on, we can speak of possible pasts that are neither necessary nor impossible, and that as such insist instead of exist and are fed back into the present exploration of a "horizon of the past." And where in earlier statements about the past, it was modeled as a closed, linear order juxtaposed with the ramified structure of statements about the future, meanwhile this network of possible worlds also reveals itself in the past and in the medium of possible narrative acts.

"In civilizations without boats," Foucault said in 1967, "dreams dry up, espionage takes the place of adventure, and the police take the place of pirates."[153] Perhaps modernity, with its apparent canceling of the elementary land-sea distinction, forfeited genuine seafaring along with the classical sea. But even if it indeed has lost that endless horizon for which the sea conceptually stands, it has only begun to explore the riddling deep dimension that the word "sea" signifies. For this reason, the last adventure at sea may lie in maritime research—all police and military security measures notwithstanding. Early modern *aventiure* was first and foremost a maritime affair. At sea it showed itself initially as a reckless, then as a risky, and finally as entrepreneurial action. Here *adventura* was always oriented toward *advenire*, toward a possible "approach" or a future. And it was carried forward by the hope for fortune's befalling, especially that a valuable, normally not buyable thing would fall to one's share. German-speaking merchants called this *Abenteuergut*, a commodity gained through adventure,

---

153 Foucault 1986, p. 27.

by which it was also possible to understand plundered jewels.[154] Or else, following the old goddess, it was simply christened *fortuna*.[155] Henceforth turned to the past, to its coincidental arrival, the passage of deep-sea archeology again leads to *fortuna di mare:* to the riches, worth more than can be expressed in money, and found objects on the seafloor. While the Middle Ages, for which the sea only appeared as a threat, could only see loss and punishment in its *fortuna*, the modern age at least conquered the sea's wild surface. Since then, not only the sea's harboring of danger and losses passed as *fortuna* but also the remains of a shipwreck, encountered as flotsam or jetsam: those found things, then, that the sea coincidentally released on shore.[156] This, precisely, is what unites the old high seas adventure with that on the seafloor: the *fortuna di mare*, in which the sea's fortune and misfortune both come to light.

154 See Welzig 1969, pp. 439 f.
155 On the concurrence between allegories of *ventura* and *fortuna*, see Patch 1967, pp. 39 f.
156 See Détrée 2005, pp. 32 f.

Cited Literature

Aczel, Amir D.: The Riddle of the Compass: The Invention that Changed the World, New York, San Diego, and London 2001.
Adam, Sophia: "Aspects de la sécurité de la navigation dans l'Antiquité grecque," in: Symposion 6 (1985), pp. 283–291.
Adamowsky, Natascha: "Annäherungen an eine Ästhetik des Geheimnisvollen. Beispiele aus der Meeresforschung des 19. Jahrhunderts," in: Wolfgang Krohn (ed.), Ästhetik in der Wissenschaft. Interdisziplinärer Diskurs über das Gestalten und Darstellen von Wissen, Hamburg 2006, pp. 219–232.
Adams, Jonathan: "Ships and Boats as Archaeological Source Material," in: World Archaeology 32/3 (2001), pp. 292–311.
Adorno, Theodor W., Prisms, transl. Samuel and Sherry Weber, Cambridge (MA) 1983.
Agamben, Giorgio: Bartleby oder die Kontingenz. Die absolute Immanenz, Berlin 1998.
Agamben, Giorgio: State of Exception, transl. Kevin Attell, Chicago and London 2005a.
Agamben, Giorgio: The Time that Remains: A Commentary on the Letter to the Romans, transl. Patricia Dailey, Stanford 2005b.
Ahlström, Christian: Looking for Leads: Shipwrecks of the Past Revealed by Contemporary Documents and the Archaeological Record, Helsinki 1997.
Akashi, Kinji: Cornelius van Bynkershoek: His Role in the History of International Law, The Hague, London, and Boston 1998.
Akinside [i. e. Akenside], Mark: The Pleasures of Imagination: A Poem. In Three Books, London 1744.
Albrecht, Michael von: Vergil. Bucolica, Georgica, Aeneis. Eine Einführung, Heidelberg 2006.
Aldane, Michael (transl.): Fortunatus. First complete translation of the editio princeps [1509] into English, with original illustrations: http://www.michaelhaldane.com/Fortunatus%20Illustrated.pdf.
Alertz, Ulrich: Vom Schiffbauhandwerk zur Schiffbautechnik. Die Entwicklung neuer Entwurfs- und Konstruktionsmethoden im italienischen Galeerenbau (1400–1700), Hamburg 1991.
Alexander, Amir R.: Geometrical Landscapes: The Voyages of Discovery and the Transformation of Mathematical Practice, Stanford 2002.
Alexander, Caroline: Die Bounty. Die wahre Geschichte der Meuterei auf der Bounty, Berlin 2004.
Alighieri, Dante: Princeton Dante Project, http://etcweb.princeton.edu/dante/pdp/
Ampère, Jean-Marie: Essai sur la philosophie des Sciences ou exposition analytique d'une classification naturelle de toutes les sciences, vol. 2, Paris 1843.
Anderson, Gregory: "Capturing Uncertainty: The Role of Insurance in the Construction of Modern Life," in: Zeitschrift für die gesamte Versicherungswissenschaft 2 (2007), pp. 129–147.
Andreae, Bernhard: Odysseus. Archäologie des europäischen Menschenbildes, Frankfurt a. M. 1982.
Andresen, Carl et al. (eds.): Lexikon der Alten Welt, 3 vol., Augsburg 1994.
Andriopoulos, Stefan: "The Invisible Hand: Supernatural Agency in Political Economy and the Gothic Novel," in ELH 66/3 (1999), pp. 739–758.
Anonymous: "Von den Gründen des Vergnügens an traurigen Gegenständen," in: Neue Bibliothek der schönen Wissenschaften und der freyen Künste 43/2 (1791), pp. 177–185

https://doi.org/10.1515/9783110610734-015

Anton, Michael: “Neuer Schutz archäologischer Kulturgüter,” in: Michael Wittinger, Rudolf Wendt, and Georg Ress (eds.): Verfassung—Volkerrecht—Kulturgüterschutz. Festschrift für Wilhelm Fiedler zum 70. Geburtstag, Berlin 2011, pp. 319–351.
Apelt, Ursula: “Der Text in der Karte. Hydrographie und Globalisierung in der Frühen Neuzeit,” in: Bernhard Klein and Gesa Mackenthun (eds.): Das Meer als kulturelle Kontaktzone. Räume, Reisen, Repräsentationen, Constance 2003, pp. 95–114.
Apollonius of Rhodes: Argonautica, transl. R.C. Seaton, London and Cambridge (MA) 1912.
Arata, Stephen: “Stevenson’s Careful Observances,” in: Romanticism and Victorianism on the Net 47 (August 2007): www.erudit.org/revue/ravon/2007/v/n47/016704ar.html.
Arendt, Hannah: Origins of Totalitarianism, New York 1976.
Aristotle: The Basic Works of Aristotle, ed. Richard McKeon, New York 1941.
Armstrong, Tim: “Slavery, Insurance, and Sacrifice in the Black Atlantic,” in: Bernhard Klein and Gesa Mackenthun (eds.): Sea Changes: Historicizing the Ocean, London 2004, pp. 167–185.
Asals, Heather: “The Voices of Silence and Underwater Experience,” in: Anna-Teresa Tymieniecka (ed.): Poetics of the Elements in the Human Condition. The Sea. From Elemental Stirrings to Symbolic Inspiration, Language, and Life-Significance in Literary Interpretation and Theory, Dordrecht, Boston, and Lancaster 1985, pp. 299–308.
Ashburner, Walter: Nómos Rhodion nautikós. The Rhodian Sea-Law. Edited from the Manuscripts, Oxford 1909.
Asher, Eugene L.: The Resistance to the Maritime Classes: The Survival of Feudalism in the France of Colbert, Berkeley and Los Angeles 1960.
Assmann, Aleida: “Fest und flüssig: Anmerkungen zu einer Denkfigur,” in: Aleida Assmann and Dietrich Harth (eds.): Kultur als Lebenswelt und Monument, Frankfurt a.M. 1991, pp. 181–199.
Assmann, Jan: “Das gerettete Wissen. Flutkatastrophen und geheime Archive,” in: Jan Assmann and Martin Mulsow (eds.): Sintflut und Gedächtnis. Erinnern und Vergessen des Ursprungs, Munich 2006, pp. 291–301.
Aubin, Nicolas: Dictionnaire de marine, contenant les termes de la navigation et de l’architecture navale avec les règles & proportions qui doivent y être observées, Amsterdam 1702.
Auden, Wystan Hugh: The Enchafèd Flood: or, The Romantic Iconography of the Sea, New York 1967.
Auerbach, Erich: Mimesis: The Representation of Reality in Western Literature, transl. Willard R. Trask, Princeton 1974.
Austin, John Langshaw: How to Do Things with Words: The William James Lectures delivered at Harvard University in 1955, 2nd edn., Oxford and New York 1975.
B. Traven: “Mein Roman ‘Das Totenschiff,’” in: Die Büchergilde, 3 (1926), pp. 34–38.
B. Traven: “In Memoriam ‘Yorikke,’” in: Johannes Beck, Klaus Bergmann, and Heiner Boehncke (eds.): Das B. Traven-Buch, Reinbek 1976, pp. 160–163.
B. Traven: Das Totenschiff, Frankfurt a.M., Vienna, and Zurich 1978.
Baas, Bernard: Das reine Begehren, Vienna 1995.
Babits, Lawrence E. and Hans van Tilburg (eds.): Maritime Archaeology: A Reader of Substantive and Theoretical Contributions, New York and London 1998.
Bachelard, Gaston: Water and Dreams: An Essay on the Imagination of Matter, transl. Edith R. Farrell, Dallas 1983.

Bachelard, Gaston: The Poetics of Space, transl. Maria Jolas, Boston 1994.
Bacon, Francis: Works, 10 vols., London 1826 ff.
Bacon, Francis: "The New Atlantis," in: Charles W. Eliot (ed.): Essays, Civil and Moral & The New Atlantis by Francis Bacon: Areopagitica & Tractate on Education by John Milton. Religio Medici by Sir Thomas Browne, New York 1965, pp. 145–183.
Bacon, Francis: The New Organon, ed. Lisa Jardine and Michael Silverthorne, Cambridge (UK) 2000.
Bacon, Francis: Of the Dignity and Advancement of Learning, in: The Works of Francis Bacon, ed. James Spedding, Robert Leslie Ellis, and Douglas Denon Heath, Cambridge (UK) 2011.
Badiou, Alain: Saint Paul: The Foundation of Universalism, transl. Ray Brassier, Stanford 2003.
Baecker, Dirk: Womit handeln Banken? Eine Untersuchung zur Risikoverarbeitung in der Wirtschaft, 2nd edn., Frankfurt a. M. 2008.
Bailey, Geoff: "The Wider Significance of Submerged Archaeological Sites and Their Relevance to World Prehistory," in: Nicholas C. Flemming (ed.): Submarine Prehistoric Archaeology of the North Sea: Research Priorities and Collaboration with Industry, York 2004, pp. 3–10.
Bakhtin, Mikhail: "Forms of Time and of the Chronotope in the Novel," in: Mikhail Bakhtin, The Dialogic Imagination: Four Essays, Austin and London 1981, pp. 84–259.
Balke, Friedrich: Der Staat nach seinem Ende. Die Versuchung Carl Schmitts, Munich 1996.
Ballabriga, Alain: Les fictions d'Homère. L'invention mythologique et cosmographique dans l'Odyssée, Paris 1998.
Balmer, Hans: Die Romfahrt des Apostels Paulus und die Seefahrtkunde im römischen Kaiserzeitalter, Bern 1905.
Balmer, Heinz: Beiträge zur Geschichte der Erkenntnis des Erdmagnetismus, Aarau 1956.
Baltsavias, Emmanuel (ed.): Recording, Modeling and Visualization of Cultural Heritage, London 2006.
Barber, Peter: "England II: Monarchs, Ministers, and Maps 1550–1625," in: David Buisseret (ed.): Monarchs, Ministers and Maps: The Emergence of Cartography as a Tool of Government in Early Modern Europe, Chicago and London 1992, pp. 57–98.
Barbour, Violet: "Marine Risks and Insurance in the Seventeenth Century," in: Journal of Economic and Business History 1 (1929), pp. 561–596.
Barbour, Violet: Capitalism in Amsterdam in the 17th Century, Baltimore 1950.
Barolini, Theodolinda: "Dante's Ulysses. Narrative and Transgression," in: Amilcare A. Jannucci (ed.): Dante: Contemporary Perspectives, Toronto 1997, pp. 113–132.
Barth, Johannes: "Neues zum Fliegenden Holländer. Die bislang unbekannte erste Mitteilung der Sage in deutscher Sprache und Wilhelm Hauffs Geschichte von dem Gespensterschiff," in: Fabula. Zeitschrift für Erzählforschung 35/3–4 (1994), pp. 311–315.
Barth, Karl: Epistle to the Romans, transl. Edwyn Hosykns, Oxford 1968.
Barthes, Roland: The Preparation of the Novel: Lecture Courses and Seminars at the Collège de France (1978–1979 and 1979–1980), transl. Nathalie Léger, New York 2011.
Barthold, Wassili Wladimirowitsch: "Der Koran und das Meer," in: Zeitschrift der deutschen Morgendländischen Gesellschaft 1929, pp. 37–43.
Baruzzi, Arno: Mensch und Maschine. Das Denken sub specie machinae, Munich 1973.

Baslar, Kemal: The Concept of the Common Heritage of Mankind in International Law, The Hague, Boston, and London 1998.
Bass, George: Archaeology under Water, Bristol 1966.
Bassett, Fletcher S.: Legends and Superstitions of the Sea and of Sailors, London 1885.
Bathurst, Bela: The Lighthouse Stevensons, London 2000.
Bathurst, Bela: The Wreckers: A Story of Killing Seas, False Lights and Plundered Ships, London 2005.
Battesti, Michèle: "Le sauvetage maritime en France au XIXe siècle ou la mer apprivoisée?," in: Christian Buchet and Claude Thomasset (eds.): Le naufrage. Actes du Colloque tenu à l'Institut Catholique de Paris (28–30 janvier 1998), Paris 1999, pp. 111–138.
Baur, Christine O'Connell: Dante's Hermeneutics of Salvation: Passages to Freedom in the Divine Comedy, Toronto 2007.
Beattie, James: Dissertations Moral and Critical, London 1783.
Beck, Horace: Folklore of the Sea, Connecticut 1985.
Beckert, Erwin and Gerhard Breuer: Öffentliches Seerecht, Berlin and New York 1991.
Beebe, William: Half Mile Down, New York 1934.
Begemann, Christian: Furcht und Angst im Prozeß der Aufklärung. Zu Literatur und Bewußtseinsgeschichte des 18. Jahrhunderts, Frankfurt a. M. 1987.
Behme, Thomas: Samuel von Pufendorf: Naturrecht und Staat, Göttingen 1995.
Behrmann, Walter: Über die niederdeutschen Seebücher des 15. und 16. Jahrhunderts, Amsterdam 1978.
Bene, F. Alexander: Der Begriff des Versicherungsfalles in der Seeversicherung, Hamburg 1928.
Beniger, James R.: The Control Revolution: Technological and Economic Origins of the Information Society, Cambridge (MA) and London 1986.
Benjamin, Walter: The Arcades Project, transl. Howard Eiland and Kevin McLaughlin, Cambridge (MA) and London 1999a.
Benjamin, Walter: Selected Writings, transl. Rodney Livingstone et al., vol. 2, part 2, 1931–1934, Cambridge (MA) 1999b.
Bense, Max: "Kybernetik oder Die Metatechnik einer neuen Maschine" (1951), in Max Bense: Ausgewählte Schriften, Bd. 2: Philosophie der Mathematik, Naturwissenschaft und Technik, Stuttgart / Weimar 1998, pp. 429–446.
Benton, Lauren: A Search for Sovereignty: Law and Geography in European Empires, 1400–1900, Cambridge (UK) 2010.
Bérard, Victor: Les Phéniciens et l'Odyssée, vol. 1., Paris 1902.
Bérard, Victor: Dans le sillage d'Ulysse (Album odysséen), avec des photos de Frédéric Boissonnas, Paris 1933.
Bérard, Victor: Les navigations d'Ulysse, 4 vols., Paris 1927–1929, reprint Paris 1971.
Bernoulli, Jakob: Wahrscheinlichkeitsrechnung (Ars conjectandi), ed. Robert Haussner, Leipzig 1899.
Bernstein, Karl Hamilkar: Die Schiffsgewalt, Göttingen 1904 (diss.).
Berressem, Hanjo: "Sailing with Serres, Deleuze and Guattari: Writing Nautical Space," in: Klaus Benesch, Jon-K Adams, and Kerstin Schmidt (eds.): The Sea and the American Imagination, Tübingen 2004, pp. 215–250.
Beurdeley, Cécile: L'archéologie sous-marine. L'Odyssée des trésors, Paris 1999.

Bietz, Hauke: Die Entwicklung der deutschen Wattforschung. Ein wissenschaftshistorischer Beitrag zur Bedeutung der Pionierarbeiten, Bremerhaven 2004.
Billecocq, Xavier Beguin: Des voyageurs français au Cap de Bonne-Espérance, Paris 1996.
Binder, Hartmut: "Der Jäger Gracchus. Zu Kafkas Schaffensweise und poetischer Topographie," in: Jahrbuch der Schiller-Gesellschaft 15 (1971), pp. 375–440.
Binggeli, Bruno: Primum mobile. Dantes Jenseitsreise und die moderne Kosmologie, Zurich 2006.
Blackbourn, David: The Conquest of Nature: Water, Landscape and the Making of Modern Germany, London 2006.
Blackmore, Josiah: Manifest perdition: Shipwreck Narrative and the Disruption of Empire, Minneapolis 2002.
Blake, John: The Sea Chart: The Illustrated History of Nautical Maps and Navigational Charts, London 2004.
Blanchot, Maurice: "The Essential Solitude," in: Maurice Blanchot, The Space of Literature, transl. Ann Smock, Lincoln and London 1982, pp. 19–34.
Blanchot, Maurice: "The Song of the Sirens," in: Maurice Blanchot, The Book to Come, transl. Charlotte Mandell, Stanford 2003, pp. 3–26.
Blasius, Julia: Schiffsarztjournale der Royal Navy als medizinhistorische Quellen. Die Überseereise der Fregatte "Bristol" in den Aufzeichnungen der Schiffsärzte Marcus Allen und Alexander Rattray (1871), Frankfurt a.M. et al. 2001.
Bligh, William: The Bounty Mutiny, ed. R. D. Madison, New York et al., 2001.
Bloch, Ernst: Avicenna und die Aristotelische Linke, Frankfurt a.M. 1963.
Blot, Jean-Yves: Archéologie sous-marine, Paris 1988.
Blume, Bernhard: "Das Bild des Schiffbruchs in der Romantik," in: Jahrbuch der deutschen Schillergesellschaft 2 (1958), pp. 145–161.
Blume, Dieter: "Schiffe und Bilder in der Wissenskultur des 15. Jahrhunderts," in: Hannah Baader and Gerhard Wolf (eds.): Das Meer, der Tausch und die Grenzen der Repräsentation, Zurich and Berlin 2010, pp. 295–316.
Blumenberg, Hans: "The Concept of Reality and the Possibility of the Novel," in: Richard Amacher and Victor Lange (eds.), New Perspectives in German Literary Criticism: A Collection of Essays, transl. David Henry Wilson et al., Princeton 1979, pp. 29–48.
Blumenberg, Hans: The Legitimacy of the Modern Age, transl. Robert M. Wallace, Cambridge (MA) 1985a.
Blumenberg, Hans: Work on Myth, transl. Robert M. Wallace, Cambridge (MA) and London 1985b.
Blumenberg, Hans: The Genesis of the Copernican World, transl. Robert B. Wallace, Cambridge (MA) and London 1987.
Blumenberg, Hans: Shipwreck with Spectator: Paradigm of a Metaphor for Existence, transl. Steven Rendall, Cambridge (MA) 1996.
Blumenberg, Hans: Die Vollzähligkeit der Sterne, Frankfurt a.M. 2000.
Blumenberg, Hans: "Anthropologische Annäherung an die Aktualität der Rhetorik," in: Hans Blumenberg, Ästhetische und metaphorologische Schriften, ed. Anselm Haverkamp, Frankfurt a.M. 2001, pp. 406–431.
Blumenberg, Hans, "Sokrates und das 'objet ambigu.' Paul Valérys Auseinandersetzung mit der Tradition der Ontologie des ästhetischen Gegenstandes," in: Hans Blumenberg,

Ästhetische und metaphorologische Schriften, ed. Anselm Haverkamp, Frankfurt a. M. 2001, pp. 74 – 111.
Blumenberg, Hans: Beschreibung des Menschen, ed. Manfred Sommer, Frankfurt a. M. 2006.
Blumenberg, Hans: Der Mann vom Mond. Über Ernst Jünger, ed. Alexander Schmitz and Marcel Lepper, Frankfurt a. M. 2007.
Blumenberg, Hans: Geistesgeschichte der Technik, ed. Alexander Schmitz and Bernd Stiegler, Frankfurt a. M. 2009.
Blumenberg, Hans: Care Crosses the River, transl. Paul Fleming, Stanford 2010.
Blumenberg, Hans: Quellen, Ströme, Eisberge, ed. Ulrich von Bülow and Dorit Krusche, Berlin 2012.
Blumhardt: "Der Einfluß des kanonischen Wucherverbots auf die Entwicklung der Assekuranz," in: Zeitschrift für die gesamte Versicherungswissenschaft 11 (1911), pp. 66 – 80.
Blunck, Hans Friedrich: Nordseesagen, Bayreuth 1982.
Boeckmann, Kurt von: Vom Kulturreich des Meeres, Berlin 1924.
Bodin, Jean: Six Books of the Commonwealth, abridged and transl. M.J. Tooley, Oxford 1955.
Bodin, Jean: On Sovereignty: Four Chapters from the Six Books of the Commonwealth, transl. Julian H. Franklin, Cambridge (UK) 1992.
Bodmer, Johann Jakob: Critische Betrachtungen über die poetischen Gemählde der Dichter, Zurich 1741, reprint Frankfurt a. M. 1971.
Boehncke, Heiner: "Arbeit und Abenteuer," in: Heinz Ludwig Arnold (ed.): B. Traven. Text und Kritik 102 (1989), pp. 6 – 16.
Böhme, Gernot and Hartmut: Feuer, Wasser, Erde, Luft. Eine Kulturgeschichte der Elemente, Munich 1996.
Böhme, Hartmut: "Umriß einer Kulturgeschichte des Wassers. Eine Einleitung," in: Hartmut Böhme (ed.), Kulturgeschichte des Wassers, Frankfurt a. M. 1988, pp. 7 – 42.
Böhme, Hartmut: "Vom phobos zur Angst. Zur Transformation und Kulturgeschichte der Angst," in: Martin Harbsmeier and Sebastian Möckel (ed.), Pathos, Affekt, Emotion. Transformationen der Antike, Frankfurt a. M. 2009, pp. 154 – 184.
Boesten, Eke: Archaeological and/or Historic Valuable Shipwrecks in International Waters: Public International Law and What it Offers, The Hague 2002.
Boitani, Pietro: The Shadow of Ulysses: Figures of a Myth, Oxford 1994.
Boiteux, Lucas Alexandre: La fortune de mer. Le besoin de sécurité et les débuts de l'assurance maritime, Paris 1968.
Bonß, Wolfgang: Vom Risiko. Unsicherheit und Ungewißheit in der Moderne, Hamburg 1995.
Bosse, Heinrich: Autorschaft ist Werkherrschaft. Über die Entstehung des Urheberrechts aus dem Geist der Goethezeit, Paderborn et al. 1981.
Bourne, William: A Regiment of the Sea and Other Writings on Navigation, Cambridge (UK) 1963.
Bowden, Hugh: Classical Athens and the Delphic Oracle: Divination and Democracy, Cambridge (UK) 2005.
Bowden, Hugh (ed.): "Misadventures of the Santo Antônio and Jorge d'Albuquerque. By Alfonso Luis," in: Charles R. Boxer (ed.), Further Selections from the Tragic History of the Sea: 1559 – 1565, Cambridge (UK) 1968, pp. 109 – 157.
Boxer, Charles R. (ed.), The Tragic History of the Sea, 1589 – 1622 : Narratives of the Shipwrecks of the Portuguese East Indiamen São Thomé (1589), Santo Alberto (1593),

São João Baptista (1622), and the Journeys of the Survivors in South East Africa, Cambridge 1959.
Boxer, Charles R.: From Lisbon to Goa, 1500–1750: Studies in Portuguese Maritime Enterprise, London 1984.
Boyle, Robert: Some Considerations Touching the Vsefvlnesse of Experimental Naturall Philosophy, 2nd edn., London 1664.
Boyle, Robert: Two Essays Concerning the Unsuccessfulness of Experiments, containing divers Admonitions and Observations (chiefly Chymical) touching that Subject, London 1669.
Boyle, Robert: Experiments and Notes About the Mechanical Production of Magnetism, London 1676.
Boyle, Robert: The Philosophical Works, ed. Peter Shaw, London 1725.
Brannigan, Fionán: "Nautisches über die Romfahrt des hl. Paulus," in: Theologie und Glaube 25 (1933), pp. 170–186.
Brant, Sebastian: Das Narrenschiff, ed. Hans-Joachim Mähl, Stuttgart 2002.
Braudel, Fernand: The Mediterranean and the Mediterranean World in the Age of Philip II, vol. 1, transl. Siân Reynolds, London 1972.
Braudel, Fernand: La dynamique du capitalisme, Paris 1985.
Braudel, Fernand: "'Frontière'—Wort und Bedeutung," in: Fernand Braudel, Das Gewissen des Historikers, ed. Ulrich Raulff, Berlin 1988.
Braudel, Fernand: The Mediterranean in the Ancient World, London 2001.
Braun, Heinrich: Geschichte der Lebensversicherung und der Lebensversicherungstechnik, 2nd edn., Berlin 1963.
Brecht, Christoph Heinrich: Zur Haftung der Schiffer im antiken Recht, Munich 1962.
Bremmer, Jan N.: "Greek Normative Animal Sacrifice," in: Daniel Ogden (ed.): A Companion to Greek Religion, Malden (MA), Oxford, and Victoria 2007.
Brenk, Frederick E.: "'Vunum pro multis caput.' Myth, History, and Symbolic Imagery in Vergil's Palinurus Incident," in: Latomus 43 (1984), pp. 776–801.
Brentano, Franz: Wahrheit und Evidenz. Erkenntnistheoretische Abhandlungen und Briefe, ed. Oskar Kraus, Leipzig 1930.
Breusing, Arthur: Die Nautik der Alten, Bremen 1886.
Breusing, Arthur: "Flavio Gioja und der Schiffskompaß," in: Wolfgang Köberer (ed.): Das rechte Fundament der Seefahrt. Deutsche Beiträge zur Geschichte der Navigation, Berlin 1982, pp. 79–95.
"British Wreck Commissioner's Inquiry" (n.a.), in: Titanic Inquiry Project, online über: http://www.titanicinquiry.org/BOTInq/BOTInq24Young03.php (accessed: 26 Nov. 2013).
Brito, Bernardo Gomes de: História trágico-marítima. Berichte aus der großen Zeit der portugiesischen Seefahrt 1552–1602, ed. Johannes Pögl, Nördlingen 1987.
Broch, Hermann: "James Joyce und die Gegenwart. Rede zu Joyces 50. Geburtstag," in: Schriften zur Literatur, vol. 1: Kritik, Frankfurt 1975, pp. 63–94.
Brockes, Barthold Heinrich: Irdisches Vergnügen in Gott. Naturlyrik und Lehrdichtung, ed. Hans-Georg Kemper, Stuttgart 1999.
Brunner-Taut, Emma (ed.): "Der Schiffbrüchige," in: Altägyptische Märchen. Mythen und andere volkstümliche Erzählungen, Munich 1990, pp. 34–39.
Brunotte, Ulrike: Hinab in den Maelstrom. Das Mysterium der Katastrophe im Werk Edgar Allan Poes, Stuttgart and Weimar 1993.

Brunotte, Ulrike: “Wasserkatastrophen—Katastrophenwasser. Von der Urflut bis zum Maelstrom,” in: Barbara Wally (ed.): Aquaria. Über die außergewöhnliche Beziehung von Wasser und Mensch, Linz and Chemnitz 2002, pp. 97–116.
Brustat-Naval, Fritz: Lichter über dem Meer. Leuchttürme, Feuerschiffe und ihre Männer, Wiesbaden 1969.
Budgen, Frank: James Joyce and the Making of Ulysses and Other Writings, London, Oxford, and Melbourne 1972.
Budisavljevic, Bojan: “D’Annunzios Torpedowesen. Instrumente der Vorsehung und Geschicke des Meeres im Seekrieg um Fiume herum,” in: Hans-Ulrich Gumbrecht, Friedrich Kittler, and Bernhard Siegert (eds.): Der Dichter als Kommandant. D’Annunzio erobert Fiume, Munich 1996, pp. 227–260.
Bueb, Volkmar: Die ‘Junge Schule’ der französischen Marine. Strategie und Politik 1875–1900, Boppard am Rhein 1971.
Büchner, Franz: “Zum Begriff und zum Beginn der Versicherung,” in: Horst Baumann, Helmut Schirmer, and Reimer Schmidt (eds.): Festschrift für Karl Sieg, Karlsruhe 1976, pp. 111–125.
Bürgel, Johann Christian and Marianne Chenou (eds.): Sindbad der Seefahrer. Eine Geschichte aus Tausendundeiner Nacht, Stuttgart 1995.
Buffon, Georges Louis Leclerc Comte de: Les Époques de la Nature, Paris 1780.
Burckhardt, Jacob: “Weltgeschichtliche Betrachtungen”, in: Das Geschichtswerk, vol. 2, Frankfurt a. M. 2007, pp. 763–972.
Burke, Edmund: The Writing and Speeches of Edmund Burke, New York 2008.
Burkert, Walter: Homo necans. Interpretationen altgriechischer Opferriten und Mythen, Berlin 1997.
Burkert, Walter: “‘Irrevocabile verbum’: Spuren mündlichen Erzählens in der Odyssee,” in: Kleine Schriften I: Homerica, Göttingen 2001, pp. 117–126.
Burkert, Walter: “The Making of Homer in the Sixth Century B.C.: Rhapsodes versus Stesichoros,” in: Kleine Schriften I: Homerica, Göttingen 2001, pp. 198–217.
Burkert, Walter: “Odysseen. Phantasien, Realitäten und Homer,” in: Elisabeth Wagner and Burkhardt Wolf (eds.): Odysseen. Mosse-Lectures 2007, Berlin 2008, pp. 14–36.
Burnet, Thomas: The Sacred Theory of the Earth, 2 vols., London 1719.
Buschmeier, Matthias: Poesie und Philologie in der Goethe-Zeit. Studien zum Verhältnis der Literatur mit ihrer Wissenschaft, Tübingen 2008.
Busmann, Hendrik: Sovereign of the Seas. Die Skulpturen des britischen Königsschiffes von 1637, Hamburg 2002.
Buß, Hedwig: Was die Alten einst erzählten… Vom ewigen Jäger, Moospfaff und vielen anderen Geistern. Totensagen aus dem Harmersbachtal und dem mittleren Schwarzwald aufgeschrieben und erläutert, Waldkirch 1996.
Cabantous, Alain: La Vergue et les Fers. Mutins et Déserteurs dans la marine de l’ancienne France (XVIIe–XVIIIe siècles), Paris 1984.
Cabantous, Alain: Le ciel dans la mer. Christianisme et civilisation maritime (XVe–XIXe siècle), Paris 1990.
Cabantous, Alain: Les côtes barbares. Pilleurs d’épaves et sociétés littorales en France (1680–1830), Paris 1993.
Cacciari, Massimo: Der Archipel Europa, Cologne 1998.

Cahagnet, Louis Alphonse: The Celestial Telegraph: or, Secrets of the Life to Come Revealed Through Magnetism, New York 1851.
Calcoen, Roger: Le cartographe Gerard Mercator 1512–1594, Brussels 1994.
Caldwell, Charles: Facts in Mesmerism, and Thoughts on Its Causes and Uses, Louisville 1842.
Calvin, Jean: Institutes of the Christian Religion, vol. 2, 4th edn., Philadelphia 1843.
Calvin, Jean: Lectures or Daily Sermons, of that Reuerend Diuine, D. Iohn Caluine, Pastor of the Church of God in Geneua, upon the Prophet Jonas, by N.B. student in Diuinitie, London 1578.
Cameron, Alan and Roy Farndon, Scenes from Sea and City. Lloyds List 1734–1984, Essex 1984.
Camões, Luís de: Os Lusíadas (The Lusiads), transl. Richard Francis Burton, 2 vols., London 1880.
Campe, Rüdiger: "Die Schreibszene, Schreiben," in: Hans Ulrich Gumbrecht and K. Ludwig Pfeiffer (eds.): Paradoxien, Dissonanzen, Zusammenbrüche. Situationen offener Epistemologie, Frankfurt a. M. 1991, pp. 759–772.
Campe, Rüdiger: "Kafkas Institutionenroman. Der Proceß, Das Schloß," in: Rüdiger Campe and Michael Niehaus (eds.): Gesetz und Ironie. Festschrift für Manfred Schneider, Heidelberg 2004, pp. 197–208.
Campe, Rüdiger, "Robert Walsers Institutionenroman Jakob von Gunten," in: Rudolf Behrens and Jörn Steigerwald (eds.): Die Macht und das Imaginäre. Eine kulturelle Verwandschaft in der Literatur zwischen Früher Neuzeit und Moderne, Würzburg 2005, pp. 235–250.
Campe, Rüdiger: "Die Sorge der Prinzessin und die Zukunft des Ereignisses," in: Arne Höcker, Jeannie Moser and Philippe Weber (eds.): Wissen. Erzählen. Narrative der Humanwissenschaften, Bielefeld 2006a, pp. 65–84.
Campe, Rüdiger: "Epoche der Evidenz. Knoten in einem terminologischen Netzwerk," in: Sibylle Peters and Martin Jörg Schäfer (eds.): 'Intellektuelle Anschauung'. Figurationen von Evidenz zwischen Kunst und Wissen, Bielefeld 2006b, pp. 25–43.
Campe, Rüdiger: "Kafkas Fürsprache," in: Oliver Simon and Arne Höcker (eds.): Kafkas Institutionen, Bielefeld 2007, pp. 188–212.
Campe, Rüdiger: The Game of Probability: Literature and Calculation from Pascal to Kleist, transl. Ellwood H. Wiggins, Jr., Stanford 2012.
Canetti, Elias: Crowds and Power, transl. Carol Stewart, New York 1978.
Carey, Daniel: "Compiling Nature's History: Travellers and Travel Narratives in the Early Royal Society," in: Annals of Science 54 (1997), pp. 269–292.
Carolis, Massimo de: Das Leben im Zeitalter seiner technischen Reproduzierbarkeit, Zurich and Berlin 2009.
Carpano, Claudio Mecchegiani, Luigui Fozzati, and Edoardo Tortorici: "The ArcheoMar Project: A Register of Underwater Archaeological Sites in For Regions of Southern Italy," in: Albert Hafner (ed.): Die neue Sicht. Unterwasserarchäologie und Geschichtsbild. Akten des 2. Internationalen Kongresses für Unterwasserarchäologie, Rüschlikon bei Zürich, 21.–24. Oktober 2004, Basel 2006, pp. 246–249.
Carrer, Olivier Le: Die Vermessung der Ozeane. Welt- und Seekarten von der Antike bis zur Neuzeit, Bielefeld 2009.
Casarino, Cesare: Modernity at Sea. Melville, Marx, Conrad in Crisis, Minneapolis et al. 2002.

Casey, Edward S.: Representing Place. Landscape Painting and Maps, Minneapolis and London 2002.
Cassirer, Ernst: The Philosophy of Symbolic Form, vol. 3, The Phenomenology of Knowledge, transl. Ralph Mannheim, New Haven, and London 1957.
Casson, Lionel: Die Seefahrer der Antike, Munich 1979.
Castel, Robert: "Von der Gefährlichkeit zum Risiko," in: Manfred Max Wambach (ed.): Der Mensch als Risiko. Zur Logik von Prävention und Früherkennung, Frankfurt a. M. 1983, pp. 51–74.
Castex, Raoul: "Mer-terre-air," in: Revue Militaire Générale (Juli 1937), pp. 13–28.
Castoriadis, Cornelius: The Imaginary Institution of Society, transl. Kathleen Blamey, Cambridge (UK) 1987.
Ceccarelli, Giovanni: "Risky Business. Theological and Canonical Thought on Insurance from the Thirteenth to the Seventeenth Century," in: Journal of Medieval and Early Modern Studies 31/3 (2001), pp. 607–658.
Cerasano, Susan P.: William Shakespeare's 'The Merchant of Venice.' A Sourcebook, New York et al. 2004.
Certeau, Michel de: "'Räume' und 'Orte,'" in: Jörg Dünne and Stephan Günzel (eds.): Raumtheorie. Grundlagentexte aus Philosophie und Kulturwissenschaften, Frankfurt a. M. 2006, pp. 345–352.
Certeau, Michel de: The Practice of Everyday Life, transl. Steven Rendall, Berkeley et al. 2006.
Chalmers, George: Another Account of the Incidents, from which the Title, and a Part of the Story of Shakespeare's Tempest, Were Derived; And the True Era of it Ascertained, evincing the Original Connexion of the Royal Family with the Poet's Drama, London 1815, reprint New York 1973.
Chankin, Donald O.: Anonymity and Death. The Fiction of B. Traven, Pennsylvania 1975.
Chaunu, Pierre: Séville et l'Atlantique (1504–1650). Partie interprétative. Les Structures, vol. 8 of Pierre Chaunu, Séville et l'Amérique aux XVIe et XVIIe siècles, Paris 1959.
Childers, Erskine: The Riddle of the Sands: A Record of Secret Service, New York 1915.
Clarke, Arthur C.: The Exploration of Space, New York 1951.
Clarke, Arthur C.: The Challenge of the Spaceship. Previews of Tomorrow's World, New York 1959.
Clarke, Arthur C.: Profiles of the Future: An Enquiry into the Limits of the Possible, London 1962.
Clarke, Arthur C.: By Space Possessed. Essays on the Exploration of Space, London 1993.
Clarke, Arthur C.: "A Recursion in Metastories," in: The Collected Stories, London 2000, pp. 854.
Clarke, Arthur C.: 2001. A Space Odyssey, London 2006.
Classen, Peter: "Romanum gubernans imperium," in: Gunther Wolf (ed.): Zum Kaisertum Karls des Großen. Beiträge und Aufsätze, Darmstadt 1972, pp. 4–29.
Claus, Felix: "25 Jahre Sozialversicherung," in: Soziale Praxis und Archiv für Volkswohlfahrt 20/1 (1910/1911), pp. 10–14.
Cockrell, Wilburn A.: "Why Dr. Bass Couldn't Convince Dr. Gumbel. The Trouble With Treasure Revisited, Again," in: Lawrence E. Babits and Hans van Tilburg (eds.): Maritime Archaeology. A Reader of Substantive and Theoretical Contributions, New York and London 1998, pp. 85–96.

Cohen, Margaret: The Novel and the Sea, Princeton 2010.
Columbus, Christopher: The Voyage of Christopher Columbus: Columbus' Own Journal of Discovery, transl. and ed. John Cummins, London 1992.
Columbus, Christopher: The Book of Prophecies, ed. Roberto Rusconi, Berkeley, Los Angeles, and London 1997.
Columbus, Christopher: Der erste Brief aus der Neuen Welt, lateinisch–deutsch, ed. Robert Wallisch, Stuttgart 2006.
Conrad, Joseph: Last Essays, London and Toronto 1926.
Conrad, Joseph: Collected Works. The Medallion Edition, 22 vols., London 1995.
Cooper, James Fenimore: Sea Tales. The Pilot. The Rover, ed. Kay Seymour House und Thomas L. Philbrick, New York 1991.
Corbin, Alain: The Lure of the Sea: The Discovery of the Seaside in the Western World 1750–1840, transl. Jocelyn Phelps, Berkeley and Los Angeles 1994.
Cormack, Bradin: "Marginal Waters. Pericles and the Idea of Jurisdiction," in: Andrew Gordon and Bernhard Klein (eds.): Literature, Mapping, and the Politics of Space in Early Modern England, Cambridge (UK) 2001, pp. 155–180.
Cornils, Anja: Vom Geist Gottes erzählen. Analysen zur Apostelgeschichte, Tübingen 2006.
Correa, Gaspar: The Three Voyages of Vasco da Gama and His Viceroyalty, ed. Henry E. J. Stanley, reprint New York 1869.
Cortés, Martin: The Arte of Nauigation Conteynyng a Compendious Description of the Sphere, with the Makyng of Certen Instrumentes and Rules for Nauigations: and Exemplified by Manye Demonstrations. Wrytten in the Spanyshe Tongue by Martin Cortes, and Directed to the Emperour Charles the Fyfte. Translated out of Spanyshe into Englyshe by Richard Eden, London 1561.
Corvisier, Jean-Nicolas: "Le naufrage en Grèce ancienne," in: Christian Buchet and Claude Thomasset (eds.): Le naufrage. Actes du Colloque tenu à l'Institut Catholique de Paris (28–30 janvier 1998), Paris 1999, pp. 17–32.
Cotter, Charles H.: "The Early History of Ship Magnetism: The Airy-Scoresby Controversy," in: Annals of Science 34/6 (1977), pp. 589–599.
Craig, Kenneth M.: A Poetics of Jonah. Art in the Service of Ideology, Columbia (SC) 1994.
Craig, William Lane: The Problem of Divine Foreknowledge and Future Contingents from Aristotle to Suarez, Leiden et al. 1988.
Crane, Nicholas: Mercator: The Man who Mapped the Planet, London 2002.
Crumlin-Pedersen, Ole: "Experimental Archaeology and Ships—Principles, Problems and Examples," in: Lucy Katherine Blue, Frederick M. Hocker, and Anton Englert (eds.): Connected by the Sea. Proceedings of the Tenth International Symposium on Boat and Ship Archaeology, Roskilde 2003, Oxford 2006, pp. 1–7.
Crump, Charles: The Morte Stone. A Tale of the Coast, Based on Fact. With a Remonstrance Against the System of Wrecking; and an Appeal fort Shipwrecked Seamen, London 1850.
Curtius, Ernst Robert: European Literature and the Latin Middle Ages, transl. Willard R. Trask, Princeton 1953.
Curtius, Ernst Robert: "The Ship of the Argonauts," in: Ernst Robert Curtius, Essays on European Literature, transl. Michael Kowal, Princeton 1973, pp. 465–496.
Daiber, Jürgen: Experimentalphysik des Geistes. Novalis und das romantische Experiment, Göttingen 2001.

Damlos-Kinzel, Christiane: "The Tension between Oeconomics and Chrematistics in William Shakespeare's 'The Merchant of Venice,'" in: Christian Bode and Wolfgang Klooss (eds.): Historicizing / Contemporizing Shakespeare. Essays in Honour of Rudolf Böhm, Trier 2000, pp. 115–132.

Dammann, Günter: "Die Rückübersetzung von Traven-Romanen ins Deutsche während des Exils der Büchergilde Gutenberg in Zürich," in: Günter Dammann (ed.): B. Traven. Autor—Werk—Werkgeschichte, Würzburg 2012, pp. 13–28.

Dana, Richard Henry: "Cruelty to Seamen: Being the Case of Nichols & Couch," in: The American Jurist (October 1839), pp. 92–107.

Dana, Richard Henry: The Seaman's Friend: Containing a Treatise on Practical Seamanship, with Plates; A Dictionary of Sea Terms; Customs and Usages of the Merchant Service; Laws Relating to the Practical Duties of Master and Mariners (1841), 2nd edn., London 1856.

Dana, Richard Henry: Two Years before the Mast, Hertfordshire 1996.

Danek, Georg: Epos und Zitat. Studien zu den Quellen der Odyssee, Vienna 1998.

Daniel, Stephen H.: "Myth and the Grammar of Discovery in Francis Bacon," in: Philosophy and Rhetoric 15/4 (1982), pp. 219–237.

Darcy, Robert F.: "Freeing Daughters on Open Markets: The Incest Clause in The Merchant of Venice," in: Linda Woodbridge (ed.): Money and the Age of Shakespeare. Essays in New Economic Criticism, New York et al. 2003, pp. 189–200.

Day, Archibald: The Admiralty Hydrographic Service 1795–1919, London 1967.

Deacon, Margaret: Scientists and the Sea 1650–1900. A Study of Marine Science, London 1971.

Dear, Peter: "Totius in Verba: Rhetoric and Authority in the Early Royal Society," in: Isis 76/2 (1985), pp. 145–161.

Defoe, Daniel: An Essay upon Public Credit Being an Enquiry how the Publick Credit Comes to Depend Upon the Change of the Ministry, Or the Dissolutions of Parliaments, and Whether it Does So Or No?: with an Argument, Proving that the Public Credit May be Upheld and Maintained in this Nation, and Perhaps Brought Toa Greater Height Than it Ever Yet Arrived At, Though All the Changes Or Dissolutions Already Made, Pretended To, and Now Discoursed Of, Should Come to Pass in the World, London 1710.

Defoe, Daniel: A Tour Through the Island of Great Britain: Divided into Circuits or Journeys, vol. 1, London 1778.

Defoe, Daniel: Roxana: The Fortunate Mistress or, a History of the Life and Vast Variety of Fortunes of Mademoiselle de Beleau, London 1840.

Defoe, Daniel: "The Anatomy of Exchange Alley; or, a System of Stock-Jobbing: Proving that Scandalous Trade, as it is Now Carried on, to be Knavish in Its Private Practice, and Treason in Its Public," in: John Francis (ed.): Chronicles and Characters of the Stock Exchange, Boston 1850.

Defoe, Daniel: "The Storm, or, a Collection of the Most Remarkable Casualties and Disasters which Happened in the Late Dreadful Tempest, Both by Sea and Land," in: The Novels and Miscellaneous Works of Daniel Defoe, London 1855, pp. 249–412.

Defoe, Daniel: Serious Reflections during the Life and Surprising Adventures of Robinson, London 1899.

Defoe, Daniel: Roxana: The Fortunate Mistress or, a History of the Life and Vast Variety of Fortunes of Mademoiselle de Beleau, ed. David Blewett, London et al. 1982.

Defoe, Daniel: The Fortunes and Misfortunes of the Famous Moll Flanders, Toronto et al. 1989.
Defoe, Daniel: An Essay upon Projects, ed. Joyce D. Kennedy, Michael Seidel und Maximillian E. Nowack, New York 1999.
Defoe, Daniel: Robinson Crusoe, London 1994.
Defoe, Daniel: The Further Adventures of Robinson Crusoe, Doylestown 2004.
Deist, Wilhelm: Flottenpolitik und Flottenpropaganda. Das Nachrichtenbureau des Reichsmarineamtes 1897–1914, Stuttgart 1979.
Deleuze, Gilles: The Logic of Sense, transl. Mark Lester with Charles Stivale, New York 1990.
Deleuze, Gilles: Difference and Repetition, transl. Paul Patton, London 1994.
Deleuze, Gilles: Cinema 1: The Movement-Image, transl. Hugh Tomilinson and Barbara Habberjam, Minneapolis 1996.
Deleuze, Gilles: Cinema 2: The Time-Image, transl. Hugh Tomlinson and Robert Galeta, Minneapolis 1997.
Deleuze, Gilles: L'île déserte et autres textes 1953–1974, ed. David Lapoujade, Paris 2002.
Deleuze, Gilles: The Fold: Leibniz and the Baroque, transl. Tom Conley, London and New York 2006.
Deleuze, Gilles: "Bartleby; or the Formula," trans. Daniel W. Smith and Michael A. Greco, in: Oncurating 40 (Sept. 2018), pp. 49–64.
Deleuze, Gilles and Guattari, Félix: A Thousand Plateaus: Capitalism and Schizophrenia, transl. Brian Massumi, Minneapolis 1980.
Deleuze, Joseph P. F.: Practical Instruction in Animal Magnetism, Providence 1837.
Delumeau, Jean: Rassurer et protéger. Le sentiment de sécurité dans l'Occident d'autrefois, Paris 1989.
Delumeau, Jean: "Die Angst der Seefahrer vor dem Meer," in: Klaus Bergmann and Solveig Ockenfuß (eds.): Neue Horizonte. Eine Reise durch die Reisen, Reinbek 1984, pp. 51–68.
Dening, Greg: Mr. Bligh's Bad Languages. Passion, Power and Theatre on the Bounty, Cambridge (UK), New York, and Melbourne 2002.
Derrida, Jacques: Given Time: 1. Counterfeit Money, transl. Peggy Kamuf, Chicago and London 1992a.
Derrida, Jacques: The Other Heading: Reflections on Today's Europe, transl. Pascale Anne Brault and Michael B. Naas, Bloomington 1992b.
Derrida, Jacques: "Pace Not(s)," transl. John P. Leavey, in: Jacques Derrida, Parages, transl. John P. Leavey et al., Stanford 1994.
Derrida, Jacques: "Archive Fever: A Freudian Impression," transl. Eric Prenowitz, in: Diacritics 25/2 (summer, 1995), pp. 9–63.
DeRoover, F. E.: "Early Examples of Marine Insurance," in: Journal of Economic History 5 (1945), pp. 172–200.
Despoix, Philippe: Die Welt vermessen. Dispositive der Entdeckungsreise im Zeitalter der Aufklärung, Göttingen 2009.
Desrosières, Alain: The Politics of Large Numbers: A History of Statistical Reasoning, transl. Camille Naish, Cambridge (MA) 1998.
Detel, Wolfgang: "Das Prinzip des Wassers bei Thales," in: Hartmut Böhme (ed.): Kulturgeschichte des Wassers, Frankfurt a. M. 1988, pp. 43–64.
Detienne, Marcel und Pierre Vernant: La cuisine du sacrifice en pays grec, Paris 1983.

Détrée, Jean-François: “Fortunes de mer,” in: Michel L’Hour and Elisabeth Veyrat (eds.): La mer pour mémoire. Archéologie sous-marine des épaves atlantiques, Paris 2005, pp. 32–33.
Dictionnaire de marine contenant les termes de la Navigation et de L’architecture navale. Avec les Règles & Proportions qui doivent y être observées (n.a.), Amsterdam 1702.
“Die Unfall- und Krankenversicherung der Seeleute in Österreich” (n.a.), in: Soziale Praxis 20/20 (1911), pp. 630f.
Dilke, Oswald A. W.: Greek and Roman Maps, London 1985.
Dircksen, Rolf: Das Wattenmeer. Landschaft ewigen Wandels, Munich 1959.
Domingues, Francisco Contente: “Neue Meere befahren und Neue Welten darstellen. Nautik und Kartographie in der portugiesischen Seefahrt,” in: Michael Kraus and Hans Ottomeyer (eds.): Novos Mundos—Neue Welten. Portugal und das Zeitalter der Entdeckungen, Dresden 2007, pp. 63–72.
Dölger, Franz Joseph: “Die Eucharistie als Reiseschutz,” in: Antike und Christentum 5 (1976), pp. 232–247.
Döring, Jörg: “Rhetorik des Schweigens. Zu Franz Kafkas Odysseus-Travestie im dritten Oktavheft,” in: Iris Hermann and Anne Maximiliane Jäger-Gogoll (eds.): Durchquerungen. Für Ralf Schnell zum 65. Geburtstag, Heidelberg 2008, pp. 55–67.
Döring, Tobias: “The Sea Is History. Historicizing the Homeric Sea in Victorian Passages,” in: Bernhard Klein (ed.): Fictions of the Sea: Critical Perspectives on the Ocean in British Literature and Culture, Burlington 2002, pp. 121–140.
Dolzer, Rudolf: “Die Deklaration des Kulturguts zum ‘common heritage of mankind,’” in: Rudolf Dolzer et al. (eds.): Rechtsfragen des internationalen Kulturgüterschutzes. Symposion vom 22./23. Juni 1990 im Internationalen Wissenschaftsforum Heidelberg, Heidelberg 1994.
Dotson, John E.: Merchant Culture in Fourteenth Century Venice: The Zibaldone da Canal, Binghamton 1994.
Dotzler, Bernhard (ed.): Babbages Rechen-Automate. Ausgewählte Schriften, Vienna and New York 1996.
Dröge, Karl-Eckhard: “Ein neues Zeitalter bricht an. Die deutsche Marine auf dem Weg ins nächste Jahrhundert,” in: Truppenpraxis. Wehrausbildung. Zeitschrift für Führung, Ausbildung und Erziehung 42/2 (1998), pp. 112–117.
Drummond, Maldwin: The Riddle, London 1985.
Drux, Rudolf: “Des Dichters Schiffahrt. Struktur und Pragmatik einer poetologischen Allegorie,” in: Walter Haug (ed.): Formen und Funktionen der Allegorie. Symposium Wolfenbüttel 1978, Stuttgart 1979, pp. 38–51.
Dünne, Jörg: Die kartographische Imagination. Erinnern, Erzählen und Fingieren in der Frühen Neuzeit, Munich 2011.
Ducoin, Jacques: Naufrages, conditions de navigation et assurances dans la marine de commerce du XVIIIe siècle, 2 vols., Paris 1989.
Ducoin, Jacques: “Naufrageurs et pilleurs d’épaves sur les côtes bretonnes au XVIIIe siècle,” in: Christian Buchet and Claude Thomasset (eds.): Le naufrage. Actes du Colloque tenu à l’Institut Catholique de Paris (28–30 janvier 1998), Paris 1999, pp. 311–325.
Duffy, James: Shipwreck and Empire: Being an Account of Portuguese Maritime Disasters in a Century of Decline, Cambridge (MA) 1955.
Durkheim, Émile: Soziologie und Philosophie, Frankfurt a. M. 1976.

Ebel, Wilhelm: “Über Sklaven- und Sklavereiversicherung,” in: ZVersWiss 3 (1963), pp. 207–230.
Ebeling, Knut: Wilde Archäologien 1. Theorien der materiellen Kultur von Kant bis Kittler, Berlin 2012.
Eco, Umberto: The Aesthetics of Chaosmos: The Middle Ages of James Joyce, Cambridge (MA) 1989.
Eder, Markus: Crime and punishment in the Royal Navy of the Seven Years’ War, 1755–1763, Aldershot et al. 2004.
Edson, Evelyn, Emilie Savage-Smith, and Anna-Dorothee von den Brincken: Der mittelalterliche Kosmos. Karten der christlichen und islamischen Welt, Darmstadt 2005.
Eidelmann, David J.: SOS. Schiff und Mensch in Not, Rostock 1985.
Eisen, Ute E.: Die Poetik der Apostelgeschichte. Eine narratologische Studie, Fribourg and Göttingen 2006.
Ellmann, Richard: James Joyce, Oxford 1959.
Ellmers, Detlef: “Schiffsarchäologie in Deutschland,” in: Berichte zur Denkmalpflege in Niedersachsen 17/1 (1997), pp. 23–26.
Elster, Walter: Die Anzeigepflicht bei der Seeversicherung, Leipzig 1931 (diss.).
Engel, Christoph: Wissen, Nichtwissen, unsicheres Wissen, Baden-Baden 2002.
Engert, Rolf: “Die Sage vom fliegenden Holländer,” in: Meereskunde. Sammlung volkstümlicher Vorträge zum Verständnis der nationalen Bedeutung von Meer und Seewesen, ed. Institut für Meereskunde Berlin, Berlin 1927.
Ernst, Wolfgang: “Das Archiv als Gedächtnisort,” in: Knut Ebeling and Stephan Günzel (eds.): Archivologie. Theorien des Archivs in Philosophie, Medien und Künsten, Berlin 2009, pp. 177–200.
Ernst, Wolfgang: “Der Appell der Medien: Wissensgeschichte und ihr Anderes,” in: Ana Ofak and Philipp von Hilgers (eds.): Rekursionen. Von Faltungen des Wissens, Munich 2010, pp. 177–197.
Ernst, Wolfgang: “Troja zwischen Medien und Archäologie,” in: Knut Ebeling and Stefan Altekamp (eds.): Die Aktualität des Archäologischen in Wissenschaft, Medien und Künsten, Frankfurt a.M. 2004, pp. 233–251.
Esménard, Jean Alphonse: La Navigation, 2nd edn., Paris 1806.
Esposito, Elena: Soziales Vergessen. Formen und Medien des Gedächtnisses der Gesellschaft, Frankfurt a.M. 2002.
Esposito, Elena: Die Fiktion der wahrscheinlichen Realität, Frankfurt a.M. 2007.
Ewald, François: The Birth of Solidarity, rev. edition, transl. Timothy Scott Johnson, Duke UP 2020.
Fabre, David: Seafaring in ancient Egypt, London 2005.
Fabricius, Johann Albert: Hydrotheologie oder Versuch, durch aufmerksame Betrachtung der Eigenschaften, reichen Austheilung und Bewegung der Wasser, die Menschen zur Liebe und Bewunderung Ihres Gütigsten, Weisesten, Mächtigsten Schöpfers zu ermuntern, Hamburg 1734.
Fahlke, Eberhard: “Auf der Suche nach ‘Inselgeschichten’” in: Uwe Johnson: Inselgeschichten, ed. Eberhard Fahlke, Frankfurt a.M. 1995, pp. 169–198.
Falconer, William: An Universal Dictionary of the Marine, or, A Copious Explanation of the Technical Terms and Phrases Employed in the Construction, Equipment, Furniture, Machinery, Movements, and Military Operations of a Ship, London 1780.

Falconer, William: A Critical Edition of the Poetical Works of William Falconer, ed. William R. Jones, Lewiston et al. 2003.
Febvre, Lucien: "Pour l'histoire d'un sentiment: le besoin de sécurité," in: Annales. Économies, Sociétés, Civilisation (1956), pp. 244–247.
Feldbusch, Thorsten: Zwischen Land und Meer. Schreiben auf den Grenzen, Würzburg 2003.
Feneberg, Wolfgang: Paulus der Weltbürger. Eine Biographie, Munich 1992.
Ferreiro, Larrie D.: Ships and Science: The Birth of Naval Architecture in the Scientific Revolution, 1600–1800, Cambridge (MA) and London 2007.
Feshbach, Sidney: "Literal/Littoral/Littorananima: The Figure on the Shore in the Works of James Joyce," in: Anna-Teresa Tymieniecka (ed.): Poetics of the Elements in the Human Condition: The Sea. From Elemental Stirrings to Symbolic Inspiration, Language, and Life-Significance in Literary Interpretation and Theory, Dordrecht et al. 1985, pp. 325–342.
Fichte, Jörg O.: "Providentia—Fatum—Fortuna," in: Das Mittelalter 1 (1996), pp. 5–19.
Finsler, Georg: Homer in der Neuzeit von Dante bis Goethe, Leipzig and Berlin 1912.
Firth, Antony: Managing Archaeology Underwater: A Theoretical, Historical and Comparative Perspective on Society and Its Submerged Past, Oxford 2002.
Fitger, Emil: Die wirtschaftliche und technische Entwicklung der Seeschiffahrt von der Mitte des 19. Jahrhunderts bis auf die Gegenwart, Leipzig 1902.
Flasche, Hans: Die Struktur des Auto Sacramental "Los Encantos de la Culpa" von Calderón. Antiker Mythos in christlicher Umprägung, Cologne and Opladen 1968.
Flatman, Joe und Mark Staniforth: "Historical Maritime Archaeology," in: Dan Hicks and Mary C. Beaudry (eds.): The Cambridge Companion to Historical Archaeology, Cambridge (UK) 2006, pp. 168–188.
Flinders Petrie, W. K, transl.: "The Tale of the Shipwrecked Sailor," in: Eva March Tappan (ed.), The World's Story: A History of the World in Story, Song and Art, vol. 3, Egypt, Africa, and Arabia, Boston 1914, pp. 41–46.
Flohr, Dieter: Der Fliegende Holländer. Legenden und Tatsachen, Rostock 2002.
Florschütz, Georg: "Die Versicherung der Abgelehnten," in: Zeitschrift für die gesamte Versicherungswissenschaft 16 (1916), pp. 431–435.
Fludernik, Monika: Erzähltheorie. Eine Einführung, 2nd edn., Darmstadt 2008.
Flusser, Vilém: "Räume," in: Jörg Dünne and Stephan Günzel (eds.): Raumtheorie. Grundlagentexte aus Philosophie und Kulturwissenschaften, Frankfurt a. M. 2006, pp. 274–284.
Fock, Gorch: "Etwas über Sagenbildung," in: Heimatbuch für unser hamburgisches Wandergebiet. Mit 19 Autotypien, ed. Pädagogische Vereinigung von 1905 in Hamburg, Hamburg 1914, pp. 46–51.
Förster, Heinz von: "Cybernetics of Epistemology," in: Understanding Understanding. Essays on Cybernetics and Cognition, New York et al. 2003, pp. 229–246.
Fohrmann, Jürgen: Abenteuer und Bürgertum. Zur Geschichte der deutschen Robinsonaden im 18. Jahrhundert, Stuttgart 1981.
Foucault, Michel: Madness and Civilization: A History of Insanity in the Age of Reason, transl. Richard Howard, New York 1965.
Foucault, Michel: The Order of Things: An Archeology of the Human Sciences, New York 1970.
Foucault, Michel: The Archeology of Knowledge, transl. A.M. Sheridan Smith, New York 1972.

Foucault, Michel: "Language to Infinity," in: Michel Foucault, Language, Counter-Memory, Practice: Selected Interviews, ed. Donald F. Bouchard, transl. Donald F. Bouchard and Sherry Simon, Ithaca and New York 1977, pp. 53–67.
Foucault, Michel: "What is an Author?," in: Michel Foucault, Language, Counter-Memory, Practice: Selected Interviews, ed. Donald F. Bouchard, transl. Donald F. Bouchard and Sherry Simon, Ithaca and New York 1977, pp. 113–138.
Foucault, Michel: "Omnes et Singulatim: Towards a Criticism of 'Political Reason,'" in: Sterling M. McMurrin (ed.), The Tanner Lectures on Human Values II, Cambridge (UK) 1981, pp. 223–255.
Foucault, Michel: "Space, Knowledge, and Power," in: Michel Foucault, The Foucault Reader, ed. Paul Rabinow, New York 1984, pp. 239–256.
Foucault, Michel: "Of Other Spaces: Utopias and Heterotopias," transl. Jay Miskowiec, in: Diacritics 16/1 (1986), pp. 22–27.
Foucault, Michel: "The Thought from Outside," in: Michel Foucault, Maurice Blanchot: The Thought from Outside, transl. Brian Massumi, Cambridge (MA) and London 1987, pp. 7–60.
Foucault, Michel: Dits et écrits 1954–1988, vol. 1, Paris 1994.
Foucault, Michel: "Nietzsche, Freud, Marx," transl. Jon Anderson and Gary Hentzi, in: Michel Foucault, Aesthetics, Method, and Epistemology, New York 1998, pp. 269–278.
Foucault, Michel: Security, Territory, Population: Lectures at the Collège de France 1977–1978, transl. Graham Burchell, New York 2007.
Foulke, Robert: The Sea Voyage Narrative, New York 1997.
Fourier, Charles: The theory of the Four Movements, ed. Gareth Stedman Jones and Ian Patterson, transl. Ian Patterson, Cambridge (UK) 1996.
Fournier, Georges: Hydrographie contenant la Théorie et la Pratique de toutes les Parties de la Navigation, 2nd edn., Paris 1667.
Fowler, Elizabeth: "The Ship Adrift," in: Peter Hulme and William H. Sherman (eds.): 'The Tempest' and Its Travels, London 2000, pp. 37–40.
Fowles, John: "Shipwreck," in: Jonathan Raban (ed.): The Oxford Book of the Sea, New York 1992, pp. 494 f.
Francke, Ernst: "Erster Verhandlungstag, Montag, 14. September 1903: Die Lage der in der Seeschiffahrt beschäftigten Arbeiter," in: Verhandlungen des Vereins für Socialpolitik über die Lage der in der Seeschiffahrt beschäftigten Arbeiter und über die Störungen im deutschen Wirtschaftsleben während der Jahre 1900 ff., vol. 113, Leipzig 1904, pp. 11–44.
Frank, Manfred: Die unendliche Fahrt. Die Geschichte des Fliegenden Holländers und verwandter Motive, Leipzig 1995.
Frank, Michael C. und Kirsten Mahlke: "Nachwort," in: Michail M. Bachtin: Chronotopos, Frankfurt a. M. 2008, pp. 201–242.
Frazer, James George: The Golden Bough. A Study in Magic and Religion, abridged edition, New York 1944.
Frede, Dorothea: Aristoteles und die 'Seeschlacht.' Das Problem der Contingentia Futura in De Interpretatione 9, Göttingen 1970.
Frege, Gottlob: "On Sense and Reference," in: Gottlob Frege, Translations from the Philosophical Writings of Gottlob Frege, transl. Peter Geach and Max Black, Oxford 1960.

Freitas, William: Camõens and His Epic. A Historic, Geographic, and Cultural Survey, Stanford 1963.
French, Gregory A.: Der Tiefseebergbau. Eine interdisziplinäre Untersuchung der völkerrechtlichen Problematik, Cologne 1990.
Freud, Sigmund: Delusion & Dream: An Interpretation in the Light of Psychoanalysis of Gradiva, A Novel, by Wilhelm Jensen, transl. Helen M. Downey, New York 1922.
Freud, Sigmund: Civilization ad ist Discontents, transl. James Strachey, New York 1962.
Freud, Sigmund: "The Theme of the Three Caskets," in: James Strachey (ed.), The Standard Edition of the Complete Psychological Works of Sigmund Freud, London 1953 – 1974a, vol. 12, pp. 291 – 301.
Freud, Sigmund: "The Dissection of the Psychical Personality," in: James Strachey (ed.), The Standard Edition of the Complete Psychological Works of Sigmund Freud, London 1953 – 1974b, vol. 22, pp. 56 – 79.
Fricke, Beate: "Schaumgeburten. Zur Topologie der Creatio ex Nihilo bei Albrecht Dürer und ihrer Vorgeschichte," in: Hannah Baader and Gerhard Wolf (eds.): Das Meer, der Tausch und die Grenzen der Repräsentation, Zurich and Berlin 2010, pp. 41 – 66.
Friedl, Herwig: "Edgar Allan Poe: 'Ms. Found in a Bottle,'" in: Klaus Lubbers (ed.): Die englische und amerikanische Kurzgeschichte, Darmstadt 1990, pp. 40 – 51.
Friedland, Klaus: "Schiff und Besatzung. Seemännische Berufsgemeinschaften im spätmittelalterlichen Nordeuropa," in: Klaus Friedland, Mensch und Seefahrt zur Hansezeit, Cologne, Weimar, and Vienna 1995, pp. 256 – 267.
Friedmann, John Block and Kristen M. Figg (eds.): Trade, Travel, and Exploration in the Middle Ages: An Encyclopedia, New York 2000.
Friedrich, Hugo: "Odysseus in der Hölle (Dante, Inferno XXVI)," in: Ernesto Grassi (ed.): Geistige Überlieferung. Das zweite Jahrbuch, Berlin 1942, pp. 154 – 200.
Friedrich, Lars: Der Achill-Komplex. Versuch einer dekonstruktiven Gattungspoetik, Paderborn and Munich 2009.
Friedrich, Rainer: "The Hybris of Odysseus," in: The Journal of Hellenic Studies 111 (1991), pp. 16 – 28.
Friel, Ian: The Good Ship: Ships, Shipbuilding and Technology in England 1200 – 1520, London 1995.
Fritscher, Bernhard: "'Archiv der Erde'. Zur Codierung von Erdgeschichte um 1800," in: Knut Ebeling and Stephan Günzel (eds.): Archivologie. Theorien des Archivs in Philosophie, Medien und Künsten, Berlin 2009, pp. 201 – 219.
Fugger, Rose Marie Gräfin: "The Commercial Activities of the Fugger Family in the Sixteenth Century," in: K. P. Mathew (ed.): Mariners, Merchants and Oceans. Studies in Maritime History, New Delhi 1995, pp. 111 – 120.
Full, Patrick: Der Abgesang der Imagination. Edgar Allen Poes Neubestimmung der menschlichen Kreativität, Trier 2007.
Fulton, Thomas W.: The Sovereignty of the Sea: An Historical Account of the Claims of England to the Dominion of the British Seas, and of the Evolution of the Territorial Waters, with Special reference to the Rights of Fishing and the Naval Salute, Edinburgh and London 1911.
Furbank, P.N. und W.R. Owens: "The 'Lost' Continuation of Defoe's Roxana," in: Eighteenth-Century Fiction 9/3 (1997), pp. 299 – 308.

Furetière, Antoine: Dictionnaire universel. Contenant généralement tous les mots françois, tant vieux que modernes, & les termes de toutes les sciences et des arts. Compilé par feu Messire Antoine Furetière, La Haye and Rotterdam 1690.
Fyfe, Herbert C.: Submarine Warfare. Past, Present, and Future, London 1903.
Galison, Peter: "Die Ontologie des Feindes. Norbert Wiener und die Vision der Kybernetik," in: Hans-Jörg Rheinberger, Michael Hagner, and Bettina Wahrig-Schmidt (eds.): Räume des Wissens. Repräsentation, Codierung, Spur, Berlin 1997, pp. 281–324.
Galison, Peter: How Experiments End, Chicago 1987.
Gamper, Michael: Elektropoetologie. Fiktionen der Elektrizität 1740–1870, Göttingen 2009.
Garabello, Roberta and Tullio Scovazzi (eds.): The Protection of the Underwater Cultural Heritage: Before and After the 2001 UNESCO Convention, Leiden et al. 2003.
Garnsey, Peter: "Grain for Rome," in: Peter Garnsey, Keith Hopkins, and C. R. Whittaker (eds.): Trade in the Ancient Economy, Berkeley and Los Angeles 1983, pp. 118–130.
Gaukroger, Stephen: Francis Bacon and the Transformation of Early-Modern Philosophy, Cambridge (UK) 2001.
Gay, Franco and Cesare Ciano: The Ships of Christopher Columbus, Rome 1996.
Geduld, Carolyn: Filmguide to 2001: A Space Odyssey, Bloomington 1973.
Gehlen, Arnold: Der Mensch. Seine Natur und seine Stellung in der Welt, ed. Karl-Siegbert Rehberg, Frankfurt a. M. 1993.
Geneste, Pascal: "The Archivist and the Hydrographer: The Origins of the French Hydrographic Service in the Dépôt des Cartes et Plans de la Marine, 1720–1850," in: Pieter van der Merwe (ed.): Science and the French and British Navies, 1700–1850, London 2003, pp. 34–47.
Genette, Gérard: The Architext: An Introduction, transl. Jane E. Lewin, Berkeley et al. 1992.
Genette, Gérard: Fiction and Diction, transl. Katharine Porter, Ithaca and London 1993.
Gennep, Arnold van: The Rites of Passage, transl. Monika B. Vizedom and Gabrielle L. Caffee, London and Henley 1960.
Gerds, Peter: Glaube, Seefahrt, Christentum. Religiöse Symbolik im maritimen Leben, Rostock 2009.
Gerhards, Meik: "Zum motivgeschichtlichen Hintergrund der Verschlingung des Jona," in: Theologische Zeitschrift 59 (2003), pp. 222–247.
Gerndt, Helge: Fliegender Holländer und Klabautermann, Göttingen 1971.
Gerndt, Helge: Kultur als Forschungsfeld, Munich 1986.
Gerndt, Helge: "Sagen und Sagenforschung im Spannungsfeld von Mündlichkeit und Schriftlichkeit," in: Fabula. Zeitschrift für Erzählforschung 29/1–2 (1988), pp. 1–20.
Gerstenberger, Heide und Ulrich Welke: Vom Wind zum Dampf. Sozialgeschichte der deutschen Handelsschiffahrt im Zeitalter der Industrialisierung, Münster 1996.
Gerstenberger, Heide und Ulrich Welke: Arbeit auf See. Zur Ökonomie und Ethnologie der Globalisierung, Münster 2002.
Gibb, David E. W.: Lloyd's of London: A Study in Individualism, London and New York 1957.
Gibbins, David and Jonathan Adams: "Shipwrecks and Maritime Archaeology," in: World Archaeology 32/3 (2001), pp. 279–292.
Giertz, Gernot (ed.): Vasco da Gama. Die Entdeckung des Seewegs nach Indien. Ein Augenzeugenbericht 1497–1499, Berlin 1986.

Gil, Isabel Capeloa: "Hydrography, or the Anxiety of the Sea. The Nautical Chart as a Cultural Model," in: Isabel Capeloa Gil (ed.): Fleeting, Floating, Flowing. Water Writing and Modernity, Würzburg 2008, pp. 103–120.
Gilchrist, J. M.: "The Historicity of Paul's Shipwreck," in: Journal for the Study of the New Testament 61 (1996), pp. 29–51.
Gitlitz, David: "Inquisition Confessions and Lazarillo de Tormes," in: Hispanic Review 68/1 (2000), pp. 53–74.
Gladstone, William Ewart: Studies on Homer and the Homeric Age, 3 vols., Oxford 1858.
Glanvill, Joseph: Plus Ultra, or, The Progress and Advancement of Knowledge Since the Days of Aristotle in an Account of Some of the Most Remarkable Late Improvements of Practical, Useful Learning, to Encourage Philosophical Endeavours, London 1668.
Glanvill, Joseph: Scepsis Scientifica, or, Confest Ignorance, the Way to Science; in an Essay of the Vanity of Dogmatizing and Confident Opinion, ed. John Owen, reprint London 2005.
Goedde, Lawrence Otto: Tempest and Shipwreck in Dutch and Flemish Art: Convention, Rhetoric, and Interpretation, University Park et al. 1989.
Goetz, Hans-Werner: "Fortuna in der hochmittelalterlichen Geschichsschreibung," in: Das Mittelalter 1 (1996), pp. 75–89.
Goldammer, Kurt: "Das Schiff der Kirche. Ein antiker Symbolbegriff aus der politischen Metaphorik in eschatologischer und ekklesiologischer Umdeutung," in: Theologische Zeitschrift der Universität Basel 6 (1950), pp. 233–237.
Goldast, Melchior: Von der Bestendigkeit [De Constantia], Faksimiliedruck der deutschen Übersetzung des Andreas Viritius nach der zweiten Auflage von 1601, mit den wichtigsten Lesarten der ersten Auflage von 1599, ed. Leonard Forster, Stuttgart 1965.
Goldschmidt, Jean-Arthur: Als Freud das Meer sah. Freud und die deutsche Sprache, Zurich 1999.
Goldstücker, Eduard: "Kafkas 'Der Heizer'. Versuch einer Interpretation," in: Acta Universitas Carolinae—Philologica 1. Germanistica Pragensia, vol. 3 (1964), pp. 49–64.
Gondek, Hans-Dieter: Angst—Einbildungskraft—Sprache. Ein verbindender Aufriß zwischen Freud—Kant—Lacan, Munich 1990.
Gonzalez, Ariel W.: "Negotiating the Convention on Underwater Cultural Heritage: Myths and Reality," in: Roberta Garabello and Tullio Scovazzi (eds.), The Protection of the Underwater Cultural Heritage: Before and After the 2001 UNESCO Convention, Leiden et al. 2003, pp. 83–88.
Göttlicher, Arvid: Materialien für ein Corpus der Schiffsmodelle im Altertum, Mainz 1978.
Göttlicher, Arvid: "Fortuna Gubernatrix—Das Steuerruder als römisches Glückssymbol," in: Antike Welt 12/4 (1981a), pp. 27–33.
Göttlicher, Arvid: Nautische Attribute römischer Gottheiten, Bremen 1981b.
Göttlicher, Arvid: Die Schiffe der Antike. Eine Einführung in die Archäologie der Wasserfahrzeuge, Darmstadt 1985.
Göttlicher, Arvid: Kultschiffe und Schiffskulte im Altertum, Berlin 1992.
Göttlicher, Arvid: Die Schiffe im Neuen Testament, Berlin 1999.
Göttlicher, Arvid: Seefahrt in der Antike. Das Schiffswesen bei Herodot, Darmstadt 2006.
Golther, Wolfgang: "Der Fliegende Holländer in Sage und Dichtung," in: Zur deutschen Sage und Dichtung. Gesammelte Aufsätze, Leipzig 1911, pp. 7–16.

Gorman, Michael John: "The Angel and the Compass. Athanasius Kircher's Magnetic Geography," in: Paula Findlen (ed.): Athanasius Kircher. The Last Man Who Knew Everything, New York and London 2004, pp. 239–259.
Gould, Richard A.: Shipwreck Anthropology, Albuquerque (New Mexico) 1983.
Gould, Richard A.: Archaeology and Social History of Ships, Cambridge (UK) 2000.
Gould, Richard A.: "An Underwater Crime Scene? The Wreck of the North Carolina," in: Disaster Archaeology, Salt Lake City 2007, pp. 168–191.
Graeber, David: Debt: The First 5,000 Years, New York 2011.
Gragnolati, Manuele: Experiencing the Afterlife—Soul and Body in Dante and Medieval Culture, Notre Dame 2005.
Grampp, William D.: "What Did Smith Mean by the Invisible Hand?," in: The Journal of Political Economy 108/3 (2000), pp. 441–465.
Grant, Michael and John Hazel: Lexikon der antiken Mythen und Gestalten, 5th edn., Munich 1987.
Green, Jeremy: Maritime Archaeology: A Technical Handbook, Amsterdam et al. 2004.
Greenblatt, Stephen: Shakespearean Negotiations: The Circulation of Social Energy in Renaissance England, Berkeley and Los Angeles 1988.
Greenfield, Bruce Robert: Narrating Discovery, New York 1992.
Gregersen, Niels Henrik: "Risk and Religion. Towards a Theology of Risk Taking (Beck, Luhmann)," in: Zygon. Journal of Religion and Science 38/2 (2003), pp. 355–377.
Grethlein, Jonas und Antonios Rengakos: Narratology and Interpretation: The Content of Narrative Form in Ancient Literature, Berlin et al. 2009.
Griffiths, Denis: Steam at Sea: Two Centuries of Steam-powered Ships, London 1997.
Grimm, Jacob and Wilhelm (eds.): Deutsche Sagen. Herausgegeben von den Brüdern Grimm, ed. Heinz Rölleke, Frankfurt a. M. 1994.
Grotius, Hugo: Von der Freiheit des Meeres (1609), Leipzig 1919.
Groys, Boris: "Der submediale Raum des Archivs," in: Knut Ebeling and Stephan Günzel (eds.): Archivologie. Theorien des Archivs in Philosophie, Medien und Künsten, Berlin 2009, pp. 139–151.
Gruenter, Rainer: "Das Schiff. Ein Beitrag zur historischen Metaphorik," in: Werner Kohlschmidt and Hermann Meyer (eds.): Tradition und Ursprünglichkeit. Akten des III. Internatioalen Germanistenkongresses 1965 in Amsterdam, Bern and Munich 1966, pp. 86–101.
Güngerich, Rudolf: Die Küstenbeschreibung in der griechischen Literatur, Münster 1950.
Günther, Gotthard: Das Bewußtsein der Maschinen. Eine Metaphysik der Kybernetik, Krefeld and Baden-Baden 1954.
Günther, Gotthard: "Die philosophische Idee einer nicht-aristotelischen Logik," in: Beiträge zur Grundlegung einer operationsfähigen Dialektik, vol. 1, Hamburg 1976, pp. 24–30.
Günther, Gotthard: Beiträge zur Grundlegung einer operationsfähigen Dialektik, vol. 2., Hamburg 1979.
Günther, Siegmund: "Über die Genesis der nautischen Kartenprojektion G. Mercators," in: Wolfgang Köberer (ed.): Das rechte Fundament der Seefahrt. Deutsche Beiträge zur Geschichte der Navigation, Berlin 1982, pp. 273–282.
Guillaumier, Paul: "New Perspective on the Historicity of St. Paul's Shipwreck on Melite," in: Michael Galea and John Ciarlò (eds.): St. Paul in Malta: A Compendium of Pauline Studies, Malta 1992, pp. 53–114.

Guillemot, Vanessa und Laurent Champonnois: Vaisseaux-phantômes et navires errants, Grenoble 1995.
Gurney, Alan: Compass: A Story of Exploration and Innovation, New York et al. 2004.
Guthke, Karl S.: B. Traven. Biographie eines Rätsels, Zurich 1990.
Guttridge, Leonard F.: Mutiny: A History of Naval Insurrection, Anapolis 1992.
Gutzkow, Karl: "Julius Mosens Ahasver," in: Karl Gutzkow: Vermittelungen. Kritiken und Charakteristiken, vol. 2, Leipzig 1842. pp. 154–177.
Habermann, Ina: "Death by Water: The Theory and Practice of Shipwrecking," in: Bernhard Klein (ed.): Fictions of the Sea: Critical Perspectives in English Literature and Culture, Aldershot 2002, pp. 104–120.
Häberle, Peter: Die Erinnerungskultur im Verfassungsstaat. 'Denk-Mal'-Themen, Geschichtsorte, Museen, nationaler und internationaler Kulturgüterschutz, Berlin 2011.
Hagen, Otto: "Begriff und Rechtsgrundlagen der Rückversicherung," in: Zeitschrift für die gesamte Versicherungswissenschaft 20 (1920), pp. 137–160.
Hagena, Katharina: Was die wilden Wellen sagen. Der Seeweg durch den Ulysses, Hamburg 2006.
Hager, Fritz-Peter: "Empeiria," in: Historisches Wörterbuch der Philosophie, vol. 2, ed. Joachim Ritter, Basel 1972, pp. 453 f.
Hahn, Hans, "The Crisis in Intuition," in: Hans Hahn, Empiricism, Logic and Mathematics: Philosophical Papers, ed. Brian McGuinness (no translator indicated), Dordrecht 1980, pp. 73–102.
Hale, Edward Everett: "The Man without a Country," in: The Man without a Country and other Tales, New York 1968, pp. 7–48.
Halley, Edmond: "An Estimate of the Degrees of the Mortality of Mankind, drawn from curious Tables of the Births and Funerals at the City of Breslaw; with an Attempt to ascertain the Price of Annuities upon Lives," in: Philosophical Transactions (1692/1693), pp. 596–656.
Halley, Edmond: Curieuse Erzehlungen von denen Winden / Nach der heutigen Seefahrer und Anwohner des Meers beständigen Anmerckungen, in: Kornelius Drebbel, Tractat oder Abhandlung von Natur und Eigenschafft der Elementen, Hoff 1723, pp. 67–106.
Hamann, Günther: Der Eintritt der südlichen Hemisphäre in die europäische Geschichte. Die Erschließung des Afrikaweges nach Asien vom Zeitalter Heinrichs des Seefahrers bis zu Vasco da Gama, Vienna et al. 1968.
Hamburger, Käte: The Logic of Literature, transl. M.J. Rose, Bloomington 1993.
Hamilton-Paterson, James: The Great Deep: The Sea and its Thresholds, New York 1992.
Hamilton-Paterson, James: Three Miles Down, London 1998.
Hamlin, Cyrus: "Die Poetik des Gedächtnisses. Aus einem Gesprach über Hölderlins 'Andenken,'" in: Hölderlin-Jahrbuch 24 (1984/85), pp. 119–138.
Hansen, Christian Peter: Das Schleswig'sche Wattenmeer und die friesischen Inseln, Glogau 1865, reprint Wiesbaden 1972.
Hanses, Dieter: Die rechtliche Stellung des Kapitäns auf deutschen Seeschiffen unter besonderer Berücksichtigung der historischen Entwicklung, Berlin 1983.
Hanway, Jonas: An Account of the Marine Society, Recommending the Piety and Policy of the Institution, and Pointing out the Advantages Accruing to the Nation. With the Motives for Establishing it. Also a Full Detail of Their Rules and Forms of Business. With a List of the Subscribers, From the Commencement in July 1756, to September 30. 1759. Also a

Proposal for Accommodating the Boys, Equipped by Them, in the Merchants Service, When the War Is Finished. The Sixth Edition, Adapted to the Present Time, London 1759.
Hanway, Jonas: Serious Considerations on the Salutary Design of the Act of Parliament for a Regular, Uniform Register of the Parishpoor in all the Parishes within the Bills of Mortality. To which are Added, Some Thoughts on the Usefulness of Ventilation [...] in Two Letters Addressed to a Churchwarden, London 1762.
Hanway, Jonas: Proposal for County Naval Free-Schools, to Be Built on Waste Lands. Giving Such Effectual Instructions to Poor Boys, as May Nurse Them for the Sea Service. Teaching Them also to Cultivate the Earth, that in due Time They May Furnish Their Own Food; and to Spin, Knit, Weave Make Shoes &c., London 1783.
Harding, Richard: Seapower and Naval Warfare. 1650 – 1830, London 1999.
Harley, J. Brian und David Woodward: The History of Cartography: Cartography in Prehistoric, Ancient, and Medieval Europe and the Mediterranean, vol. 1, Chicago and London 1987.
Harrison, Jane: Myths of the Odyssey in Art and Literature, London 1882.
Harrison, Robin: "Das 'Rettende' oder 'Gefahr'? Die Bedeutung des Gedächtnisses in Hölderlins Hymne 'Mnemosyne,'" in: Hölderlin-Jahrbuch 24 (1984/85), pp. 195 – 206.
Hart, Henry H.: Luis de Camoëns and the Epic of the Lusiads, Norman 1962.
Hartmann, Albert: Untersuchungen über die Sagen vom Tod des Odysseus, Munich 1917.
Hartmann, Martin: Die Praxis des Vertrauens, Berlin 2011.
Harvey, Ronald C.: The Critical History of Edgar Allan Poe's "The Narrative of Arthur Gordon Pym". A Dialogue with Unreason, New York et al. 1998.
Hauer, Adolf: Beweisrecht im Seeprozeß, Rostock 1990.
Hausen, Jürgen: Schiffbau in der Antike, Herford 1979.
Hauser, Walter: Die Wurzeln der Wahrscheinlichkeitsrechnung. Die Verbindung von Glücksspieltheorie und statistischer Praxis vor Laplace, Stuttgart 1997.
Hauss, Jon: "Manuscripts of the Maelstrom," in: Western Humanities Review 55 (2001), pp. 139 – 157.
Hausschild, Jan-Christoph: B. Traven—Die unbekannten Jahre, Zurich, Vienna, and New York 2012.
Hazlitt, William (ed.): The Romanticist and Novelist's Library, London 1841.
Hegel, Georg Wilhelm Friedrich: Lectures on the Philosophy of Religion, transl. E.B. Speirs and J. Burdon Sanderson, vol. 2, London 1895.
Hegel, Georg Wilhelm Friedrich: Lectures on the Philosophy of History, transl. J. Sibree, London 1914.
Hegel, Georg Wilhelm Friedrich: Philosophy of Nature, transl. A.V. Miller, Oxford 1970.
Hegel, Georg Wilhelm Friedrich: Lectures on Fine Art, vol. 1, transl. T.M. Knox, Oxford 1975a.
Hegel, Georg Wilhelm Friedrich: "The Spirit of Judaism," in: G.W.F. Hegel, On Christianity: Early Theological Writings, transl. T.M. Knox, Philadelphia 1975b, pp. 182 – 204.
Hegel, Georg Wilhelm Friedrich: Hegel's Philosophy of Subjective Spirit, vol. 2, Anthropology, transl. M.J. Petry, Dordrecht and Boston 1978.
Hegel, Georg Wilhelm Friedrich: Elements of the Philosophy of Right, transl. H. B.Nisbet, Cambridge (UK) 1991.
Hegel, Georg Wilhelm Friedrich: "The German Constitution (1798 – 1802)," in: G.W.F. Hegel, Political Writings, transl. H.B. Nisbet, Cambridge (UK) 1999, pp. 6 – 102.
Heidegger, Martin: Being and Time, transl. John Macquarrie and Edward Robinson, Oxford 1962.

Heidegger, Martin: "The Nature of Language," in: Martin Heidegger, On the Way to Language, transl. Peter D. Hertz, New York 1971a, pp. 57–110.
Heidegger, Martin: "The Way to Language," in: Martin Heidegger, On the Way to Language, transl. Peter D. Hertz, New York 1971b, pp. 111–138.
Heidegger, Martin: "The Question Concerning Technology," in: Martin Heidegger, The Question Concerning Technology and Other Essays, transl. William Lovitt, New York and London 1977a, pp. 3–35.
Heidegger, Martin: "The Turning," in: Martin Heidegger, The Question Concerning Technology and Other Essays, transl. William Lovitt, New York and London 1977b, pp. 36–49.
Heidegger, Martin: "Das Ende der Philosophie und die Aufgabe des Denkens," in: Zur Sache des Denkens, 3rd edn., Tübingen 1988, pp. 61–80.
Heidegger, Martin: "Remembrance," in: Martin Heidegger, Elucidations of Hölderlin's Poetry, transl. Keith Hoeller, Amherst 2000, pp. 101–174.
Heidegger, Martin: "The Age of the World Picture," in: Martin Heidegger, Off the Beaten Track, ed. and transl. Julian Young and Kenneth Haynes, Cambridge (UK) 2002.
Heike, Heinz: Der Versicherungsbetrug im geltenden Recht und in den Entwürfen zu einem neuen Strafgesetzbuch, Göttingen 1925 (diss.).
Heimerdinger, Timo: Der Seemann. Ein Berufsstand und seine Inszenierung (1844–2003), Cologne, Weimar, and Vienna 2005.
Heims, Paul Gerhard: Seespuk. Aberglauben, Märchen und Schnurren in Seemannskreisen gesammelt und bearbeitet, Leipzig 1888.
Hein, Günter W.: "Satellite Navigation," in: Berndt Feuerbacher and Heinz Stoewer (eds.): Utilization of Space: Today and Tomorrow, Berlin et al. 2006, pp. 251–273.
Heinrich, Klaus: Das Floß der Medusa. 3 Studien zur Faszinationsgeschichte mit mehreren Beilagen und einem Anhang, Basel, Frankfurt a. M. 1995.
Heisenberg, Werner: "The Idea of Nature in Contemporary Physics," in: Werner Heisenberg, The Physicist's Conception of Nature, transl. Arnold J. Pomerans, London 1958, pp. 7–31.
Hellenkemper, Hansgerd: "Der Weg in die Katastrophe. Mutmaßungen über die letzte Fahrt des Mahdia-Schiffes," in: Gisela Hellenkemper Salies et al. (eds.): Das Wrack. Der antike Schiffsfund von Mahdia, vol. 1, Bonn 1994, pp. 153–165.
Heller-Roazen, Daniel: The Enemy of All: Piracy and the Law of Nations, New York 2009.
Helmes, Günter: "Die erzählstrategische Vermittlung von Diskursen in B. Travens Das Totenschiff," in: Günter Dammann (ed.): B. Traven. Autor—Werk—Werkgeschichte, Würzburg 2012, pp. 183–193.
Henkel, Arthur and Albrecht Schöne (eds.): Emblemata. Handbuch zur Sinnbildkunst des XVI. und XVII. Jahrhunderts, Stuttgart 1967.
Henning, Eckart: "Einleitung," in: Friedrich Beck and Eckart Henning (eds.): Die archivalischen Quellen. Mit einer Einführung in die Historischen Hilfswissenschaften, Neustadt a. d. Aisch 2000, pp. 1–6.
Hennig, Richard: "Die Kenntnis des Sternenhimmels in ihrer Bedeutung für die Nautik vor der Einführung des Kompasses," in: Marine-Rundschau 35 (1930), pp. 500–509, 546–555.
Henrich, Dieter: "Angedenken, Erinnerung, Gedächtnis," in: Sinn und Form. Beiträge zur Literatur 42/2 (1990), pp. 379–384.
Herdendorf, Charles E.: "Science on a Deep-Ocean Shipwreck," in: The Ohio Journal of Science, Special Issue 95/1 (1995), pp. 1–224.

Herder, Johann Gottfried: Journal meiner Reise im Jahr 1769, historisch-kritische Ausg., ed. Katharina Mommsen, Stuttgart 2002.
Hergt, Gustav: Die Nordlandfahrt des Pytheas, Halle an der Saale 1893 (diss.).
Hertel, Peter: Das Geheimnis der alten Seefahrer. Aus der Geschichte der Navigation, Gotha 1990.
Herwig, Holger H.: "Der Einfluß von Alfred Th. Mahan auf die deutsche Seemacht," in: Werner Rahn, Deutsche Marinen im Wandel. Vom Symbol nationaler Einheit zum Instrument internationaler Sicherheit, Munich 2005, pp. 127–142.
Hertzberg, Wilhelm (ed.): The Libell of English Policye (1436), Text u. metrische Übers., Leipzig 1878.
Herzog, Ronald: Kulturgut unter Wasser. Schatztaucher, das Seevölkerrecht und der Schutz des kulturellen Erbes, New York et al. 2002.
Hesiod: Theogony & Works and Days, transl. M.L. West, Oxford 1988.
Hesse, Eva: New Approaches to Ezra Pound. A Co-ordinated Investigation of Pound's Poetry and Ideas, London 1969.
Heubeck, Alfred, Joseph Louis Russo, and Manuel Fernández-Galiano: A Commentary on Homer's Odyssey, 4 vols., Oxford 1988–1992.
Heydenreich, Titus: Tadel und Lob der Seefahrt. Das Nachleben eines antiken Themas in den romanischen Literaturen, Heidelberg 1970.
Hickman, William: A Treatise on the Law and Practice of Naval Courts-Martial, London 1851.
Hill, John R.: The Oxford Illustrated History of the Royal Navy, Oxford 1995.
Hintermeyer, Hellmut: Schiffskatastrophen. Von der spanischen Armada bis zum Untergang der Pamir, Stuttgart 1998.
Hirdt, Willi: Wie Dante das Jenseits erfährt. Zur Erkenntnistheorie des Dichters der Göttlichen Komödie, Bonn 1989.
Hirsch, Gordon: "The Commercial World of The Wrecker," in: Journal of Stevenson Studies 2 (2005), pp. 70–97.
Hirschfeld, Nicolle: "The Ship of Saint Paul: Historical Background," in: Biblical Archaeologist 53/1 (1990), pp. 25–30.
Hoar, Allen: The Submarine Torpedo Boat, New York 1916.
Hobe, Stephan: "Die staatsfreien Räume—insbesondere der Weltraum," in: Karl Schmitt (ed.): Politik und Raum, Baden-Baden 2002, pp. 79–88.
Hobson, Rolf: Maritimer Imperialismus. Seemachtideologie, seestrategisches Denken und der Tirpitzplan 1876 bis 1914, Munich 2004.
Hochart, Patrick: "Wilhelm von Ockham. Das Zeichen und seine Duplizität," in: François Châtelet (ed.): Die Philosophie des Mittelalters, vol. 2 of Geschichte der Philosophie, Frankfurt, Berlin, and Vienna 1973, pp. 175–194.
Hocker, Frederick M.: "Maritime Archaeology and the ISBSA—Where to in the 21st Century?," in: Carlo Beltrame (ed.): Boats, Ships and Shipyards. Proceedings of the Ninth International Symposium on Boat and Ship Archaeology, Venice 2000, pp. 1–6.
Höckmann, Olaf: Antike Seefahrt, Munich 1985.
Hölderlin, Friedrich: Sämtliche Werke, Briefe und Dokumente. Bremer Ausgabe, ed. D. E. Sattler, 12 vols., Munich 2004.
Holländer, Hans: "Die Kugel der Fortuna," in: Das Mittelalter 1 (1996), pp. 149–167.
Hölscher, Uvo: Die Odyssee. Epos zwischen Märchen und Roman, Munich 1990.
Homer: The Odyssey, transl. Emily Wilson, New York 2018.

Homer: Werke. Ilias. Odyssee, transl. Johann Heinrich Voß, 4 vol., 2nd. ed., Königsberg 1802.
Hönig, Christoph: Die Lebensfahrt auf dem Meer der Welt. Der Topos. Texte und Interpretationen, Würzburg 2000.
Hörisch, Jochen: Kopf oder Zahl. Die Poesie des Geldes, Frankfurt a. M. 1998.
Holl, Ute: Cinema, Trance and Cybernetics, transl. Daniel Hendrickson, Amsterdam 2017.
Homburg, Walter: Leitfaden der Nautik, 3rd edn., Berlin 1962.
Hooke, Robert: The Posthumous Works, London 1705.
Hooke, Robert and Robert Moray: "Directions for Observations and Experiments to Be Made by Masters of Ships, Pilots, and Other Fit Persons in Their Sea-Voyages," in: Philosophical Transactions 2 (1666/1667), pp. 433–448.
Horden, Peregrine and Nicholas Purcell: The Corrupting Sea: A Study of Mediterranean History, Oxford 2000.
Horkheimer, Max and Theodor W. Adorno: Dialectic of Enlightenment: Philosophical Fragments, transl. Edmund Jephcott, Stanford 2002.
Horn, Eva: The Secret War: Treason, Espionage, and Modern Fiction, transl. Geoffrey Winthrop-Young, Evanston 2013.
Hoste, P. Paul: Théorie de la construction des vaisseaux, qui contient plusieurs traitez de mathématique sur des matières nouvelles et curieuses, Lyon 1697.
Hourani, George Fadlo: Arab Seafaring in the Indian Ocean in Ancient and Early Medieval Times, New York 1975.
Howgego, Raymond John: "Vivaldi," in: Encyclopedia of Exploration to 1800, Potts Point (NSW) 2003.
Huber, Ryk: "Auf der Spurensuche in der Antike. Annona-Segler," in: Das Logbuch 38 (2002), pp. 172–179.
Hubert, Henri und Marcel Mauss: "Essai sur la nature et la fonction du sacrifice," in: Marcel Mauss, Œuvres I. Les fonctions sociales du sacré. Le sens commun, Paris 1968, pp. 193–307.
Hücking, Günter: Gefahr, gegenwärtige Gefahr, Dauergefahr, Gefährlichkeit: Erscheinungsformen des Gefahrbegriffs und ihre Bedeutung im geltenden Strafrecht (unter besonderer Berücksichtigung der Rechtsprechung des R.G.), Düsseldorf 1935 (diss.).
Huemmerich, Franz: Vasco da Gama und die Entdeckung des Seeweges nach Ostindien, reprint Munich edn. 1898, Hildesheim 1977.
Hufnagel, Henning: "'Nun, Schifflein! Sieh dich vor!'—Meerfahrt mit Nietzsche. Zu einem Motiv der fröhlichen Wissenschaft," in: Nietzsche-Studien 37 (2008), pp. 143–159.
Hugo, Victor: The Toilers of the Sea, transl. James Hogarth, New York 2002.
Huler, Scott: Defining the Wind: The Beaufort Scale and how a Nineteenth-century Admiral Turned Science into Poetry, New York 2004
Humboldt, Alexander: Kosmos. Entwurf einer physischen Weltbeschreibung, ed. Ottmar Ette und Oliver Lubrich, Frankfurt a. M. 2004.
Hume, David: "Idea of a Perfect Commonwealth," in: Essays Moral, Political and Literary, London 1875, pp. 480–492.
Huntress, Keith: A Checklist of Narratives of Shipwrecks and Disasters, Ames (IA) 1979.
Husserl, Edmund: The Crisis of European Sciences and Transcendental Phenomenology: An Introduction to Phenomenological Philosophy, transl. David Carr, Evanston 1970.

Huxley, H. H.: "The Perils and Penalties of Seafaring. A Stock Theme in Latin Poetry," in: Proceedings of the Leeds Philosophical and Literary Society: Literary and Historical Section 6 / Part 8, Leeds 1951, pp. 576–582.
Iacomella, Lucia: "'Glauben sie nicht [...], dass ich klage um zu klagen'. Kafkas Beschwerdeordnungen," in: off topic. Zeitschrift der Kunsthochschule für Medien Köln 2 (2010), pp. 18–26.
Ifrah, Georges: Universalgeschichte der Zahlen, Frankfurt a. M. 1993.
Imbach, Rüedi: "Experiens Ulixes. Hinweise zur Figur des Odysseus im Denken der Patristik, des Mittelalters und bei Dante," in: Gotthard Fuchs (ed.): Lange Irrfahrt—grosse Heimkehr. Odysseus als Archetyp—zur Aktualität des Mythos, Frankfurt a. M. 1994, pp. 59–80.
Iser, Wolfgang: Die Appellstruktur der Texte. Unbestimmtheit als Wirkungsbedingung literarischer Prosa, Constance 1970.
Jackson, Richard P.: "From Profit-Sailing to Wage-Sailing: Mediterranean Owner-Captains and their Crews during the Medieval Commercial Revolution," in: Journal of European Economic History 18 (1989), pp. 605–628.
Jäger, Ernst Ludwig: Lucas Paccioli und Simon Stevin, nebst einigen juengeren Schriftstellern ueber Buchhaltung. Skizzen zur Geschichte der kaufmännischen, staatlichen und landwirthschaftlichen Buchführung, Stuttgart 1876.
Jakobson, Roman: "Linguistics and Poetics," in: Thomas A. Sebeok (ed.), Style in Language, Cambridge (MA) 1960, pp. 1–26.
Jal, Augustin: Scènes de la vie maritime, vol. 1, Paris 1832.
Jal, Augustin: Archéologie navale, vol. 1, Paris 1840.
Jal, Augustin: Glossaire nautique. Répertoire polyglotte de termes de marine anciens et modernes, vol. 1, Paris 1848.
Janeway, James: Mr. James Janeway's Legacy to his Friends, Containing Twenty-Seven Famous Instances of God's Providence In and About Sea-Dangers and Deliverances, 2nd edn., London 1680.
Johnson, Uwe: "Ein unergründliches Schiff," in: Merkur 33/6 (1979), pp. 537–550.
Jonkers, Art Roeland Theo: Earth's Magnetism in the Age of Sail, Baltimore 2003.
Jonson, Ben, George Chapman, and John Marston: Eastward Ho!, ed. Christopher G. Petter, London 1994.
Jourdin, Michel Mollat du: Europe and the Sea, Oxford and Cambridge (MA) 1993.
Jourdan, Silvester: A Plaine Description of the Barmvdas, Now Called Sommer Ilands: With Their Manner of Their Discouerie Anno 1609: By the Shipwrack and Admirable Deliuerance of Sir Thomas Gates, and Sir George Sommers, London 1613, reprint Amsterdam and New York 1971.
Journal de la Société générale des naufrages et de l'union des nations, sous les rapports du sauvetage, des sciences et du commerce [n.a.], vol. 3, Paris 1836.
Joyce, James: Ulysses, London 2000.
Jünger, Ernst: "Der Waldgang," in: Sämtliche Werke, vol. 7.: Essays I, Stuttgart 1980, pp. 281–375.
Jünger, Ernst: "Total Mobilisation", trans. Joel Golb, in: Richard Wolin (ed.), The Heidegger Controversy: A Critical Reader, New York 1991, pp. 119–139.
Jünger, Ernst: "On Danger," transl. Donald Reneau, in: New German Critique, spring/summer 1993, pp. 27–32.

Jünger, Ernst: The Worker, transl. Bogdan Costea, Evanston 2017.
Jünger, Ernst: A German Officer in Occupied Paris: The War Journals 1941–1945, transl. Thomas S. Hansen and Abby J. Hansen, New York 2019.
Jürgens, Thomas: Diplomatischer Schutz und Staatenlose, Berlin 1986.
Justi, Johann Gottlob Heinrich: Die Chimäre des Gleichgewichts der Handlung und Schiffahrt, oder: Ungrund und Nichtigkeit einiger neuerlich geäußerten Meynungen von denen Maassregeln der freyen Mächte gegen die zu befürchtende Herrschaft und Obermacht zur See, Altona 1759.
Kafka, Franz: The Castle, transl. Willa and Edwin Muir, New York 1946.
Kafka, Franz: Diaries, transl. Martin Greenberg, with cooperation of Hannah Arendt, New York 1949.
Kafka, Franz: "Description of a Struggle," transl. James and Tania Stern, in: Franz Kafka, The Complete Stories, ed. Nahum Glatzer, New York 1971, pp. 9–51.
Kafka, Franz: "The Great Wall of China" transl. Willa and Edwin Muir, in: Franz Kafka, The Complete Stories, ed. Nahum Glatzer, New York 1971, pp. 235–248.
Kafka, Franz: "The Hunter Gracchus," transl. Willa and Edwin Muir, in: Franz Kafka, The Complete Stories, ed. Nahum Glatzer, New York 1971, pp. 226–231.
Kafka, Franz: "The Hunter Gracchus: A Fragment," transl. Tania and James Stern, in: Franz Kafka, The Complete Stories, ed. Nahum Glatzer, New York 1971, pp. 231–234.
Kafka, Franz: "The News of the Building of the Wall: A Fragment," in: Franz Kafka, The Complete Stories, ed. Nahum Glatzer, New York 1971, pp. 248 f.
Kafka, Franz: "A Report to an Academy," transl. Willa and Edwn Muir, in: Franz Kafka, The Complete Stories, ed. Nahum Glatzer, New York 1971, pp. 250–259.
Kafka, Franz: "A Report to an Academy: Two Fragments," transl. Tania and James Stern, in: Franz Kafka, The Complete Stories, ed. Nahum Glatzer, New York 1971, pp. 259–262.
Kafka, Franz: Letters to Felice, ed. Erich Heller and Jürgen Born, transl. James Stern and Elisabeth Duckworth, New York 1973.
Kafka, Franz: Briefe 1902–1924, ed. Max Brod Frankfurt a. M. 1975.
Kafka, Franz: Briefe an Felice und andere Korrespondenz aus der Verlobungszeit, ed. Erich Heller und Jürgen Born, Frankfurt a. M. 1976.
Kafka, Franz: Letters to Friends, Family, and Editors, transl. Richard and Clara Winston, New York 1977.
Kafka, Franz: Letters to Milena, transl. Philip Boehm, New York 1990.
Kafka, Franz: Nachgelassene Schriften und Fragmente, vol. 2, ed. Jost Schillemeit, Frankfurt a. M. 1992.
Kafka, Franz: Nachgelassene Schriften, vol. 1, Apparatband, ed. Malcolm Pasley, Frankfurt a. M. 1993.
Kafka, Franz: The Man Who Disappeared (Amerika), transl. Michael Hofmann, London and New York 1996.
Kafka, Franz: The Trial, transl. Breon Mitchell, New York 1998.
Kafka, Franz: Amtliche Schriften, ed. Klaus Hermsdorf und Benno Wagner, Frankfurt a. M. 2004.
Kafka, Franz: The Office Writings, ed. Stanley Corngold et al., transl. Eric Patton with Ruth Hein, Princeton and Oxford 2009.
Kahlmeyer, Johannes: Seesturm und Schiffbruch als Bild im antiken Schrifttum, Greifswald and Hildesheim 1934 (diss.).

Kant, Immanuel: "On the Common Saying: That May be Correct in Theory, But it is of no Use in Practice," in: Immanuel Kant, Practical Philosophy," transl. Mary Gregor, Cambridge (UK) 1966, pp. 277 – 309.
Kant, Immanuel: The Metaphysics of Morals, transl. Mary Gregor, Cambridge (UK) 1996a.
Kant, Immanuel: "What Does it Mean to Orient Oneself in Thinking?" transl. Allen Wood, in: Immanuel Kant, Religion and Rational Theology, ed. Allen W. Wood and George Di Giovanni, Cambridge (UK) 1996b.
Kant, Immanuel: Critique of Pure Reason, transl. and ed. Paul Guyer and Allen W. Wood, Cambridge (UK) 1998.
Kant, Immanuel: Critique of the Power of Judgment, transl. Paul Guyer and Eric Matthews, Cambridge (UK) 2000.
Kant, Immanuel: Religion within the Bounds of Bare Reason, transl. Werner S. Pluhar, Indianapolis and Cambridge (UK) 2009.
Kanthak, Gerhard: Der Akademiegedanke zwischen utopischem Entwurf und barocker Projektemacherei, Berlin 1987.
Kapp, Ernst: Vergleichende Allgemeine Erdkunde in wissenschaftlicher Darstellung, 2nd edn., Braunschweig 1868.
Karcev, Vladimir Petrovic and Petr Michaijlovic Chazanovskij: Warum irrten die Experten? Unglücksfälle und Katastrophen aus der Sicht technischer Zuverlässigkeit, Moscow and East Berlin 1984.
Karten, Walter: "Leibniz' versicherungswissenschaftliche Schriften aus ökonomischer Sicht," in: Zeitschrift für die gesamte Versicherungswissenschaft 90 (2001), pp. 327 – 337.
Kastely, James L.: "Understanding the 'work' of Literature: B. Travens 'The Death Ship,'" in: Mosaic 18 (1985), pp. 79 – 96.
Katsos, Charalampos: Nachhaltiger Schutz des kulturellen Erbes. Zur ökologischen Dimension des Kulturgüterschutzes, Baden-Baden 2011.
Kaufmanns, Eitel-Friedrich von: Die öffentlich-rechtlichen Machtbefugnisse des Kapitäns gegenüber der Schiffsmannschaft und den Passagieren unter Berücksichtigung des in England, Frankreich und den Vereinigten Staaten von Amerika geltenden Seerechts, Hamburg 1931.
Kay, Bernhard: "Ans Ende der Welt und darüber hinaus". Das Abenteuer, die Welt mit dem Schiff zu entdecken, Frankfurt a. M. 1995.
Keller, Hildegard Elisabeth: "'Die Kartoffel des Lebens'. Metaphysische Aspekte des Risikos," in: Renate Schlesier and Ulrike Zellmann (eds.): Ritual als provoziertes Risiko, Würzburg 2009, pp. 123 – 135.
Keller, Ludwig: Die nichtkriegerische militärische Gewaltmassnahme. Eine Untersuchung über ihre Abgrenzung vom Krieg und ihre Zulässigkeit als Repressalie und Notwehrakt, Berlin 1934.
Kelletat, Dieter: Physische Geographie der Meere und Küsten, Leipzig 1999.
Kemper, Raimund: Il était un petit navire… Zur Archäologie der Narrenschiff-Phantasien Michel Foucaults, Frankfurt a. M. 1996.
Kennedy, Paul M.: The Rise and Fall of British Naval Mastery, London 1983.
Kepler, J. S.: "The Operating Potential of London Marine Insurance in the 1570s: Some Evidence from 'A Booke of Orders of Assurances within the Royall Exchange,'" in: Business History 17/1 (1975), pp. 44 – 55.

Kettenbach, Günter: Einführung in die Schiffahrtsmetaphorik der Bibel, Frankfurt a. M. et al. 1994.
Kettenbach, Günter: Das Logbuch des Lukas. Das antike Schiff in Fahrt und vor Anker, Frankfurt a. M. et al. 1997.
Kierkegaard, Søren: The Concept of Dread, transl. Walter Lowrie, Princeton 1957.
Kierkegaard, Søren: Fear and Trembling and The Book on Adler, transl. Walter Lowrie, New York et al., 1994.
Kindervater, Christian Victor: "Ueber das Wohlgefallen an traurigen Vorstellungen," in: Denkwürdigkeiten aus der philosophischen Welt 5 (1787), pp. 95–135.
Kingsley, Sean: "Introduction," in: Greg Stemm and Sean Kingsley (eds.): Oceans Odyssey: Deep-Sea Shipwrecks in the English Channel, Straits of Gibraltar & Atlantic Ocean: Odyssey Marine Exploration Reports 1, Oxford and Oakville 2010, pp. ix–xii.
Kinzel, Ulrich: "Bergungen. Das Wrack, die Wahrheit und die Kunst in Uwe Johnsons 'Ein unergründliches Schiff,'" in: Johnson-Jahrbuch 8 (2001), pp. 134–151.
Kinzel, Ulrich: "Orientation as a Paradigm of Maritime Modernity," in: Bernhard Klein (ed.): Fictions of the Sea: Critical Perspectives in English Literature and Culture, Aldershot 2002, pp. 28–48.
Kisch, Egon Erwin: Reportagen von der Seefahrt 1914–1924, Oldenburg, Munich, and Hamburg 1979.
Kittler, Friedrich: "In the Wake of the Odyssey," in: Cultural Politics, vol. 8, issue 3 (2012), pp. 413–427.
Kittler, Friedrich: Musik und Mathematik I. Hellas 1: Aphrodite, Munich 2006.
Klein, Bernhard: Maps and the writing of space in early modern England and Ireland, Basingstoke et al. 2001.
Kleinschmidt, Erich: "Denkform im geschichtlichen Prozeß. Zum Funktionswandel der Allegorie in der frühen Neuzeit," in: Walter Haug (ed.): Formen und Funktionen der Allegorie. Symposium Wolfenbüttel 1978, Stuttgart 1979, pp. 388–404.
Klesczewski, Reinhard: "Dantes Odysseus-Gesang," in: R. Baum and W. Hirdt (eds.): Dante Alighieri, Tübingen 1985, pp. 17–30.
Klose, Alexander: Das Container-Prinzip. Wie eine Box unser Denken verändert, Hamburg 2009.
Knight, Frank: Risk, Uncertainty and Profit, Boston and New York 1921, reprint London 1933.
Knox-Shaw, Peter: The Explorer in English Fiction, Houndsmills et al. 1987.
Koch, Peter: Pioniere des Versicherungsgedankens. 300 Jahre Versicherungsgeschichte in Lebensbildern 1550–1850, Wiesbaden 1968.
Koch, Peter: "Der geistesgeschichtliche Hintergrund der Versicherungswirtschaft," in: Heinz Leo Müller-Lutz and Karl-Heinz Rehnert (eds.): Beiträge zur Geschichte des deutschen Versicherungswesens. Aus Anlaß des 60. Geburtstages von Peter Koch, Karlsruhe 1995, pp. 109–116, pp. 151–163.
Koch, Peter: "Der Versicherungsgedanke bei Adam Smith, dem Begründer der klassischen Nationalökonomie," in: Heinz Leo Müller-Lutz and Karl-Heinz Rehnert (eds.): Beiträge zur Geschichte des deutschen Versicherungswesens. Aus Anlaß des 60. Geburtstages von Peter Koch, Karlsruhe 1995, pp. 109–116.
Kocziszky, Eva: Hölderlins Orient, Würzburg 2009.
Körber, Hans-Günther: Zur Geschichte der Konstruktion von Sonnenuhren und Kompassen des 16. bis 18. Jahrhunderts (unter besonderer Berücksichtigung der im

Geomagnetischen Institut Potsdam und im Staatlichen Mathematisch-physikalischen Salon Dresden vorhandenen Instrumente), Berlin 1965.
Körte, Mona und Robert Stockhammer: Ahasvers Spur. Dichtungen und Dokumente vom 'Ewigen Juden', Leipzig 1995.
Köster, August: Das antike Seewesen, Berlin 1923.
Köster, Werner and Thomas Lischeid (eds.), Titanic. Ein Medienmythos, Leipzig 2000.
Kohler, Alfred: Columbus und seine Zeit, Munich 2006.
Koldau, Linda Maria: Titanic. Das Schiff—der Untergang—die Legenden, Munich 2012.
Konersmann, Ralf: "Die Umdeutung des Unsichtbaren," in: Die Schleier des Timanthes. Perspektiven der historischen Semantik, Frankfurt a. M. 1994, pp. 56–83.
Konersmann, Ralf: "Die Philosophen und das Meer," in: Akzente 50 (2003), pp. 218–233.
Konstam, Angus: Atlas versunkener Schiffe, Augsburg 1999.
Kopley, Richard: Poe's Pym: Critical Explorations, Durham et al. 1992.
Koschorke, Albrecht: Geschichte des Horizonts. Grenze und Grenzüberschreitung in literarischen Landschaftsbildern, Frankfurt a. M. 1990.
Koschorke, Albrecht: Fact and Fiction: Elements of a General Theory of Narrative, transl. Joel Golb, Berlin and Boston 2018.
Koyré, Alexandre: From the Closed World to the Infinite Universe, Baltimore 1957.
Krah, Hans: "Räume, Grenzen, Grenzüberschreitungen: Einführende Überlegungen," in: Kodikas/Code 22/1–2 (1999), pp. 3–12.
Kramer, Jürgen: "Conrad's Crews Revisited," in: Bernhard Klein (ed.): Fictions of the Sea: Critical Perspectives in English Literature and Culture, Aldershot 2002, pp. 121–140.
Kratz, Reinhard: Rettungswunder. Motiv-, traditions- und formkritische Aufarbeitung einer biblischen Gattung, Frankfurt a. M. 1979.
Kretschmer, Konrad: Die italienischen Portolane des Mittelalters. Ein Beitrag zur Geschichte der Kartographie und Nautik, Berlin 1909.
Kreutziger, Günter: Seerechtliches für Nautiker, Karlsruhe 1966.
Krohn, Wolfgang: Francis Bacon, Munich 1987.
Kullmann, Wolfgang: Homerische Motive. Beiträge zur Entstehung, Eigenart und Wirkung von Ilias und Odyssee, ed. Roland J. Müller, Stuttgart 1992.
Kurson, Robert: Shadow Divers: The True Adventure of Two Americans Who Risked Everything to Solve One of the Last Mysteries of World War II, New York 2005.
Kuske, Bruno: "Die Begriffe Angst und Abenteuer in der deutschen Wirtschaft des Mittelalters," in: Zeitschrift für handelswissenschaftliche Forschung N.F. 1 (1949), pp. 547–550.
Lacan, Jacques: Freud's Papers on Technique, 1953–54: The Seminar of Jacques Lacan, Book I, transl. John Forrester, New York 1988.
Lacan, Jacques: The Ethics of Psychoanalysis, 1959–1960: The Seminar of Jacques Lacan, Book VII, transl. Dennis Porter, New York 1992.
Lacan, Jacques: "Science and Truth," in: Jacques Lacan, Écrits, transl. Bruce Fink, New York and London 2006, pp. 726–745.
Lacan, Jacques: Anxiety: The Seminar of Jacques Lacan, Book X, transl. A.R. Price, Cambridge (UK) and Malden 2014.
Lachs, Johannes: "Verhaltensspiele von Küstenbewohnern bei der Rettung Schiffbrüchiger," in: Jahrbuch für Volkskunde und Kulturgeschichte 32 (1989), pp. 116–121.
Lambert, Andrew: War at Sea in the Age of Sail 1650–1850, London 2000.

Lamont-Brown, Raymond: Phantoms, Legends, Customs and Superstitions of the Sea, London 1972.
Lanczkowski, Günter: "Die Geschichte des Schiffbrüchigen. Versuch einer religionsgeschichtlichen Interpretation," in: Zeitschrift der Deutschen Morgenländischen Gesellschaft 103/NF 28 (1953), pp. 360–371.
Landy, Marcia: "The Cinematographic Brain in *2001. A Space Odyssey,*" in: Robert Kolker (ed.), Stanley Kubrick's 2001: A Space Odyssey. New Essays, New York 2006, pp. 87–104.
Lane, Frederic C.: Andrea Barbarigo: Merchant of Venice, Baltimore 1944.
Lane, Frederic C.: "The Economic Meaning of the Invention of the Compass," in: The American Historical Review 68/3 (1963), pp. 605–617.
Lane, Frederic C.: Studies in Venetian Social and Economic History, ed. Benjamin G. Kohl und Reinhold C. Mueller, London 1987.
Lane, Frederic C.: Venice. A Maritime Republic, Baltimore 1973.
Larson, Jennifer: "A Land Full of Gods: Nature Deities in Greek Religion," in: Daniel Ogden (ed.): A Companion to Greek Religion, Oxford 2007, pp. 56–70.
Latacz, Joachim: Homer. Der erste Dichter des Abendlandes, Düsseldorf and Zurich 2003.
Latour, Bruno: Reassembling the Social: An Introduction to Actor-Network-Theory, Oxford 2005.
Lavery, Brian: Shipboard Life and Organisation 1731–1815, Aldershot et al. 1998.
Le Boulicaut, Yannick: "Shores in Joseph Conrad's Works," in: Conradiana: A Journal of Joseph Conrad Studies 37/2 (2005), pp. 233–244.
Leer, David van: "Poe's Cosmology: The World of the Mind," in: Hermann Josef Schnackertz (ed.): POEtic Effect and Cultural Discourses, Heidelberg 2003, pp. 189–207.
Lefebvre, Jean-Pierre: "Auch die Stege sind Holzwege," in: Hölderlin-Jahrbuch 26 (1988/89), pp. 202–223.
Le Goff, Jacques: Your Money or Your Life: Economy and Religion in the Middle Ages, transl. Patricia Ranum, New York 1988.
Le Goff, Jacques: Money and the Middle Ages, transl. Jean Birrell, Cambridge (MA) and Malden 2012.
Lehmann-Brune, Marlies: Die Story von Lloyd's of London. Glanz und Tragödien des legendären Versicherungshauses, Düsseldorf 1999.
Leibniz, Gottfried Wilhelm: "Öffentliche Assekuranzen," in: Sämtliche Schriften und Briefe. Akademieausgabe, vol. 4. 3., Berlin 1923ff., pp. 421–432.
Leidinger, Hans: Kapitalismus, Vienna et al. 2008.
Lem, Stanislaw: Philosophie des Zufalls. Zu einer empirischen Theorie der Literatur, Frankfurt a. M. 1983.
Leonhard, Karin: "Was ist Raum im 17. Jahrhundert? Die Raumfrage des Barocks: Von Descartes zu Newton und Leibniz," in: Horst Bredekamp (ed.): Visuelle Argumentationen. Die Mysterien der Repräsentation und die Berechenbarkeit der Welt, Munich 2006, pp. 11–34.
Lesky, Albin: Thalatta. Der Weg der Griechen zum Meer, Vienna 1947.
Levi, Primo: If This be a Man, transl. Stuart Woolf, New York 1959.
Lewin, Kurt: "Kriegslandschaft" (1917), in: Jörg Dünne and Stephan Günzel (eds.): Raumtheorie. Grundlagentexte aus Philosophie und Kulturwissenschaften, Frankfurt a. M. 2006, pp. 129–140.

Libera, Alain de: La querelle des universaux De Platon à la fin di Moyen Age, Paris 1996.
Lichtenstein, Heinrich: "Geschichte der Entdeckung des Vorgebirges der Guten Hoffnung," in: Vaterländisches Museum, 4 (Oktober 1810), pp. 369–415.
Liebig, Eugen von: Die Seeversicherung. Systematische Darstellung des Seeversicherungswesens, Berlin 1914.
Limburg, James: Jonah: A Commentary, Louisville (KY) 1993.
Linebaugh, Peter und Markus Rediker: Die vielköpfige Hydra. Die verborgene Geschichte des revolutionären Atlantiks, Berlin and Hamburg 2008.
Link, Jürgen: Versuch über den Normalismus. Wie Normalität produziert wird, 2nd edn., Opladen 1999.
Linke, Ute: Beobachtungen von Schiffsärzten der Royal Navy über die häufigsten Erkrankungen zur See, dargestellt an ausgewählten Bordjournalen des beginnenden 19. Jahrhunderts, Frankfurt a.M. et al. 1987.
Linschoten, Jan Huygen van: Discours of Voyages into y East & West Indies, Amsterdam 1974.
Lisch, Ralf: Totale Institution Schiff, Berlin 1976.
Lohenstein, Daniel Casper von: Grossmüthiger Feldherr Arminius oder Herrmann, Ein tapfferer Beschirmer der deutschen Freyheit / nebst seiner Durchlauchtigen Thußnelda / In einer sinnreichen Staats=Liebes= und Helden=Geschichte / Dem Vaterlande zu Liebe / Dem deutschen Adel aber zu Ehren und rühnlichen Nachfolge, 2 vols., ed. Elida Maria Szarota, Bern and Frankfurt a.M. 1973.
London, Jack: The Mutiny of the Elsinore, Rockville (MD) 2009.
London, Jack: The Sea Wolf, Mineola (NY) 2000.
Lotman, Jurij M.: The Structure of the Artistic Text, transl. Ronald Vroon, Anne Arbor 1977.
Lotman, Jurij M.: Universe of the Mind: A Semiotic Theory of Culture, transl. Ann Shukman, London and New York 1990.
Lübeck, Maria Holmberg: Iphigeneia, Agamemnon's Daughter: A Study of Ancient Conceptions in Greek Myth and Literature Associated with the Atrides, Stockholm 1993.
Luhmann, Niklas: "Klassische Theorie der Macht: Kritik ihrer Prämissen," in: Zeitschrift für Politik 16 (1969), pp. 149–170.
Luhmann, Niklas: Love as Passion: The Codification of Intimacy, transl. Jeremy Gaines and Doris L. Jones, Cambridge (MA) 1986.
Luhmann, Niklas: Risk: A Sociological Theory, transl. Rhodes Barrett, Berlin and New York 1993.
Luhmann, Niklas: Die Wirtschaft der Gesellschaft, 2nd edn., Frankfurt a.M. 1996.
Luhmann, Niklas: "Kultur als historischer Begriff," in: Niklas Luhmann, Gesellschaftsstruktur und Semantik, Studien zur Wissenssoziologie der modernen Gesellschaft, vol. 4, Frankfurt a.M. 1999, pp. 31–54.
Luhmann, Niklas: "Literatur als fiktionale Realität," in: Niklas Luhmann, Schriften zur Kunst und Literatur, ed. Niels Werber, Frankfurt a.M. 2008, pp. 276–291.
Luhmann, Niklas: Theory of Society, vol. 1, transl. Rhodes Barrett, Stanford 2012.
Luhmann, Niklas: A Systems Theory of Religion, transl. David A. Brenner, Stanford 2013.
Lukács, Georg: The Theory of the Novel: A Historico-Philosophical Essay on the Forms of Great Epic Literature, transl. Anna Bostock, Cambridge (MA) 1971.
Lutun, Bernard: La marine de Colbert. Études d'organisation, Paris 2003.
Luzzana, Ilaria Caraci: The History of the Life and Deeds of the Admiral Don Christopher Columbus: Attributed to His Son Fernando Colón, Turnhout 2004.

Lyell, Charles: Principles of Geology, 4 vols., London 1830–35.
Macdonald, Dennis R.: "The Shipwrecks of Odysseus and Paul," in: New Testament Studies 45/1 (1999), pp. 88–107.
Macfie, Alec: "The Invisible Hand of Jupiter," in: Journal of the History of Ideas 32 (1971), pp. 595–599.
Mahan, Alfred Thayer: On Naval Warfare, ed. Allan Westcott, Boston 1941.
Maier, Walter: Die Meuterei im geltenden Strafrecht unter besonderer Berücksichtigung der neueren und neuesten Entwürfe, Aschaffenburg 1932.
Malinowski, Bronislaw: Argonauts of the Western Pacific, New York 2014.
Malkin, Irad: The Returns of Odysseus: Colonization and Ethnicity, Berkeley 1998.
Mandelstam, Ossip: "Conversations about Dante," transl. Jane Gary Harris and Constance Link, in: Peter Hawkins and Rachel Jacoff (eds.), The Poet's Dante, New York 2001, pp. 40–93.
Manes, Alfred: Das Recht des Pseudonyms unter besonderer Beachtung des Bürgerlichen Gesetzbuchs und ausländischen Rechts, Göttingen 1899.
Manes, Alfred: Englands Seeherrschaft im Wanken. Ein Vortrag nach tausend Tagen Weltkrieg, Bonn 1917.
Manes, Alfred: Grundzüge des Versicherungswesens (Privatversicherung), Leipzig and Berlin 1918.
Marin, Louis: Portrait of the King, transl. Martha M. Houle, Minneapolis 1988.
Mark, Samuel: Homeric Seafaring, College Station 2005.
Marryat, Frederick: The Phantom Ship, Ithaca 2000.
Martel, Christoph: "'Noch denket das mir wohl…'. Poetik der Erinnerung in Hölderlins Hymnen Andenken und Mnemosyne," in: Euphorion. Zeitschrift für Literaturgeschichte 98 (2004), pp. 385–406.
Martin, Colin, "De-particularizing the Particular. Approaches to the Investigation of Well-documented Post-medieval Shipwrecks," in: World Archaeology, 32/3 (2001), pp. 383–400.
Martus, Steffen: Werkpolitik. Zur Literaturgeschichte kritischer Kommunikation vom 17. bis ins 20. Jahrhundert, Berlin and New York 2007.
Marx, Karl: Capital, vol. 1, fourth edition (1890), transl. Hans. G. Ehrbar (annotated German/English edition), http://content.csbs.utah.edu/~ehrbar/cap1.pdf
Marx, Karl: A Contribution to the Critique of Political Economy, transl, N. I. Stone, Chicago 1904.
Mateas, Michael: "Reading HAL. Representation and Artificial Intelligence," in: Robert Kolker (ed.): Stanley Kubrick's 2001: A Space Odyssey. New Essays, New York 2006, pp. 105–126.
Mathewson, R. Duncan III: "Archaeology on Trial," in: Babits, Lawrence E. and Hans van Tilburg (eds.): Maritime Archaeology: A Reader of Substantive and Theoretical Contributions, New York and London 1998, pp. 97–104.
Matschke, Reinhard G.: Überleben auf See. Medizinische Aspekte des Schiffbrüchigen in historischen Darstellungen, Düsseldorf 1975.
Matthews, Colonel Lloyd J.: General Henry Lockwood of Delaware: Shipmate of Melville, Co-builder of the Naval Academy, Civil War Commander, Newark 2014.
Maurer, Karl: "Topographie des Unbetretbaren. Der Raum in Dantes Divina Commedia," in: Rudolf Behrens and Rainer Stillers (eds.): Orientierungen im Raum. Darstellung

räumlichen Sinns in der italienischen Literatur von Dante bis zur Postmoderne, Heidelberg 2008, pp. 11–26.
Maurier, Daphne du: Jamaica Inn, London 1977.
Maury, Matthew Fontaine: The Physical Geography of the Sea, New York 1855.
Maury, Matthew Fontaine: The Physical Geography of the Sea and Its Metereology, 10th edn., London 1861.
Mauss, Marcel: The Gift: Forms and Functions of Exchange in Archaic Societies, transl. Ian Cunnison, London 1966.
Maxwell, James Clark: "On Governors," in: Proceedings of the Royal Society 16 (1868), pp. 270–283.
May, W. E.: "The Birth of the Compass," in: Journal of the Institute of Navigation 2 (1949), pp. 259–263.
Maye, Harun: "Die Paradoxie der Billigkeit in Recht und Hermeneutik," in: Cornelia Vismann and Thomas Weitin (eds.): Urteilen / Entscheiden, Munich 2006, pp. 56–71.
Mayr, Otto: Zur Frühgeschichte der technischen Regelungen, Munich and Vienna 1969.
Mazzotta, Giuseppe: "Canto XXVI—Ulysses: Persuasion versus Prophecy," in: Giuseppe Mazzotta: (ed.): Inferno: A New Verse Translation, Backgrounds and Contexts, Criticism, New York et al. 2008, pp. 326–334.
Meder, Stephan: Schuld, Zufall, Risiko. Untersuchungen struktureller Probleme privatrechtlicher Zurechnung, Frankfurt a. M. 1993.
Meinecke, Friedrich: Die Idee der Staatsräson in der neueren Geschichte, Werke, ed. Walther Hofer, 4th edn., Munich and Vienna 1976.
McConnell, Anita: No Sea too Deep: The History of Oceanographic Instruments, Bristol 1982.
McGrail, Séan: "The Ship: Carrier of Goods, People and Ideas," in: Ellen E. Rice (ed.): The Sea and History, Stroud (Gloucestershire) 1996, pp. 67–96.
McGrail, Séan: Studies in Maritime Archaeology, Oxford 1997.
McLuhan, Marshall: Understanding Media: The Extensions of Man, New York 1964.
McLuhan, Marshall and Quentin Fiore: The Medium is the Message, Corte Madera 1967.
Meinardus, Otto F.: Die Reisen des Apostel Paulus. Nachvollzogen im 20. Jahrhundert, Kiel 1988.
Melville, Herman: Mardi and A Voyage Tither, ed. Harrison Hayford, Hershel Parker und G. Thomas Tanselle, Evanston and Chicago 1970.
Melville, Herman: The Confidence-Man: His Masquerade, ed. Hershel Parker, New York and London 1971.
Melville, Herman: Complete Shorter Fiction, ed. John Updike, New York and Toronto 1997.
Melville, Herman: Moby-Dick or The Whale, ed. Harrison Hayford, Hershel Parker und G. Thomas Tanselle, 4th edn., Evanston and Chicago 2000.
Melville, Herman: White-Jacket or, The World in a Man-of-War, ed. Elizabeth Hardwick, New York 2002.
Menke, Bettine: "Das Schweigen der Sirenen: die Rhetorik und das Schweigen," in: Johannes Janota (ed.): Vorträge des Augsburger Germanistentags 1991, Tübingen 1993, pp. 134–162.
Menke, Bettine: Prosopopoiia. Stimme und Text bei Brentano, Hoffmann, Kleist und Kafka, Munich 2000.
Menke, Bettine: "Das Polargebiet als Bibliotheksphänomen und die Polargebiete der Bibliothek. Nachfahren Petrarcas und Dantes im Eis und in den Texten," in: Annelore

Engel-Braunschmidt et al. (eds.): Ultima Thule. Bilder des Nordens von der Antike bis zur Gegenwart, Frankfurt a. M. et al. 2001, pp. 145–172.
Mertens, Sabine: Seesturm und Schiffbruch. Eine motivgeschichtliche Studie, Rostock 1987.
Metzeltin, Michael: Die Terminologie des Seekompasses in Italien und auf der iberischen Halbinsel bis 1600, Basel 1970 (diss.).
Metzner, Joachim: Persönlichkeitszerstörung und Weltuntergang. Das Verhältnis von Wahnbildung und literarischer Imagination, Tübingen 1976.
Meuli, Karl: Odyssee und Argonautika, Berlin 1921.
Mevert, Hans Erich: Zum ursächlichen Zusammenhang in der Seeversicherung, Marburg 1921 (diss.).
Meyer-Abich, Klaus: "Laborforschung im Erkenntnishandeln der Experimentiergesellschaft. Eine holistisch-pragmatische Perspektive für die Wissenschaftstheorie," in: Philosophia naturalis 43/1 (2006), pp. 168–191.
Meyer-Haßfurther, Monika und Ingo: 500 Jahre Navigation. Navigationsinstrumente vom 15. bis zum 19. Jahrhundert, Hamburg 2005.
Meyn, Matthias et al. (eds.): Die großen Entdeckungen, vol. 2 of Dokumente zur Geschichte der europäischen Expansion, Munich 1984.
Michelet, Jules: History of France, from the Earliest Period to the Present Time, vol. 1, transl. G.H. Smith, New York 1845.
Michelet, Jules: The Sea, New York 1861.
Mignon, Molly Raymond: Dictionary of Concepts in Archaeology, Westport (CN) and London 1993.
Millett, Paul: "Maritime Loans and the Structure of Credit in Fourth-Century Athens," in: Peter Garnsey, Keith Hopkins, and C. R. Whittaker (eds.): Trade in the Ancient Economy, Berkeley and Los Angeles 1983, pp. 36–52.
Milton, John: The Complete Poetical Works of John Milton, Boston etc. 1924.
Mindell, David A.: Between Human and Machine: Feedback, Control, and Computing Before Cybernetics, Baltimore 2004.
Mindell, David A.: Digital Apollo: Human and Machine in Spaceflight, Cambridge (UK) and London 2008.
Minois, Georges: Geschichte der Zukunft. Orakel, Prophezeiungen, Utopien, Prognosen, Düsseldorf and Zurich 1998.
Mödersheim, Sabine: "Materiale und mediale Aspekte der Emblematik," in: Eva Horn and Manfred Weinberg (eds.): Allegorie. Konfiguration von Text, Bild und Lektüre, Opladen 1998, pp. 201–220.
Möhring, Christa: "Robert Louis Stevenson und das schottische Leuchtturmwesen," in: Lorenz Engell, Bernhard Siegert, and Joseph Vogl (eds.): Archiv für Mediengeschichte, vol. 2: Licht und Leitung, Weimar 2002, pp. 229–242.
Mokyr, Joel (ed.): The Oxford Encyclopedia of Economic History, vol. 3, Oxford 2003.
Mommsen, Theodor: Heortologie. Antiquarische Untersuchungen über die städtischen Feste der Athener, Leipzig 1864.
Monceau, Henri Louis Duhamel du: Anfangsgründe der Schiffbaukunst oder Praktische Abhandlung über den Schiffbau, ed. and transl. Christian G. Müller, Berlin 1791, reprint Kassel 1973.

Montanari, Franco: "Episches Meer, Epos des Meeres," in: Hannah Baader and Gerhard Wolf (eds.): Das Meer, der Tausch und die Grenzen der Repräsentation, Zurich and Berlin 2010, pp. 193–206.
Moog-Grünewald, Maria: "Zwischen Kontingenz und Ordo. Das Emblem in Renaissance und Barock," in: Joachim Küpper and Friedrich Wolfzettell (eds.): Diskurse des Barock. Dezentrierte oder rezentrierte Welt?, Munich 2000, pp. 187–216.
Mosès, Stéphane: "Franz Kafka: 'Das Schweigen der Sirenen,'" in: Spuren der Schrift. Von Goethe bis Celan, Frankfurt a. M. 1987, pp. 52–72.
Moss, David A.: When all Else Fails: Government as the Ultimate Risk Manager, Cambridge (MA) et al. 2002.
Mostert, Noël: Supership, Harmondsworth and New York 1976.
Muckelroy, Keith: "The Archaeology of Shipwrecks," in: Lawrence E. Babits and Hans van Tilburg (eds.): Maritime Archaeology: A Reader of Substantive and Theoretical Contributions, New York and London 1998, pp. 267–290.
Müller, Jan-Dirk: "Literarische und andere Spiele. Zum Fiktionalitätsproblem in vormoderner Literatur," in: Poetica 36 (2004), pp. 281–311.
Müller, Jan-Dirk: "Rationalisierung und Mythisierung in Erzähltexten der Frühen Neuzeit," in: Klaus Ridder (ed.): Reflexion und Inszenierung von Rationalität in der mittelalterlichen Literatur. Blaubeurer Kolloquium 2006, vol. 20 of Wolfram-Studien, Berlin 2008, pp. 435–456.
Münchberg, Katharina: "Im Gespräch mit Odysseus. Zum Verhältnis von Allegorie und Dialog in Dantes Commedia, Inferno XXVI," in: Deutsches Dante-Jahrbuch 72 (1997), pp. 83–96.
Münchberg, Katharina: Dante. Die Möglichkeit der Kunst, Heidelberg 2005.
Muldoon, James: "Who Owns the Sea?," in: Bernhard Klein (ed.): Fictions of the Sea: Critical Perspectives in English Literature and Culture, Aldershot 2002, pp. 13–27.
Mumford, Lewis: Herman Melville: A Study of His Life and Vision, London 1963.
Munk, Walter: "The Circulation of the Oceans" (1955), in: Jonathan Raban (ed.): The Oxford Book of the Sea, New York 1992, pp. 443–446.
Nägele, Rainer: "Andenken an Hyperion," in: Hansjörg Bay (ed.): Hyperion—terra incognita. Expeditionen in Hölderlins Roman, Wiesbaden 1998, pp. 17–37.
Nagel, Bert: Kafka und die Weltliteratur. Zusammenhänge und Wechselwirkungen, Munich 1983.
Najder, Zdzislaw: Conrad in Perspective: Essays on Art and Fidelity, Cambridge (UK) 1997.
Naremore, James: On Kubrick, London 2007.
Neckham, Alexander: De naturis rerum libri duo; with the poem of the same author De Laudibus divinae sapientiae, ed. Thomas Wright, London 1863.
Nehlsen-von Stryk, Karin: Die venezianische Seeversicherung im 15. Jahrhundert, Ebelsbach 1986.
Nehlsen-von Stryk, Karin: "Kalkül und Hazard in der spätmittelalterlichen Seeversicherungspraxis," in: Rechtshistorisches Journal 8 (1989), pp. 195–208.
Neitzel, Britta and Rolf F. Nohr: "Einleitung," in: Neitzel/Nohr (eds.): Das Spiel mit dem Medium. Partizipation—Immersion—Interaktion. Zur Teilhabe an den Medien von Kunst bis Computerspiel, Marburg 2006, pp. 9–17.
Nelis, Damien: Vergil's Aeneid and the Argonautica of Apollonius Rhodius, Leeds 2001.
Nelson, Curtis L.: Hunters in the Shallows: A History of the PT Boat, Washington et al. 1998.

Nerlich, Michael: Abenteuer oder das verlorene Selbstverständnis der Moderne. Von der Unaufhebbarkeit experimentellen Handelns, Munich 1997.
Nes, Dirk van: Die maritime Bildersprache des Aischylos, Groningen 1963.
Neumann, Gerhard: "Umkehrung und Ablenkung: Franz Kafkas 'Gleitendes Paradox,'" in: Heinz Politzer (ed.): Franz Kafka, Darmstadt 1973, pp. 459–515.
Nietzsche, Friedrich: Briefwechsel. Kritische Gesamtausgabe, ed. Giorgio Colli and Mazzino Montinari, vol. 3.1, Berlin 1981.
Nietzsche, Friedrich: "On Truth and Lying in an Extra-Moral Sense," in: Friedrich Nietzsche, Friedrich Nietzsche on Rhetoric and Language, transl. Sander Gilman, Carol Blair, and David J. Parent, New York and Oxford 1989.
Nietzsche, Friedrich: Beyond Good and Evil, transl. Judith Norman, Cambridge (UK) 2001a.
Nietzsche, Friedrich: The Gay Science, transl. Josefine Nauckhoff, Cambridge (UK) 2001b.
Nitemaa, Vilho: Das Strandrecht in Nordeuropa im Mittelalter, Helsinki 1955.
Noonan, John T. Jr.: The Scholastic Analysis of Usury, Cambridge (UK) 1957.
Nora, Pierre: Zwischen Geschichte und Gedächtnis, Frankfurt a. M. 1998.
Nordenskiöld, Adolf Erik: Periplus: An Essay on the Early History of Charts and Sailing-Directions, Stockholm 1897.
Nork, F. [alias Friedrich Korn]: Mythologie der Volkssagen und Volksmärchen; eine Darstellung ihrer genetischen Entwicklung mit vorzugsweise Berücksichtigung jener durch Deutungs-Versuche von Naturerscheinungen, Lokaleigenthümlichkeiten, Orts- und Personennamen, Wahrzeichen von Städten, Wappenbildern sc. erzeugten Sagenbildungen, Stuttgart 1848.
Norris, Margot: "Possible-Worlds Theory and Joyce's 'Wandering Rocks': The Case of Father Conmee," in: Joyce Studies Annual (2007), pp. 21–43.
North, Michael: Kleine Geschichte des Geldes. Vom Mittelalter bis heute, Munich 2009.
Nova, Alessandro: "Kirche, Nation, Individuum. Das stürmische Meer als Allegorie, Metapher und Seelenzustand," in: Hannah Baader and Gerhard Wolf (eds.): Das Meer, der Tausch und die Grenzen der Repräsentation, Zurich and Berlin 2010, pp. 67–94.
Novak, Maximilian E.: Economics and the Fiction of Daniel Defoe, Berkeley and Los Angeles 1962.
Nowacki, Horst: "Archimedes und die Stabilität von Schiffen," in: Deutsches Schiffahrtsarchiv 24 (2001), pp. 7–37.
Nüstedt, Holger: Joseph Conrads Seegeschichten. Variationen eines anthropologischen Initiationskonzepts, Frankfurt a. M. 1998.
O'Keefe, Patrick J.: "International Waters," in: Sarah Dromgoole (ed.): Legal Protection of the Underwater Cultural Heritage: National and International Perspectives, The Hague, London, and Boston 1999, pp. 223–235.
Obrist, Barbara: "Wind diagrams and medieval cosmology," in: Speculum 72 (1997), pp. 33–84.
Oduntan, Gbenga: Sovereignty and Jurisdiction in the Airspace and Outerspace. Legal Criteria for Spatial Delimitation, New York 2012.
Oldelehr, Herrmann: Seekrankheit in der Antike, Düsseldorf 1977.
Ordonnance de Louis XIV Roi de France et de Navarre. Donné à Fontainebleau au mois d'Août 1681. Touchant la Marine (n.a.), Paris 1687.
Ortlieb, Cornelia: "Eismeere. Zur Geschichte eines modernen Phantasmas," in: Hannah Baader and Gerhard Wolf (eds.): Das Meer, der Tausch und die Grenzen der Repräsentation, Zurich and Berlin 2010, pp. 123–145.

Orsenna, Erik: Lob des Golfstroms, Munich 2008.
Oschmann, Dirk: "Anonymität als Symptom in der Literatur der Weimarer Republik," in: Stefan Pabst (ed.): Anonymität und Autorschaft. Zur Literatur- und Rechtsgeschichte der Namenlosigkeit, Berlin and Boston 2011, pp. 289–306.
Oswald, Renate: Das Ende der Odyssee. Studien zu Strukturen epischen Gestaltens, Graz 1993.
Owen, William: Captain Owen's Narrative of Voyages to Explore the Shores of Africa, Arabia, and Madagascar, 2 vols., London 1833.
Pacioli, Luca: Abhandlung über die Buchhaltung 1494, ed. Rudolf Seyffert, Stuttgart 1997.
Pabst, Stefan: "Einleitung," in: Stefan Pabst (ed.): Anonymität und Autorschaft. Zur Literatur- und Rechtsgeschichte der Namenlosigkeit, Berlin and Boston 2011, pp. 1–34.
Pallas, Nadine Christina: Maritimer Kulturgüterschutz, Berlin 2004.
Palma, Paola: "Monitoring of Shipwreck Sites," in: The International Journal of Nautical Archaeology 34/2 (2005), pp. 323–332.
Palmer, Michael A.: Command at Sea: Naval Command and Control Since the Sixteenth Century, Cambridge (MA) and London 2005.
Panofsky, Erwin: "What is Baroque," in: Erwin Panofky, Three Essays on Style," Cambridge (MA) 1995, pp. 17–80.
Parker, A. J.: Ancient Shipwrecks of the Mediterranean & the Roman Provinces, Oxford 1992.
Parker, Hershel: Herman Melville: A Biography, vol. 1: 1819–1851, Baltimore and London 1996.
Parry, John Horace: The Discovery of the Sea, New York 1974.
Pascal, Blaise: "Discours sur la condition des grands," in: Blaise Pascal: Œuvres, ed. Léon Brunschvicg, Pierre Boutroux, and Félix Gazier, vol. 9, Paris 1914, pp. 359–374.
Pascal, Blaise: Thoughts, Letters, and Minor Works, transl. W.F. Trotter et al., New York 1910.
Pasolini, Pier Paolo: Barbarische Erinnerungen. La Divina Mimesis, Berlin 1983.
Pasolini, Pier Paolo: "Il genocidio," in: Pier Paolo Pasolini, "Scritti corsari," in: Walter Siti and Silvia De Laude (eds.): Pier Paolo Pasolini. Saggi sulla politica e sulla società, Milano 1999, pp. 511–517.
Patch, Howard Rollin "The Goddess Fortuna in the Divine Comedy," in: Annual Reports of the Dante Society 33 (1914), pp. 13–28.
Patch, Howard Rollin: The Goddess Fortuna in Medieval Literature, reprint, London 1967.
Patroni, Giovanni: "Intorno al mito delle Sirene," in: Rivista di Filologia e d'Istruzione Classica, 19 (1891), pp. 321–340.
Pechmann, Alexander: Herman Melville. Leben & Werk, Vienna, Cologne, and Weimar 2003.
Peck, John: Maritime Fiction: Sailors and the Sea in British and American Novels, 1719–1917, Houndsmills 2001.
Peil, Dietmar: Untersuchungen zur Staats- und Herrschaftsmetaphorik in literarischen Zeugnissen von der Antike bis zur Gegenwart, Munich 1983.
Pepys, Samuel: Pepys' Memoires of the Royal Navy 1679–1688, Oxford 1906.
Pepys, Samuel: Samuel Pepys and the Second Dutch War: Pepys's Navy White Book and Brooke House Papers, Aldershot 1995.
Perdikas, Panayotis: "Die Entstehung der Seeversicherung im Mittelalter—Geschichtliche Grundlagen im Verhältnis zu Seedarlehen, Überseekauf, commenda und Bodmereien," in: Zeitschrift für Versicherungswissenschaft (1966), pp. 425–509.

Pérez-Mallaína, Pablo E.: Spain's Men of the Sea: Daily Life on the Indies Fleets in the Sixteenth Century, Baltimore 1998.
Person, Leland S. Jr.: "Trusting the Tellers. Paradoxes of Narrative Authority in Poes 'A Descent into the Maelstrom,'" in: Journal of Narrative Technique 23/1 (Winter 1993), pp. 45–56.
Peters, Sibylle and Martin Jörg Schäfer (eds.), "Intellektuelle Anschauung." Figurationen von Evidenz zwischen Kunst und Wissen, Bielefeld 2006.
Peterson, Erik: "Das Schiff der Kirche. Die Tat des Messias im eschatologischen Meeressturm in der jüdischen und altchristlichen Überlieferung," in: Theologische Zeitschrift 6 (1950), pp. 77–79.
Phillips, Gervase: "To Cry 'Home! Home!': Mutiny, Morale, and Indiscipline in Tudor Armies," in: The Journal of Military History 65/2 (2001), pp. 313–332.
Pias, Claus: Computer Spiel Welten, Munich 2002.
Pias, Claus (ed.): Cybernetics—Kybernetik. The Macy-Conferences 1946–1953, vol. 1: Protokolle, Zurich and Berlin 2003a.
Pias, Claus: "Zeit der Kybernetik—eine Einstimmung," in: Claus Pias (ed.): Cybernetics—vol. 2: Essays & Dokumente, Zurich and Berlin 2003b, pp. 9–41.
Pierce, Frank: "Camões' Adamastor," in: D. M. Atkinson and A. H. Clarke (eds.): Hispanic Studies in Honour of Joseph Manson, Oxford 1972.
Pietrzik, Dominik: Die Brandan-Legende. Ausgewählte Motive in der frühneuhochdeutschen sogenannten 'Reise'-Version, Frankfurt a.M. et al. 1999.
Piper, Leonard: Dangerous Waters: The Life and Death of Erskine Childers, London and New York 2003.
Pircher, Wolfgang: "Im Schatten der Kybernetik. Rückkopplung im operativen Einsatz: operational research," in: Michael Hagner and Erich Hörl (eds.): Die Transformation des Humanen. Beiträge zur Kulturgeschichte der Kybernetik, Frankfurt a.M. 2008, pp. 348–376.
Plato: The Collected Dialogues of Plato, ed. Edith Hamilton and Huntington Cairns, New York 1961.
Plotinus: The Enneads, transl. Stephan MacKenna, London 1956.
Poe, Edgar Allan: Essays and Reviews. Theory of Poetry. Reviews of British and Continental Authors. Reviews of American Authors and American Literature. Magazines and Criticism. The Literary and Social Scene. Articles and Marginalia, ed. Gary R. Thompson, New York 1984a.
Poe, Edgar Allan: Poetry and Tales, ed. Patrick F. Quinn, New York 1984b.
Polonsky, Rachel: "Poe's aesthetic theory," in: Kevin J. Hayes (ed.): The Cambridge Companion to Edgar Allan Poe, Cambridge (UK) 2002, pp. 42–56.
Pope, Alexander: Works, vol. 3: Moral Essays: With His Last Corrections, London 1751.
Popitz, Heinrich: Phänomene der Macht, Tübingen 1992.
Postyshev, Vladimir: The Concept of the Common Heritage of Mankind: From New Thinking to New Practice, Moscow 1990.
Potapova, Galina: "B. Travens Roman Das Totenschiff. Fassungsgeschichte eines Werks zwischen der Erstausgabe und dem Edierten Text der Gesamtausgabe," in: Günter Dammann (ed.): B. Traven. Autor—Werk—Werkgeschichte, Würzburg 2012, pp. 57–94.
Pound, Ezra: The Cantos, New York 1996.
Powell, Barry: Homer and the Origin of the Greek Alphabet, Cambridge (UK) 1991.

Powell, Barry: "Homer and Writing," in: Ian Morris and Barry Powell (eds.): A New Companion to Homer, Leiden, New York and Cologne 1997, pp. 3–32.
Powell, Barry: Writing and the Origins of Greek Literature, Cambridge (UK) 2002.
Power, Arthur: Conversations with James Joyce, ed. Clive Hart, Chicago 1974.
Praeder, Susan Marie: "Acts 27:1–28:16: Sea Voyages in Ancient Literature and the Theology of Luke-Act," in: The Catholic Biblical Quarterly 46/4 (1984), pp. 683–706.
Preller, Ludwig: Griechische Mythologie, vol. 1, Berlin 1860.
Priestley, Joseph: The History and Present State of Electricity: With Original Experiments, 2 vols., New York and London 1966.
Pucci, Pietro: Odysseus Polutropos: Intertextual Readings in the Odyssey and the Iliad, Ithaca and London 1987.
Pucci, Pietro: "The Song of the Sirens," in: Arethusa, 12/2 (1979), pp. 121–129.
Pufendorf, Samuel: De Officio Hominis Et Civis Iuxta Legem Naturalem Libri Duo, Erfurt, Rudolstadt 1678.
Pufendorf, Samuel: Acht Bücher vom Natur- und Völcker-Rechte, vol. 1, Frankfurt a. M. 1711.
Pufendorf, Samuel: On the Natural State of Men: The 1678 Latin Edition and English Translation, ed. Michael Seidler, Lewiston (NY) et al. 1990.
Pufendorf, Samuel: On the Duty of Man and Citizen According to Natural Law, ed. James Tully, Cambridge et al. 1991.
Purchas, Samuel: Hakluytus Posthumus or Purchas His Pilgrimes: Contayning a History of the World in Sea Voyages and Lande Travells by Englishmen and others, vol. 1, Glasgow 1905.
Pytheas of Massalia: On the Ocean, ed. and transl. Christina Horst Roseman, Chicago 1994.
Quaritsch, Helmut: "Das Schiff als Gleichnis," in: Hans Peter Ipsen and Karl-Hartmann Necker (eds.): Recht über See. Festschrift. Rolf Stödter zum 70. Geburtstag am 22. April 1979, Hamburg and Heidelberg 1979, pp. 251–286.
Quételet, Adolphe: Sur l'homme et le développement de ses facultés, 2 vols, vol. 1, Brussels 1836; vol 2., Brussels 1835 (sic).
Quint, David: Epic and Empire: Politics and Generic Form from Virgil to Milton, Princeton 1993.
Rachewiltz, Siegfried von: De Sirenibus: An Inquiry into Sirens from Homer to Shakespeare, New York and London 1987.
Radermacher, Ludwig: Mythos und Sage bei den Griechen, 2nd edn., Brünn, Munich and Vienna 1938.
Radloff, Bernhard: Will and Representation: The Philosophical Foundations of Melville's Theatrum Mundi, New York et al. 1996.
Raether, Martin: "Dantis Ulixes figura poetae. Eine typologische Interpretation von Inferno 26 als Poetik der Divina Commedia," in: Poetica 13 (1981), pp. 280–308.
Rahner, Hugo: Symbole der Kirche. Die Ekklesiologie der Väter, Salzburg 1964.
Raitz, Walter: 'Fortunatus', Munich 1984.
Rammstedt, Ottheim: "Risiko," in: Joachim Ritter, Karlfried Gründer, and Gottfried Gabriel (eds.): Historisches Wörterbuch der Philosophie, vol. 8, Basel, Stuttgart 1992, col. 1045–1050.
Ramras-Rauch, Gila: "The Response of Biblical Man to the Challenge of the Sea," in: Anna-Teresa Tymieniecka (ed.): Poetics of the Elements in the Human Condition: The Sea:

From Elemental Stirrings to Symbolic Inspiration, Language, and Life-Significance in Literary Interpretation and Theory, Dordrecht, Boston, and Lancaster 1985, pp. 139 – 148.
Randles, William Graham L.: "The Alleged Nautical School Founded in the Fifteenth Century at Sagres by Prince Henry of Portugal, Called the 'Navigator,'" in: Imago Mundi 45 (1993), pp. 20 – 28.
Randles, William Graham L.: Geography, Cartography and Nautical Science in the Renaissance: The Impact of the Great Discoveries, Aldershot 2000.
Rapp, Hans Reinhard: Mensch, Gott und Zahl. Kybernetik im Horizont der Theologie, Hamburg 1967.
Rapske, Brian M.: "Acts, Travel, and Shipwreck," in: David William J. Gill and Conrad Gempf (eds.): The Book of Acts in Its Greco-Roman Setting, Carlisle 1994, pp. 1 – 47.
Rasor, Eugene L.: The Titanic: Historiography and Annotated Bibliography, Westport and London 2001.
Rath, Jürgen: "...doch stehlen können sie meisterlich!" Über Strandungen, Strandräuber und Strandvögte, Hamburg 2007.
Ratzel, Friedrich: Das Meer als Quelle der Völkergröße. Eine politisch-geographische Studie, 2nd edn., ed. Hans Helmolt, Munich and Berlin 1911.
Ratzenhofer, Gustav: Die Staatswehr. Wissenschaftliche Untersuchung der öffentlichen Wehrangelegenheiten, Stuttgart 1881.
Ravenstein, Ernest George: A Journal of the First Voyage of Vasco da Gama 1497 – 99, reprint, New York 1963.
Ray, G. Carleton and Bruce P. Hayden: "Coastal Zone Ecotones," in: Andrew J. Hansen and Francesco di Castri (eds.): Landscape Boundaries: Consequences for Biotic Diversity and Ecological Flows, New York et al. 1992, pp. 403 – 420.
Regn, Gerhard: "Gott als Dichter. Die Wirklichkeit der Fiktion in Dantes Paradiso," in: Ursula Peters and Rainer Warning (eds.): Fiktion und Fiktionalität in den Literaturen des Mittelalters. Jan-Dirk Müller zum 65. Geburtstag, Munich 2009, pp. 365 – 386.
Reichert, Klaus: Fortuna oder Die Beständigkeit des Wechsels, Frankfurt a. M. 1985.
Reichert, Klaus: "Joyces Memoria," in: Welt-Alltag der Epoche. Essays zum Werk von James Joyce, Frankfurt a. M. 1994, pp. 77 – 106.
Reichert, Klaus: "Odysseus und Ulysses. Irrfahrten und Wanderungen von Wörtern und Helden," in: Elisabeth Wagner and Burkhardt Wolf (eds.): Odysseen. Mosse-Lectures 2007, Berlin 2008, pp. 64 – 92.
Reinitzer, Heimo: "Wasser des Todes und Wasser des Lebens. Über den geistigen Sinn des Wassers im Mittelalter," in: Hartmut Böhme (ed.): Kulturgeschichte des Wassers, Frankfurt a. M. 1988, pp. 99 – 144.
Reinke-Kunze, Christine: Schiffe der Welt. Welt der Forschungsschiffe, Hamburg 1994.
Rengakos, Antonios: Apollonios Rhodios und die antike Homererklärung, Munich 1994.
Reynolds, Mary T.: Joyce and Dante: The Shaping Imagination, Princeton 1981.
Rheinberger, Hans-Jörg: Experimentalsysteme und epistemische Dinge. Eine Geschichte der Proteinsynthese im Reagenzglas, Göttingen 2001.
Ricœur, Paul: Time and Narrative, vol. 3, transl. Kathleen Blamey and David Palauer, Chicago and London 1988.
Ricœur, Paul: "Archiv, Dokument, Spur," in: Knut Ebeling and Stephan Günzel (eds.): Archivologie. Theorien des Archivs in Philosophie, Medien und Künsten, Berlin 2009, pp. 123 – 138.

Riehle, Wolfgang: Daniel Defoe, Reinbek 2002.
Rimpau, Laetitia: "Räume des Übergangs: Brunetto Latini, Dante Alighieri, Francesco Colonna," in: Rudolf Behrens and Rainer Stillers (eds.): Orientierungen im Raum. Darstellung räumlichen Sinns in der italienischen Literatur von Dante bis zur Postmoderne, Heidelberg 2008, pp. 27–44.
Ring, Jim: Erskine Childers, London 1996.
Ripa, Cesare: Iconologie ov explication novvelle de plusievrs images, reprint, Dijon 1999.
Ritter, C[arl]: "Vom Grundsatze der allgemeinen Gefahrendeckung im Seeversicherungsrechte, insbesondere die Gefahr des Verschuldens der Schiffsbesatzung," in: ZfgV 14 (1914), pp. 28–66.
Röller, Nils: Ahabs Steuer. Navigationen zwischen Kunst und Wissenschaft, Berlin 2005.
Röller, Nils: Magnetismus. Eine Geschichte der Orientierung, Munich 2010.
Röpling, Gerhard: Die Struktur der englischen Seeversicherung und ihr Einfluß auf die Zahlungsbilanz Großbritanniens, Berlin 1960.
Rösch, Gerhard: Venedig. Geschichte einer Seerepublik, Stuttgart, Berlin, and Cologne 2000.
Rösler, Wolfgang: Dichter und Gruppe. Eine Untersuchung zu den Bedingungen und zur historischen Funktion früher griechischer Lyrik am Beispiel Alkaios, Munich 1980.
Robertson, Noel: "Sacrifice to the Sea: a Custom Prior to the 'Olympian' and 'Chthonian' Categories?," in: Robin Hägg and Brita Alroth (eds.): Greek Sacrificial Ritual, Olympian and Chthonian, Proceedings of the Sixth International Seminar on Ancient Greek Cult, Stockholm 2005, pp. 85–98.
Rogers, Francis M.: "The Vivaldi Expedition," in: 73rd Annual Report of the Dante Society, Cambridge (MA) 1955, pp. 31–45.
Rohland, Woldemar von: Die Gefahr im Strafrecht, 2nd edn., Dorpat 1888.
Roller, Duane W.: Through the Pillars of Herakles: Greco-Roman Exploration of the Atlantic, New York and London 2006.
Roloff, Hans-Gert (ed.): Fortunatus, Studienausgabe nach der Editio Princeps von 1509, bibliographisch ergänzte Ausgabe, Stuttgart 2007.
Roloff, Jürgen: Die Apostelgeschichte, Göttingen 1988.
Rooke, Laurence: "Directions for Sea-men, Bound for Voyages," in: Philosophical Transactions 1 (1665/1666), pp. 140–143.
Rosenblueth, Arturo, Norbert Wiener, and Julian Bigelow: "Behavior, Purpose and Teleology," in: Claus Pias (ed.): Cybernetics—Kybernetik: The Macy-Conferences 1946–1953, vol. 2: Essays & Dokumente, Zurich and Berlin 2004, pp. 327–332.
Rosenne, Shabtai: The Perplexities of Modern International Law, Leiden 2004.
Rost, Georg Alexander: Vom Seewesen und Seehandel in der Antike. Eine Studie aus maritim-militärischer Sicht, Amsterdam 1968.
Rüegg, August: "Zum Danteschen Ulisse-Problem," in: Deutsches Dante-Jahrbuch 43 (1965), pp. 172–198.
Rüger, Jan: The Great Naval Game: Britain and Germany in the Age of Empire, Cambridge (UK) 2007.
Ruge, Sophus: Die Entdeckung des Seewegs nach Ostindien durch Vasco da Gama 1497/8, Dresden 1898.
Rule, John G.: "Wrecking and Coastal Plunder," in: Douglas Hay et al. (eds.): Albion's Fatal Tree: Crime and Society in Eighteenth-Century England, London 1975.

Ruud, Jay: Critical Companion to Dante: A literary Reference to His Life and Work, New York 2008.
Ryan, Marie-Laure: Possible Worlds, Artificial Intelligence, and Narrative Theory, Indiana 1991.
Ryan, Marie-Laure: Narrative as Virtual Reality: Immersion and Interactivity in Literature and Electronic Media, Baltimore et al. 2001.
Ryle, Gilbert: Dilemmas: The Tarner Lectures 1953, Cambridge (UK) 1954.
Salewski, Michael: "Die Wilhelminischen Flottengesetze. Realität und Illusion," in: Die Deutschen und die See. Studien zur deutschen Marinegeschichte des 19. und 20. Jahrhunderts, ed. Jürgen Elvert and Stefan Lippert, Stuttgart 1998, pp. 119–125.
Salingar, Leo: "The New World in The Tempest," in: Jean-Pierre Maquerlot and Michèle Willems (eds.): Travel and Drama in Shakespeare's Time, Cambridge (UK) 1996, pp. 209–222.
Salters, Robert B.: Jonah & Lamentations, Sheffield 1994.
Sandau, Paul: Das Recht des Pseudonyms, Marburg 1932 (diss.).
Sanford, Rhonda Lemke: Maps and Memory in Early Modern England: A Sense of Place, New York and Basingstoke 2002.
Sartre, Jean-Paul: Being and Nothingness, transl. Hazel E. Barnes, New York 1966.
Sasson, Jack M.: Jonah: A New Translation with Introduction, Commentary, and Interpretation, New York et al. 1990.
Saussure, Ferdinand de: Course in General Linguistics, transl. Wade Baskin, New York 1959.
Schadewaldt, Hans: "Zur Geschichte der Seekrankheit," in: Medizinische Welt 38 (1967), pp. 2258–2265.
Schäfer, Eckart: "Das Staatsschiff. Zur Präzision eines Topos," in: Peter Jehn (ed.): Toposforschung. Eine Dokumentation, Frankfurt a. M. 1972, pp. 259–392.
Schäffner, Wolfgang: "Zur Navigation der Kunst," in: Dieter Mammel: Space becomes a Whirlpool, Berlin 1996, pp. 9–23.
Schäffner, Wolfgang: "Das Trauma der Versicherung. Das Ereignis im Zeitalter der Wahrscheinlichkeit," in: Inka Mülder-Bach (ed.): Modernität und Trauma. Beiträge zum Zeitenbruch des Ersten Weltkriegs, Vienna 2000, pp. 104–120.
Schäffner, Wolfgang: "Die Verwaltung der Endlichkeit. Zur Geburt des neuzeitlichen Romans in Spanien," in: Eckart Goebel and Martin von Koppenfels (eds.): Die Endlichkeit der Literatur, Berlin 2002, pp. 1–12.
Schäffner, Wolfgang, "Raster-Orte," in: Annett Zinsmeister (ed.): Constructing Utopia: Konstruktionen künstlicher Welten, Zurich and Berlin 2005, pp. 47–56.
Schäffner, Wolfgang: "Das Ei des Brunelleschi. Projekte, Fiktionen und die Erfindung des Neuen," in: Daniel Weidner (ed.): Figuren des Europäischen. Kulturgeschichtliche Perspektiven, Paderborn 2006, pp. 43–58.
Scherer, Burkhard: Mythos, Katalog und Prophezeiung. Studien zu den 'Argonautika' des Apollonios Rhodios, Stuttgart 2006.
Schestag, Thomas: "'das kind beim rechten namen nennen'. ... Goethe ... Auerbach ... Homer...," in: Tatjana Petzer, Sylvia Sasse, Franziska Thun-Hohenstein, and Sandro Zanetti (eds.): Namen. Benennung—Verehrung—Wirkung. Positionen der europäischen Moderne, Berlin 2009, pp. 273–294.
Schiebe, August: Das französische Seerecht (based on Pardessus, Cours de Droit Commercial), Bremen 1840.

Schlaffer, Heinz: Poesie und Wissen. Die Entstehung des ästhetischen Bewußtseins und der philologischen Erkenntnis, Frankfurt a. M. 2005.
Schlag, Richard: Beitrag zum Capitel Seekrankheit, Greifswald 1899 (diss.).
Schlee, Susan: A History of Oceanography: The Edge of an Unfamiliar World, London 1975.
Schlegel, Friedrich: Lectures on the History of Literature, Ancient and Modern, [transl. J.G. Lockhart.], New York 1841.
Schlesier, Renate: "Transgressionen des Odysseus," in: Andreas Luther (ed.): Epos und Zitat. Geschichte und Fiktion in der homerischen Odyssee, Munich 2006, pp. 107–116.
Schmidgen, Wolfram: "Robinson Crusoe, Enumeration and the Mercantile Fetish," in: Eighteenth-Century Studies 35/1 (2001), pp. 19–39.
Schmidt, Arnold: "Walter Scott's The Pirate: Imperialism, Nationalism and Bourgeois Values," in: Bernhard Klein (ed.): Fictions of the Sea: Critical Perspectives in English Literature and Culture, Aldershot 2002, pp. 89–103.
Schmidt, Jochen: Hölderlins letzte Hymnen: 'Andenken' und 'Mnemosyne', Tübingen 1970.
Schmitt, Carl: "Nehmen / Teilen / Weiden," in: Carl Schmitt, Verfassungsrechtliche Aufsätze aus den Jahren 1924–1954. Materialien zu einer Verfassungslehre, 2nd edn., Berlin 1973, pp. 489–504.
Schmitt, Carl: "The Turn to the Discriminating Concept of War," in: Carl Schmitt, Writings on War, transl. Timothy Nunan, Cambridge (UK) and Malden 1988, pp. 30–74.
Schmitt, Carl: "Die letzte globale Linie," in: Carl Schmitt, Staat, Großraum, Nomos. Arbeiten aus den Jahren 1916–1969, ed. Günter Maschke, Berlin 1995a, pp. 441–452.
Schmitt, Carl: "Das Meer gegen das Land," in: Carl Schmitt, Staat, Großraum, Nomos. Arbeiten aus den Jahren 1916–1969, ed. Günter Maschke, Berlin 1995b, pp. 395–400.
Schmitt, Carl: "Der neue Nomos der Erde," in: Carl Schmitt, Staat, Großraum, Nomos. Arbeiten aus den Jahren 1916–1969, ed. Günter Maschke, Berlin 1995c, pp. 518–522.
Schmitt, Carl: "Nomos—Nahme—Name," in: Carl Schmitt, Staat, Großraum, Nomos. Arbeiten aus den Jahren 1916–1969, ed. Günter Maschke, Berlin 1995d, pp. 573–591.
Schmitt, Carl: "Die Raumrevolution. Durch den totalen Krieg zu einem totalen Frieden," in: Carl Schmitt, Staat, Großraum, Nomos. Arbeiten aus den Jahren 1916–1969, ed. Günter Maschke, Berlin 1995e, pp. 388–394.
Schmitt, Carl: "Gespräch über den Neuen Raum (1955/58)," in: Carl Schmitt, Staat, Großraum, Nomos. Arbeitenaus den Jahren 1916–1969, ed. Günter Maschke, Berlin 1995, pp. 552–572.
Schmitt, Carl: Der Begriff des Politischen. Text von 1932 mit einem Vorwort und drei Corollarien, Berlin 1996.
Schmitt, Carl: The Nomos of the Earth in the International Law of the Jus Publicum Europaeum (1950), transl. G.L. Ulmen, Candor (NY) 2003.
Schmitt, Carl: Dictatorship: From the Origin of the Modern Concept of Sovereignty to Proletarian Class Struggle, transl. Michael Hoelzl and Graham Ward, Cambridge (UK) and Malden (MA) 2014.
Schmitt, Carl: Land and Sea: A World-Historical Meditation (1942), transl. Samuel Garrett Zeitlin, Candor (NY) 2015.
Schmitt, Thomas M.: Cultural Governance. Zur Kulturgeographie des UNESCO-Welterberegimes, Stuttgart 2011.
Schmitt-Lermann, Hans: Der Versicherungsgedanke im deutschen Geistesleben des Barock und der Aufklärung, Munich 1954.

Schmitt-Neuerburg, Tilman: Vergils Aeneis und die antike Homerexegese. Untersuchungen zum Einfluß ethischer und kritischer Homerrezeption auf 'imitatio' und 'aemulatio' Vergils, Berlin et al. 1999.
Schmitz-Emans, Monika: Seetiefen und Seelentiefen. Literarische Spiegelungen innerer und äußerer Fremde, Würzburg 2003.
Schmitzer, Ulrich: "Odysseus—ein griechischer Held im kaiserzeitlichen Rom," in: Andreas Luther (ed.): Odyssee-Rezeptionen, Frankfurt a. M. 2005, pp. 33–53.
Schnall, Uwe: "Practical Navigation in the Late Middle Ages. Some Remarks on the Transfer of Knowledge from the Mediterranean to the Northern Seas," in: Christiane Villain-Gandossi (ed.): Medieval Ships and the Birth of Technological Societies, vol. 2: The Mediterranean Area and European integration, Vienna 1991, pp. 271–279.
Schnapp, Alain: Die Entdeckung der Vergangenheit. Ursprünge und Abenteuer der Archäologie, 3rd ed., Stuttgart 2011.
Schneider, Gerhard: Die Apostelgeschichte. II. Teil, Freiburg, Basel and Vienna 1982, pp. 379–408.
Schneider, Ivo: Die Entwicklung der Wahrscheinlichkeitstheorie von den Anfängen bis 1933. Einführungen und Texte, Darmstadt 1988.
Schneider, Manfred: Liebe und Betrug. Die Sprachen des Verlangens, Munich 1994.
Schneider, Manfred: Der Barbar. Endzeitstimmung und Kulturrecycling. Munich 1997.
Schneider, Ulrich: "'An Actuality of the Possible as Possible.' Reflections on the Theme of History in 'Nestor,'" in: Gottlieb Gaiser (ed.): International Perspectives on James Joyce, Troy (NY) and New York 1986, pp. 44–58.
Schnobel, Martin: Die rechtliche Stellung des Kapitäns zur Schiffsmannschaft, Würzburg 1926 (diss.).
Schoeck, R. J.: "Shakespeare and the Law: An Overview," in: W.R. Elton and John M. Mucciolo (eds.): The Shakespearean International Yearbook, vol. 1: Where are We Now on Shakespeare Studies?, Aldershot 1999, pp. 219–239.
Schopenhauer, Arthur: "Animal Magnetism and Magic," in: Arthur Schopenhauer, Two Essays by Arthur Schopenhauer: I. On the Fourfold Root of the Principle of Sufficient Reason. II. On the Will to Reason, transl, Mme. Karl Hillebrand, London 1891, pp. 326–358.
Schopenhauer, Arthur: The World as Will and Representation, 2 vols., transl. E.F.J Pyne, New York 1969.
Schück, Albert: Der Kompass, 3 vols., Hamburg 1911–1918.
Schücking, Levin: "Die drei Freier" (1848), in: Der Bauernfürst, Leipzig 1851, pp. 32–103.
Schücking, Walther: Der Bund der Völker. Studien und Vorträge zum organisatorischen Pazifismus, Leipzig 1918.
Schüttpelz, Erhard: "Eine Berichtigung für eine Akademie," in: Oliver Simon and Arne Höcker (eds.): Kafkas Institutionen, Bielefeld 2007, pp. 91–118.
Schulz, Raimund: Die Antike und das Meer, Darmstadt 2005.
Schuster, Susanne: Narrative Politik. Untersuchungen zur politischen Lektüre von Daniel Defoes Robinson Crusoe-Trilogie, Würzburg 2000.
Schwarz, Arnold: Statistik durch Anschauung, mit 130 Figuren, Zurich 1945.
Schwarztrauber, Helmut: Fiktion der Fiktion. Begründung und Bewahrung des Erzählens durch theoretische Selbstreflexion im Werk N. Hawthornes und E. A. Poes, Heidelberg 2000.
Schwerdt, Carl: Eine akute durch Traumen bedingte Stoffwechselstörung und ihre Verhütung, Jena 1914.

Schwingeler, Stephan: Die Raummaschine. Raum und Perspektive im Computerspiel, Boizenburg 2008.
Scott, Sir Walter: Memoirs of the life of Sir Walter Scott, ed. John Gibson Lockhart, vol. 1, Philadelphia 1837.
Searle, John R.: Speech Acts: An Essay in the Philosophy of Language, Cambridge (UK) 1969.
Seeßlen, Georg and Fernand Jung: Stanley Kubrick und seine Filme, Marburg 1999.
Segal, Allan F.: Life after Death: A History of the Afterlife in Western Religion, New York et al. 2004.
Seger, Tobias: "Stürze in den Malstrom. Edgar Allan Poes 'A Descent into the Maelström' im Horizont von Kants 'Analytik des Erhabenen,'" in: Literaturwissenschaftliches Jahrbuch 42 (2001), pp. 225–243.
Seidensticker, Bernd: "Irrfahrten des Odysseus?," in: Bernd Blaschke et al. (eds.): Umwege. Ästhetik und Poetik exzentrischer Reisen, Bielefeld 2008, pp. 17–32.
Seife, Charles: Zwilling der Unendlichkeit. Eine Biographie der Zahl Null, Berlin 2000.
Seifert, Arno: "Staatenkunde—eine neue Disziplin und ihr wissenschaftstheoretischer Ort," in: Mohammed Rassem and Justin Stagl (eds.): Statistik und Staatsbeschreibung in der Neuzeit, vornehmlich im 16.–18. Jahrhundert, Paderborn u. a. 1980, pp. 217–248.
Selden, John: Of the Dominion, or, Ownership of the Sea. [...] Written at First in Latin, and Entituled, Mare Clausum seu De Dominio Maris, London 1652 (Lat. edn. of 1635), reprint New York 1972.
Seligmann, Matthew S.: Spies in Uniform: British Military and Naval Intelligence on the Eve of the First World War, Oxford et al. 2006.
Sellert, Wolfgang: "Einführung," in: Okko Behrends and Wolfgang Sellert (eds.): Nomos und Gesetz. Ursprung und Wirkungen des griechischen Gesetzesdenkens, 6. Symposion der Kommission 'Die Funktion des Gesetzes in Geschichte und Gegenwart', Göttingen 1995, pp. 7–11.
Séris, Jean-Pierre: Machine et communication. Du théâtre des machines à la mécanique industrielle, Paris 1987.
Serres, Michel: with Bruno Latour: Conversations on Science, Culture, and Time, transl. Roxanne Lapidus, Anne Arbor 1995.
Shell, Marc: Money, Language, and Thought: Literary and Philosophical Economics from the Medieval to the Modern Era, Berkeley et al. 1982.
Sherman, Sandra: Finance and Fictionality in the Early Eighteenth Century: Accounting for Defoe, Cambridge (UK) 1996.
Sherwood, Yvonne: A biblical text and its afterlives: the survival of Jonah in western culture, Cambridge Univ. Press 2000.
Siegert, Bernhard: "Die Verortung Amerikas im Nachrichtendispositiv um 1500 oder: Die Neue Welt der Casa de la Contratación," in: Horst Wenzel (ed.): Gutenberg und die Neue Welt, Munich 1994, pp. 307–325.
Siegert, Bernhard: Passage des Digitalen. Zeichenpraktiken der neuzeitlichen Wissenschaften 1500–1900, Berlin 2003.
Siegert, Bernhard: "Translatio Imperii: Der cursus publicus im römischen Kaiserreich," in: Lorenz Engell, Bernhard Siegert and Joseph Vogl (eds.): Medien der Antike, Mediale Historiographien 3 (2003), pp. 41–59.
Siegert, Bernhard: "Pasajeros a Indias. 'Auto'-biographische Schrift zwischen Alter und Neuer Welt im 16. Jahrhundert," in: Anja K. Maier and Burkhardt Wolf (eds.): Wege des

Kybernetes. Schreibpraktiken und Steuerungsmodelle von Politik, Reise, Migration, Münster 2004, 260–276.
Siegert, Bernhard: Passagiere und Papiere. Schreibakte auf der Schwelle zwischen Spanien und Amerika, Munich 2006.
Siegert, Bernhard: "Die Geburt der Literatur aus dem Rauschen der Kanäle. Zum poetologischen Status der phatischen Funktion des Zeichens in der Literatur des 20. Jahrhunderts," in: Michael Franz, Wolfgang Schäffner and Bernhard Siegert (eds.): Electric Laokoon. Zeichen und Medien, von der Lochkarte zur Grammatologie, Berlin 2007, pp. 5–41.
Simek, Rudolf: Heaven and Earth in the Middle Ages: The Physical World Before Columbus, transl. Angela Hall, Woodbridge (EN) 1996.
Simmel, Georg: "The Sociology of Space," in: Georg Simmel, Simmel on Culture: Selected Writings, ed. David Fisby and Mike Featherstone, London et al. 1997.
Simmel, Georg: The Philosophy of Money, transl. Tom Bottomore and David Frisby, London and New York 2004.
Simon, Uriel: Jona. Ein jüdischer Kommentar, Stuttgart 1994.
Simpson, W. K.: "Schiffbrüchiger," in: Wolfgang Helck and Wolfhart Westendorf (eds.): Lexikon der Ägyptologie, vol. 5, Wiesbaden 1984, col. 619–622.
Sinz, Herbert: Das geteilte Risiko—The shared risk. Geschichten zur Geschichte der Rückversicherung, Munich 1979.
Slaby, Adolf: Entdeckungsfahrten in den elektrischen Ozean, Berlin 1922.
Sloterdijk, Peter: Spheres, vol. 2, Globes: Macrospherology, transl. Wieland Hoban, South Pasadena 2014.
Smidt, Heinrich: "Der ewige Segler," in: Poetische Versuche, Altona 1825, pp. 95–99.
Smidt, Heinrich: Seemanns-Sagen und Schiffer-Märchen, 2nd edn., 2 vols., Berlin 1849.
Smith, Adam: "The Principles Which Lead and Direct Philosophical Enquiries: Illustrated by the History of Astronomy," in: The Early Writings of Adam Smith, ed. J. Ralph Lindgren, London 1967, pp. 30–109.
Smith, Adam: An Enquiry into the Nature and Causes of the Wealth of Nations, ed. Edwin Cannan, Chicago 1976.
Smith, Adam: The Theory of Moral Sentiments, ed. Knud Haakonssen, Cambridge (UK) 2002.
Smith, John: Sea Grammar, Amsterdam and New York 1968.
Smith, M.A.: "The Limitations of Inference in Archaeology," in: Lawrence E. Babits and Hans van Tilburg (eds.): Maritime Archaeology: A Reader of Substantive and Theoretical Contributions, New York and London 1998, pp. 167–174.
Smuda, Manfred: "Variation und Innovation. Modelle literarischer Möglichkeiten der Prosa in der Nachfolge Edgar Allan Poes," in: Jochen Vogt (ed.): Der Kriminalroman. Poetik, Theorie, Geschichte, Munich 1998, pp. 121–141.
Snow, C.P.: The Two Cultures, London 1953.
Snyder, John P.: Flattening the Earth: Two Thousand Years of Map Projections, Chicago and London 1993.
Söhn, Gerhard: Literaten hinter Masken. Eine Betrachtung über das Pseudonym in der Literatur, Berlin 1974.
Sohn-Rethel, Alfred: Warenform und Denkform. Aufsätze, Frankfurt a. M. 1971.
Sohn-Rethel, Alfred: Das Geld, die bare Münze des Apriori, Berlin 1990.

Sokol, B. J.: A Brave New World of Knowledge: Shakespeare's 'The Tempest' and Early Modern Epistemology, Madison (NJ) et al. 2003.
Solove, Daniel J.: "Melville's Billy Budd and Security in Times of Crisis," in: Cardozo Law Review 26/6 (2005), pp. 2443 – 2470.
Sombart, Werner: Der moderne Kapitalismus, vols. 1 – 3, reprint, Munich 1987.
Sowinski, Bernhard: Herzog Ernst, in der mittelhochdeutschen Fassung B nach der Ausgabe von Karl Bartsch mit den Bruchstücken der Fassung A, durchges. und verb. Ausgabe, Stuttgart 1979.
Spencer, Eric: "Taking Excess, Exceeding Account. Aristotle Meets 'The Merchant of Venice,'" in: Linda Woodbridge (ed.): Money and the Age of Shakespeare: Essays in New Economic Criticism, New York et al. 2003, pp. 143 – 158.
Sperl, Stephan: Die Semantisierung der Musik im filmischen Werk Stanley Kubricks, Würzburg 2006.
Spooner, Frank: Risks at Sea: Amsterdam Insurance and Maritime Europe 1766 – 1780, Cambridge (UK) 1983.
Sprat, Thomas: The History of the Royal Society of London: For the Improving of Natural Knowledge, 4th edn, London 1734.
Stammler, Wolfgang: "Seemanns Brauch und Glaube," in: Deutsche Philologie im Aufriß, 2nd edn., Berlin 1967, pp. 2901 – 2972.
Stanford, W. B.: The Ulysses Theme: A Study in the Adaptability of a Traditional Hero, Oxford 1954.
Starr, Chester G.: The Influence of Sea Power on Ancient History, New York 1989.
Staudacher, Hermann: Die Friedensblockade. Ein Beitrag zur Theorie und Praxis der nichtkriegerischen Selbsthilfe, Leipzig 1909.
Stefani, Giuseppe: Insurance in Venice from the Origins to the End of the Serenissima, 2 vols., Triest 1958.
Steffen, Uwe: Das Mysterium von Tod und Auferstehung. Formen und Wandlungen des Jona-Motivs, Göttingen 1963.
Steffen, Uwe: Die Jona-Geschichte. Ihre Auslegung und Darstellung im Judentum, Christentum und Islam, Neukirchen-Vluyn 1994.
Steffy, Richard J.: Wooden Ship Building and the Interpretation of Shipwrecks, College Station 1994.
Stein, Gustav: "Die Invaliden- und Altersversicherung in der Schiffahrt auf internationalen Strömen," in: Soziale Praxis 20/13 (1910), pp. 403 – 406.
Steinbuch, Karl: Automat und Mensch. Auf dem Weg zu einer kybernetischen Anthropologie, Berlin et al. 1971.
Stephenson, Neal: Cryptonomicon, New York 2002.
Sternberger, Dolf: "Hohe See und Schiffbruch—Zur Geschichte einer Allegorie (Erstversion 1935)," in: Schriften, vol. 4.: Vexierbilder des Menschen, Frankfurt a. M. 1981, pp. 227 – 246.
Stevenson, Robert Louis: Virginibus Puerisque and Other Papers, New York 1901.
Stevenson, Robert Louis: Works, vol. 12: Records of a Family of Engineers, London and New York 1922.
Stevenson, Robert Louis and Osbourne, Lloyd: The Wrecker, in: Robert Louis Stevenson: Tales of the South Seas, ed. Jenni Calder, Edinburgh 1996.

Stevenson, Sarah York: "Insurance and Business Adventure in the Days of Shakespeare and in Those of William Penn," in: The Proceedings of the Numismatic and Antiquarian Society of Philadelphia, Philadelphia 1913, pp. 241–268.
Stierle, Karlheinz: "Der Schrecken der Kontingenz—Ein verborgenes Thema in Dantes Commedia," in: Bernhard Greiner and Maria Moog-Grünewald (eds.): Kontingenz und Ordo. Selbstbegründung des Erzählens in der Neuzeit, Heidelberg 2000, pp. 29–46.
Stierle, Karlheinz: Das große Meer des Sinns. Hermenautische Erkundungen in Dantes 'Commedia', Munich 2007.
Stierle, Karlheinz: "Fiktion," in: Karlheinz Barck et al. (eds.): Ästhetische Grundbegriffe. Historisches Wörterbuch in sieben Bänden, vol. 2, Stuttgart 2010, pp. 380–428.
Stocker, Werner: Das Prinzip des Common Heritage of Mankind als Ausdruck des Staatengemeinschaftsinteresses im Völkerrecht, Zurich 1993.
Stockhammer, Robert: Die Kartierung der Erde. Macht und Lust in Karten und Literatur, Munich 2007.
Stoker, Bram, The Mystery of the Sea, ed. Carol A. Senf, Kansas City (MO) 2007.
Stollberg-Rilinger, Babara: Der Staat als Maschine. Zur politischen Metaphorik des absoluten Fürstenstaats, Berlin 1986.
Storm, Theodor: "Eine Halligfahrt," in: Werke. Dritter Teil: Novellen II (1867–1875), ed. Felix Lorenz, Berlin and Leipzig 1927, pp. 40–63.
Straßburg, Gottfried von: Tristan, mittelhochdeutsch—neuhochdeutsch, 3 vols., ed. Rüdiger Krohn, Stuttgart 1993.
Strijbosch, Clara: "Searching for a Versatile Saint: Introduction," in: Glynn S. Burgess and Clara Strijbosch (eds.): The Brendan Legend. Texts and Versions, Leiden 2006, pp. 1–33.
Styliaras, Georgios, Dimitrios Koukopoulos, and Fotis Lazarinis: Handbook of research on technologies and cultural heritage: Applications and environments, Hershey 2011.
Subrahmanyam, Sanjay: The Career and Legend of Vasco da Gama, Cambridge (UK) 1997.
Sulivan, Richard Joseph: Thoughts on Martial Law, London 1784.
Supple, Barry: "Insurance in British Industry," in: Oliver M. Westall (ed.): The Historian and the Business of Insurance, Manchester 1984, pp. 1–8.
Suvin, Darko: Poetik der Science Fiction. Zur Theorie und Geschichte einer literarischen Gattung, Frankfurt a. M. 1979.
Talanga, Jospi: Zukunftsurteile und Fatum. Eine Untersuchung über Aristoteles 'De interpretatione 9' und Ciceros 'De fato', mit einem Überblick über die spätantiken Heimarmene-Lehren, Bonn 1986.
Tang, Chenxi: The Geographic Imagination of Modernity: Geography, Literature, and Philosophy in German Romanticism, Stanford 2008.
Taubes, Jacob: The Political Theology of Paul, transl. Dana Hollander, Stanford 2003.
Taylor, Eva G. R. and Michael William Richey: The Geometrical Seaman: A Book of Early Nautical Instruments, London 1962.
Taylor, Eva G. R.: The Haven-Finding Art. A History of Navigation from Odysseus to Captain Cook, London 1971.
Taylor, James S.: Jonas Hanway: Founder of the Marine Society. Charity and Policy in Eighteenth-Century Britain, London and Berkeley 1985.
Telotte, J. P.: "The Gravity of *2001: A Space Odyssey,*" in: Robert Kolker (ed.): Stanley Kubrick's 2001: A Space Odyssey. New Essays, New York 2006, pp. 43–54.

Tenenti, Alberto: "Der Kaufmann und der Bankier," in: Eugenio Garin (ed.): Der Mensch der Renaissance, Frankfurt a. M. and New York 1990, pp. 215–250.
Teschke, Henning: "An den Grenzen der geglaubten Welt: Dantes Odysseus," in: Italienisch 59 (2008), pp. 2–19.
Thompson, Carl: The Suffering Traveller and the Romantic imagination, Oxford 2007.
Throckmorton, Peter (ed.): The Sea Remembers: Shipwrecks and Archaeology from Homer's Greece to the Rediscovery of the Titanic, London 1996.
Till, Geoffrey: Seapower: A Guide for the Twenty-First Century, London et al. 2004.
Timmermann, Gerhard: Die Suche nach der günstigsten Schiffsform, Oldenburg and Hamburg 1977.
Tomaševsij, Boris: "Literature and Biography," transl. Herbert Eagle, in: Ladislav Matejka and Krysyna Pomorska (eds.), Readings in Russian Poetics: Formalist and Structuralist Views, Cambridge (MA) 1971, pp. 47–55.
Torpey, John: "The Great War and the Birth of the Modern Passport System," in: Jane Caplan and John Torpey (eds.): Documenting Individual Identity: The Development of State Practices in the Modern World, Princeton 2001, pp. 256–270.
Trenerry, Charles Farley: The Origin and Early History of Insurance, London 1926.
Tuerk, Helmut: Reflections on the Contemporary Law of the Sea, Leiden, Boston 2012.
Turner, Victor: "Liminality and Communitas," in: Michael Lambek (ed.): A Reader in the Anthropology of Religion, Malden (MA) 2009, pp. 358–374.
Tuschling, Anna: "Affektive Risikoverarbeitung. Mediengeschichtliche Anmerkungen zu Gefahr bei Goldstein und Lacan," in: Lorenz Engell, Bernhard Siegert, and Joseph Vogl (eds.): Gefahrensinn. Archiv für Mediengeschichte 9, Munich 2009, pp. 115–126.
"Über den Ursprung der Sage von dem fliegenden Holländer oder Gespensterschiff[i]" (n.a.), in: Das Ausland. Ein Tagblatt für Kunde des geistigen und sittlichen Lebens der Völker 14 (25. August 1841), pp. 945 f.
Übereinkommen über die Rechtsstellung der Staatenlosen, (n.a.), agreement on stateless persons, 28. September 1954 (BGBl. 1976 II pp. 474), ch. 1: "Allgemeine Bestimmungen," par. 11: "Staatenlose Seeleute," online at: http://beck-online.beck.de/default.aspx?vpath=bibdata%2Fges%2FUNRechtStaatenlosUeb%2Fcont%2FUNRechtStaatenlosUeb%2EA11%2Ehtm (accessed: 26 Nov. 2019).
Ulrich, R.: "Die Titanic-Katastrophe," in: ZfgV 12 (1912), 1042–1059.
Vaihinger, Hans: Die Philosophie des Als Ob. System der theoretischen, praktischen und religiösen Fiktionen der Menschheit auf Grund eines idealistischen Positivismus. Mit einem Anhang über Kant und Nietzsche, Berlin 1911, reprint Saarbrücken 2007.
Valéry, Paul: Eupalinos, or the Architect, in: Paul Valéry, Dialogues, transl. William McCausland Stewart, New York 1956, pp. 65–152.
Valin, René-Josué: Nouveau commentaire sur l'ordonnance de la Marine du mois d'août 1681 […], vol. 2., La Rochelle 1760.
Vance, W. R.: "The Early History of Insurance Law," in: Columbia Law Review 8/1 (1908), pp. 1–17.
"Vanderdecken's Message Home; Or, the Tenacity of Natural Affection" (n.a.), in: Blackwood's Edinburgh Magazine 50 (Mai 1821), pp. 127–131.
Van D'Elden, Karl H.: "The Development of the Insurance Concept and Insurance Law in the Middle Ages," in: Harold J. Johnson (ed.): The Medieval Tradition of Natural Law, Kalamazoo 1987, pp. 191–200.

Vanoli, Alessandro: "Bild und Geographie. Das Mittelmeer denken," in: Hannah Baader and Gerhard Wolf (eds.): Das Meer, der Tausch und die Grenzen der Repräsentation, Zurich and Berlin 2010, pp. 207–238.
Vattel, Emer de: The Law of Nations, or, Principles of the Law of Nature, Applied to the Conduct and Affairs of Nations and Sovereigns, with Three Early Essays on the Origin and Nature of Natural Law and on Luxury (1758), transl. Thomas Nugent, Indianapolis 2008.
Vaz d'Almada, Francisco: "Treatise of the Misfortunes that Befell the Great Ship São João Baptista," in: Charles R. Boxer (ed.), The Tragic History of the Sea: 1589–1622, Cambridge (UK) 1959.
Vega, Joseph de la: Die Verwirrung der Verwirrungen. Vier Dialoge über die Börse, Breslau 1919.
Verlinden, Charles and Eberhard Schmitt (eds.): Die mittelalterlichen Ursprünge der europäischen Expansion, Munich 1986.
Verne, Jules: An Antarctic Mystery: The Sphinx of the Ice Fields, The Floating Press (e-book; n.p., no translator indicated) 2010.
Vickers, Ilse: Defoe and the New Sciences, Cambridge (UK) 1996.
Vico, Giambattista: The New Science, transl. Thomas Goddard Bergin and Max Harold Fisch, Ithaca 1948.
Vidal-Naquet, Pierre: The Atlantis Story: A Short History of Plato's Myth, transl. Janet Lloyd, Exeter 2007.
Villain-Gandossi, Christiane: Pour une histoire du 'fait maritime'. Sources et champs de recherches, Paris 2001.
Vinci, Leonardo da: Das Wasserbuch. Schriften und Zeichnungen, ed. Marianne Schneider, Munich 1996.
Vinken, Barbara: Unentrinnbare Neugierde. Die Weltverfallenheit des Romans. Richardsons "Clarissa," Laclos' "Liaisons dangereuses," Freiburg 1991.
Virgil: Eclogues, Georgics, Aeneid, transl. H.R. Fairclough, Cambridge (MA) 1916.
Vismann, Cornelia: Akten. Medientechnik und Recht, Frankfurt a. M. 2000.
Vismann, Cornelia: "Quellen," in: Cornelia Vismann. Das Recht und seine Mittel. Ausgewählte Schriften, ed. Markus Krajewski and Fabian Steinhauer, Frankfurt a. M. 2012.
Vogl, Joseph: Kalkül und Leidenschaft. Poetik des ökonomischen Menschen, Munich 2002.
Vogl, Joseph: "Asyl des Politischen. Zur Topologie politischer Gelegenheiten," in: Uwe Hebekus, Ethel Matala de Mazza, and Albrecht Koschorke (eds.): Das Politische. Figurenlehren des sozialen Körpers nach der Romantik, Munich 2003, pp. 23–38.
Vogl, Joseph: "Regierung und Regelkreis. Historisches Vorspiel," in: Claus Pias (ed.): Cybernetics–Kybernetik: The Macy-Conferences 1946–1953. Essays & Documents, Zurich and Berlin 2004, pp. 67–79.
Vogl, Joseph: "Das charismatische Tier," in: Anne von der Heiden and Joseph Vogl (eds.): Politische Zoologie, Zurich and Berlin 2007, pp. 119–130.
Vogl, Joseph: "Lebende Anstalt," in: Friedrich Balke, Joseph Vogl, and Benno Wagner (eds.), Für Alle und Keinen. Lektüre, Schrift und Leben bei Nietzsche und Kafka, Zurich and Berlin 2008, pp. 21–33.
Vogl, Joseph: Spectre of Capital, transl. Joachim Redner and Robert Savage, Stanford 2015.
Vogt, Peter: Kontingenz und Zufall. Begriffsgeschichtliche Präzisierungen, Ideengeschichtliche Skizzen, philosophische Schlussfolgerungen, Berlin 2011.

Volkert, Klaus Thomas: Die Krise der Anschauung, Göttingen 1986.
Waage, Frederick O.: "Touching the Compass. Empiricism in Popular Scientific Writing of Bacon's Time," in: Huntington Library Quarterly: A Journal for the History and Interpretation of English and American Civilization 41 (1978), pp. 201–216.
Wachsmuth, Dietrich: Πομπιμοσ ο Δαιμον. Untersuchungen zu den antiken Sakralhandlungen bei Seereisen, Berlin 1967 (diss.).
Wade-Gery, Henry Theodore: The Poet of the Iliad, Cambridge (UK) 1952.
Wagner, Benno: Der Unversicherbare. Kafkas Protokolle, unveröffentl. Habil., Siegen 1998.
Wagner, Benno: "Beim Bau der chinesischen Mauer," in: Manfred Engel and Bernd Auerochs (eds.): Kafka-Handbuch, Stuttgart 2010, pp. 250–259.
Wagner, Rudolf: Handbuch des Seerechts, vol. 1, Leipzig 1906.
Waldenfels, Bernhard: "Das Ordentliche und das Außer-ordentliche," in: Bernhard Greiner and Maria Moog-Grünewald (eds.): Kontingenz und Ordo. Selbstbegründung des Erzählens in der Neuzeit, Heidelberg 2000, pp. 1–13.
Wallerstein, Immanuel: Das moderne Weltsystem, vol. 2. Der Merkantilismus. Europa zwischen 1600 und 1750, Vienna 1998.
Warburg, Aby: "Francesco Sassetti's Last Injunctions to his Sons," in: Aby Warburg, The Renewal of Pagan Antiquity: Contributions to the Cultural History of the European Renaissance, transl. David Britt, Los Angeles 1999.
Warnecke, Heinz: Paulus im Sturm. Über den Schiffbruch der Exegese und die Rettung des Apostels auf Kephallenia, Nuremberg 2000.
Warning, Rainer: "Fiktion und Transgression," in: Ursula Peters and Rainer Warning (eds.): Fiktion und Fiktionalität in den Literaturen des Mittelalters. Jan-Dirk Müller zum 65. Geburtstag, Munich 2009, pp. 31–56.
Waswo, Richard: "Crises of Credit. Monetary and Erotic Economies in the Jacobean Theatre," in: Dieter Mehl, Angela Stock, and Anne-Julia Zwierlein (eds.): Plotting Early Modern London: New Essays on Jacobean City Comedy, Aldershot 2004, pp. 55–74.
Watson, Roderick: "'The unrest and movement of our century'. The universe of The Wrecker," in: The Journal of Stevenson Studies 4 (2007), pp. 114–128.
Weber, Max: Economy and Society: An Outline of Interpretive Sociology, transl. Ephraim Fischoff et al., Berkeley, Los Angeles, and London 1978.
Weber, Otto: Die Schiffsgewalt des Kapitäns und ihre rechtlichen Grundlagen, Tübingen 1907.
Weeks, Andrew: "Kafka und die Zeugnisse vom versunkenen Kakanien," in: Sprache im technischen Zeitalter 88/15 (1983), pp. 320–337.
Wegmann, Thomas: Tauschverhältnisse. Zur Ökonomie des Literarischen und zum Ökonomischen in der Literatur von Gellert bis Goethe, Würzburg 2002.
Wehnert, Jochen: "Gestrandet. Zu einer neuen These über den Schiffbruch des Apostels Paulus auf dem Weg nach Rom," in: Zeitschrift für Theologie und Kirche 87 (1990), pp. 67–99.
Weibust, Knut: Deep Sea Sailors: A Study in Maritime Ethnology, Stockholm 1969.
Weigel, Sigrid: "Das Gedankenexperiment. Nagelprobe auf die facultas fingendi in Wissenschaft und Literatur," in: Thomas Macho and Annette Wunschel (eds.): Science & Fiction. Über Gedankenexperimente in Wissenschaft, Philosophie und Literatur, Frankfurt a. M. 2004a, pp. 183–205.
Weigel, Sigrid: Literatur als Voraussetzung der Kulturgeschichte. Schauplätze von Shakespeare bis Benjamin, Munich 2004b.

Weisberg, Richard: The Failure of the World: The Protagonist as Lawyer in Modern Fiction, New Haven and London 1984.
Weitin, Thomas: Zeugenschaft. Das Recht der Literatur, Munich 2009.
Welke, Ulrich: Der Kapitän. Die Erfindung einer Herrschaftsform, Münster 1997.
Welsch, Caroline: "Die Sirenen und das Klavier. Vom Mythos der Sphärenharmonie zur experimentellen Sinnesphysiologie," in: Bernhard Dotzler, Henning Schmidgen, and Cornelia Weber (eds.): Parasiten und Sirenen: Zwei ZwischenRäume, MPI-Preprint 253, Berlin 2004, pp. 57–85.
Welzig, Werner: "Der Wandel des Abenteurertums," in: Helmut Heidenreich (ed.), Pikarische Welt. Schriften zum europäischen Schelmenroman, Darmstadt 1969, pp. 438–454.
Werber, Niels: "Krieg und Nicht-Krieg. Anmerkungen zur militärischen Weltraumordnung," Rudolf Maresch and Niels Werber (eds.): Raum—Wissen—Macht, Frankfurt a.M. 2002, pp. 287–305.
Werber, Niels: "Ahab's Charisma. Captains, Kings, and Prophets," in: New German Critique 114 (2011), pp. 51–62.
Wertheimer, Eric: Underwriting: The Poetics of Insurance in America 1722–1872, Stanford 2006.
Weski, Tim: "The Value of Experimental Archaeology for Reconstructing Ancient Seafaring," in: Lucy Katherine Blue, Frederick M. Hocker, and Anton Englert (eds.): Connected by the Sea: Proceedings of the Tenth International Symposium on Boat and Ship Archaeology, Roskilde 2003, Oxford 2006, pp. 63–67.
Wheeler, R.E. Mortimer: Archaeology from the Earth, Oxford 1954.
Wiedemann, Peter M.: "Tabu, Sünde, Risiko. Veränderungen der gesellschaftlichen Wahrnehmung von Gefährdungen," in: Ulrike Becker et al. (eds.): Risiko ist ein Konstrukt. Wahrnehmungen zur Risikowahrnehmung, Munich 1993, pp. 43–67.
Wiemann, Renate: Die Erzählstruktur im Volksbuch Fortunatus, Hildesheim and New York 1970.
Wiener, Norbert: Cybernetics or Control and Communication in the Animal and in the Machine, Cambridge (MA) 1962.
Williams, Dennis: "Filling the Void. A Lacanian Angle of Vision on Moby-Dick," in: John Bryant, Mary K. Brecaw Edwards, and Timothy Marr (eds.): "Ungraspable Phantom": Essays on Moby-Dick, Kent (OH) 2006, pp. 61–80.
Wilson, Luke: "Monetary Compensation for Injuries to the Body," in: Linda Woodbridge (ed.): Money and the Age of Shakespeare: Essays in New Economic Criticism, New York 2003, pp. 19–37.
Wilson, Luke: "Drama and Marine Insurance in Shakespeare's London," in: Constance Jordan and Karen Cunningham (eds.): The Law in Shakespeare, Basingstoke et al. 2007, pp. 127–142.
Wirz, Christian: Der gekreuzigte Odysseus. 'Umbesetzung' als Form des christlichen Verhältnisses zur Welt als dem Anderen, Regensburg 2005.
Wislicenus, Georg: Deutschlands Seemacht sonst und jetzt, 3rd edn., Leipzig 1909.
Witt, Jann Markus: Master next God? Der nordeuropäische Handelsschiffskapitän vom 17. bis zum 19. Jahrhundert, Hamburg 2001a.
Witt, Jann Markus: "'Master Next to God?'. The Merchant Captain from the 17th to the 19th century," in: Luc François and Ann Katherine Isaacs (eds.): The Sea in European History, Pisa 2001b, pp. 237–249.

Witte, Bernd: "Homerische Schatten. Ein historisch-kritischer Kommentar zu Holderlins 'Andenken,'" in: Johannes Janota (ed.): Kultureller Wandel und die Germanistik in der Bundesrepublik, vol. 3: Methodenkonkurrenz in der germanistischen Praxis, Tübingen 1993, pp. 181–192.
Witthaus, Jan-Henrik: "Kompass, Pulver, Presse. Zur technischen Ausrichtung des europäischen Raums in französischen Zeitschriften des 18. Jahrhunderts," in: Angela Oster (ed.): Europe en mouvement. Mobilisierungen von Europa-Konzepten im Spiegel der Technik, Berlin 2009, pp. 60–83.
Wittschier, Heinz Willi: Dantes Divina Commedia. Einführung und Handbuch, Frankfurt a. M. 2004.
Woeller, Waltraud: "Die Sage vom fliegenden Holländer," in: Deutsches Jahrbuch für Volkskunde, Berlin 1968, pp. 292–314.
Wolf, Armin: Hatte Homer eine Karte? Beobachtungen über die Anfänge der europäischen Kartographie, Karlsruhe 1997.
Wolf, Armin and Hans Helmut Wolf: Die wirkliche Reise des Odysseus. Zur Rekonstruktion des Homerischen Weltbildes, Munich and Vienna 1983.
Wolf, Burkhardt: "Kafkas Seekrankheit," in: Klaus R. Scherpe and Elisabeth Wagner (eds.): Kontinent Kafka. Mosse-Lectures an der Humboldt-Universität zu Berlin, Berlin 2006, pp. 120–134.
Wolf, Burkhardt: "Kap der Stürme. Der Fliegende Holländer und die Irrfahrten maritimer Globalisierung," in: Hannah Baader and Gerhard Wolf (eds.): Das Meer, der Tausch und die Grenzen der Repräsentation, Zurich and Berlin 2010, pp. 357–377.
Wolf, Burkhardt: "Homo gubernator. Defoe und das postsouveräne Erzählen um 1700," in: Lili —Zeitschrift für Literaturwissenschaft und Linguistik, 42 (2012), no. 165, pp. 35–54.
Wolf, Friedrich August: Darstellung der Altertumswissenschaft nach Begriff, Umfang und Wert, Berlin 1807.
Wolf, Friedrich August: Prolegomena zu Homer, transl. Hermann Muchau, Leipzig o. J. [1908].
Wolf, Ursula: "Einleitung," in: Ursula Wolf (ed.): Eigennamen. Dokumentation einer Kontroverse, Frankfurt a. M. 1985, pp. 9–41.
Wolff, Hans Wolter: Studien zum Jonabuch. Mit einem Anhang von Jörg Jeremias: Das Jonabuch in der Forschung seit Hans Walter Wolff, Neukirchen 2003.
Wolff, Hermann: Die akademischen Berufe, vol. 6: Der Techniker, Berlin 1919.
Wolter, Detlev: Grundlagen "Gemeinsamer Sicherheit" im Weltraum nach universellem Völkerrecht. Der Grundsatz der friedlichen Nutzung des Weltraums im Lichte des völkerrechtlichen Strukturprinzips vom 'Gemeinsamen Erbe der Menschheit', Berlin 2003.
Wood, Lorraine: "Joyce's Ineluctable Modality: (Re)Reading the Structure of 'Sirens,'" in: Joyce Studies Annual (Winter 2007), pp. 67–91.
Wossidlo, Richard: Reise, Quartier, in Gottesnaam. Das Seemannsleben auf den alten Segelschiffen im Munde alter Fahrensleute, im Auftrag des Kuratoriums der Wossidlo-Stiftung aus dem Nachlaß Richard Wossidlos, ed. Paul Beckmann, Rostock 1951.
Wright, Edward: Certaine Errors in Navigation Detected and Corrected, with Many Additions that were not in the Former Edition, 3rd edn., London 1657.
Wright, Georg Henrik von: "Determinismus, Wahrheit und Zeitlichkeit," in: Bertram Kienzle (ed.): Zustand und Ereignis, Frankfurt a. M. 1994, pp. 171–189.

Zedler, Johann Heinrich: Grosses vollständiges Universal-Lexicon Aller Wissenschafften und Künste, vol. 40, Leipzig and Halle 1744.
Zelle, Carsten: "Erhabene Weltuntergänge im Kleinen. Über Schiffbrüche und Schlachten vor Zuschauer—Bemerkungen zur Krise der Aufklärungsästhetik im Anschluß auf Lukrez' De rerum natura," in: Emilio Bonfatti (ed.): Il gesto, il bello, il sublime. Arte e letteratura in Germania tra Sette e Ottocento, Rome 1997, pp. 77–111.
Zelle, Carsten: "Schiffbruch vor Zuschauer. Über einige popularphilosophische Parallelschriften zu Schillers Abhandlungen über den Grund des Vergnügens an tragischen Gegenständen in den neunziger Jahren des 18. Jahrhunderts," in: Jahrbuch der deutschen Schillergesellschaft 34 (1990), pp. 289–316.
Zielinski, Tadeusz Stefan: "Die Behandlung gleichzeitiger Ereignisse im antiken Epos," Philologus Suppl. 8 (1899–1901) 407–49.
Zirus, Werner: Ahasverus. Der ewige Jude, Berlin and Leipzig 1930.
Žižek, Slavoj: "Grimaces of the Real, or When the Phallus Appears," in: October 58 (1991), pp. 44–68.
Zmijewski, Josef: Die Apostelgeschichte, Regensburg 1994.
Zorn, Philipp: Das Staatsrecht des deutschen Reichs, vol. 2, Berlin 1897.
Zwierlein, Anne-Julia: "Shipwrecks in the City: Commercial Risk as Romance in Early Modern City Comedy," in: Dieter Mehl, Angela Stock, and Anne-Julia Zwierlein (eds.): Plotting Early Modern London: New Essays on Jacobean City Comedy, Aldershot 2004, pp. 75–94.
Zysberg, André: "Naufrages, épaves et sauvetages maritimes au miroir des titres d'imprimés français (1600–1969)," in: Christian Buchet and Claude Thomasset (eds.): Le naufrage. Actes du Colloque tenu à l'Institut Catholique de Paris (28–30 janvier 1998), Paris 1999, pp. 179–192.

List of Figures

https://doi.org/10.1515/9783110610734-016

Index of Names

https://doi.org/10.1515/9783110610734-017

Index of Subjects

https://doi.org/10.1515/9783110610734-018

www.ingramcontent.com/pod-product-compliance
Ingram Content Group UK Ltd.
Pitfield, Milton Keynes, MK11 3LW, UK
UKHW041453071026
15587UKWH00010B/1003

9 783110 608953